Graduate Texts in Physics

Series Editors

Kurt H. Becker, NYU Polytechnic School of Engineering, Brooklyn, USA

Jean-Marc Di Meglio, Matière et Systèmes Complexes, Bâtiment Condorcet, Université Paris Diderot, Paris, France

Sadri Hassani, Department of Physics, Illinois State University, Normal, USA

Morten Hjorth-Jensen, Department of Physics, Blindern, University of Oslo, Oslo, Norway

Bill Munro, Graduate University, Okinawa Institute of Science and Technology, Onna-son, Japan

Richard Needs, Cavendish Laboratory, University of Cambridge, Cambridge, UK

William T. Rhodes, Department of Computer and Electrical Engineering and Computer Science, Florida Atlantic University, Boca Raton, USA

Susan Scott, Australian National University, Acton, Australia

Jonathan V. Selinger, Liquid Crystal Institute, Kent State University, Kent, OH, USA

H. Eugene Stanley, Department of Physics, Center for Polymer Studies, Boston University, Boston, MA, USA

Andreas Wipf, Institute of Theoretical Physics, Friedrich-Schiller-University Jena, Jena, Germany

Graduate Texts in Physics publishes core learning/teaching material for graduate- and advanced-level undergraduate courses on topics of current and emerging fields within physics, both pure and applied. These textbooks serve students at the MS- or PhD-level and their instructors as comprehensive sources of principles, definitions, derivations, experiments and applications (as relevant) for their mastery and teaching, respectively. International in scope and relevance, the textbooks correspond to course syllabi sufficiently to serve as required reading. Their didactic style, comprehensiveness and coverage of fundamental material also make them suitable as introductions or references for scientists entering, or requiring timely knowledge of, a research field.

Mustafa A. G. Abushagur

Applied Photonics
An Introduction for Physicists
and Engineers

 Springer

Mustafa A. G. Abushagur
Department of Electrical and
Microelectronic Engineering
Rochester Institute of Technology
Rochester, NY, USA

ISSN 1868-4513　　　　　　　ISSN 1868-4521　(electronic)
Graduate Texts in Physics
ISBN 978-3-031-86456-8　　　ISBN 978-3-031-86457-5　(eBook)
https://doi.org/10.1007/978-3-031-86457-5

© The Editor(s) (if applicable) and The Author(s), under exclusive license to Springer Nature Switzerland AG 2025

This work is subject to copyright. All rights are solely and exclusively licensed by the Publisher, whether the whole or part of the material is concerned, specifically the rights of translation, reprinting, reuse of illustrations, recitation, broadcasting, reproduction on microfilms or in any other physical way, and transmission or information storage and retrieval, electronic adaptation, computer software, or by similar or dissimilar methodology now known or hereafter developed.
The use of general descriptive names, registered names, trademarks, service marks, etc. in this publication does not imply, even in the absence of a specific statement, that such names are exempt from the relevant protective laws and regulations and therefore free for general use.
The publisher, the authors and the editors are safe to assume that the advice and information in this book are believed to be true and accurate at the date of publication. Neither the publisher nor the authors or the editors give a warranty, expressed or implied, with respect to the material contained herein or for any errors or omissions that may have been made. The publisher remains neutral with regard to jurisdictional claims in published maps and institutional affiliations.

This Springer imprint is published by the registered company Springer Nature Switzerland AG
The registered company address is: Gewerbestrasse 11, 6330 Cham, Switzerland

If disposing of this product, please recycle the paper.

To my beloved wife, Fatma, whose unwavering support, love, and encouragement have been my constant source of strength and inspiration.

To our wonderful children—Ousama, Soumiea, Asmah, Sarrah, and Noor—who fill my life with joy and pride through their brilliance, resilience, and love.

And to my cherished grandchildren—Ahmed, Tasneem, Ilyas, Zakaria, Yousef, Malik, Mariam, and Fajr—who embody hope and promise for the future.

This book is for you.

Preface

This book is the culmination of over four decades of teaching courses and research in the broad fields of optics, optoelectronics, and photonics. During my academic career, I had the privilege of teaching these subjects at esteemed institutions, including the University of Alabama in Huntsville, the Rochester Institute of Technology, and others. The material in this book has been carefully refined and expanded through years of engaging with students, addressing their questions, and witnessing the evolution of the field. Photonics, as a discipline, has seen exponential growth over the past few decades, becoming a cornerstone of technological innovations. Its applications span diverse industries, from telecommunications and healthcare to energy and manufacturing, demonstrating the profound impact light-based technologies have on society. This book seeks to capture that dynamism, offering readers both the theoretical foundations and practical insights necessary to navigate and contribute to this rapidly advancing field. Beyond academia, my experience extends into entrepreneurship in the photonics industry. Over the years, I have founded and led companies that bridged the gap between research and commercialization, translating innovative ideas into impactful technologies. One such endeavor was the founding of Photronix (M) Sdn. Bhd. in Malaysia, where we developed, manufactured, and marketed advanced photonics products such as fiber Bragg gratings-based devices, optical couplers, and optical amplifiers. Another venture was LiquidLight Inc., based in Atlanta, Georgia, where we pioneered optical networking hardware, including the groundbreaking triple-play technology. This innovation represented a significant leap in telecommunications by enabling the seamless integration of voice, video, and data over a single platform. My entrepreneurial journey has allowed me to witness firsthand how photonics technologies evolve from conceptual ideas to transformative solutions that reshape industries and improve lives. The content of this book reflects not only my academic expertise but also the practical knowledge and insights gained from these ventures. The chapters are organized to provide a progressive journey through the subject, beginning with fundamental concepts before delving into advanced topics. Each chapter is supplemented with problems and exercises to ensure that readers can connect theoretical knowledge to practical applications. This book is designed for graduate and advanced undergraduate students in engineering and physics, as well as researchers and professionals seeking a comprehensive understanding of photonics and its diverse applications. Whether you are new to the field or looking to deepen your knowledge, I hope the material presented here serves as a valuable resource and source of inspiration. Writing this book has been both a professional and personal journey, reflecting my enduring fascination with the science of light and its transformative potential. I am deeply grateful to my teachers, students, colleagues, and family, whose support and encouragement have made this endeavor possible. I invite you to explore the fascinating world of photonics and hope that this book sparks the same passion for the field that I have carried throughout my career.

Rochester, NY, USA Mustafa A. G. Abushagur
November 2024

Declarations

Competing Interests The author has no competing interests to declare that are relevant to the content of this manuscript.

Introduction

Photonics has transformed the way we understand, utilize, and interact with light. From enabling high-speed communication over optical fibers to paving the way for advanced imaging technologies and beyond, photonics is central to some of the most groundbreaking innovations of the twenty-first century. This book, *Applied Photonics*, aims to provide a thorough exploration of the field, blending fundamental principles with practical applications, to equip readers with the knowledge and tools to contribute to this rapidly evolving domain.

Chapter 1, "What Is Light?," serves as a foundational introduction to the nature of light, tracing its conceptual evolution through various theories, including the corpuscular, wave, and quantum perspectives. This chapter lays the groundwork for understanding the dual nature of light and its behavior across different models.

Chapter 2, "Geometrical Optics," we delve into the principles of ray optics, exploring ray tracing, reflection, refraction, and the use of lenses and mirrors. The mathematical frameworks and examples provided enable readers to analyze and design basic optical systems.

Chapter 3, "Physical Optics," moves into the wave nature of light, covering topics such as electromagnetic theory, Maxwell's equations, and the wave equation. This chapter discusses interference, diffraction, and polarization, offering insights into the phenomena that arise from light's wave-like properties.

Chapter 4, "Fourier Optics," introduces the mathematical tools of Fourier analysis in the context of optics. Topics include diffraction theory, image formation, and optical signal processing, showcasing the relevance of Fourier transforms in analyzing and designing optical systems.

Chapter 5, "Light Polarization," the focus is on the states and representation of polarization. Readers will learn about polarization by reflection, birefringence, and the use of polarizers and wave plates, essential for many optical applications.

Chapter 6, "Optical Beams and Resonators," examines Gaussian beam propagation and the design of optical resonators, such as Fabry-Perot and spherical mirror resonators. The discussion extends to beam stability and resonator losses, providing tools for understanding laser systems.

Chapter 7, "Optical Waveguides," explores the principles of light confinement in waveguides. Topics include slab waveguides, mode propagation, and coupled-mode theory, offering a deep dive into the technology underlying modern integrated photonics.

Chapter 8, "Optical Fibers," focuses on fiber optics, including step-index and graded-index fibers, signal degradation, and dispersion management. The material emphasizes the role of fibers in high-speed, long-distance optical communication.

Chapter 9, "Introduction to Quantum Mechanics," provides a primer on quantum principles, from wave-particle duality to Schrödinger's equation. This chapter bridges the gap between classical and quantum theories, paving the way for understanding quantum photonics.

Chapter 10, "Semiconductor Energy Bands," introduces the electronic properties of semiconductors, discussing energy bands, density of states, and carrier densities. These concepts are critical for understanding optoelectronic devices.

Chapter 11, "Optoelectronic Semiconductors," the focus shifts to materials for photonics applications. Topics include direct and indirect bandgap semiconductors, emission and absorption probabilities, and semiconductor gain media.

Chapter 12, "Semiconductor Light Sources," examines the design and operation of devices such as LEDs, laser diodes, and optical amplifiers. The chapter covers critical concepts, including light emission mechanisms and device characteristics.

Chapter 13, "Optical Detectors," explores the principles of light detection, from photodiodes to avalanche photodiodes. The discussion includes quantum efficiency, responsivity, and optical receiver design, integral to photonic communication systems.

Chapter 14, "Optical Modulators," delves into the critical components used to modulate light for communication and processing applications. This chapter explores the principles and applications of electro-optic and acousto-optic modulators, discussing their underlying physics and practical implementations. Readers will gain insight into how these devices manipulate light to encode information, control intensity, and change wavelength, making them essential in modern photonic systems.

Finally, Chap. 15, "Optical Fiber Communication Systems," concludes the book by integrating concepts from previous chapters to examine the design and operation of optical communication networks. Topics include wavelength-division multiplexing (WDM), optical switching, and network architectures. This chapter highlights the transformative role of photonics in enabling high-speed, large-capacity data transmission over global communication infrastructures.

This book is structured to guide readers progressively from fundamental concepts to advanced applications, making it an invaluable resource for students, researchers, and professionals in photonics. It is my hope that this text not only deepens your understanding of photonics but also inspires further exploration and innovation in this dynamic field.

Contents

1. What Is Light? 1
2. Geometrical Optics 9
3. Physical Optics 43
4. Fourier Optics 105
5. Light Polarization 155
6. Optical Beams and Resonators 189
7. Optical Waveguides 225
8. Optical Fibers 259
9. Introduction to Quantum Mechanics 295
10. Semiconductor Energy Bands 339
11. Optoelectronic Semiconductors 379
12. Semiconductor Light Sources 403
13. Optical Detectors 461
14. Optical Modulators 497
15. Optical Fiber Communication Systems 519

List of Figures

Fig. 2.1	Path of a light ray between points A and B in inhomogeneous media according to Fermat's principle	11
Fig. 2.2	Reflection of ray AO from a mirror	11
Fig. 2.3	The refraction of a light ray through different media	12
Fig. 2.4	Sign convention	13
Fig. 2.5	Paraxial rays..	14
Fig. 2.6	Reflection of light from a plane mirror	15
Fig. 2.7	Focusing of parallel light rays by a paraboloidal mirror, where C is the center of the mirror, and F is its focal point.......	16
Fig. 2.8	Imaging by a spherical concave mirror	16
Fig. 2.9	Image formation by a spherical mirror. In this case, the image is real, h_2 tall and inverted	18
Fig. 2.10	A biconvex spherical lens formed by the intersection of two spherical surfaces	19
Fig. 2.11	Image formation by a thin lens......................................	20
Fig. 2.12	Light rays propagating through an optical system	23
Fig. 2.13	The output of the system based on setting the ray matrix coefficients in turn to be zero......................................	25
Fig. 2.14	A translation of a ray for a distance L in a homogeneous media .	26
Fig. 2.15	A transition of a ray between two mediums	27
Fig. 2.16	A transition of a ray between two spherical dielectric media.....	27
Fig. 2.17	Reflection of rays from a spherical mirror........................	29
Fig. 2.18	Ray matrix for cascaded interfaces	30
Fig. 2.19	A thick lens	31
Fig. 2.20	A single lens imaging system	32
Fig. 2.21	A two-lens imaging system..	34
Fig. 2.22	Ray matrix representation of a periodic optical system	36
Fig. 3.1	Propagation the wavefronts of planewave	51
Fig. 3.2	Propagation of planewave wavefronts	51
Fig. 3.3	A planewave traveling in the z direction is a periodic function in z with spatial period λ	52
Fig. 3.4	A spherical emitted from a point source	56
Fig. 3.5	Reflection of a planewave off a mirror surface	60
Fig. 3.6	Refraction of a planewave through different media	61

Fig. 3.7	Reflection and transmission of an electric field polarized perpendicular to the plane of incidence. This symbol ⊙ denotes that the field is pointing out of the plane of incidence	62
Fig. 3.8	Reflection and transmission of an electric field polarized in the the plane of incidence	62
Fig. 3.9	The amplitude coefficients of reflection and transmission as a function of incident angle. These correspond to external reflection $n_t > n_i$ at an air–glass interface ($\frac{n_t}{n_i} = 1.5$). The parallel component of the incidence field becomes zero at $\theta_i = \theta_p$ which is known as the polarization angle	63
Fig. 3.10	The amplitude coefficients of reflection and transmission as a function of incident angle. These correspond to internal reflection $n_t < n_i$ at an air–glass interface ($\frac{n_i}{n_t} = 1.5$). Total internal reflection occurs at a critical angle $\theta_c = 41.8°$	64
Fig. 3.11	Reflectance and transmittance as a function of the angle of incidence for external reflection, $n_t > n_i$, at an air–glass interface where $\frac{nt}{n_i} = 1.5$	64
Fig. 3.12	Planewave passing through a rectangular glass plate of thickness d. (**a**) oblique incidence and (**b**) $\bar{k}$ is making and angle θ	66
Fig. 3.13	A plate of a varying thickness $d(x, y)$	67
Fig. 3.14	(**a**) A planewave passing through a convex lens emerges as a converging paraboloidal wave focused at the back focal plane, and (**b**) a diverging paraboloidal wave originating at the front focal plane of a convex lens emerges as a planewave	69
Fig. 3.15	The intensity pattern for the interference of two waves as given by Eq. (3.57)	71
Fig. 3.16	Interference of two planewaves	72
Fig. 3.17	Interference of two spherical waves emitted by two point sources located at $(d, 0, 0)$ and $(-d, 0, 0)$	75
Fig. 3.18	The fringe pattern of the interference of two spherical waves emitted by two point sources	77
Fig. 3.19	MATLAB plots for (**a**) the intensity of the interference of two spherical waves for: $d = 1$ mm, $z = 10$ m, $\lambda = 1000$ nm, and (**b**) the fringe pattern as seen on an observation screen	78
Fig. 3.20	MATLAB plots for the intensity of the interference of M=5 spherical waves	81
Fig. 3.21	The intensity pattern, Eq. (3.77) is plotted as a function of $\phi[radians]$ for the finesse, F=1, 5, and 50. The intensity peaks are sharper and narrower for larger finesse values	83
Fig. 3.22	General representation of incident, reflected, and transmitted beams	83

Fig. 5.9	A dichroic crystal allows the E-field component parallel to its optic axis to pass through without any absorption, while the E-field component perpendicular to the optic axis is absorbed as the light propagates through the crystal	167
Fig. 5.10	One layer of a cubic crystalline structure of sodium chloride. Each chloride ion (red spheres) is surrounded by six individual sodium ions (blue spheres) and vice versa for the sodium ions	167
Fig. 5.11	Index ellipsoids of Isotropic and anisotropic crystals classifications	168
Fig. 5.12	The ordinary and extraordinary waves in an uniaxial crystal	169
Fig. 5.13	Incident and refracted waves' relations for isotropic and anisotropic materials	169
Fig. 5.14	Propagation of a light wave through a calcite crystal with various optical axis orientations. Ordinary waves, which are polarized perpendicular to the plane of the paper, and extraordinary waves, which are polarized parallel to the plane of the paper, are depicted for three different orientations of the optical axis. Double refraction occurs when the optical axis is neither parallel nor perpendicular to the wave's propagation direction	170
Fig. 5.15	Polarizing prisms: (**a**) Glan–Foucault, (**b**) Glan-Taylot, and (**c**) Wollaston prism	171
Fig. 5.16	Polarization by reflection	173
Fig. 5.17	Wave retardation for positive birefringence $n_e > n_o$ and negative birefringence, $n_e < n_o$	174
Fig. 5.18	The state of polarization as E_x leads E_y by φ in increments of $\pi/4$	174
Fig. 5.19	A half-wave plate rotates light initially linearly polarized at an angle θ through a total angle of 2θ. Here light was incident oscillating in the first and third quadrants, and it emerged oscillating in the second and fourth quadrants	175
Fig. 5.20	A quarter-wave plate transforms a linearly polarized light at 45^0 into a circular polarized state	176
Fig. 5.21	Representation of the relation between the input and output of an optical polarizing device in terms of Jones calculus	177
Fig. 5.22	Polarization states for the output of a quarter-wave plate for an incident light wave linearly polarized at angle α	182
Fig. 5.23	A system of cascaded Jones matrices for polarizing optical components	183
Fig. 5.24	Unpolarized light passing through two linear polarizers	185
Fig. 6.1	The cross section of the Gaussian profile of Eq. (6.20) is plotted along the radial distance r_t showing the the radius ω_o at amplitude $1/e$	194

Fig. 6.2	The size of the Gaussian beam along the direction of propagation	195
Fig. 6.3	The Gaussian beam wavefront at a distance z from the beam waist	196
Fig. 6.4	The Gaussian beam radius of curvature $R(z)$	200
Fig. 6.5	Gaussian beam intensity plot	201
Fig. 6.6	The Gaussian beam intensity, $I(0, z)$ along the z-axis for $r_t = 0$	201
Fig. 6.7	A Gaussian beam propagating along the z-axis of a complex radius of curvature q_1 at distance z_1 from the beam waist and q_2 at distance z_2 from the beam waist	203
Fig. 6.8	Gaussian beam propagation through a thin lens. The complex radius of curvature, q_1, represents the beam at the input (left surface) of the lens, while q_2 represents the beam at the output of the lens	204
Fig. 6.9	A Gaussian beam with complex radius of curvature of q_{in} at the input plane of an optical system that can be described with the paraxial ABCD matrix. The Gaussian beam exiting the optical system has an output complex radius of curvature of q_{out}	205
Fig. 6.10	Fabry–Perot resonator comprised two parallel planar mirrors separated by a distance d. (**a**) Light rays perpendicular to the mirrors reflect back and forth without escaping the mirrors. (**b**) Adjacent resonant frequencies, ν_i, separated by frequency ν_F	207
Fig. 6.11	Intensity of the output beam for a Fabry–Perot resonator as a function of mirror reflectance R around a resonant frequency. As the reflectance R increases, the peak intensity becomes sharper and more pronounced, indicating higher finesse and narrower bandwidth of the resonator	211
Fig. 6.12	(**a**) A light beam reflected from a mirror of $R = 0.999$. (**b**) The same beam as it passes through a resonator	212
Fig. 6.13	Geometry for spherical mirror resonator constructed with two concave mirrors	215
Fig. 6.14	A spherical-mirror resonator with a stable Gaussian mode	216
Fig. 6.15	An example of a simple optical resonator	218
Fig. 6.16	Optical directional coupler	221
Fig. 6.17	A typical single ring resonator	221
Fig. 6.18	Typical characteristic of a ring resonator	222
Fig. 7.1	Cross section of several standard waveguide structures	226
Fig. 7.2	Basic structure of an optical slab waveguide	227
Fig. 7.3	Propagation of a wave within the core of a waveguide	228
Fig. 7.4	Transverse electric (TE) and transverse magnetic polarization (TM)	230

Fig. 7.5	K-vector triangles at the core–cladding interface and the boundary conditions that govern them	231
Fig. 7.6	The electric field profile for the first three guided modes for $m = 1, 2$ and 3	235
Fig. 7.7	A cross section of an asymmetric slab waveguide	236
Fig. 7.8	Sketch of the electric field distribution as a function of β for a slab waveguide	239
Fig. 7.9	Graphical solution for the transcendental equation of asymmetric slab waveguide	242
Fig. 7.10	Solution for Eq. (7.69) for a systemic slab waveguide where the core thickness (**a**) $d = 2\,\mu$m and (**b**) $d = 10\,\mu$m	245
Fig. 7.11	The normalized index b as a function of the normalized frequency V for three mode orders m and two asymmetry coefficients $a = 0$ and $a = 5$. The inset provides a detailed view of the plots in the range $0 \leq V \leq 20$	248
Fig. 7.12	Schematic diagram of (**a**) a two-channel directional coupler of a length l consisting of two parallel waveguides and (**b**) its index profile assuming two step-index waveguides on the same substrate. The coupler is symmetric if $n_a = n_b = n_1$ and $d_a. = d = d$	254
Fig. 7.13	Periodic amplitude and power of the guided modes exchange between two codirectionally coupled modes for (**a**) power exchange a function of z, and (**b**) mode amplitude exchange a function of z. The blue solid line is for waveguid 1, and dotted red line is for waveguide 2	256
Fig. 8.1	Typical construction of an optical fiber	261
Fig. 8.2	Types of optical fibers and their refractive index profiles	262
Fig. 8.3	Light confinement through total internal reflection in step-index fibers. Rays for which $\phi < \phi_c$ are refracted out of the core	262
Fig. 8.4	Ray trajectories in a graded-index fiber	263
Fig. 8.5	(**a**) A meridional ray zig-zags down the fiber, passing through the origin. There is no angular rotation of the ray path as it propagates. (**b**) A skew ray travels in a spiral path down the fiber. The ray does not go through the origin	267
Fig. 8.6	Cylindrical coordinate system where any point in space can be given by $P(r, \theta, z)$ coordinates	267
Fig. 8.7	Plots for Bessel functions 1st and 2nd kind and modified Bessel 1st and 2nd kin	270
Fig. 8.8	Electric field as a function of the radial distance r for the first two modes (m=0 and m=1) for the parameters given in the example	272

Fig. 8.9	Plotting the transcendental equation Eq. (8.36) as a function of κ. There are two roots of the equation, which are marked with circles. Ignore the vertical lines, since they are artifacts of the calculations	275
Fig. 8.10	The allowed regions for the LP modes of order $m = 0, 1$ against normalized frequency (V) for a circular optical waveguide with a constant refractive index core (step index fiber). [D. Gloge. Appl. Opt., 10, p. 2552, 1971]	275
Fig. 8.11	The normalized propagation constant b as a function of normalized frequency V for a number of LP modes. [D. Gloge, Appl. Opt., 10, p. 2552, 1971]	276
Fig. 8.12	Spectral dependence of loss mechanisms and total attenuation in a fiber	282
Fig. 8.13	Digital bit pattern 1011 broadening of light pulses, due to dispersion, as they propagate along an optical fiber: (**a**) pulses at the fiber input, (**b**) pulses at distance L_1 alone the fiber, and (**c**) pulses at a distance $L_2 > L_1$ along the fiber	283
Fig. 8.14	Optical axial and meridian rays paths for the lowest-order and highest-order modes in step-index multimode fiber	284
Fig. 8.15	The refractive index n and group index n_g for pure silica as a function of wavelength	288
Fig. 8.16	Schematic of the material and waveguide dispersions for a fused silica fiber as a function of the wavelength. The chromatic dispersion is the sum of both dispersions (plot is not to scale)	290
Fig. 9.1	A three-dimensional cavity emitting electromagnetic radiation from a tiny hole	297
Fig. 9.2	Resonant frequencies of the cavity that exist between frequencies v and $v + dv$	298
Fig. 9.3	Blackbody radiation as a function of wavelengths for observed radiation (red and blue lines) and the Rayleigh-Jeans radiation law (black line). The plot for the classical method given by the Rayleigh-Jeans law prediction fails at short wavelengths to match the experimental results	299
Fig. 9.4	Kinetic energy of a photoelectron extracted from a material surface illuminated with a light beam of frequency v. The minimum frequency of the photon needed to extract a photoelectron is v_o, which is a function of the work function of the material	300
Fig. 9.5	Blackbody radiation energy density emitted as a function of wavelength and temperature	303
Fig. 9.6	The probability of finding a quantum particle between x_1 and x_2	317

Fig. 9.7	Electrons double slit experiment showing that electrons interfere like wave	318
Fig. 9.8	The boundary conditions for an infinite potential well	322
Fig. 9.9	Electron in an infinite potential well: (**a**) four lowest discrete energy levels, (**b**) corresponding wave functions, and (**c**) corresponding probability density functions	325
Fig. 9.10	The quantized energy levels for a hydrogen atom	329
Fig. 9.11	The boundary conditions for a particle in a finite potential well	330
Fig. 9.12	Graphical solution for the transcendental equation (**a**) for even-parity quantized bound states, and (**b**) for odd quantized bound states	333
Fig. 9.13	The wave functions for even- and odd-parity solutions in a finite potential well	335
Fig. 10.1	Energy level splitting: (**a**) two atoms far from each other have $n = 1$ energy levels, (**b**) the atoms brought closer so that their $n = 1$ energy level splits into two levels, and (**c**) three atoms are brought closer so their $n = 1$ energy level splits into three levels	341
Fig. 10.2	Schematic of energy level splitting for three energy levels $n = 1, 2,$ and 3.	342
Fig. 10.3	Potential and electron energy functions of a single, noninteracting, one-electron atom	343
Fig. 10.4	Periodic potential for a one-dimensional single crystal	343
Fig. 10.5	The one-dimensional periodic potential function of the Kronig-Penney model approximating the potential well-shown in Fig. 10.4	343
Fig. 10.6	Plot of the function $f(\alpha a)$. The shaded areas show the allowed values of (αa) corresponding to real values of ka	346
Fig. 10.7	The $E - k$ diagram generated from Fig. 10.6. The allowed energy bands and forbidden energy bandgaps are indicated	347
Fig. 10.8	The plot for $f(\alpha a)$ to determine the energy bandgap when $ka = \pi$ and the parameter $p = 10$ used in the example	348
Fig. 10.9	The E-k diagram for the extended region including the energy bands and bandgaps. The reduced (Brillouin) region is shaded	349
Fig. 10.10	(**a**) Conduction and valence bands, (**b**) the $E - k$ diagram of the conduction and valence bands of a semiconductor at $T = 0\,K°$, and (**c**) at $T > 0\,K°$	350
Fig. 10.11	(**a**) A two-dimensional array of allowed quantum states in k space (indicated by blue circles). (**b**) Three-dimensional *k-space* showing the volume of each quantum state by a blue box	355
Fig. 10.12	(**a**) The density of states in the conduction and in the valence bands as a function of energy and (**b**) the $E - k$ diagram for a bulk semiconductor material	357

Fig. 10.13	The allowed quantum states (denoted by blue circles) for a quantum well exist in a disk at $k_z = q\pi/L_z$	359
Fig. 10.14	The $E - k$ diagram and density of states for a quantum well structure	360
Fig. 10.15	A two-dimensional array of allowed quantum states in the k_T-$space$ showing the states in the annular representing the quantum states between k_T and $k_T + dk_T$	361
Fig. 10.16	$E - K$ diagram used to calculate N_c between two energy levels E_c and E_2	363
Fig. 10.17	The Fermi-Dirac distribution function at $0\,K$ and $300\,K$	365
Fig. 10.18	Carrier density as a function of energy for (**a**) intrinsic semiconductors, (**b**) n-type doped semiconductors, and (**c**) p-type doped semiconductors	367
Fig. 10.19	Fermi-Dirac function $f(E)$ for quasi-Fermi levels. In this plot, the following values were used: $E_c = 0.7$, $E_{fn}0.68$, $E_{fv}0.52$, $and\ E_v = 0.5$ eVs	373
Fig. 10.20	pn-junctions under several biasing conditions	375
Fig. 10.21	Fermi energy levels for (**a**) no current injection and (**b**) current injection	376
Fig. 11.1	Direct and indirect bandgap semiconductors: (**a**) GaAs direct bandgap ($E_g = 1.4\,\text{eV}$) and (**b**) Si indirect bandgap ($E_g = 1.1\,\text{eV}$)	381
Fig. 11.2	Sketch of the $E - k$ diagrams for GaAs and AlAs semiconductors showing their direct and indirect bandgaps	383
Fig. 11.3	Dependence of the formation of the ternary semiconductor $AlGaAs$ on the composition factor x	383
Fig. 11.4	Interband transition resulting in the emission or absorption of a photon	385
Fig. 11.5	The optical joint density of states ($\rho_{opt}(\nu)$) as a function of photon energy	387
Fig. 11.6	The rate of spontaneous emission is plotted as a function of the photon energy ($h\nu$)	393
Fig. 11.7	Spontaneous emission rate as a function of photon energy at various temperatures	394
Fig. 11.8	Schematic of a box of semiconductor material showing the input and output beams of light as it goes through absorption and stimulated emissions	395
Fig. 11.9	The input-output relationship for the beam intensity in cases of both gain and attenuation	397
Fig. 11.10	The gain and attenuation coefficients for a semiconductor medium	400
Fig. 11.11	The gain coefficient ($\gamma(\nu)$) for a range of photon energies $h\nu > E_{fc} - E_{fv}$ where the gain medium becomes an absorber following the attenuation coefficient shown as a blue dashed line	400

Fig. 11.12	Effect of temperature on the gain coefficient	401
Fig. 12.1	The unbiased *pn-junction* (**a**) charge carriers, (**b**) energy band structure, (**c**) built-in electric field, and (**d**) built-in potential	404
Fig. 12.2	A *pn-junction* under forward-bias condition	406
Fig. 12.3	A *pn-junction* under reverse-biased conditions	407
Fig. 12.4	The applied bias potential V reduces V_o and thereby allows electrons to diffuse and be injected, into the *p-side*. Recombination around the junction of the electrons on the *p-side* leads to spontaneous photon emission	409
Fig. 12.5	A heterojunction light-emitting diode: (**a**) device structure, (**b**) energy band diagram for an unbiased device, (**c**) energy band diagram for a forward biased device, and (**d**) electron-hole recombination and emission of photons in the narrow bandgap	410
Fig. 12.6	Current-optic power characteristic for a typical light-emitting diode	413
Fig. 12.7	Rate of spontaneous emission as a function of wavelength at various temperatures. Units are arbitrary where (**a**) assuming E_g is assumed to be independent of temperature and (**b**) applied values for (E_g) as a function of temperature, as given by Eq. (12.41). Amplitudes are normalized	418
Fig. 12.8	Schematic of the experimental results for the emitted optical power as a function of wavelength ay various operating temperatures, not to scale	419
Fig. 12.9	(**a**) Energy distribution of electrons in the CB and holes in the VB. (**b**) A simplified energy-momentum ($E - k$) diagram showing direct recombination transitions where momentum (k) is conserved. (**c**) Relative emitted intensity as a function of photon energy based on transitions depicted in (**b**)	420
Fig. 12.10	Illustrating the loss mechanisms: (**a**) a schematic of an *LED* showing emitted photons traversing the device, (**b**) losses due to device surface reflection, and (**c**) losses due to total internal reflection	422
Fig. 12.11	Schematic of a typical structure of a surface-emitting LED. In this case, the output is coupled directly into an optical fiber	425
Fig. 12.12	Double heterojunction edge-emitting light-emitting diode (ELED)	428
Fig. 12.13	Optical fiber pigtailed ELED	428
Fig. 12.14	The white light spectrum generated by (**a**) red, green, and blue LEDs and (**b**) blue LED and yellow phosphorus emulsions	429
Fig. 12.15	A schematic for an optical amplifier	431
Fig. 12.16	The $E - k$ diagram for a gain medium satisfying the conditions for net stimulated emission	431

Fig. 12.17	(**a**) The optical amplifier peak gain coefficient as a function of excess carrier density and (**b**) the gain versus photon energy curves for a variety of carrier injections...........	432
Fig. 12.18	The gain medium volume (active region).........................	433
Fig. 12.19	Semiconductor optical amplifier structure for a heterojunction pn-diode illustrating the optical confinement within the fundamental mode of the optical field ...	437
Fig. 12.20	Packaged semiconductor optical amplifier illustrating the electric contacts and input and output optical fibers, made by InPhenix Inc.	437
Fig. 12.21	Schematic of a cross section of a double heterostructure semiconductor optical amplifier	439
Fig. 12.22	The cavity gain is plotted as a function of frequency below and above the threshold...	442
Fig. 12.23	The plots illustrate how the output optical power of a laser diode is influenced by the net gain above the threshold condition combined with the response of the cavity modes. Panels (**a**) and (**d**) show the gain spectrum of the active region of the LD. Panel (**b**) depicts the Fabry-Perot longitudinal modes for a reflectivity $R = 0.9$, whereas panel (**e**) presents these modes for $R = 0.32$. The corresponding output spectra of the laser diode are represented by the red line in panels (**c**) and (**f**) for $R = 0.9$ and $R = 0.32$, respectively. The $x\text{-}axis$ scale is arbitrary, and the amplitudes of the curves are normalized	443
Fig. 12.24	The optical power output as a function of current injection in a semiconductor laser. Above the threshold, the presence of a high photon density causes stimulated emission to dominate	446
Fig. 12.25	Laser diode threshold current dependence on temperature where $T_3 > T_2 > T_1$	447
Fig. 12.26	Angular profile of the laser diode beam	447
Fig. 12.27	A single-mode selection in which the gain is less than the losses except at one resonant mode.	448
Fig. 12.28	External cavity laser utilizing an Fabry-Perot etalon..............	449
Fig. 12.29	Schematic of a distributed feedback (DFB) laser utilizing a Bragg grating along the active regions...........................	450
Fig. 12.30	Schematic of a distributed Bragg reflector (DBR) laser diode....	451
Fig. 12.31	Schematic of a double heterostructure vertical cavity surface-emitting laser (VCSEL) utilizing Bragg reflectors	454
Fig. 12.32	(**a**) Bragg reflector stacks and (**b**) the reflectances of the stacks as a function of wavelength plotted for various number of periods..	455
Fig. 12.33	Tunable laser diodes based on a distributed Bragg reflector	456

Fig. 12.34	Quantum well structure (**a**) $E - k$ diagram, (**b**) and (**c**) the density of states as a function of energy	457
Fig. 12.35	(**a**) Energy band diagram for a quantum well structure illustrating the available quantum states and (**b**) the effect of the width d_x on the energy of the available quantum states	457
Fig. 12.36	Gain coefficient for bulk and quantum well semiconductors as a function of energy	459
Fig. 13.1	Schematic of a photoconductor detector	462
Fig. 13.2	(**a**) A schematic diagram of a reverse biased pn-junction photodiode. (**b**) Energy band diagram for a reverse biased pn-junction, (**c**) net space charge across the diode in the depletion region, and (**d**) the electric field in the depletion region	466
Fig. 13.3	(**a**) Photodiode circuit diagram and (**b**) the I–V characteristic curves for a dark (blue line) and illuminated (red line) photodiode	467
Fig. 13.4	Absorption coefficient (α) versus wavelength for some of the semiconductors listed in Table 13.1	469
Fig. 13.5	Absorption in Ge for light beams with wavelength and absorption coefficients given un the example	470
Fig. 13.6	Typical responsivity vs. wavelength for Si, Ge, and $InGaAs$	472
Fig. 13.7	A schematic of a reverse biased photodiode	472
Fig. 13.8	(**a**) A photodetector illuminated with a single photon at $x = L$, the electron-hole pair generated drifts across the detector to reach the respected collecting electrodes, (**b**) a plot of the electron and hole currents, and (**c**) a plot of the total photocurrent generated by the electron and the hole drift currents	475
Fig. 13.9	(**a**) A photodetector illuminated with a very large number of photons across the device surface, the electron-hole pairs generated drift across the detector to reach the respected collecting electrodes, and (**b**) a plot of the electron and hole currents as a function of time	476
Fig. 13.10	(**a**) A typical photodiode receiver circuit and (**b**) photodiode equivalent circuit	478
Fig. 13.11	(**a**) p-i-n photodiode in reverse bias, (**b**) space charge density across the device, and (**c**) induced electric field	479
Fig. 13.12	(**a**) Schematic illustration of an avalanche photodiode (APD) structure and the identification of the absorption and multiplication regions. (**b**) Net space charge density across the photodiode. (**c**) Electric field across the diode	481
Fig. 13.13	The process of impact ionization	481
Fig. 13.14	Gain across the APD	482
Fig. 13.15	Digital signal detection and noise distribution	491

Fig. 13.16	BER plotted as a function of the Q-factor using Eq. (13.90)	493
Fig. 14.1	(a) Laser diode modulation drive circuit including the DC bias current I_1 and the modulating drive current i_{in} and (b) the input current and output optic power characteristics	499
Fig. 14.2	The frequency dependence of $P_{mo}(2\pi f)/P_{mo}(0)$	500
Fig. 14.3	Laser diode direct modulation of a digital signal	501
Fig. 14.4	(b) Laser diode emitted optical power for (a) an input step function drive current	501
Fig. 14.5	Dependence of the refractive index on the electric field: (a) Pockels medium and (b) Kerr medium	503
Fig. 14.6	Phase shift of a Pockels cell as a function of the voltage applied across the cell	504
Fig. 14.7	Configuration of (a) the longitudinal phase modulator and (b) the transverse phase modulator. The voltage is applied across the Pockels cell via transparent electrodes	505
Fig. 14.8	Integrated optical waveguide phase modulator	506
Fig. 14.9	(a) A Mach-Zehnder intensity modulator and (b) the transmittance of the modulator as a function of the applied voltage	508
Fig. 14.10	Integrated Mach-Zehnder intensity modulator	508
Fig. 14.11	A dual-drive push-pull optical amplitude modulator	509
Fig. 14.12	Franz-Keldysh effect	510
Fig. 14.13	Energy bands tilt caused by the application of an electric field in a quantum well structure	510
Fig. 14.14	Electro-absorption modulator integrated waveguide device	511
Fig. 14.15	Liquid crystal molecules tilt angle θ as a function of the applied voltage	514
Fig. 14.16	Liquid crystal intensity modulator: (a) no electric field applied and light beam passes through the "ON" state, and (b) electric field E applied across the device and the light beam does not pass through the "OFF" state	514
Fig. 14.17	(a) Schematic layout of the ring resonator-based modulator and (b) the cross section of the ring	516
Fig. 15.1	Sketch of the fiber attenuation and optical fiber communication system wavelength transmission bands	520
Fig. 15.2	Schematic block diagram of (a) an electrical communication system and (b) an optical fiber communication system	521
Fig. 15.3	Point-to-point optical fiber network	522
Fig. 15.4	Schematic of communication networks topologies	523
Fig. 15.5	Schematic of wavelength-division multiplexer and demultiplexer	524
Fig. 15.6	(a) Reflective diffraction grating based wavelength-division multiplexer and (b) demultiplexer	525

Fig. 15.7	Schematic of an arrayed-waveguide-grating multiplexer/demultiplexer [John Senior,...]	526
Fig. 15.8	An optical add/drop multiplexer (OADM) using an FBG reflective filter	527
Fig. 15.9	Schematic of an erbium-doped fiber (EDF) amplifier	530
Fig. 15.10	Schematic for an erbium-doped amplifier (EDFA) spectrum	531
Fig. 15.11	Stokes shift and the resulting Raman gain spectrum produced by a pump laser operating at 1453 nm. In this example, a signal at 1550 nm	532
Fig. 15.12	Typical configuration of a Raman optical amplifier with co- and counter-directional coupling	532
Fig. 15.13	Raman amplifier with four pumping laser sources in counter-directional pumping. WSC is a wavelength selective filter	533
Fig. 15.14	A block diagram of a point-to-point optical link, where OTx represents the optical transmitter and ORx represents the optical receiver	534
Fig. 15.15	Block diagrams for (**a**) an optical transmitter and (**b**) and optical receiver	534
Fig. 15.16	Optical fiber bus-network with N-taps	535
Fig. 15.17	Schematic of a passive optical network (PON) system architecture	536
Fig. 15.18	Block diagram for an example of a wavelength-division multiplexed network identifying key optical components. The spacing between the components may be in tens of kilometers	537
Fig. 15.19	Schematic for a DWDM multiplexing and channel bandwidth and channel spacing for the 25 GHz standard	539

What Is Light?

1.1 Introduction

The question of what light has fascinated people since ancient times. To understand the different responses of science throughout the evolution of physics, we must look back in history. Scientists have searched for logical answers to the mysteries of nature for generations. This involves explaining the main features of the physicist's work, along with the pure thinking of the researcher. The role of thought and ideas is essential in the pursuit of knowledge about the material world. We gather cues from the books of nature and use our brains to understand how humans discover the laws that govern them. In this introduction, we focus on the last 300 years of physics evolution.

When researchers studied the properties of light, they noted that some light phenomena could be explained via current research methodologies and logical conclusions. The fundamental issue of mobility was one of the first cues noted. Nature's movements, such as a stone hurling into the air, a ship sailing in the ocean, or a cart being pushed down the street, are quite complex. To move a body at rest, we have to push it, lift it, or pull it, i.e., exert a force on it. This insight led to the emergence of Newton's mechanical theory of nature, which allowed for the idea that everything in the universe is composed of particles that are moved by forces.

On the basis of this view, physicists have attempted to explain electric current, and this model has also been used to explain the phenomenon of light as a stream of particles.

1.2 Light as a Substance

To better understand the nature of light, let us examine some of the clues provided by its behavior on the basis of the mechanical nature of the world. Light moves through space, as evidenced by our ability to see the stars. Optical tests can also be conducted by observing how light behaves when it passes through various substances, such as air or glass. One of the fundamental observations we make about light is that it propagates in a straight line. When a light source illuminates an aperture in an opaque screen, a bright and dark zone of light falls on the observation screen. This demonstrates that light moves in a straight line and does not bend or curve around obstacles. This phenomenon is known as the rectilinear propagation of light, and it is a crucial factor in our understanding of the behavior of light.

Another important optical phenomenon to consider is the propagation of light through different media, known as refraction. When a light beam travels through a vacuum and encounters a glass plate, instead of continuing in a straight line, it bends.

Any valid theory of light must be able to explain these phenomena, including both the rectilinear propagation of light and its refraction as it passes through different media. These observations and experiments provide valuable clues to the nature of light and help us develop a deeper understanding of the behavior of this fascinating phenomenon.

1.3   Corpuscular Theory of Light

The earliest attempt to explain the properties of light was based on corpuscular theory, which assumes that light is made of particles. According to this theory, all illuminated objects emit light corpuscles that strike our eyes and create the sensation of light. This mechanical explanation of light can account for the reflection of light by mirrors, similar to bouncing an elastic ball from a wall.

However, the corpuscular theory has difficulty explaining the phenomenon of refraction. According to this theory, when light corpuscles fall on a glass surface, the particles of the medium exert a force on them. Curiously, these forces act only in the vicinity of the glass surface. Any force that acts on a moving particle will change its velocity. Therefore, if the net force acting on the light corpuscles is an attraction parallel to the glass's surface. The new motion of the corpuscles occurs between the initial path line and the line perpendicular to the glass surface.

It is assumed that the net force acting on the light corpuscles is an attraction parallel to the glass surface. The new motion in that situation will be positioned between the initial path line and the line perpendicular to the surface. The color paradox, also known as dispersion, needs to be addressed by this theory. Sunlight that has been refracted by a prism appears in a spectrum of every color in the rainbow. All other colors are positioned between the red and violet edges of this spectrum. Newton explained this phenomenon in the following manner: Every color was already present in white light. White light is a mixture of corpuscles of different colors, and the prism separates them in space. According to mechanical theory, refraction is due to forces acting on light particles and originating from glass particles. These forces differ for corpuscles of various colors, with violet having the strongest effects and red having the weakest effects. As a result, as the light exits the prism, each color will be refracted along a separate path. This justification does not make sense; hence, another theory and additional research are needed.

As a result, a new theory of light, known as the wave theory of light, was developed. This theory proposes that light travels as a wave, similar to sound waves, and could explain a wider range of optical phenomena, including the bending of light during refraction and the color paradox. This theory is still widely accepted today and has been developed further through additional research and experimentation.

1.4 Light as a Wave

To understand the wave theory of light, we first need to understand what a wave is. In a wave, there is a distinction between the motion of the wave itself and the motion of the particles of the medium through which the wave is traveling. For example, if a stone is thrown into the water, it creates a circular wave that spreads outward, whereas the water particles oscillate only up and down.

Another type of wave can be generated by a sphere that expands and contracts rhythmically, causing changes in density that propagate throughout the entire medium. The particles of the medium themselves only perform small vibrations, but the overall motion is a progressive wave.

Importantly, the motion of a wave is not a motion of matter but rather a propagation of energy through matter. This concept is central to the wave theory of light, which proposes that light travels as a wave, with oscillations of electric and magnetic fields propagating through a medium (such as air or a vacuum).

The wave theory of light can explain many optical phenomena, including the bending of light during refraction and the color paradox. It remains a fundamental concept in the study of light and its behavior.

In an experiment with a pulsating sphere, we can introduce two important physical concepts: the wave length and the wave velocity. The wavelength is the distance between two successive troughs or crests of a wave, and it is dependent on the frequency of the pulsation of the sphere. On the other hand, the wave's propagation velocity is dependent on the properties of the medium through which it travels.

In some types of waves, such as sound waves, the motion of the wave and that of the particles of the medium are in the same direction. These types of waves are called longitudinal waves. In a longitudinal wave, the particles of the medium oscillate parallel to the direction of wave propagation.

In contrast, in transverse waves, such as those that propagate on a string, the motion of the wave and the particles of the medium are perpendicular to each other. In a transverse wave, the particles of the medium oscillate up and down or side to side, perpendicular to the direction of wave propagation.

The distinction between longitudinal and transverse waves is important in understanding the behavior of waves and how they interact with the medium through which they propagate.

To better understand transverse waves, let us consider an experiment where a sphere is placed in a jelly-like medium, such as gelatin, and rotated around a specific axis with a rhythmic motion in one direction through a small angle and then back again. This oscillation generates a wave that propagates outward from the sphere. In this case, the motion of the particles is perpendicular to that of the wave, and these types of waves are called transverse waves.

The wave theory of light proposes that light travels as a transverse wave, with oscillations of electric and magnetic fields propagating through a medium or a vacuum. This theory can explain a wide range of optical phenomena, including refraction, reflection, and diffraction.

1.5 Wave Theory of Light

In 1678, Huygens proposed the wave theory of light, which holds that light is an energy transmission in the form of a wave rather than a physical entity. However, wave theory faces challenges in explaining certain physical phenomena, such as how light can propagate in a vacuum where sound waves cannot.

One solution proposed by wave theory involves the existence of ether, a hypothetical weightless substance that is believed to fill all of space and through which light waves can propagate. However, later experiments and observations revealed that ether does not exist, and the wave theory of light was modified to eliminate this concept.

(a) **Rectilinear Propagation of Light**: The wave theory of light can explain the rectilinear propagation of light, as light waves do not bend around corners in the same way that larger waves in water do. However, it can also account for the refraction of light as it passes through a medium with a different refractive index, such as a glass plate.
(b) **Refraction of Light**: Suppose that we imagine the case of two men holding a rigid pole and moving with the same velocity. It is obvious that the stick at this point will turn and no longer be moved parallel to the original position if for a brief fraction of a second, the motions of the two men are not the same. During the interval of time when the two men's speeds were different, the direction changed. Consider the case in which a planewave traveling through the ether strikes a glass plate. Since the speed of light depends on the medium through which it passes, it is different in glass than in air. During the short period in which the wave front enters the glass, other parts of the wave front have different velocities. The direction of the wave itself changes as a result of this variation in speed along the wave front during the period of "immersion" in the glass. This justification for the occurrence of refraction is shown considerably more clearly and simply than by the corpuscular hypothesis.
(c) **Color Dispersion**: The wave theory of light can also explain the phenomenon of color dispersion, as different wavelengths of light correspond to different colors.

1.5 Wave Theory of Light

When white light passes through a prism, each wavelength refracts at a slightly different angle, resulting in separation of the colors of the spectrum.

(d) **Diffraction**: The phenomenon of diffraction is best explained by the wave theory of light. When light passes through a small aperture or encounters an obstacle, it diffracts and produces a pattern of bright and dark fringes on a screen. The diffraction of light was first observed by Young in his famous double-slit experiment, where he passed a beam of light through two small slits and observed the resulting interference pattern on a screen.

The interference pattern arises from the constructive and destructive interference of waves from the two slits, as they interfere with each other. When the crest of a wave from one slit meets the crest of a wave from the other slit, they reinforce each other and produce a bright fringe. Conversely, when the crest of a wave from one slit meets the trough of a wave from the other slit, they cancel each other out and produce a dark fringe.

This diffraction pattern provides evidence for the wave nature of light, as it cannot be explained by the corpuscular theory of light. The wave theory of light also provides a more comprehensive explanation for a wide range of optical phenomena, including reflection, refraction, and polarization.

(e) **Are light Waves Longitudinal or Transverse?** Does light spread like sound does? Ether is an air-like or elastic (jelly-like) medium. To answer these questions, we will run the following experiment. Two thin plates of tourmaline crystals (polarizers) were used. Gradually rotate one of the crystals around the axis of the incoming ray of light. A strange thing will happen. The light becomes increasingly weaker until it vanishes completely. It reappears as the rotation continues. Can this phenomenon be explained if light waves are longitudinal? The ether particles move along the axis like the beam does in the case of longitudinal waves. Therefore, the notion that light waves are transverse rather than longitudinal is the only one that can account for this phenomenon. Furthermore, it is necessary to believe that ether has a "jelly-like" consistency.

In summary, while the corpuscular theory of light was an early attempt to explain the nature of light, it was ultimately superseded by the wave theory of light, which has proven to be a more accurate and comprehensive explanation for a wide range of optical phenomena.

One of the fundamental ideas in the wave theory of light is that the speed of light is constant in a vacuum, regardless of the motion of the observer or the source of light. This principle, known as the principle of relativity, was first proposed by Albert Einstein in his theory of special relativity and has been confirmed through numerous experiments.

The wave theory of light has developed further over time, with new discoveries and experimental evidence providing a deeper understanding of the nature of light and its behavior.

Overall, the wave theory of light has been refined and developed over time, providing a deeper understanding of the nature of light and its behavior. It remains a fundamental concept in the study of optics and has led to numerous technological advancements in fields such as telecommunications and medicine.

1.6 Quantum Theory of Light

The wave theory dominated scientific thought until the end of the 19th century. In 1887, Heinrich Hertz discovered the photoelectric effect, where ultraviolet light shone on a metal plate caused it to emit sparks. This phenomenon was initially puzzling. Although metals easily conduct electricity because their electrons are loosely bound, increasing the light's brightness produced more electrons without increasing their energy. Additionally, each metal required a different minimum light frequency to emit electrons.

This behavior can be compared to a wall being eroded by waves versus being struck by bullets. While both processes reduce the wall's mass, their effects can be clearly distinguished. Similarly, when light impinges on a metal surface, it extracts electrons and accelerates them. This process, called the photoelectric effect, involves the transfer of the light's energy into the kinetic energy required to free the electrons.

In a typical photoelectric effect experiment, a cathode and an anode are placed inside a vacuum tube, with light of varying wavelengths shone on the cathode while adjusting the voltage and measuring the current. The results showed that current would only flow if the light's frequency exceeded a threshold specific to the material. For instance, ultraviolet light on a copper cathode generated current, while red, orange, or green light did not. Increasing light intensity raised the current but did not affect the energy of the ejected electrons. This demonstrated that the photoelectric effect depended on light's frequency, not its intensity, contradicting predictions made by the wave theory of light.

The photoelectric effect led to the concept of photons—discrete energy packets that travel at the speed of light. Planck's introduction of energy quanta helped explain this effect and marked the birth of quantum theory. In quantum theory, when a photon strikes an atom with sufficient energy, it ejects an electron, with the energy of the ejected electron being directly related to the photon's energy. The energy of a photon is proportional to its frequency, expressed as $E = h\nu$, where h is Planck's constant, and ν is the photon's frequency.

Further support for quantum theory came from the Compton effect, where X-rays collided with electrons, similar to how billiard balls transfer energy in collisions. Compton's experiment demonstrated that photons could behave like particles, further validating the quantum theory of light.

One of the first major successes of quantum theory came when Max Planck explained blackbody radiation by proposing that electromagnetic energy is radiated in discrete quanta, proportional to its frequency. This led to the well-known relationship $E = h\nu$. Energy quantization also plays a critical role in atomic systems, as described by the Schrödinger equation. For example, electrons in atoms can only occupy discrete energy levels, and when they transition between these levels, they emit photons with energy determined by Planck's equation. This principle underpins the operation of lasers, where electrons are excited to higher energy levels and, when they return to lower levels, emit coherent light with a frequency determined by the energy difference.

This dual nature of light—its ability to behave as both a wave and a particle—is a fundamental concept in physics. The wave–particle duality of light is not unique; quantum theory also shows that particles, such as electrons, exhibit wave-like properties. According to the de Broglie equation, $\lambda = h/p$, where λ is the wavelength, and p is the momentum, even particles like electrons can display wave-like behavior, as verified experimentally.

The dual nature of light is complementary, not contradictory. High-energy waves, such as gamma rays, exhibit more particle-like behavior, while low-energy waves, like radio waves, act more like waves. This duality is practically important in engineering: light behaves as particles when generated or detected (e.g., in semiconductor lasers or detectors) but propagates as waves, as described by Maxwell's equations. This view is known as the semiclassical theory of light. When studying light's generation and detection, a quantum approach is used; for light propagation, classical field theory applies.

1.7 Summary

In summary, the quantum theory of light can explain certain phenomena that wave theory cannot explain (e.g., the photoelectric effect), whereas wave theory can explain some events that quantum theory cannot explain (e.g., diffraction). Furthermore, there are phenomena, such as the rectilinear propagation of light, that can be explained by both theories. Therefore, what is light? Is it a wave or a stream of photons?

In all these experiments, light manifests itself as an electromagnetic wave. This is not the case for two separate phenomena. A signal travels across vast regions of space without knowing which experiment we will perform. However, when we observe it one way, we find it arriving continuously, and when we observe it another way, it appears to come in compact bursts of energy. Nature responds to questions concerning interference in wave form, while it responds as a stream of photons when we attempt to increase the current in a photoelectric cell.

Thus, light has a dual nature as both a wave and a stream of photons. This duality is a fundamental aspect of the behavior of light and other quantum particles, reflecting the underlying principles of quantum mechanics .

Bibliography

1. Albert Einstein and Leopold Infeld (1942), The Evolution of Physics, Simon and Schuster.
2. Huygens, Christiaan. (1690), Traité de la Lumière (Treatise on Light). Leyden: Pieter van der Aa.

3. Young, Thomas (1807), A Course of Lectures on Natural Philosophy and the Mechanical Arts. London: Taylor and Walton.
4. Feynman, Richard P. (1985), QED: The Strange Theory of Light and Matter. Princeton University Press,.
5. Planck, Max. (1901), On the Law of Distribution of Energy in the Normal Spectrum. Annalen der Physik.
6. Compton, Arthur H. A. (1923), Quantum Theory of the Scattering of X-Rays by Light Elements. Physical Review.
7. Dirac, Paul A. M. (1930), The Principles of Quantum Mechanics. Oxford University Press.
8. Einstein, Albert (1905), On a Heuristic Viewpoint Concerning the Production and Transformation of Light. Annalen der Physik.

Geometrical Optics

2.1 Ray Tracing

As discussed in the previous chapter, light exhibits dual behavior, as a particle or a wave, depending on the experiment performed. In this chapter, we explore the wave nature of light in the analysis and design of optical systems. This chapter focuses on the use approximate methods to analyze optical systems by employing geometrical properties and laws governing the propagation of light rays. Through this approach, we trace light rays through space and optical components used in imaging systems.

In the following sections, we will represent the propagation of optical rays using matrix methods. This technique can be applied to analyze and design complex optical systems, such as optical resonators used in laser systems. By understanding the wave nature of light and utilizing these methods, we can gain valuable insights into the functioning and optimization of various optical devices and systems.

2.1.1 Light Rays

Light, as an electromagnetic wave, carries energy along its path in the direction of its propagation. The straight line paths followed by a narrow beam of light are known as light rays. Thus, light rays trace the direction of propagation of light energy in a medium. In optical systems, light rays are represented by straight arrows.

In a homogeneous medium, light rays travel in a straight line and change their direction when they transition between different media, following the laws of refraction, which will be discussed later. The propagation of light rays is governed by three laws:

1. Rectilinear propagation: Light propagates in a straight line.
2. Refraction: Light rays change direction as they cross different media.
3. Reflection: Light rays change direction at reflecting surfaces.

By understanding and applying these laws, we can analyze and design various optical systems and predict the behavior of light in different scenarios.

2.1.2 Refractive Index

Light travels at different speeds in different media. It reaches its maximum speed in a vacuum, approximately 300,000 kilometers per second or 3×10^8 m/s. This speed is represented by the lowercase "c" throughout this book. The speed of light in optical media, denoted by "v," is slower than its speed in a vacuum.

The ratio of the speed of light in a vacuum to that in a medium is called the refractive index or index of refraction. It is represented by the lowercase "n":

$$n = \frac{c}{v} = \frac{speed\ of\ light\ in\ free\ space}{speed\ of\ light\ in\ the\ medium} \tag{2.1}$$

The refractive index is a fundamental property of the medium and affects the behavior of light as it propagates through it. Higher refractive indices correspond to slower light speeds within the medium, which influences phenomena such as reflection, refraction, and dispersion.

2.1.3 Postulates of Ray Optics

These are the fundamental postulates from which the basic principles governing the propagation of light rays through optical media are derived:

(a) Light travels in the form of rays, which are emitted from light sources and can be observed when they reach a detector.
(b) An optical medium is characterized by a refractive index $n \geq 1$. Consequently, the time it takes for light to travel a distance d in the medium is given by the following formula:

$$t = \frac{d}{v} = \frac{nd}{c}, \tag{2.2}$$

where t is the time taken, d is the distance traveled, c is the speed of light in a vacuum, and n is the refractive index of the medium. The quantity nd is called the optical path length (OPL)
(c) In inhomogeneous media, the refractive index n(r) is a function of the position $r = (x, y, z)$. The optical path length along a given path between points A and B, as shown in Fig. 2.1, is given by

$$OPL = \int_A^B n(r)ds \tag{2.3}$$

2.1 Ray Tracing

Fig. 2.1 Path of a light ray between points A and B in inhomogeneous media according to Fermat's principle

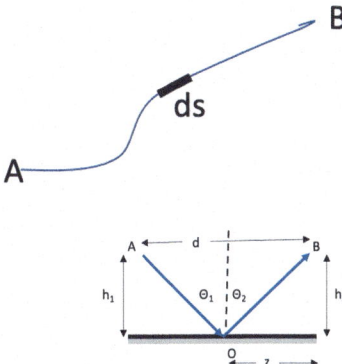

Fig. 2.2 Reflection of ray AO from a mirror

which is dependent on the refractive index, Where ds is the length of a differential element along the path from A to B.

(d) Fermat's Principle: Light rays travel along the path of least time. Fermat's principle states that the time a ray takes to travel from one point to another along its actual path is stationary, i.e., the light follows the path that takes the shortest time. The time it takes to traverse the geometric path ds in a medium with refractive index $n(r)$

$$\frac{ds}{c/n} = \frac{nds}{c} \qquad (2.4)$$

According to Fermat, the ray following the OPL is an extremum. In other words,

$$\delta(OPL) = \delta\left(\int_A^B n(r)\,ds\right) = 0 \qquad (2.5)$$

2.1.4 Reflection Law

One of the fundamental laws of geometrical optics is the law of reflection. It is known that the angle of incidence is equal to the angle of reflection. We use Fermat's principle to prove this law. Let the incident ray be noted by AO and reflected ray be noted by OB, as shown in Fig. 2.2. Fermat's principle states that the time taken by the ray to travel from point A to O and then to B is stationary.

The time requited for the light to travel from point A to point B is given by

$$t(z) = \frac{AO + OB}{v} \qquad (2.6)$$

where v is the speed of light in the medium. Using the geometry in Fig. 2.2, we obtain

$$t(z) = \frac{1}{v}\left[\left[h_1^2 + (d-z)^2\right]^{1/2} + \left[h_2^2 + z^2\right]^{1/2}\right] \tag{2.7}$$

Applying Fermat's principle to the time t(z)

$$\frac{d(t(z))}{dz} = 0 \tag{2.8}$$

Therefore, by taking the derivative of T(z) in terms of z, we obtain

$$\frac{dt(z)}{dz} = \frac{d-z}{\left[h_1^2 + (d-z)^2\right]^{1/2}} - \frac{z}{\left[h_2^2 + d^2\right]^{\frac{1}{2}}} = 0 \tag{2.9}$$

$$\frac{(d-z)}{\left[h_1^2 + (d-z)^2\right]^{\frac{1}{2}}} = \frac{z}{\left[h_2^2 + d^2\right]^{\frac{1}{2}}} \tag{2.10}$$

Therefore,

$$sin(\theta_1) = sin(\theta_2) \tag{2.11}$$

i.e., $\theta_1 = \theta_2$, which is the reflection law.

2.1.5 Law of Refraction

We use the geometry of the refraction of a ray of light, as shown in Fig. 2.3, using similar procedure to that in the previous section.

The time it takes for the light ray to travel from point A to point O in medium 1, and then from point O to point B in medium 2, is given by

$$t(z) = \frac{[h_1^2 + z^2]^{\frac{1}{2}}}{v_1} + \frac{[h_2^2 + (d-z)^2]^{\frac{1}{2}}}{v_2} \tag{2.12}$$

Fig. 2.3 The refraction of a light ray through different media

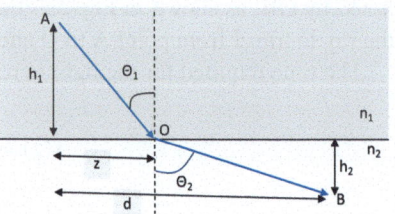

2.1 Ray Tracing

According to Fermat's principle,

$$\frac{dt}{dz} = \frac{z}{v_1 \left[h_1^2 + z^2\right]^{\frac{1}{2}}} - \frac{d-z}{v_2 \left[h_2^2 + (d-z)^2\right]^{\frac{1}{2}}} = 0 \tag{2.13}$$

$$\frac{z}{v_1 \left[h_1^2 + z^2\right]^{\frac{1}{2}}} = \frac{d-z}{v_2 \left[h_2^2 + (d-z)^2\right]^{\frac{1}{2}}} \tag{2.14}$$

$$\frac{sin\theta_1}{\frac{c}{n_1}} = \frac{sin\theta_2}{\frac{c}{n_2}}. \tag{2.15}$$

Therefore, $n_1 sin\theta_1 = n_2 sin\theta_2$, which is Snell's law. Since,

$$\frac{sin\theta_1}{sin\theta_2} = \frac{n_2}{n_1} \tag{2.16}$$

If $n_2 > n_1$ then the refracted angle, θ_2, will be larger than the incident angle, θ_1. As θ_1 increases θ_2 will eventually reach 90° before θ_1 does. This condition leads to Total Internal Reflection (TIR), which occurs when the incident angle equals the critical angle, θ_c. At the critical angle, all of the incident light is reflected back into the medium, and no refraction occurs. This phenomenon is known as Total Internal Reflection (TIR).

2.1.6 Sign Convention

Analyzing optical systems requires following the trajectory of light rays as they propagate through the optical elements. This requires a sign convention to determine the position with respect to the optical elements. The following are the sign conventions used, as shown in Fig. 2.4:

1. Light initially travels from left to right in a positive direction.
2. The distances to the right of and above (left of and below) a reference point are positive (negative).
3. The radius of curvature of a surface is treated as the distance of its center of curvature from its vertex. Thus, it is positive (negative) when the center of curvature lies to the right (left) of the vertex.

Fig. 2.4 Sign convention

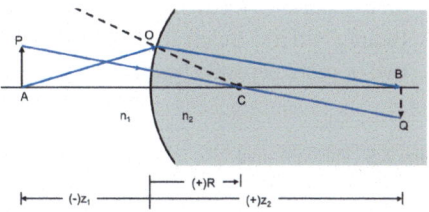

4. The acute angle of a ray from the optical axis or from the surface normal is positive (negative) if it is counterclockwise (clockwise).
The sign convention is demonstrated in Fig. 2.4.

2.2 Paraxial Rays

Optical rays that make very small angles with the optic axis are called paraxial rays as long as they fulfill the following relations for a ray making an angle θ with the optic axis as shown in Fig. 2.5:

$$sin(\theta) \cong \theta,$$

$$tan(\theta) \cong \theta \text{ and}$$

$$cos(\theta) \cong 1.$$

2.3 Optical Components

Optical systems consist of a combination of components that manipulate light rays as they traverse the system to achieve a specific function. These components usually either reflect or refract light. In this section, we present some of the essential components and their operations on the basis of ray optics:

1. Mirrors: Mirrors are reflective surfaces that change the direction of incoming light rays according to the law of reflection. Mirrors can be planar or curved (concave or convex), affecting the reflected rays differently. Concave mirrors converge light rays, whereas convex mirrors diverge them.
2. Lenses: Lenses are refractive components made of transparent materials, such as glass or plastic, which are shaped to focus, disperse, or otherwise manipulate incoming light rays. Lenses can be converging (convex) or diverging (concave). Converging lenses bring parallel light rays to a single focal point, whereas diverging lenses spread the rays apart.
3. Prisms: Prisms are transparent optical elements with flat, polished surfaces that refract light. They are often used to disperse light into its constituent colors (as in a rainbow) or to deviate the path of light. The refraction at each surface depends on the angle of incidence and the refractive index of the prism material.
4. Beam splitters: Beam splitters are optical devices that divide incoming light into two or more separate beams. They can be made from partially reflective materials

Fig. 2.5 Paraxial rays

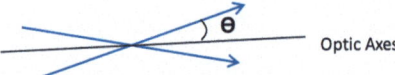

2.3 Optical Components

or use a combination of mirrors and prisms. Beam splitters are often used in interferometers, spectrometers, and other optical instruments.

5. Gratings: Gratings are optical elements with a periodic structure that diffracts light into various orders. They are commonly used for dispersing light, measuring wavelengths, and filtering specific wavelengths. Gratings can be either reflective or transmissive.

These components, along with other components such as filters, polarizers, and waveplates, form the basis of a wide range of optical systems, from microscopes and telescopes to cameras and optical communication systems.

In the following subsections, some of these components are analyzed in the context of geometrical optics, whereas others are addressed in subsequent chapters. Geometrical optics provide a useful framework for understanding the basic principles governing the behavior of light in these optical components and offers insights into their design and function. By studying these components and their interactions with light rays, we can gain a deeper understanding of how complex optical systems are constructed and optimized to perform specific tasks.

2.3.1 Plane Mirrors

A plane mirror reflects light rays off its surface such that their angle of reflection is equal to the angle of incidence with respect to the normal to its surface. When light rays originate from point P_1, as shown in Fig. 2.6, they are reflected as if they were originating from point P_2 at the back of the mirror. This is how plane mirrors create images of objects. The image appears to be the same distance behind the mirror as the object is in front of it, resulting in a virtual and upright image.

2.3.2 Paraboloidal and Spherical Mirrors

Paraboloidal mirrors have surfaces that are shaped like parabolas. In the case of the paraxial approximation, spherical mirrors can approximate paraboloidal mirrors. Figure 2.7 shows the focusing of a parallel light beam by a paraboloidal mirror or,

Fig. 2.6 Reflection of light from a plane mirror

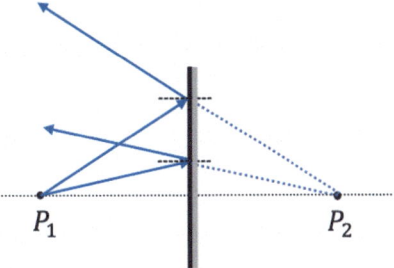

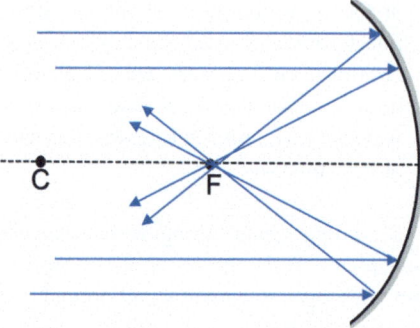

Fig. 2.7 Focusing of parallel light rays by a paraboloidal mirror, where C is the center of the mirror, and F is its focal point

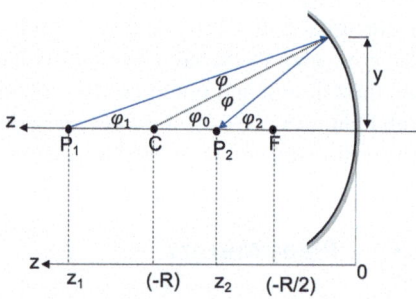

Fig. 2.8 Imaging by a spherical concave mirror

in the case of paraxial rays, the focusing of light by a spherical mirror. The focal length of the mirror is determined by the shape of the parabola or, in the case of a spherical mirror, the radius of curvature. In both cases, the focal length represents the distance from the mirror's surface (vertex) to the point where incoming parallel light rays converge after reflection.

For a spherical mirror, the focal length (f) is related to the radius of curvature (R) as follows:

$$f = (-)R/2.$$

2.3.2.1 Reflection by a Spherical Mirrors

A spherical mirror of radius R acts like a paraboloidal mirror with focal length $f = \frac{R}{2}$. All the paraxial rays originating from a point on the optic axis are reflected and focused in a point on the axis. Therefore, a ray originates from point P_1 at a distance z_1, as shown in Fig. 2.8, with angle φ_1 being reflected from the concave mirror at an angle $-\varphi_2$ to meet the optic axis at point P_2 at a distance z_2 from the mirror. Using the trigonometry of the figure

$$\varphi_o = \varphi_1 + \varphi \quad so \ \varphi_1 = \varphi_o - \varphi$$

2.3 Optical Components

$$\varphi_2 = \varphi_o + \varphi = 2\varphi_o - \varphi_1$$

$$\varphi_1 + \varphi_2 = 2\varphi_0 \tag{2.17}$$

For paraxial rays

$$\varphi_o \cong \frac{y}{-R}$$

$$\varphi_1 + \varphi_2 \cong \frac{2y}{-R}$$

Also

$$\varphi_1 \cong \frac{y}{z_1}$$

and

$$\varphi_2 \cong \frac{y}{z_2}$$

R is negative because the center of curvature of the mirror is to the left. Therefore, using the previous paraxial approximations

$$\frac{y}{z_1} + \frac{y}{z_2} = \frac{2y}{-R}$$

$$\frac{1}{z_1} + \frac{1}{z_2} = \frac{2}{-R}$$

Let

$$f = \frac{-R}{2}$$

Equations (2.20) and (2.28) can be rewritten in the following form:

$$\frac{1}{z_1} + \frac{1}{z_2} = \frac{1}{f} \tag{2.18}$$

This equation is known as the imaging equation for paraxial rays.

2.3.2.2 Image Formation by a Spherical Mirror

Paraboloidal mirrors have a wide range of applications, from flashlights to telescopes. Compared to lenses, their lightweight nature makes them the ultimate choice for large-diameter telescopes. Spherical and paraboloidal mirrors can be made from

Fig. 2.9 Image formation by a spherical mirror. In this case, the image is real, h_2 tall and inverted

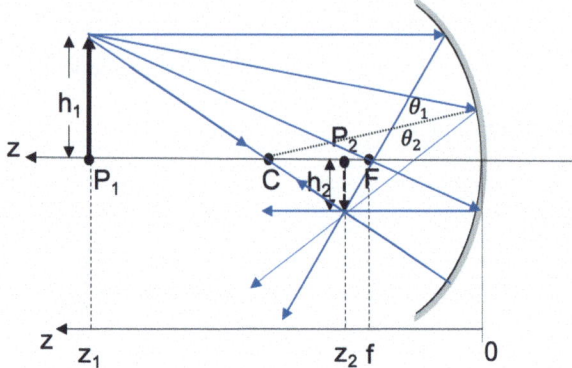

a large array of small square planar mirrors that can be shaped to form such mirrors. Figure 2.9 shows the image of an h_1 tall object at point P_1.

To determine the location and height of the image, four rays originating from the tip of the object are shown as they reflect from the mirror:

1. a ray parallel to the optic axis reflects through the focal point F,
2. a ray passes through the center of the circle, C, reflects back on itself,
3. a ray is reflected from a point on the mirror such that $\theta_1 = \theta_2$ and
4. a ray passing through the focal point, F, reflects parallel to the optic axis.

The intersection of the four rays forms the tip of the image. In this case, the image is real, h_2 tall, and inverted. By analyzing the geometric relationships between the object and its image, it is possible to calculate the image's location and properties, providing valuable insights into the behavior of paraboloidal and spherical mirrors in various optical systems.

Example A converging spherical mirror has a radius of curvature of 60 cm. An object is placed 60 cm away from the mirror.

Determine the position of the image formed by the mirror, the magnification, and the nature of the image (i.e., virtual or real, and inverted or erect).

Solution First, we use the mirror equation to find the position of the image:

$$\frac{1}{f} = \frac{1}{d_o} + \frac{1}{d_i}$$

where f is the focal length, d_o is the object distance, and d_i is the image distance.

(continued)

Calculate the focal length of the converging spherical mirror:

$$f = \frac{R}{2} = \frac{60\,\text{cm}}{2} = 30\,\text{cm}$$

The image position is calculated using the mirror equation:

$$\frac{1}{30\,\text{cm}} = \frac{1}{60\,\text{cm}} + \frac{1}{d_i}$$

Solving for d_i, we obtain $d_i = 60\,\text{cm}$.

The magnification is calculated using the magnification formula:

$$M = -\frac{d_i}{d_o} = -\frac{60\,\text{cm}}{60\,\text{cm}} = -1$$

Determine the nature of the image:

Since d_i is positive, the image is real, and since the magnification is negative, therefore, image is inverted.

In conclusion, the image is formed at a distance of 60 cm from the converging spherical mirror, with a magnification of -1, and it is real and inverted.

2.3.3 Lenses

Spherical lenses are formed by two spherical surfaces as shown in Fig. 2.10. Its properties are defined by the radii of the two surfaces R_1 and R_2, the refractive indices of the material, and its thickness Δ.

Fig. 2.10 A biconvex spherical lens formed by the intersection of two spherical surfaces

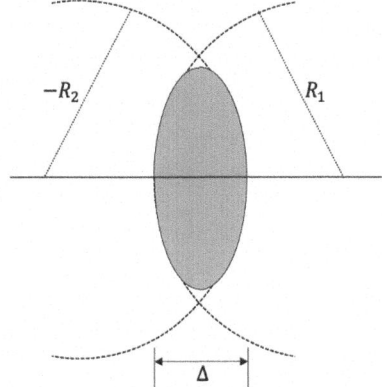

Fig. 2.11 Image formation by a thin lens

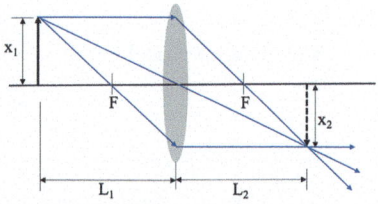

2.3.3.1 Image Formation by a Convex Lens

Figure 2.11 demonstrates imaging by a lens, showing an x_1 tall object at a distance L_1 in front of a lens with a focal length f. To determine the location and height of the image, three rays originating from the tip of the object are shown as they traverse from the object through the lens:

1. A ray parallel to the optic axis passes through the focal point, F, behind the lens.
2. A ray passes through the center of the lens without deviation.
3. A ray passes through the front focal point, F, and emerges parallel to the optic axis.

The intersection of the three rays forms the tip of the image. In this case, the image is real, x_2 tall, inverted, and located at L_2 behind the lens. The relationship between these distances is given by:

$$\frac{1}{L_1} + \frac{1}{L_2} = \frac{1}{f}. \qquad (2.19)$$

This equation is known as the lens imaging equation. It is a fundamental formula in geometrical optics that relates the object distance, image distance, and focal length of a lens, allowing for the calculation of image properties and the design of various optical systems that utilize lenses.

In the following section, we introduce a method to analyze optical systems using a set of ray matrices that can represent various optical components. The entire system is represented by a ray matrix known as the ABCD matrix. This method can be used to analyze complex systems, such as optical resonators.

Example An object is placed 30 cm away from a convex lens with a focal length of 10 cm.

Determine the position, nature, and magnification of the image formed.

(continued)

2.3 Optical Components

Solution To find the position of the image, we can use the thin lens formula:

$$\frac{1}{f} = \frac{1}{d_o} + \frac{1}{d_i}$$

where f is the focal length, d_o is the object distance, and d_i is the image distance.

Plug in the given values in the previous equation we obtain:

$$\frac{1}{10\,\text{cm}} = \frac{1}{30\,\text{cm}} + \frac{1}{d_i}$$

Solve for d_i :

$$\frac{1}{d_i} = \frac{1}{10\,\text{cm}} - \frac{1}{30\,\text{cm}} = \frac{2}{30\,\text{cm}} = \frac{1}{15\,\text{cm}}$$

$$d_i = 15\,\text{cm}$$

The image distance is positive, which means that the image is real and formed on the opposite side of the lens.

To find the magnification (m), we can use the following formula:

$$m = -\frac{d_i}{d_o} = -\frac{15\,\text{cm}}{30\,\text{cm}} = -0.5$$

The magnification is -0.5, which means that the image is 0.5 times the size of the object and inverted.

Therefore, the image is formed 15 cm away from the lens and is real, inverted, and 0.5 times the size of the object.

Example A converging lens with a focal length of 20 cm is used to focus light of two different wavelengths: 400 nm (blue) and 700 nm (red).

Determine the difference in focal lengths for these two wavelengths due to chromatic aberration. Assume the refractive index difference for blue and red wavelengths is 0.01.

(continued)

Solution The lens maker's formula relates the focal length (f), the refractive index of the lens material (n), and the radii of curvature of the two lens surfaces (R1 and R2):

$$\frac{1}{f} = (n-1)\left(\frac{1}{R1} - \frac{1}{R2}\right)$$

Since the lens is symmetrical, $R1 = -R2$. We can rewrite the lens maker's formula as:

$$\frac{1}{f} = \frac{2(n-1)}{R1}$$

We need to find the difference in focal lengths for the two wavelengths. First, we calculate the difference in refractive indices for the two wavelengths:

$$\Delta n = n_{blue} - n_{red}$$

We do not have the actual values for n_{blue} and n_{red}, but we can assume that the difference is small, around 0.01 (which is a rough estimate; the actual value depends on the material of the lens).

Now, we can find the difference in focal lengths via the lens maker's formula:

$$\Delta f = f_{blue} - f_{red}$$

Since Δn is small, we can use the following approximation:

$$\Delta f \approx -f^2 \frac{\Delta n}{2(n-1)R1}$$

We are given the focal length $f = 20$ cm. To proceed, we need the value of R1. If we assume that the lens is thin, we can approximate the radius of curvature R1 as the focal length divided by the difference in refractive indices:

$$R1 \approx \frac{f}{\Delta n} = \frac{20 \,\text{cm}}{0.01} = 2000 \,\text{cm}$$

Now, we can calculate the difference in focal lengths:

$$\Delta f \approx -\frac{(20 \,\text{cm})^2 \cdot 0.01}{2(1.5-1) \cdot 2000 \,\text{cm}} = -\frac{400}{1000} = -0.4 \,\text{cm}$$

(continued)

> The difference in focal lengths due to chromatic aberration is 0.4 cm.
> Therefore, the focal length of the lens for blue light is 0.4 cm shorter than that for red light, causing the blue light to focus closer to the lens than the red light. This chromatic aberration can lead to blurring and color fringing in the image formed by the lens.

2.4 Ray Matrices

Ray matrices, also known as transfer matrices, are efficient tools for analyzing paraxial light rays in optical systems. By applying matrix multiplication, the behavior of light rays as they pass through different components of an optical system can be easily determined, simplifying the analysis of even the most complicated setups.

The ABCD matrix method allows for the representation of different optical elements, such as lenses, mirrors, and free-space propagation, as individual matrices. The overall matrix for the entire optical system is found by multiplying the individual matrices in the order in which the light rays encounter each element.

Once the overall ABCD matrix is obtained, it can be used to calculate various properties of the optical system, such as the focal length, magnification, and image position. The ABCD matrix method is particularly useful for analyzing complex systems such optical resonators, where multiple reflections and transmissions occur within the system. By employing this method, the design and optimization of optical systems become more efficient and accessible.

Analyzing complex systems using basic component equations, such as the lens equation, can become highly challenging. As has been done in some linear systems analysis, matrices can be implemented in optical systems. These matrices are referred to as ray matrices. The ray matrix or ABCD matrix is used in ray tracing through optical systems. A complete system can be represented by a 2×2 matrix relating the output of the system to the input. Any ray of light at a specific plane can be represented by two quantities: its position with respect to the optic axis and its slope. In Fig. 2.12, the optical system is represented by a box.

A ray of light emanating from the tip of the arrow (the object) intersects with reference plane I (the input plane); after being traced through the optical system, this ray of light emerges on the right side of the system. The output ray passes

Fig. 2.12 Light rays propagating through an optical system

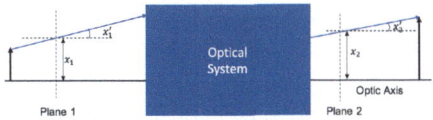

through reference plane II (the output plane) and then reaches the tip of the image of the arrow. The symbols x_1, x'_1, x_2, x'_2 are defined as:

x_1 and x_2 = distances of the input and output rays from the optical axis, respectively, and
x'_1 and x'_2 = slopes of the input and output rays, respectively.

The optical system relates the quantities x_1, x'_1, x_2, and x'_2 by a transformation law given by

$$x_2 = f(x_1, x'_1) \quad (2.20)$$

$$x'_2 = g(x_1, x'_1) \quad (2.21)$$

If the optical system forms a perfect image in the image plane, the transformation laws should be linear. Therefore, Eqs. (2.20) and (2.21) can take the following form:

$$x_2 = Ax_1 + Bx'_1 \quad (2.22)$$

$$x'_2 = Cx_1 + Dx'_1 \quad (2.23)$$

These equations can be written in the following matrix form:

$$\begin{bmatrix} x_2 \\ x'_2 \end{bmatrix} = \begin{bmatrix} A & B \\ C & D \end{bmatrix} \begin{bmatrix} x_1 \\ x'_1 \end{bmatrix} \quad (2.24)$$

2.4.1 The Significance of the Ray Matrix Coefficients

The coefficients of the ray matrix: A, B, C, and D, determine the relationship between the input and output of the optical system. We now determine the effect of each of these coefficients on the output as they are in turn made equal to zero on the output of the system.

1. $A = 0$. In this case, $x_2 = Bx'_1$, so x_2 is independent of the position of the input ray, and it is a function of the slope of the input ray, so that all rays departing from the input plane at the same angle, regardless of their position, arrive at the same position x_2 at the output plane. As shown in the figure, the output plane functions as the first focal plane.
2. $B = 0$, In this case $x_2 = Ax$, where x_2 depends only on the position of the input ray and is independent of its slope. Therefore, all rays departing from the input plane at the same height, regardless of their slope, arrive at the same position x_2 at the output plane. As shown in the figure, the output plane functions as an image plane, and the input plane is the object plane. Since $\frac{x_2}{x_1} = A$, then the coefficient

2.4 Ray Matrices

A is the linear magnification. On this basis, $B = 0$ is referred to as the imaging condition.

3. $C = 0$. In this case, $x'_2 = Dx'_1$, so x'_2 is independent of the position of the input ray, and it is a function of the slope of the input ray. Therefore, all rays departing the input plane regardless of their position arrive as parallel rays with slope x'_2 at the output plane. As shown in the figure, the output plane functions as a focal plane. Since $\frac{x'_2}{x'_1} = a$, then the coefficient D is the angular magnification.

4. $D = 0$. In this case, $x'_2 = Cx_1$, so x'_2 is independent of the slope of the input ray, and it is a function of its position. So all rays departing at a single point in the input plane arrive as parallel rays at the output plane with slope x'_2. Thus, the input plane is the first focal plane.

Figure 2.13 demonstrates graphically these four scenarios of setting the coefficients of the matrix to zero on at a time and the expected output of the system.

In the following sections, we derive a set of ray matrices for elementary optical components that enable us to analyze optical systems using this powerful technique.

2.4.2 The Translation Matrix

A ray of light travels in a straight line in a homogeneous medium. The ray of light at plane I in Fig. 2.14 is represented by its position x_1 and its slope x'_1 with respect to the optical axis, and the ray at the output plane II is represented by x_2 and x'_2, respectively. It is evident from the figure, as the ray propagates from plane I to pane II, the slope remains the same, i.e., $x'_2 = x'_1$ while the position increases by x'_1 so that $x_2 = x_1 + Lx'_1$. These relationships can be expressed in the following equations:

$$x_2 = x_1 + Lx'_1$$

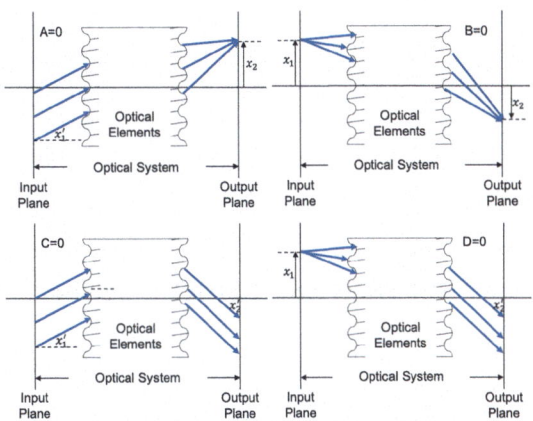

Fig. 2.13 The output of the system based on setting the ray matrix coefficients in turn to be zero

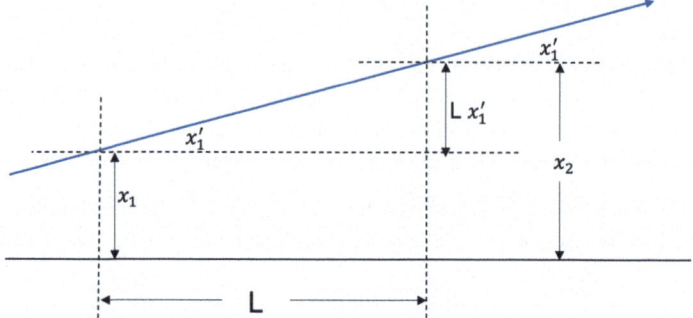

Fig. 2.14 A translation of a ray for a distance L in a homogeneous media

and

$$x_2' = x_1'.$$

Therefore, the ray translation matrix is given by

$$\begin{bmatrix} x_2 \\ x_2' \end{bmatrix} = \begin{bmatrix} 1 & L \\ 0 & 1 \end{bmatrix} \begin{bmatrix} x_1 \\ x_1' \end{bmatrix} \quad (2.25)$$

2.4.3 The Transition Matrix at a Planar Dielectric Interface

Now we consider the case at a spherical interface between two dielectric mediums. Let the two mediums have refractive indices n_1 and n_2, respectively. As shown in Fig. 2.15, as the light refracts from medium 1 to medium 2, the position (height) of the ray does not change, so $x_2 = x_1$, but the slope of the ray will change according to Snell's law $n_1 x_1' = n_2 x_2'$. Therefore, the ray matrix for such a transition is given by

$$\begin{bmatrix} x_2 \\ x_2' \end{bmatrix} = \begin{bmatrix} 1 & 0 \\ 0 & \frac{n_1}{n_2} \end{bmatrix} \begin{bmatrix} x_1 \\ x_1' \end{bmatrix} \quad (2.26)$$

2.4.4 The Transition Matrix for a Spherical Dielectric Interface

Now we consider the case at a spherical interface between two dielectric media. Let the two media have refractive indices n_1 and n_2, respectively, and let the interface have a radius of curvature R (according to our sign convention R is a negative quantity). Now we are interested in deriving the ABCD matrix of the transition

2.4 Ray Matrices

Fig. 2.15 A transition of a ray between two mediums

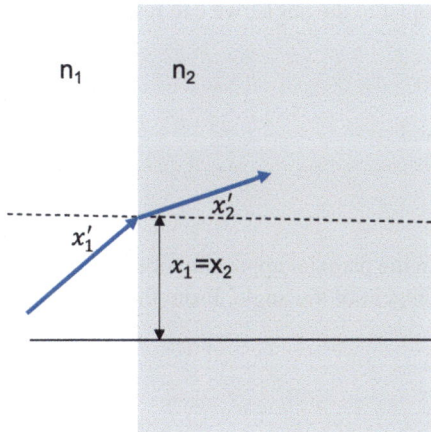

Fig. 2.16 A transition of a ray between two spherical dielectric media

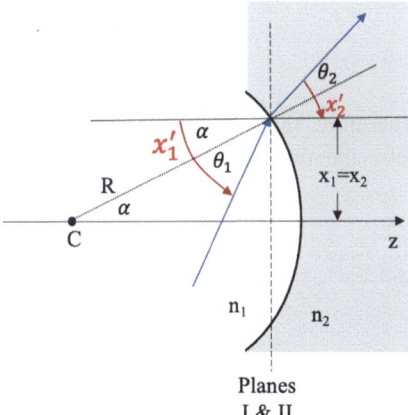

from medium 1 to medium 2 across the interface. Figure 2.16 clearly shows that both planes I and II coincide with each other because the height of the ray does not experience any change when propagating across the interface. However, the slope of the ray changes according to Snell's law. From the geometry of Fig. 2.16, we can write the following:

$$n_1 sin(\theta_1) = n_2 sin(\theta_2)$$

and

$$sin(\alpha_1) = \frac{x_1}{-R}$$

The radius of curvature of the spherical interface R is negative according to the sign convention since it falls to the left of the curvature. For the paraxial ray approximation, where the sine of the angle and the tangent of the angle are both equal to the angle, we can rewrite the equations as follows:

$$n_1 \theta_1 = n_2 \theta_2 \tag{2.27}$$

and

$$\alpha_1 = \frac{x_1}{-R} \tag{2.28}$$

In the paraxial approximation, we interchange the angle with the slope, which is the tangent of the angle. From the geometry of Fig. 2.16 and Eqs. (2.27) and (2.28)

$$\left(\frac{n_1}{n_2}\right)\theta_1 = \theta_2$$

and

$$\left(\frac{n_1}{n_2}\right)(x_1' - \alpha) = (x_2' - \alpha)$$

Therefore, the relationships between the position and the slope of the input and refractive rays are, respectively, given by

$$x_2 = x_1$$

and

$$x_2' = -\left(1 - \frac{n_1}{n_2}\right)\frac{x_1}{R} + \frac{n_1}{n_2} x_1'$$

Therefore, the transition matrix for a spherical dielectric interface is then

$$\begin{bmatrix} x_2 \\ x_2' \end{bmatrix} = \begin{bmatrix} 1 & 0 \\ \frac{(n_1 - n_2)}{n_2 R} & \frac{n_1}{n_2} \end{bmatrix} \begin{bmatrix} x_1 \\ x_1' \end{bmatrix} \tag{2.29}$$

2.5 The Ray Matrix For Reflection From a Spherical Mirror

Now we consider the reflection from a spherical mirror as shown in Fig. 2.17. At the surface of the mirror, the rays' height does not experience any change, but the slopes change according to the law of reflection. Using the geometry shown in the figure, we can derive the following formulas using the paraxial approximations:

$$x_2 = x_1$$

$$\theta_1 = x_1' + \alpha$$

2.5 The Ray Matrix For Reflection From a Spherical Mirror

Fig. 2.17 Reflection of rays from a spherical mirror

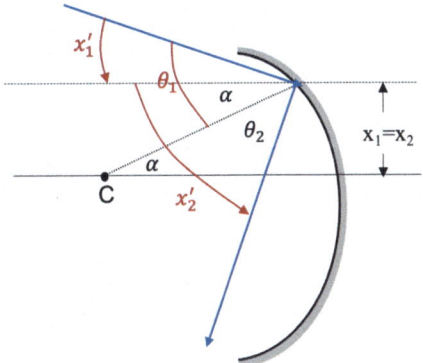

and

$$\theta_1 = \theta_2$$

therefore,

$$x'_2 - \alpha = x'_1 + \alpha$$

then

$$x'_2 = x'_1 + 2\alpha$$

but

$$\alpha = \frac{x_1}{-R}$$

from the aforementioned equation, we can obtain the following:

$$x_2 = \frac{-2}{R}x_1 + x'_1 \quad (2.30)$$

Therefore, the ray matrix for a spherical concave mirror is given by

$$\begin{bmatrix} x_2 \\ x'_2 \end{bmatrix} = \begin{bmatrix} 1 & 0 \\ \frac{-2}{R} & 0 \end{bmatrix} \begin{bmatrix} x_1 \\ x'_1 \end{bmatrix} \quad (2.31)$$

Since the focal length of a mirror is $f = \frac{R}{2}$, the ray matrix for the concave spherical mirror can be rewritten as

$$\begin{bmatrix} x_2 \\ x'_2 \end{bmatrix} = \begin{bmatrix} 1 & 0 \\ \frac{-1}{f} & 1 \end{bmatrix} \begin{bmatrix} x_1 \\ x'_1 \end{bmatrix} \quad (2.32)$$

The reflection from a plane mirror can be deduced from the ray matrix of the spherical mirror by letting $R = \infty$. Therefore, the ray matrix for a plane mirror is given by:

$$\begin{bmatrix} x_2 \\ x'_2 \end{bmatrix} = \begin{bmatrix} 1 & 0 \\ 0 & 1 \end{bmatrix} \begin{bmatrix} x_1 \\ x'_1 \end{bmatrix} \quad (2.33)$$

2.6 Ray Matrix for a Series of Interfaces

Now we consider that the ray matrix of a system consists of a series of interfaces. As shown in Fig. 2.18, there is a cascade of interfaces labeled by their ray matrices M_i, e.g., M_1 is the ray matrix between input plane I and output plane II. We can write the relationship between the input and output rays in this subsystem as follows:

$$\begin{bmatrix} x_2 \\ x'_2 \end{bmatrix} = M_1 \begin{bmatrix} x_1 \\ x'_1 \end{bmatrix}$$

Similarly,

$$\begin{bmatrix} x_3 \\ x'_3 \end{bmatrix} = M_2 \begin{bmatrix} x_2 \\ x'_2 \end{bmatrix}$$

Using the last two equations, we can write the relation between input plane I and output plane III as

$$\begin{bmatrix} x_3 \\ x'_3 \end{bmatrix} = M_2 M_1 \begin{bmatrix} x_1 \\ x'_1 \end{bmatrix}$$

Therefore, the ray matrix representing the output of the series of interfaces in terms of the input is given by

$$\begin{bmatrix} x_5 \\ x'_5 \end{bmatrix} = M_4 M_3 M_2 M_1 \begin{bmatrix} x_1 \\ x'_1 \end{bmatrix} \quad (2.34)$$

Fig. 2.18 Ray matrix for cascaded interfaces

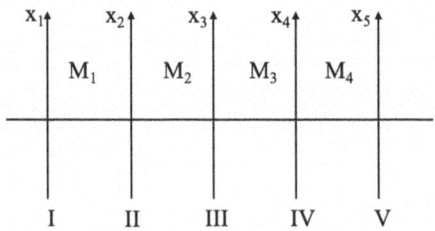

2.7 Ray Matrix for a Thick Lens

Lenses are crucial components in optical systems. A thick lens is composed of a material with a refractive index different from the surrounding media. It has two spherical surfaces separated by a distance, L, which represents its thickness. In the general case, we assume different media on either side of the lens, with the first surface having a positive radius of curvature, R_1, and the second surface having a negative radius of curvature, R_2, as illustrated in Fig. 2.19. A ray passing through the lens undergoes two refractions and a translation. We have already derived the ray matrices for these types of interfaces. By applying the method developed in the previous section for a sequence of interfaces, the ray matrix for a thick lens, as it propagates between the input plane I and output plane II, is expressed as:

$$\begin{bmatrix} x_2 \\ x_2' \end{bmatrix} = M_3 M_2 M_1 \begin{bmatrix} x_1 \\ x_1' \end{bmatrix} \tag{2.35}$$

where

$$M_1 = \begin{bmatrix} 1 & 0 \\ \frac{(n_1-n)}{nR_1} & \frac{n_1}{n} \end{bmatrix},$$

$$M_2 = \begin{bmatrix} 1 & L \\ 0 & 1 \end{bmatrix}$$

and

$$M_3 = \begin{bmatrix} 1 & 0 \\ \frac{(n-n_1)}{-nR_2} & \frac{n_1}{n} \end{bmatrix}$$

Thus, the ray matrix for a thick lens is the product of these three matrices.

$$\begin{bmatrix} x_2 \\ x_2' \end{bmatrix} = \begin{bmatrix} 1 & 0 \\ \frac{(n-n_1)}{-nR_2} & \frac{n_1}{n} \end{bmatrix} \begin{bmatrix} 1 & L \\ 0 & 1 \end{bmatrix} \begin{bmatrix} 1 & 0 \\ \frac{(n_1-n)}{nR_1} & \frac{n_1}{n} \end{bmatrix} \begin{bmatrix} x_1 \\ x_1' \end{bmatrix} \tag{2.36}$$

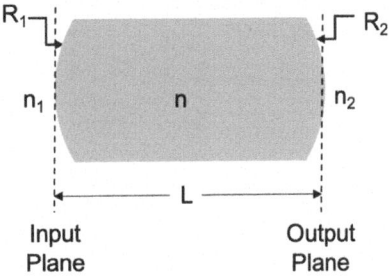

Fig. 2.19 A thick lens

For the case of a thin lens, $L = 0$ surrounded by air, $n_1 = 1$ and $n_2 = 1$, and $R_1 = R_2 = R$ the transfer matrix is reduced to

$$M = \begin{bmatrix} 1 & 0 \\ \frac{-1}{f} & 1 \end{bmatrix} \quad (2.37)$$

where the focal length of the lens

$$f = \frac{R}{2(n-1)}.$$

In order to find the position of the image, we apply the imaging condition, $B = 0$; hence,

$$L_2 - L_1 \left(\frac{L_2}{f} - 1 = 0 \right)$$

This can be reduced to

$$\frac{1}{L_1} + \frac{1}{L_2} = \frac{1}{f} \quad (2.38)$$

which is the lens equation derived in Sect. 2.1.

Example Let us consider the system in Fig. 2.20. The object of height x_1 is placed at a distance, L_1, in front of a thin lens with a focal length, f. Determine the position, height, and properties of the image. The position of the image is assumed to be a distance L_2 behind the lens.

Solution The ray matrix, also called the ABCD matrix of the system, is given by:

$$\begin{bmatrix} x_2 \\ x_2' \end{bmatrix} = \begin{bmatrix} A & B \\ C & D \end{bmatrix} \begin{bmatrix} x_1 \\ x_1' \end{bmatrix}$$

Fig. 2.20 A single lens imaging system

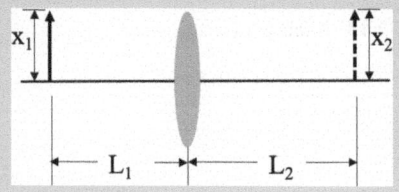

(continued)

2.7 Ray Matrix for a Thick Lens

$$\begin{bmatrix} x_2 \\ x_2' \end{bmatrix} = \begin{bmatrix} 1 & L_2 \\ 0 & 1 \end{bmatrix} \begin{bmatrix} 1 & 0 \\ \frac{-1}{f} & 1 \end{bmatrix} \begin{bmatrix} 1 & L_1 \\ 0 & 1 \end{bmatrix} \begin{bmatrix} x_1 \\ x_1' \end{bmatrix}$$

Multiplying the three matrix results in

$$\begin{bmatrix} x_2 \\ x_2' \end{bmatrix} = \begin{bmatrix} 1 - \frac{L_2}{f} & L_2 - L_1\left(\frac{L_2}{f} - 1\right) \\ \frac{-1}{f} & 1 - \frac{L_1}{f} \end{bmatrix} \begin{bmatrix} x_1 \\ x_1' \end{bmatrix}$$

To determine the position of the image, we apply the imaging condition, $B = 0$, hence,

$$L_2 - L_1\left(\frac{L_2}{f} - 1 = 0\right)$$

This can be reduced to

$$\frac{1}{L_1} + \frac{1}{L_2} = \frac{1}{f} \tag{2.39}$$

which is the lens equation shown in Sect. 2.1.

Example Let us examine the previous example further by using a lens with $f = 10$ cm, $L_1 = 15$ cm, and an object that is 5 cm tall.

Solution

$$\begin{bmatrix} x_2 \\ x_2' \end{bmatrix} = \begin{bmatrix} 1 & L_2 \\ 0 & 1 \end{bmatrix} \begin{bmatrix} 1 & 0 \\ \frac{-1}{10} & 1 \end{bmatrix} \begin{bmatrix} 1 & 15 \\ 0 & 1 \end{bmatrix} \begin{bmatrix} x_1 \\ x_1' \end{bmatrix}$$

$$= \begin{bmatrix} 1 - 0.1L_2 & 15 - 0.5L_2 \\ -0.1 & -0.5 \end{bmatrix} \begin{bmatrix} x_1 \\ x_1' \end{bmatrix}$$

Apply the imaging condition, $B = 15 - 0.5 L_2 = 0$; therefore, the position of the image is at $L_2 = 30$ cm. The magnification of the image is $A = 1 - 0.1 L_2 = -2$, i.e., the image is twice as tall and inverted. The size of the image is $-2 \times 5 = -10$ cm, so the image is 10 cm tall and is inverted.

(continued)

Let us move the object closer to the lens, $L_1 = 5$, and determine the position and orientation and size of the image. Let us move the object closer to the lens, $L_1 = 5$ cm and determine the position and orientation and size of the image.

$$\begin{bmatrix} x_2 \\ x_2' \end{bmatrix} = \begin{bmatrix} 1 & L_2 \\ 0 & 1 \end{bmatrix} \begin{bmatrix} 1 & 0 \\ \frac{-1}{10} & 1 \end{bmatrix} \begin{bmatrix} 1 & 5 \\ 0 & 1 \end{bmatrix} \begin{bmatrix} x_1 \\ x_1' \end{bmatrix}$$

$$= \begin{bmatrix} 1 - 0.1 L_2 & 5 + 0.5 L_2 \\ -0.1 & 0.5 \end{bmatrix} \begin{bmatrix} x_1 \\ x_1' \end{bmatrix}$$

Apply the imaging condition $B = 5 + 0.5\, L_2 = 0$; therefore, the position of the image is at $L_2 = -10$ cm, that is, 10 cm to the left of the lens; therefore, the image is virtual. The magnification of the image is $A = 1 - 0.1\, L_2 = +2$, i.e., the image is twice as tall and inverted. The height of the image is $-2 \times 5 = -10$ cm, so the image is 10 cm tall and is erect.

Example Consider a two lens imaging system made of a concave (negative) lens, $f_1 = -5$ cm, and a convex (positive), $f_2 = 10$ cm as shown in Fig. 2.21. If an object 20 cm tall is placed 20 cm in front of the negative lens, determine the size, position, and orientation of the image.

Solution

$$\begin{bmatrix} x_2 \\ X_2' \end{bmatrix} = \begin{bmatrix} 1 & z \\ 0 & 1 \end{bmatrix} \begin{bmatrix} 1 & 0 \\ -0.1 & 1 \end{bmatrix} \begin{bmatrix} 1 & 10 \\ 0 & 1 \end{bmatrix} \begin{bmatrix} 1 & 0 \\ .2 & 1 \end{bmatrix} \begin{bmatrix} 1 & 20 \\ 0 & 1 \end{bmatrix} \begin{bmatrix} x_1 \\ X_1' \end{bmatrix}$$

Fig. 2.21 A two-lens imaging system

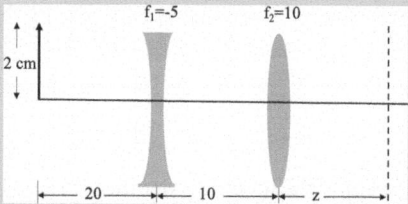

(continued)

2.7 Ray Matrix for a Thick Lens

$$\begin{bmatrix} x_2 \\ X'_2 \end{bmatrix} = \begin{bmatrix} -0.1Z + 3 & -2Z + 70 \\ -0.1 & -2 \end{bmatrix} \begin{bmatrix} x_1 \\ X'_1 \end{bmatrix}$$

By using the imaging condition $B = 70 - 2Z = 0$, thus, the image is at $Z = 35$ cm behind the positive lens. Using the magnification $A = 3 - 0.1 Z = -0.5$ cm, so the image is inverted, real, and 5 cm tall.

Example An optical system consists of a thin lens with a focal length of 20 cm, followed by a convex spherical mirror with a radius of curvature of 40 cm. The lens and the mirror are separated by a distance of 50 cm. An object is placed 30 cm away from the lens. Determine the final image position and magnification of the system.

Solution To solve this problem, we will use the ray matrix method. The light ray emanating from a point on the object will pass through the lens, then reflects from the mirror, and then passes through the lens to form the image at a distance L from the lens. Therefore, the ABCD matrix of the system is given by:

$$\begin{pmatrix} A & B \\ C & D \end{pmatrix} = \begin{pmatrix} 1 & L \\ 0 & 1 \end{pmatrix} \begin{pmatrix} 1 & 0 \\ -\frac{1}{20} & 1 \end{pmatrix} \begin{pmatrix} 1 & 50 \\ 0 & 1 \end{pmatrix} \begin{pmatrix} 1 & 0 \\ -\frac{1}{20} & 1 \end{pmatrix} \begin{pmatrix} 1 & 50 \\ 0 & 1 \end{pmatrix} \begin{pmatrix} 1 & 0 \\ -\frac{1}{20} & 1 \end{pmatrix}$$

$$\times \begin{pmatrix} 1 & 30 \\ 0 & 1 \end{pmatrix} = \begin{pmatrix} \frac{3L-20}{80} & \frac{7L-260}{8} \\ \frac{3}{80} & \frac{7}{8} \end{pmatrix}$$

The imaging condition is $B = 0$, i.e.,

$$B = \frac{7L - 260}{8} = 0,$$

therefore, the position of the image is at

$$L = \frac{260}{7} = 37.14 \, \text{cm}$$

The magnification is given by the A coefficient of the matrix

$$A = \frac{3L - 20}{80} = \frac{111.4 - 20}{80} = 1.14.$$

(continued)

Therefore, the image is real, erect, and 1.14 times the size of the object, and it is located 37.14 cm in front of the lens.

2.8 Periodic Optical Systems

In many applications, such as optical resonators, the optical system consists of cascaded identical subsystems, each characterized by the same *ABCD* matrix. These subsystems can be composed of mirrors, lenses, or any combination of mirrors, lenses, and other optical components. The block diagram in Fig. 2.22 illustrates a system that consists of m identical subsystems, each represented with identical ABCD matrix. The input ray is represented by x_0, x'_0 and an the output ray by x_m, x'_m. Since the matrices for each subsystem are identical, the overall system matrix is the product of m m identical matrices, which is equivalent to raising the subsystem matrix to the power m m. The system's output is then given by:

$$\begin{bmatrix} x_m \\ x'_m \end{bmatrix} = \begin{bmatrix} A & B \\ C & D \end{bmatrix}^m \begin{bmatrix} x_0 \\ x'_0 \end{bmatrix} \tag{2.40}$$

The general relationship between the input and output of any subsystem is given by

$$\begin{bmatrix} x_{m+1} \\ x'_{m+1} \end{bmatrix} = \begin{bmatrix} A & B \\ C & D \end{bmatrix} \begin{bmatrix} x_m \\ x'_m \end{bmatrix} \tag{2.41}$$

Now, we can rewrite these equations in the following form to derive an equation for x_{+2m}

$$x_{m+1} = A x_m + B x'_m$$

$$x'_{m+1} = C x_m + D x'_m$$

$$x'_m = \frac{x_{m+1} - A x_m}{B}$$

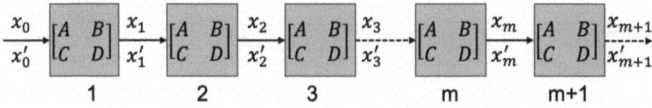

Fig. 2.22 Ray matrix representation of a periodic optical system

2.8 Periodic Optical Systems

$$x'_{m+1} = \frac{x_{m+2} - Ax_{m+1}}{B}$$

$$\frac{x_{m+2} - Ax_{m+1}}{B} = Cx_m + \frac{D}{B}(x_{m+1} - Ax_m)$$

$$x_{m+2} - Ax_{m+1} = BCx_m + Dx_{m+1} - ADy_m$$

$$x_{m+2} = Dx_{m+1} + Ax_{m+1} + (BC - AD)x_m \tag{2.42}$$

Let us change the variables so that

$$b = \frac{A+D}{2},$$

and

$$F^2 = (AD - BC) = det[M]$$

Therefore, we can write Eq. (2.42) as

$$x_{m+2} = 2bx_{m+1} - F^2 x_m \tag{2.43}$$

This is the recurrence relation for ray tracing, which is a type of difference equation. The solution to a difference equation follows a similar procedure to solving a differential equation, where we assume a solution that satisfies the given difference equation.

Let us assume a solution for the difference equation as

$$x_m = x_o h^m \tag{2.44}$$

Substituting the assumed solution in the difference Equation leads to

$$x_o h^{m+2} = 2b x_o h^{m+1} - F^2 x_o h^m,$$

divide both sides of the equation by $y_0 h^m$ results in the following:

$$h^2 - 2bh + F^2 = 0$$

which is a quadratic equation that has a solution given by

$$h = \begin{cases} b + \sqrt{b^2 - F^2} \\ b - \sqrt{b^2 - F^2} \end{cases} \tag{2.45}$$

Define a variable ϕ by the following:

$$\phi = cos^{-1}\left(\frac{b}{F}\right)$$

Therefore,

$$b = F cos\phi$$

and

$$\sqrt{F^2 - b^2} = \sqrt{F^2 - F^2 cos^2\varphi} = F sin(\phi)$$

hence,

$$h = F(cos\phi \pm j\,sin\phi)$$

$$h = F e^{\pm j\phi} \qquad (2.46)$$

Therefore, the recurrence relation solution is given by

$$x_m = x_o F^m e^{\pm j\,m\phi} \qquad (2.47)$$

Hence, the general solution of the ray recurrence relation is harmonic and can be expressed as

$$x_m = x_o F^m sin(m\phi + \phi_o) \qquad (2.48)$$

Therefore, x_m is the position of the ray with respect to the optic axis at the output of the mth subsystem as a function of the initial ray position x_0. Where x_0 and ϕ_0 are determined from the initial conditions x_0 and x'_0. The maximum position of the ray, x_{max}, is when $sin(m\phi + \phi_o) = 1$. In this case, $x_{max} = x_0 F^m$. But, $F^2 = AD - BC$, the determinant of the ABCD matrix, which is equal to unit if in the input and output media have the same refractive indices. Therefore, the ray recurrence Eq. (2.48) becomes

$$x_m = x_{max} sin(m\phi + \phi_o) \qquad (2.49)$$

For x_m to be harmonic $\phi = cos^{-1}(b)$ must be real value, so $|b| \leq 1$ or

$$\frac{|A + D|}{2} \leq 1 \qquad (2.50)$$

Equation (2.50) is known as the stability, or ray confinement, condition. As long as this condition is satisfied, the ray will stay within the periodic optical system throughout its trajectory. This is the criterion for determining the stability of optical resonators as will be discussed later in this book.

2.9 Summary

This chapter covers the fundamental concepts of geometrical optics and introduces methods for analyzing and designing optical systems using ray tracing and matrix methods. The chapter begins with an overview of light's behavior as rays, following laws of rectilinear propagation, reflection, and refraction. The concept of refractive index is explained as the ratio between the speed of light in a vacuum and in a medium, influencing how light interacts with different optical materials.

Fermat's Principle, which states that light takes the path of least time, is used to derive essential laws like reflection and Snell's Law for refraction. Various optical components such as mirrors, lenses, prisms, beam splitters, and gratings are then introduced, detailing how they manipulate light through reflection, refraction, or diffraction to achieve specific optical functions.

A key section of the chapter is the introduction of ray matrices, also known as ABCD matrices, which allow for the simplified analysis of complex optical systems. Each optical component is represented as a matrix, and by multiplying these matrices, the overall behavior of the system can be determined. This method is particularly useful for designing systems with multiple components, like optical resonators.

The chapter also explores periodic optical systems where identical subsystems, such as lenses or mirrors, are cascaded, and the system's output is analyzed through a recurrence relation. A stability criterion is introduced for ensuring that rays remain confined within such systems, crucial for applications like optical resonators.

In conclusion, this chapter provides a comprehensive framework for understanding and designing optical systems, using ray tracing and matrix methods to predict the behavior of light in various configurations.

2.10 Problems

1. A light ray in water ($n = 1.33$) strikes the boundary with air ($n = 1.00$). Calculate the critical angle for total internal reflection and determine whether a ray at 60° will undergo total internal reflection.
2. A block of flint glass, of refractive index 1.65 and of thickness of 5 cm, rests on the bottom of a water beaker. The surface of the water ($n = 1.33$) is 10 cm above the top surface of the glass block. There is a scratch on the bottom of the beaker underneath the glass block. What is the apparent depth of the scratch below the water surface?

3. For a small-angle approximation, verify the paraxial ray condition $sin(\theta) \approx \theta$ and calculate the maximum allowable angle for an error of less than 1%.
4. Derive an equation for the Numerical Aperture (NA) of a step-index optical fiber:
5. 1. Use Snell's law to show that a ray entering a planar plate of thickness d and refractive index n placed in air emerges parallel to the initial direction. The ray need not be paraxial. Derive an expression for the lateral displacement of the ray as a function of the angle of incidence θ.
6. A converging lens with a focal length of 20 cm is used to form an image of an object placed 60 cm from the lens. Use the thin lens equation to determine the position of the image and its magnification.
7. A lens with a focal length of -15 cm is used to form an image of an object placed 45 cm from the lens. Use the thin lens equation to determine the position of the image and its magnification.
8. A lens with a focal length of 15 cm for red light (700 nm) has a refractive index difference of 0.01 between red and blue light (400 nm). Calculate the focal length for blue light and the difference in focus due to chromatic aberration.
9. A biconvex lens has radii of curvature and $R_2 = -30$ cm, with a refractive index $n = 1.5$. Derive the ray matrix for this lens, assuming a thickness of 0.5 cm.
10. Derive the ray transfer matrix for a concave spherical mirror with a radius of curvature of 100 cm. Use this matrix to determine the position of the image of an object placed 150 cm in front of the mirror.
11. 2. Derive the ray matrix for an optical imaging system made of a thin convex lens of focal length f and a thin concave lens of a focal length $-f$ separated by distance f. Discuss the imaging properties of this composite lens system.
12. An optical system consists of a converging lens (focal length 20 cm) followed by a flat mirror 30 cm away. An object is placed 50 cm from the lens. Use the ABCD matrix method to find the position and nature of the final image after reflection.
13. An optical resonator consists of two mirrors separated by 50 cm. The ABCD matrix for each subsystem is the same, and the input ray is ($x_0 = 0.5$ cm, $x_0{\prime} = 0.02$ rad). Determine the output ray after three reflections using matrix multiplication.

Bibliography

1. Jenkins, F. A., & White, H. E. (2001). Fundamentals of Optics (4th ed.). McGraw-Hill.
2. Born, M., & Wolf, E. (1999). Principles of Optics (7th ed.). Cambridge University Press.
3. Hecht, E. (2017). Optics (5th ed.). Addison-Wesley.
4. Pedrotti, F. L., Pedrotti, L. S., & Pedrotti, L. M. (2017). Introduction to Optics (3rd ed.). Cambridge University Press.
5. Smith, W. J. (2007). Modern Optical Engineering (4th ed.). McGraw-Hill.

6. Meyer-Arendt, J. R. (1995). Introduction to Classical and Modern Optics (4th ed.). Prentice Hall.
7. Kingslake, R. (1994). Lens Design Fundamentals. Academic Press.
8. Longhurst, R. S. (1973). Geometrical and Physical Optics (3rd ed.). Longmans.
9. Sharma, K.K. (2006). Optics Principles and Applications, Elsecier.

Physical Optics   3

In the previous chapter, we analyzed several optical systems using geometrical laws that govern the propagation of light rays. Such analysis adheres to the corpuscular theory of light, which is based on treating the propagation of light energy as light rays. These rays are characterized by two parameters: their position and slope with respect to an optic axis. It was demonstrated that geometrical optics provides a good approximation in analyzing imaging systems. Furthermore, geometrical optics have shown excellent results in designing optical systems under the assumption that the wavelength of light is extremely small, which is the case in the visible spectrum.

As discussed in Chap. 1, Young's experiment proved without doubt that light travels as a wave when performing such an experiment. To explain this phenomenon, Huygens' principle provides a means to determine how light propagates from one point to another. Huygens asserted that every point on a wavefront, the surface of the wave's equal phase, is treated as a point source that emits light isotropically in all directions in the form of secondary spherical waves called wavelets. The sum of these wavelets, using their amplitude and phase, at a later time results in a new wavefront. In other words, the contour of the circles result in a new wavefront. This is essentially how optical waves propagate in space.

A fundamental question about the nature of light waves was raised since they can travel through vacuum, unlike sound waves, which require a medium. Optical waves are characterized by the space and time dependence of their optical field. This was explained when Maxwell developed his famous set of equations, which led to his introduction of the electromagnetic theory of light. Maxwell's equations describe the laws of electricity and magnetism. He used these equations to derive a wave equation, which he used to prove the existence of electromagnetic waves. From the wave equation, Maxwell calculated the speed of electromagnetic waves in free space. He found that the theoretical calculation of the speed of EM waves was very close to the experimentally measured value of the speed of light. On this basis, he concluded that light waves are electromagnetic waves.

Thus, according to Maxwell, light waves must be associated with time-changing electric and magnetic fields. A time-varying magnetic field generates a time and space-varying electric field, and a time-varying electric field generates a time and space-varying magnetic field. Therefore, the time-varying electric and magnetic fields result in the propagation of light waves even in vacuum.

In this chapter, we treat light as an optical wave. Optical waves are composed of coupled electric and magnetic fields governed by Maxwell's equations. Hence, they vary with time at an optical carrier frequency and propagate in a spatial direction determined by a wave vector. The behavior of an optical wave is strongly dependent on the optical properties of the medium. An optical field is a vectorial field characterized by five parameters: polarization, magnitude, phase, wave vector, and frequency. Polarization and wave vector are vectorial quantities, whereas, the magnitude, frequency, and phase are scalar quantities.

3.1 Electromagnetic Nature of Light

Wave theory describes light as an electromagnetic radiation in the frequency range of approximately 10^{15} cycles/second (Hz). In this theory, matter is treated as continuous, with the primary material response being the electric polarization. As in any other frequency range, the electromagnetic field and its interaction with matter are described by Maxwell's equations. These equations do not imply any natural time- or length scale; however, they do imply a relationship between the two scales in the form of a dispersion relation that relies on c_o, the vacuum speed of light. Electrodynamic phenomena can therefore be scaled arbitrarily if the ratio between time and length scale is conserved.

The electric and magnetic properties of matter depend strongly on the frequency. Magnetization, for example, is practically negligible at optical frequencies and is usually not considered in optics. This is because the response of the magnetic field is generally slower than the rapid oscillations of light waves in the optical frequency range. As a result, the influence of the magnetic field becomes minimal, and the primary focus in optics is on the electric field and its interaction with matter.

3.2 Definitions

In treating light as an electromagnetic wave, we use certain terms to describe these relationships. The following are definitions of these terms and their units of measurements:

Quantity	Unit
Electric field, $\bar{E}(r, t)$	V m^{-1}
Electric displacement, $\bar{D}(r, t)$	C m^{-2}
Magnetic field, $\bar{H}(r, t)$	A m^{-1}
Magnetic induction, $\bar{B}(r, t)$	T or Wb m^{-2}
Charge density, $\rho(r, t)$	C m^{-3}
Current density, J(r, t);	A m^{-2}
Polarization (electric polarization), $\bar{P}(r, t)$	C m^{-2}
Magnetization (magnetic polarization), $M(r, t)$	A m^{-1}

The response of a medium to an electromagnetic field generates the polarization and the magnetization. The relation between the electric displacement, $\bar{D}(r, t)$ and the polarization is given by:

$$\bar{D}(r, t) = \varepsilon_o \bar{E}(r, t) + \bar{P}(r, t), \tag{3.1}$$

where the free space permittivity $\epsilon_o \frac{1}{36\pi} \times 10^{-9}$ Fm^{-1} = 8.854×10^{-12} Fm^{-1}. Similarly, the relationship between the magnetic induction, $\bar{B}(r, t)$, and magnetization, $\bar{M}(r, t)$, is given by:

$$\bar{H}(r, t) = \frac{1}{\mu_o} \bar{B}(r, t) + \bar{M}(r, t), \tag{3.2}$$

where the free space permeability $\mu_o = 4\pi \times 10^{-7}$ Hm^{-1}. The polarization $P(r, t)$) and the magnetization $\bar{M}(r, t)$ are the macroscopically averaged densities of microscopic electric dipoles and magnetic dipoles that are induced by the presence of the electromagnetic field in the medium.

3.3 Maxwell's Equations

James Clerk Maxwell, in the late nineteenth century, took a number of disjointed experimental laws introduced by Faraday, Ampère, and Gauss, and due to his deep physical understanding, he was able to arrange them in a coherent new set of equations. These four equations became known as Maxwell's equations. Maxwell's ability to describe the induction current mathematically and his introduction of the displacement current allowed him to bring all these laws together in his set of equations. Furthermore, these equations led to the understanding of the relations between electric and magnetic fields and how their generation of one another causes the propagation of electromagnetic waves. The following are Maxwell's equations:

1. Gauss's Law for Electricity: $\nabla \bullet \overline{E} = \rho/\varepsilon_0$. This equation relates the electric field ($\overline{E}$) to the electric charge density (ρ) in a given region, where ε_0 is the vacuum permittivity.
2. Gauss's Law for Magnetism: $\nabla \bullet \overline{B} = 0$. This equation states that there are no magnetic monopoles, and the magnetic field ($\overline{B}$) lines are always continuous loops.
3. Faraday's Law of Electromagnetic Induction: $\nabla \times \overline{E} = -\partial \overline{B}/\partial t$. This equation describes how a changing magnetic field ($\overline{B}$) over time (t) generates an electric field ($\overline{E}$) that forms closed loops, which induces an electromotive force (EMF).
4. Ampère–Maxwell Law: $\nabla \times \overline{B} = \mu_0(\overline{J} + \varepsilon_0 \partial \overline{E}/\partial t)$. This equation relates the magnetic field ($\overline{B}$) to the electric current density ($\overline{J}$) and the rate of change of the electric field ($\overline{E}$) over time (t), where μ_0 is the vacuum permeability. The term $\varepsilon_0 \partial \overline{E}/\partial t$ represents the displacement current, introduced by Maxwell.

Therefore, the following are Maxwell's equations, the firs four equations are for free space, followed by these equations that consider polarization and magnetization:

$$\nabla \times \bar{E} = -\frac{\partial \bar{B}}{\partial t} \tag{3.3}$$

$$\nabla \times \bar{H} = \frac{\partial \bar{D}}{\partial t} + \bar{J} \tag{3.4}$$

$$\nabla \cdot \bar{D} = \rho \tag{3.5}$$

$$\nabla \cdot \bar{B} = 0 \tag{3.6}$$

$$\nabla \times \bar{E} = -\mu_o \frac{\partial \bar{H}}{\partial t} - \mu_o \frac{\partial \bar{M}}{\partial t} \tag{3.7}$$

$$\nabla \times \bar{H} = \epsilon_o \frac{\partial \bar{E}}{\partial t} + \frac{\partial \bar{P}}{\partial t} + \bar{J} \tag{3.8}$$

$$\nabla \cdot \left(\epsilon_o \bar{E}\right) = -\nabla \cdot \bar{P} + \rho \tag{3.9}$$

$$\nabla \cdot \mu_o \bar{H} = -\nabla \cdot \mu_o \bar{M} \tag{3.10}$$

$$\bar{D} = \epsilon_o \bar{E} + \bar{P} \tag{3.11}$$

$$\bar{P} = \varepsilon_o \chi \bar{E} \tag{3.12}$$

$$\varepsilon = 1 + \chi \tag{3.13}$$

$$\bar{B} = \mu_o(\bar{H} + \bar{M}) \tag{3.14}$$

where χ is the material susceptibility of the material. This chapter does not address the equations, which include polarization and magnetization, since we are mostly dealing with optical fields in free space.

3.3.1 Boundary Conditions

Boundary conditions are essential when analyzing the behavior of electromagnetic fields as they propagate from one medium to another. These conditions ensure the continuity of the electric ($\bar{E}$) and magnetic ($\bar{H}$) fields at the interface between two dielectric media. According to these boundary conditions:

1. The tangential components of the electric field ($\bar{E}$) are continuous across the boundary: $E_{1t} = E_{2t}$.
2. The tangential components of the magnetic field ($\bar{H}$) are continuous across the boundary: $H_{1t} = H_{2t}$.
3. The normal components of the magnetic flux density ($\bar{B}$) are continuous across the boundary: $B_{1n} = B_{2n}$.
4. The normal components of the electric displacement field ($\bar{D}$) are continuous across the boundary: $D_{1n} = D_{2n}$.

These boundary conditions help analyze the behavior of electromagnetic waves at the interface between different media, enabling the understanding of phenomena such as reflection, refraction, and transmission.

3.4 Wave Equation

In the remainder of this chapter, we consider Maxwell's equations in free space. In this case, there will be no current or electric charges, i.e. $\bar{J} = 0$ and $\rho = 0$. Thus, Maxwell's equations will take the following form:

$$\nabla \times \bar{E} = -\mu_o \frac{\partial \bar{H}}{\partial t} \tag{3.15}$$

$$\nabla \times \bar{E}\bar{H} = \epsilon_o \frac{\partial \bar{E}}{\partial t} \tag{3.16}$$

$$\nabla \cdot \bar{E} = 0 \tag{3.17}$$

$$\nabla \cdot \bar{H} = 0. \tag{3.18}$$

We can rearrange the first two equations to eliminate the magnetic field from the Maxwell's equations by taking the curl of Eq. (3.15) results in the following:

$$\nabla \times \nabla \times \bar{E}\bar{E} = -\mu_o \frac{\partial}{\partial t} \left(\nabla \times \bar{H} \right)$$

However,

$$\nabla \times \bar{E}\bar{H} = \epsilon_o \frac{\partial \bar{E}}{\partial t},$$

Substituting this equation in the previous one leads to the following:

$$\nabla \times \nabla \times \bar{E} = -\mu_o \epsilon_o \frac{\partial}{\partial t}\left(\frac{\partial \bar{E}}{\partial t}\right).$$

Applying the vector identity,

$$\nabla \times \nabla \times \bar{E} = \nabla\left(\nabla \cdot \bar{E}\right) - \nabla^2 \bar{E}$$

Since $\nabla \cdot \bar{E} = 0$, the last equation is reduced to:

$$\nabla^2 \bar{E} - \mu_o \epsilon_o \frac{\partial}{\partial t}\left(\frac{\partial \bar{E}}{\partial t}\right) = 0$$

$$\nabla^2 \bar{E} - \mu_o \epsilon_o \frac{\partial^2 \bar{E}}{\partial t^2} = 0 \tag{3.19}$$

However,

$$\frac{1}{\mu_o \epsilon_o} = \frac{1}{4\pi \times 10^{-7} \times \frac{1}{36\pi} \times 10^{-9}} = 9 \times 10^{16} = c^2,$$

where c is the speed of light in free space. Therefore,

$$\nabla^2 \bar{E}(r,t) - \frac{1}{c^2}\frac{\partial^2 \bar{E}(r,t)}{\partial t^2} = 0 \tag{3.20}$$

This equation will be the same for the other five components of the $\bar{E}$ and $\bar{H}$. Therefore, we can represent each of these components by the scalar wave function $u(\bar{r}, t)$. In this case, we can express the wave equation in the following format:

$$\nabla^2 u - \frac{1}{c^2}\frac{\partial^2 u}{\partial t^2} = 0, \tag{3.21}$$

where the Laplacian, ∇^2, is given by:

$$\nabla^2 = \frac{\partial^2}{\partial x^2} + \frac{\partial^2}{\partial y^2} + \frac{\partial^2}{\partial z^2}$$

Since the wave equation is linear, then the principle of superposition applies, i.e., the linear sum of solutions is also a solution.

3.4.1 Intensity, Power, and Energy

The unit of energy is the Joule, and the unit of power is Joules per second (Watts). The unit of intensity (I) is Joules per second per square centimeter (Watts/cm^2). Therefore, the optical power (in Watts) flowing through an area A A is given by:

$$P(t) = \int_A I(r,t)\, dA,$$

This equation shows the relationship between the power and intensity.

3.5 Monochromatic Waves

Monochromatic waves are electromagnetic (EM) waves with sinusoidal dependence and are characterized by angular frequency ω. These waves can be expressed in the following form:

$$u(r,t) = a(r) \cos[\omega t + \phi(r)], \tag{3.22}$$

where, $u(r)$ is the amplitude of the wave and $\phi(r)$ is the initial phase. For $u(r,t)$ to represent the scalar electromagnetic wave, it should satisfy the wave equation. Substituting $u(r,t)$ in the wave equation, and assuming the initial phase to be zero, we obtain

$$\nabla^2 u(r,t) - \frac{1}{c^2}\frac{\partial^2}{\partial t^2} u(r)\cos(\omega t) = 0$$

$$\nabla^2 u(r) + \frac{\omega^2}{c^2} u(r) = 0. \tag{3.23}$$

Thus, $u(r,t)$ is a solution of the partial differential equation. However, it is common to use the exponential form to represent the fields, as long as we keep in mind that the fields are real quantities. This representation simplifies mathematical procedures. The field is expressed as:

$$u(r,t) = Re\left[a(r)e^{j(\omega t + \phi)}\right]$$

where $Re[\ldots]$ denotes the real part, and $a(r)$ is the complex amplitude of the EM wave. A monochromatic wave is represented by a wave function with harmonic time dependence

$$u(r,t) = a(r)\cos[2\pi \nu t + \phi(r)]$$

where ν is the frequency(cycles or Hz), $\omega = 2\pi\nu$ is the angular frequency $\left(\frac{radians}{s} \text{ or } s^{-1}\right)$, and $T = 1/\nu = \frac{2\pi}{\omega}$ is the period (s).

3.6 Complex Representation of the Wave Function

It is convenient to represent the real wave function in terms of a complex function

$$U(r, t) = U(r, t) = U(r)e^{j\omega t} = a(r)e^{j\varphi(r)}e^{j2\pi\nu t} \qquad (3.24)$$

so that

$$u(r, t) = Re\{U(r, t)\} = 1/2[U(r, t) + U^*(r, t)] \qquad (3.25)$$

The '*' denotes the complex conjugate of the function U. The complex function must satisfy the same wave equation:

$$\nabla^2 U - \frac{1}{c^2}\frac{\partial^2 U}{\partial t^2} = 0,$$

$$\nabla^2[U(r, t)] - \frac{1}{c^2}\frac{\partial^2}{\partial t^2}\left[U(r)e^{j\omega t}\right] \qquad (3.26)$$

$$e^{j\omega t}\nabla^2[U(r)] - \frac{1}{c^2}(j\omega)^2 e^{j\omega t}U(r) = 0,$$

therefore, by dropping the term $e^{j\omega t}$ the wave equation will be reduced to the following depending only on the vector r:

$$\nabla^2 U(r) + \frac{\omega^2}{c^2}U(r) = 0, \qquad (3.27)$$

But

$$\frac{\omega}{c} = \frac{2\pi\nu}{c} = \frac{2\pi}{\lambda}.$$

The quantity $\frac{2\pi}{\lambda}$ is known as the wavenumber, denoted by k where $k = \frac{\omega}{c} = \frac{2\pi}{\lambda}$. Accordingly, the wave equation can be rewritten in the general form, commonly referred to as the Helmholtz wave equation:

$$\nabla^2 U + k^2 U = 0. \qquad (3.28)$$

3.7 Planewaves

Planewaves are a solution of the Helmholtz wave equation and are the simplest form of three-dimensional waves. They exist at a given time when all the surfaces upon which a field has a constant value, called wavefronts, are parallel planes. Each wavefront is perpendicular to the direction of propagation, which is identified by

3.7 Planewaves

Fig. 3.1 Propagation the wavefronts of planewave

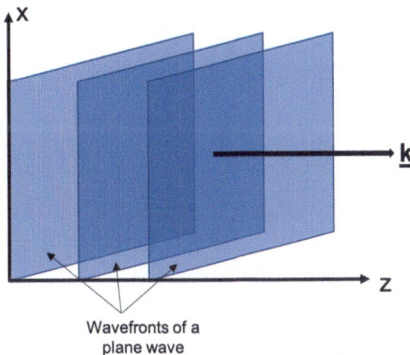

Wavefronts of a plane wave

Fig. 3.2 Propagation of planewave wavefronts

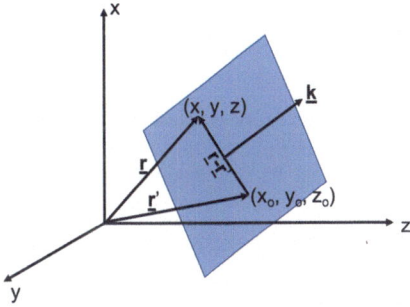

the wave vector $\bar{k}$. As the planewave propagates, it forms a series of parallel planes, as shown in Fig. 3.1. These waves are useful for understanding the behavior of electromagnetic waves in free space or uniform media.

In order to prove that the wavefront forms a plane perpendicular to the direction of propagation, consider a planewave portaging in the direction $bar k$, a plane perpendicular to $\bar{k}$ that passes through point (x_o, y_o, z_o). Let vectors $\bar{r}$ and $\bar{r}'$ be given by

$$\bar{r} = x\bar{a}_x + y\bar{a}_y + z\bar{a}_z, and$$

$$\bar{r}' = x'\bar{a}_x + y'\bar{a}_y + z'\bar{a}_z, an$$

where (x, y, z) is a point anywhere in space, and $\bar{a}_x, \bar{a}, \bar{a}_z$ are unit-vectors along x, y, and z axis, as shown in Fig. 3.2.

Let the vector $\bar{r} - \bar{r}'$ sweeps the plane of the wavefront. For the wavefront to be normal to the direction of propagation $(\bar{r} - \bar{r}') \cdot \bar{k}$ must be equal to zero. Let

$$\bar{k} = k_x\bar{a}_x + k_y\bar{a}_y + k_z\bar{a}_z$$

and substitute back into $(\bar{r} - \bar{r}') \cdot \bar{k} = 0$ we get

$$[(x - x_o)\bar{a}_x + (y - y_o)\bar{a}_y - (z - z_o)\bar{a}_z] \cdot (k_x\bar{a}_x + k_y\bar{a}_y + k_z\bar{a}_z) = 0.$$

Therefore,

$$k_x x + k_y y + k_z z = k_x x_o + k_y y_o + k_z Z_0 = constant = a$$

where a is any arbitrary value.

This leads to $\bar{k} \cdot \bar{r} = constant = a$ in the plane containing the vector $\bar{r} - \bar{r}'$. On this basis of result, the plane perpendicular to the vector $\bar{k}$ is the plane where the dot product $\bar{k} \cdot \bar{r}$ is constant, i.e., the plane of the constant phase. We can construct a set of planes over which $U(\bar{r})$ varies in space are sinusoidal as

$$u(\bar{r}) = A sin(\bar{k} \cdot \bar{r}),$$

which can be represented in a complex form by

$$U(\bar{r}) = Ae^{-j\bar{k}\cdot\bar{r}}, \qquad (3.29)$$

where $\bar{k}$ and $\bar{r}$ are vectors. Since $U(\bar{r})$ is constant over every plane defined by $\bar{k}\cdot\bar{r} = constant$, so we are dealing with harmonic functions that repeat themselves in space after a displacement of λ in the direction of $\langle k|$, as shown in Fig. 3.3.

Therefore,

$$U(\bar{r}) = U\left(\bar{r} + \lambda \frac{\bar{k}}{|k|}\right)$$

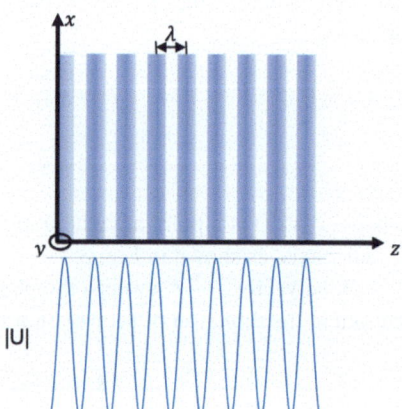

Fig. 3.3 A planewave traveling in the z direction is a periodic function in z with spatial period λ

3.7 Planewaves

where $\frac{\bar{k}}{|\bar{k}|}$ is a unit vector along $\bar{k}$, and by substituting in Eq. (3.29), we obtain

$$Ae^{-j(\bar{k}\cdot\bar{r})} = Ae^{-j\left[\bar{k}\cdot\left(\bar{r}+\lambda\frac{\bar{k}}{|\bar{k}|}\right)\right]}$$

$$= Ae^{-j(\bar{k}\cdot\bar{r})}e^{-j\lambda|\bar{k}|}$$

the second term on the right-hand side $e^{-j\lambda|\bar{k}|}$ must be equal to unity, or $\lambda|\bar{k}| = 2\pi$, and

$$|\bar{k}| = \frac{2\pi}{\lambda}.$$

The general expression for planewave in terms of space and time, using the formula for the vectors $\bar{k}$ and $\bar{r}$, is given by

$$U(\bar{r}, t) = Ae^{j(\omega t - \bar{k}\cdot\bar{r})} \qquad (3.30)$$

$$U(\bar{r}, t) = Ae^{j[\omega t - (k_x x + k_y y + k_z z)]} \qquad (3.31)$$

If the direction of the wavevector k is along the z-axis, then the wave function becomes $U(r) = Ae^{-jkz}$, and the corresponding real part of the wave function is

$$u(\bar{r}, t) = a(\bar{r})cos(2\pi \nu t - kz) = a(\bar{r})cos[2\pi \nu (t - \frac{z}{c})],$$

where ν is the frequency. The complex representation is given by

$$U(\bar{r}, t) = A(\bar{r})e^{j(\omega t - kz)}. \qquad (3.32)$$

The wave function is therefore periodic in time with period $T = \frac{1}{\nu}$, and periodic in space with period $\frac{2\pi}{k}$, which is equal to the wavelength λ. The wave function $U(r, t)$ has all the information about the wave, which will be demonstrated in the following example.

Example A planewave is described by the following equation:

$$E(x, y, z, t) = E_0 sin(kx - \omega t)$$

where E_0 is the maximum electric field strength, k is the wavenumber, ω is the angular frequency, and t is the time. The wavenumber k is given as $2\pi/\lambda$, where λ is the wavelength, and the angular frequency ω is given as $2\pi\nu$, where ν is the frequency.

(continued)

Given the following parameters:

$$E_0 = 100 \text{ V/m}, k = 0.05 \text{ rad } and \ \omega = 2*10^9 \text{ rad/s}$$

Determine the wavelength, frequency, and direction of propagation of the planewave.

Solution Calculate the wavelength: We have the wavenumber k given as 0.05 rad/cm. To convert to SI units (rad/m), we multiply by 100:

$$k = 0.05 \times 100 = 5 \text{ rad/m}$$

Now, we can determine the wavelength using the formula:

$$\lambda = 2\pi / k$$
$$\lambda = 2\pi/5 = 1.2566 \text{ m}$$

The frequency: We already have the angular frequency ω given as $2*10^9$ rad/s. We can determine the frequency using the formula:

$$f = \omega/2\pi$$
$$f = (2 \times 10^9)/2\pi = 3.1831 \times 10^8 \text{ Hz}$$

The direction of propagation: In the given wave equation:

$$E(x, y, z, t) = E_0 sin(kx - \omega t)$$

We can see that the electric field varies only with respect to x, and there is no dependence on y or z. This means that the wave is propagating in the +x direction.

Example A planewave is expressed as

$$\bar{E}(\bar{r}, t) = \left(-2\bar{a}_x + 2\sqrt{3}\bar{a}_y\right) e^{-j\left(\sqrt{3}x+y-6\times 10^8 t\right)}$$

Determine the following: (a) direction of propagation, (b) direction of polarization, (c) phase velocity, (d) frequency, (e) wavelength, and (f) amplitude

(continued)

3.7 Planewaves

Solution The wave is expressed as an electric field in vectorial form

$$\bar{E}(\bar{r},t) = \bar{E}(\bar{r})e^{-j(xk_x + yk_y - \omega)}$$

(a)
$$xk_x + yk_y = x\sqrt{3} + y,$$

therefore,

$$|\bar{k}| = \sqrt{(\sqrt{3})^2 + 1} = 2 \text{ Vm}^{-1}$$

and its direction of propagation of the wave

$$\theta = atan\left(\frac{1}{\sqrt{3}}\right) = 30°$$

with respect to the x-axis.

(b) The polarization direction is defined by the

$$\bar{E}_0 = \left(-2\bar{a}_x + 2\sqrt{3}\bar{a}_y\right).$$

Thus, its amplitude is

$$|\bar{E}_0| = \sqrt{(2\sqrt{3})^2 + 2^2} = 4 \text{ Vm}^{-1}$$

and its angle of polarization

$$\phi = atan\left(\frac{2\sqrt{3}}{-2}\right) = 120°$$

(c) the phase velocity

$$v = \frac{\omega}{k} = \frac{6 \times 10^8}{2} = 3 \times 10^8 \text{ ms}^{-1}.$$

(d) The frequency

$$\nu = \frac{\omega}{2\pi} = \frac{6 \times 10^8}{2\pi} = \frac{3}{\pi} \times 10^8 \text{ Hz}$$

(continued)

Thus, this wave's frequency is not in the optical range of the spectrum.

(e) The wavelength

$$\lambda = \frac{v}{\nu} = \frac{3 \times 10^8}{\frac{3}{\pi} \times 10^8} = \pi = 3.1416 \text{ m}.$$

(f) The amplitude is

$$\sqrt{(\sqrt{3})^2 + 1} = 2 \text{ Vm}^{-1}$$

3.8 Spherical Waves

Another simple solution of the Helmholtz eave equation is the spherical wave. These waves are generated from a point source, and the wavefronts propagate as concentric spheres emanating from the point source, as shown in Fig. 3.4. As the wavefronts expand in radius, the amplitude of the field decreases. Hence, the intensity of the optical field is inversely proportional to the square of the radius of the sphere. The wave function of the spherical wave can be expressed as:

$$U(r) = \frac{A}{r} e^{-jkr}, \tag{3.33}$$

where A is the amplitude, r is the radial distance from the point sources, and k is the wavenumber, $k = 2\pi \frac{v}{c} = \frac{\omega}{c} = \frac{2\pi}{\lambda}$. The intensity proportional to the square of the

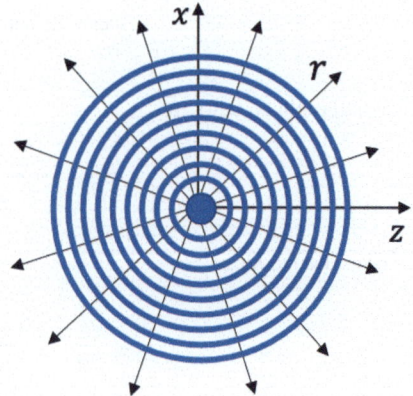

Fig. 3.4 A spherical emitted from a point source

distance r from the light source is given by

$$I(r) = |U(r)|^2 = \frac{|A|^2}{r^2},$$

For simplicity, let the $Arg\{A\} = 0$ the wavefronts are the surfaces $kr = 2\pi q$ or $r = q\lambda$, where q is an integer, q= 1, 2, 3, …,. The wavefronts, then, are concentric spheres separated by a radial distance $\lambda = \frac{2\pi}{k}$ such that it advances radically at the phase velocity c in free space.

3.9 Fresnel Approximation of the Spherical Wave

Let us examine a spherical wave originating at $r = o$, at points $r(x, y, z)$ sufficiently close to the z axis but far from the origin, so that $(x^2 + y^2)^{\frac{1}{2}} \ll z$. The paraxial approximation of ray optics would be applicable where these points are the end points of rays originating at the point source. Consider the following binomial expansion for $\sqrt{1+x}$ for $x \ll 1$

$$\sqrt{1+x} = 1 + \frac{x}{2} - \frac{x^2}{8} + \frac{x^3}{16} - \ldots, \quad (3.34)$$

The distance r for a point (x, y, z) in space is given by

$$r = \sqrt{x^2 + y^2 + z^2}$$

by taking z^2 outside the square root, we get

$$r = z\sqrt{1 + \frac{(x^2 + y^2)}{z^2}}.$$

Using the binomial expansion, we obtain

$$r \approx z + \frac{(x^2 + y^2)}{z^2} - \frac{(x^2 + y^2)^2}{8z^3} + \ldots. \quad (3.35)$$

Therefore, if $(x^2 + y^2) \ll z^2$, then the distance r can be approximated by the first two terms of the binomial expansion. Therefore, the spherical wave can be expressed as

$$U(r) \approx \frac{A_o}{z} e^{-jkz} e^{-jk\frac{x^2+y^2}{2z}}. \quad (3.36)$$

In terms of the amplitude, we can replace r with z, but in the phase term, we cannot drop r since it is multiplied by the wavenumber k, which is on the order of 10^6 m^{-1}

in the optical spectrum. In the Fresnel approximation of the spherical wave, the term e^{-jkz} represents a planewave propagating along the z-axis, while the term $e^{-jk\frac{x^2+y^2}{2z}}$ represents a paraboloidal wave. According to this approximation, the wavefronts of the spherical wave are spherical at a very close proximity to the source. As we move away along the z-axis, the wavefronts take the shape of paraboloidal, shape, and as z becomes extremely large, the wavefronts become planar. For instance, waves emanated by the sun are considered planewaves by the time they reach the Earth. This approximation holds when the third term of the binomial expansion is much less than 1, i.e.,

$$\frac{(x^2+y^2)^2}{8z^4} \ll 1.$$

But when it is multiplied by kz it can become very large. Hence, in order for this approximation to be valid, the phase

$$\frac{kz}{8}\left(\frac{x^2+y^2}{z^2}\right)^2 \ll \pi,$$

$$\left(\frac{x^2+y^2}{z^2}\right)^2 \ll \frac{8\pi}{kz} = \frac{4\lambda}{z}. \tag{3.37}$$

Therefore, the condition becomes

$$(x^2+y^2)^2 \ll 4\lambda z^3 \tag{3.38}$$

This approximation is valid for a circle with radius

$$a = \sqrt{x^2+y^2},$$

$$(a^2)^2 \ll 4\lambda z^3,$$

$$a^4 \ll 4\lambda z,^3$$

therefore, for this case

$$\frac{a^4}{4\lambda z^3} \ll 1,$$

rearranging this equations leads to the following

$$\frac{a^2}{\lambda z}\frac{a^2}{4z^3} \ll 1.$$

Therefore, a is the radius of a disc defines the Fresnel approximation validity. The term $\frac{a^2}{\lambda z}$ is known as the Fresnel number N_F,

$$N_F = \frac{a^2}{\lambda z}, \quad (3.39)$$

Example A square aperture of side length 4 mm is illuminated by a monochromatic light source with a wavelength of 600 nm. The observation screen is placed at a distance of 1.5 m from the aperture. Determine the Fresnel number for this scenario.

Solution The Fresnel number (N_F) is defined as:

$$N_F = \frac{a^2}{\lambda z}$$

where a is the size of the aperture (half of the side length for a square aperture), λ is the wavelength of the light, and z is the distance between the aperture and the observation screen.

$$a = 4 \text{ mm}/2 = 2 \text{ mm} = 2 \times 10^{-3} \text{ m}$$

Therefore, the Fresnel number is

$$N_F = (a^2)/(\lambda z)$$

$$N_F = \frac{(2 \times 10^{-3})^2}{600 \times 10^{-9} \times 1.5} = 4.44$$

3.10 Reflection and Refraction

In Chap. 2, we derived the laws of reflection and refraction using Fermat's principle. In this section, we aim to derive the same laws using wave theory. Consider a planewave shown in Fig. 3.5, propagating along $\bar{k}_1$ in the $x - z$ plane making an angle θ_1 with the z-axis. The planewave reflects from the mirror with a wave vector $\bar{k}_2$ making an angle θ_2 with the z-axis. The wave function of the incident planewave is given by $e^{-j\bar{k}_1 \cdot \bar{r}}$, and that of the reflected wave is $e^{-j\bar{k}_2 \cdot \bar{r}}$.

At the plane of the mirror, the phase of both waves must be equal, meaning $\bar{k}_1 \cdot \bar{r} = \bar{k}_2 \cdot \bar{r}$, where $\delta r = x\bar{a}_x$ lying along the surface of the mirror. The magnitudes of the wave vectors are equal, since it is the same monochromatic wave, but their

Fig. 3.5 Reflection of a planewave off a mirror surface

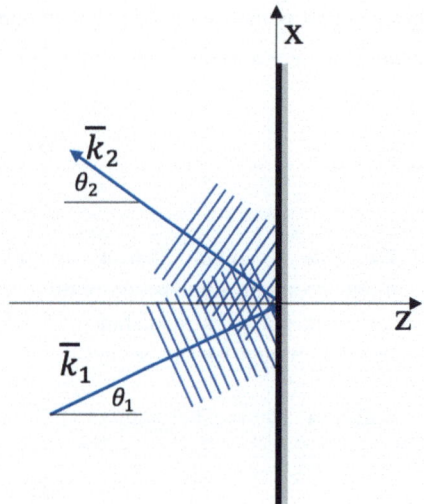

direction differ as described by

$$\bar{k}_1 = k_x a_x + k_z a_z = k\sin(\theta_1) a_x + k\cos(\theta_1) a_z,$$

$$\bar{k}_2 = k_x a_x + k_z a_z = k\sin(\theta_2) a_x - k\cos(\theta_2) a_z$$

by substituting back into $\bar{k}_1 \cdot \bar{r} = \bar{k}_2 \cdot \bar{r}$ we get:

$$k\sin(\theta_1) a_x = k\sin(\theta_2) a_x.$$

Therefore,

$$\theta_1 = \theta_2$$

which is the law of reflection.

In the case of refraction of a planewave at a plane interface between two media with refractive indices n_1 and n_2 as shown in Fig. 3.6. The phases of the incident and refracted waves must match at the interface; this condition is expressed as $\bar{k}_1 \cdot \bar{r} = \bar{k}_2 \cdot \bar{r}$ at the plane of interface between the two media. Considering that the wavenumbers in the media are functions of the free space wavenumbers, we have k_o, $\bar{k}_1 = n_1 \bar{k}_o$ and $\bar{k}_2 = n_2 \bar{k}_o$. The components of wave vectors are given by

$$\bar{k}_1 = n_1 k_o \sin\theta_1 \bar{a}_x + 0\bar{a}_y + n_1 k_o \cos\theta_1 \bar{a}_z,$$

$$\bar{k}_2 = n_2 k_o \sin\theta_2 \bar{a}_x + 0\bar{a}_y + n_2 k_o \cos\theta_2 \bar{a}_z$$

3.11 Transmission and Reflection Coefficients

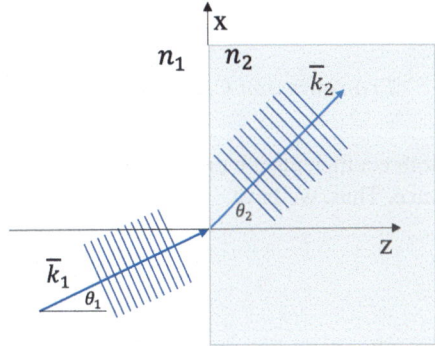

Fig. 3.6 Refraction of a planewave through different media

and a vector $\bar{r}$ anywhere is space is given by

$$\bar{r} = x\bar{a}_x + y\bar{a}_y + z\bar{a}_z$$

Therefore, the dot product between the wave vectors and the vector $\bar{r}$ are given by

$$\bar{k}_1 \cdot \bar{r} = xn_1 K_o sin\theta_1 + 0 + 0$$

and

$$\bar{k}_2 \cdot \bar{r} = xn_2 K_o sin\theta_1 + 0 + 0.$$

Since $\bar{k}_1 \cdot \bar{r} = \bar{k}_2 \cdot \bar{r}$, therefore,

$$xn_1 k_o sin\theta_1 = xn_2 k_o sin\theta_2$$

$$n_1 sin\theta_1 = n_2 sin\theta_2$$

which is Snell's law.

3.11 Transmission and Reflection Coefficients

As light waves propagate from one medium to another with different refractive indices, some of the light is reflected back in the same medium, while some of it transmitted into the other medium. The relations between the angles of incidence, reflection, and refraction are governed by the laws of reflection and refraction. The amplitudes of the reflected and refracted electric fields relative to the amplitude of the incident electric field are related through the reflection and transmission coefficient defined as follows:

$$Reflection\ Coefficient = r = \frac{amplitude\ of\ reflected\ field}{amplitude\ of\ incident\ field}$$

and

$$Transmission\ Coefficient = t = \frac{amplitude\ of\ transmitted\ field}{amplitude\ of\ incident\ field}$$

Both coefficients are expressed in terms of the electric field component of the light wave. Thus, we have:

$$r = \left|\frac{\bar{E}_r}{\bar{E}_i}\right| \text{ and } t = \left|\frac{\bar{E}_t}{\bar{E}_i}\right|$$

These coefficients depend on the polarization of the electric fields with respect to the plane of incidence, which is defined as the plane of oscillation of the electric field as it propagates. There are two possible polarizations of the electric field component of the light wave: either parallel or perpendicular to the plane of incidence. Figures 3.7 and 3.8 illustrate these two polarization states with the plane of the page considered the plane of incidence.

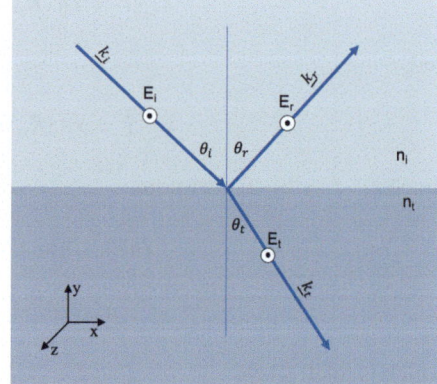

Fig. 3.7 Reflection and transmission of an electric field polarized perpendicular to the plane of incidence. This symbol ⊙ denotes that the field is pointing out of the plane of incidence

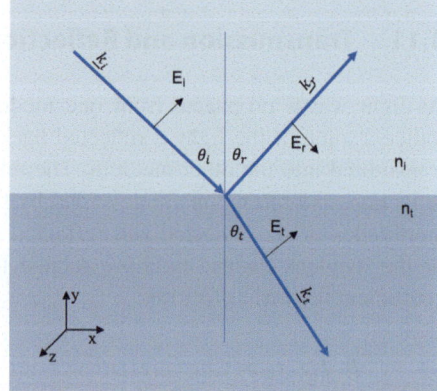

Fig. 3.8 Reflection and transmission of an electric field polarized in the the plane of incidence

3.11 Transmission and Reflection Coefficients

The propagation vectors of the incident, reflected, and transmitted planewaves are $\bar{k}_i, \bar{k}_r,$ and $\bar{k}_t$, respectively. The reflection and transmission coefficients can be derived to be

$$r_\perp = \frac{n_i \cos\theta_i - n_t \cos\theta_t}{n_i \cos\theta_i + n_t \cos\theta_t} \quad (3.40)$$

$$r_\| = \frac{n_t \cos\theta_i - n_i \cos\theta_t}{n_t \cos\theta_i + n_i \cos\theta_t} \quad (3.41)$$

$$t_\perp = \frac{2n_i \cos\theta_i}{n_i \cos\theta_i + n_t \cos\theta_t} \quad (3.42)$$

$$t_\| = \frac{2n_i \cos\theta_i}{n_t \cos\theta_i + n_i \cos\theta_t} \quad (3.43)$$

These equation are referred to as the Fresnel Equations. The physical implication of Fresnel equations is essential in understanding the relationship between the electric field components at the interface of different media. The reflection coefficient for the component of the electric field normal to the plane of incidence, $r_\perp$ as shown in Fig. 3.9, is always negative, this indicates that the normal component of the electric field undergoes a phase shift of π radians upon reflection when the incident medium has a lower refractive index than the transmitting medium.

On the other hand, the reflection coefficient of the component of the electric field, parallel to the plane of incidence $r_\|$, also shown in Fig. 3.9, remains in phase with the incident field as long as the angle of incidence is less than the polarization angle, $\theta_i \leq \theta_p$. However, for angles of incidence $\theta_i > \theta_p$, and when the incident medium has a higher refractive index than the transmitting medium, the parallel component undergoes a phase shift of π radians upon reflection.

In the case of internal reflection where $n_i > n_t$, for angles less than the critical angle, $\theta_i < \theta_c$, the reflection coefficient of the normal component of the field, $r_\perp$, as shown in Fig. 3.10, is always positive, which means that the normal component of the electric field does not undergoes a phase shift upon reflection. On the other hand, the reflection coefficient of the parallel component of the field, $r_\|$, Fig. 3.9 as, remains in phase with the incident field as long as the angle of incidence is less

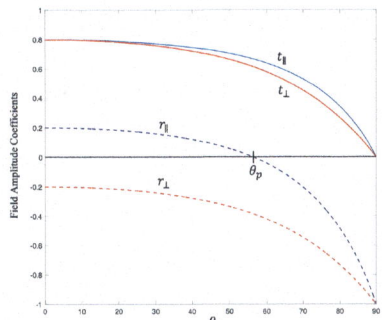

Fig. 3.9 The amplitude coefficients of reflection and transmission as a function of incident angle. These correspond to external reflection $n_t > n_i$ at an air–glass interface ($\frac{n_t}{n_i} = 1.5$). The parallel component of the incidence field becomes zero at $\theta_i = \theta_p$ which is known as the polarization angle

Fig. 3.10 The amplitude coefficients of reflection and transmission as a function of incident angle. These correspond to internal reflection $n_t < n_i$ at an air–glass interface ($\frac{n_i}{n_t} = 1.5$). Total internal reflection occurs at a critical angle $\theta_c = 41.8°$

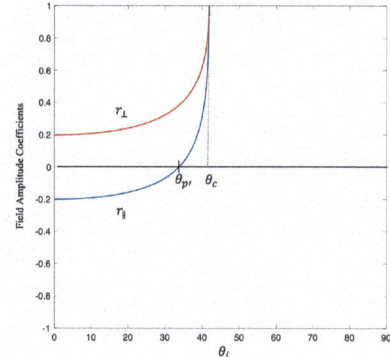

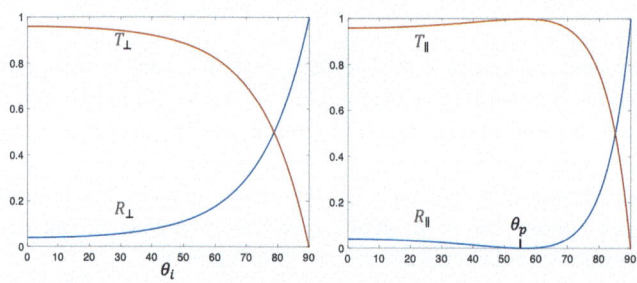

Fig. 3.11 Reflectance and transmittance as a function of the angle of incidence for external reflection, $n_t > n_i$, at an air–glass interface where $\frac{nt}{n_i} = 1.5$

than the polarization angle, $\theta_i \leq \theta_{p'}$. When the angle of incidence exceeds the polarization angle $\theta_i > \theta_{p'}$, the parallel component undergoes a phase shift of π radians upon reflection.

In the case of total internal reflection coefficients when, $\theta_i \geq \theta_c$, the reflection coefficients for both polarized fields become equal to one. In this case, all the light will be reflected back in the dense medium, and none is transmitted for all angles of incidence larger than the critical angle, $\theta_i \geq \theta_c$.

The relationship between the intensities of the incident, reflected, and transmitted fields is described by the reflectance and transmittance, as shown in Fig. 3.11, and is given by the following equations:

$$Reflectance = R = \frac{Intensity\ of\ reflected\ field}{Intensity\ of\ incident\ field|}$$

$$Transmittance = T = \frac{Intensity\ of\ rtransmitted\ field}{Intensity\ of\ incident\ field}$$

Based on conservation of power $R + T = 1$.

3.11 Transmission and Reflection Coefficients

Example Consider an interface between two transparent media, medium 1 with a refractive index n_1 and medium 2 with a refractive index n_2. A monochromatic light beam with wavelength λ is incident on the interface at an angle ϑ_1 with respect to the surface normal. Derive the Fresnel's equations for the reflection coefficients r_s and r_p, and the transmission coefficients t_s and t_p, for s-polarized (perpendicular) and p-polarized (parallel) light. Also, investigate the Brewster angle, where the reflection of p-polarized light becomes zero.

Solution Rewriting the Fresnel's equations in terms of the s and p-polarizations, we get the following:

For s-polarized light:

$$r_s = \frac{n_i \cos\theta_i - n_t \cos\theta_t}{n_i \cos\theta_i + n_t \cos\theta_t}$$

$$t_s = \frac{2n_i \cos\theta_i}{n_i \cos\theta_i + n_t \cos\theta_t}$$

For p-polarized light:

$$r_p = \frac{n_t \cos\theta_i - n_i \cos\theta_t}{n_t \cos\theta_i + n_i \cos\theta_t}$$

$$t_p = \frac{2n_i \cos\theta_i}{n_t \cos\theta_i + n_i \cos\theta_t}$$

The Brewster angle (θ_B) is the angle of incidence for which the reflection coefficient of p-polarized light becomes zero. To find this angle, we set r_p to zero and solve for θ_1:

$$\frac{n_2 \cos\theta_1 - n_1 \cos\theta_2}{n_2 \cos\theta_1 + n_1 \cos\theta_2} = 0$$

By applying Snell's law ($n_1 sin(\theta_1) = n_2 sin(\theta_2)$), we can find the Brewster angle:

$$tan(\theta_B) = n_2/n_1$$

$$\theta_B = tan^{-1}(n_2/n_1)$$

3.12 Transmission Through Optical Components

As optical waves propagate through transparent optical components, they undergo phase transformation depending on the geometry and refractive index of the specific component. Optical components are characterized by their transmission functions, which describe how the wave is modified as it passes through. The transmission function $t(x, y)$ is defined as the ratio between the output wave function U_o to the input wave function U_i, as follows:

$$t(x, y) = \frac{U_o(x, y, z)}{U_i(x, y, z)}.$$

This function determines how the optical component affects the amplitude and phase of the transmitted wave.

3.12.1 Transmission Through a Rectangular Glass Plate

The transmission function of a rectangular glass plate of thickness d and refractive index n, for a planewave propagating along the z-axis, as shown in Fig. 3.12, is

$$t(x, y) = \frac{U(x, y, d)}{U(x, y, 0)}$$

$$t(x, y) = e^{-jk_o nd} \tag{3.44}$$

Where k_o is the wavenumber in free space. The plate introduces a phase shift $nk_o d = \frac{2\pi nd}{\lambda_o}$, the wavefronts' period $\lambda = \frac{\lambda_o}{n}$, is reduced as they pass through the plate, as observed in the figure.

If the planewave is propagating along angle θ with the z-axis, as shown in the figure, the transmission function will be given by

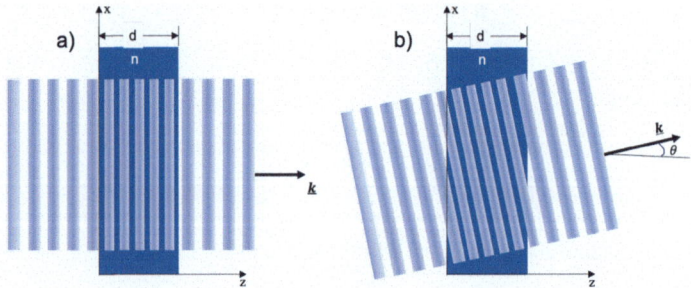

Fig. 3.12 Planewave passing through a rectangular glass plate of thickness d. (**a**) oblique incidence and (**b**) $\bar{k}$ is making and angle θ

3.12 Transmission Through Optical Components

$$t(x, y) = e^{-jk_o nd\cos(\theta)} \quad (3.45)$$

For the case, where the glass plate has a refractive index that is varying as a function of x and y as the case for a gradient index lens, where $n(x, y)$ the transmission function for an oblique planewave propagation the transmission function of such component is given by

$$t(x, y) = e^{-jk_o dn(x,y)} \quad (3.46)$$

3.12.2 Transmission Through a Plate of Varying Thickness

A planewave propagating through a plate of varying thickness $d(x, y)$, as shown Fig. 3.13, introduces a phase shift $\varphi(x, y)$ given by

$$\varphi(x, y) = k_o(d_o - d(x, y) + nd(x, y))$$

$$\varphi(x, y) = k_o d_o + (n - 1)d(x, y)$$

So the transmission function of the plate is given by

$$t(x, y) = e^{-jk_o d_o} e^{-jk_o (n-1)d(x,y)} \quad (3.47)$$

3.12.3 Transmission Through a Thin Lens

Thin lens has at least one of its surfaces convex and its thickness at its center much smaller than its diameter. Phase shift $\varphi(x, y)$ that the planewave will experience passing through the plate with varying thickness $d(x, y)$ is

$$\varphi(x, y) = k_o(d_o - d(x, y) + nd(x, y))$$

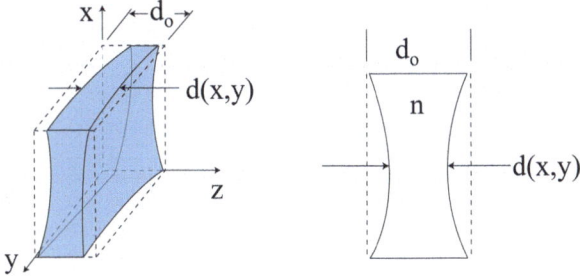

Fig. 3.13 A plate of a varying thickness $d(x, y)$

From the geometry, the varying thickness $d(x, y)$ can be expressed as

$$d(x, y) = d_o - \bar{PQ} = d_o - (R - \bar{QC})$$

where

$$QC = \sqrt{R^2 - AQ^2} = \sqrt{R^2 - (x^2 + y^2)}$$

Therefore,

$$d(x, y) = d_o - \left[R - \sqrt{R^2 - (x^2 + y^2)}\right]$$

and

$$\sqrt{R^2 - (x^2 + y^2)} = R\sqrt{1 - \frac{x^2+y^2}{R^2}} \approx R\left(1 - \frac{x^2+y^2}{2R^2}\right),$$

$$d(x, y) \approx d_o - \left[R - R\left(1 - \frac{x^2+y^2}{2R^2}\right)\right]$$

$$d(x, y) \approx d_o - \frac{x^2+y^2}{2R}$$

So the phase shift is given by

$$\varphi(x, y) = k_o d_o - d(x, y) + n d(x, y)$$

$$\varphi(x, y) = k_o d_o - (1 - n) k_o \left(d_o - \frac{x^2+y^2}{2R}\right)$$

$$= k_o d_o - k_o d_o + k_o \frac{x^2+y^2}{2R} + k_o n d_o - \frac{k_o n (x^2+y^2)}{2R}$$

$$= k_o n d_o + (1 - n)\frac{x^2+y^2}{2R}$$

Therefore,

$$\varphi(x, y) \approx k_o n d_o - \frac{(x^2+y^2)(n-1)}{2R} \tag{3.48}$$

$$= k_o n d_o - \frac{x^2+y^2}{2f} \tag{3.49}$$

3.12 Transmission Through Optical Components

Where R is the radius of the convex surface and so the focal length of the lens is

$$f = \frac{R}{n-1}.$$

So the transmission function of the thin lens is given by

$$t(x, y) = \approx e^{-jk_o nd_o} e^{jk_o \frac{x^2+y^2}{2f}} \tag{3.50}$$

The first term of the transmission function has a constant phase that has no influence on reshaping the waves passing through the lens so it can be dropped. Hence, the transmission function can be given by

$$t(x, y) = e^{jk_o \frac{x^2+y^2}{2f}} \tag{3.51}$$

For a double-convex lens, where the two surfaces having different radii of curvatures are R_1 and R_2, the focal length of the lens is given by

$$\frac{1}{f} = (n-1)\left(\frac{1}{R_1} - \frac{1}{R_2}\right). \tag{3.52}$$

The transmission function of the lens introduces a quadratic phase shift. As a result, the planewave that passes through the lens its wavefronts emerges with a quadratic phase. These wavefronts emerge as converging paraboloidal waves focused at the focal plane of the lens as shown in the Fig. 3.14. This demonstrates the focusing property of the lens. The wavefronts after the focal point propagate as diverging paraboloidal waves.

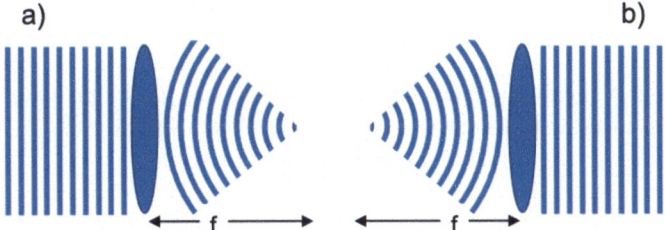

Fig. 3.14 (a) A planewave passing through a convex lens emerges as a converging paraboloidal wave focused at the back focal plane, and (b) a diverging paraboloidal wave originating at the front focal plane of a convex lens emerges as a planewave

3.13 Interference of Light Waves

When two or more light waves overlap, the resultant field is determined the superposition principle, which states the total field is the vector sum of the fields associated with the individual waves. For a particular component of the electric or magnetic fields, the resultant of adding two or more light waves is the sum of all the waves, considering their amplitude and phase. Mathematically, the interference of N waves is expressed as:

$$U(\bar{r}, t) = \sum_{i}^{N} U_i(\bar{r}, t). \qquad (3.53)$$

However, the intensity of the resultant interference does not strictly follow the superposition principle due to the loss of phase information from the different waves. That is:

$$I(\bar{r}, t) \neq \sum_{i=1}^{N} I_i(\bar{r}, t).$$

Nevertheless, the resultant intensity can be determined by:

$$I(\bar{r}, t) = |U(\bar{r}, t)|^2. \qquad (3.54)$$

For monochromatic waves, we typically drop the time dependence as all waves share the same carrier frequency represented by the exponent $e^{j\omega t}$. The superposition of N light waves will result in the interference pattern given by

$$U(\bar{r}) = \sum_{i=1}^{N} U_i(\bar{r}) = \sum_{i=1}^{N} A_i e^{-j\phi_i}, \qquad (3.55)$$

where A_i and ϕ_i are the complex amplitude and phase of ith light wave.

Now, consider the interference of two monochromatic light waves

$$U_1(\bar{r}) = A_1 e^{-j\phi_1}$$

and

$$U_2 = A_2 e^{-j\phi_2}.$$

The resultant wave is the sum of the two waves:

$$U(\bar{r}) = U_1(\bar{r}) + U_2(\bar{r}) = A_1 e^{-j\phi_1} + A_2 e^{-j\phi_2} \qquad (3.56)$$

3.13 Interference of Light Waves

The intensity of the interference pattern is given by

$$I = UU^* = (U_1 + U_2)(U_1^* + U_2^*)$$
$$= \left(A_1 e^{-j\phi_1} + A_2 e^{-j\phi_2}\right)\left(A_1 e^{j\phi_1} + A_2 e^{j\phi_2}\right)$$
$$I = |A_1|^2 + |A_2|^2 + A_1 A_2^* e^{-j(\phi_2-\phi_1)} + A_1^* A_2 e^{-j(\phi_2-\phi_1)}$$

Let $I_1 = |A_1|^2$ and $I_2 = |A_2|^2$ Then

$$I = I_1 + I_2 + 2\sqrt{I_1 I_2} \cos(\phi_2 - \phi_1).$$

$$I(\phi) = I_1 + I_2 + 2\sqrt{I_1 I_2} \cos(\phi), \qquad (3.57)$$

where $\phi = \phi_2 - \phi_1$. The interference, $I(\phi)$, varies sinusoidally as a function of the phase difference between the waves, ϕ. The intensity amplitude will be varying between $I_1 + I_2 - 2\sqrt{I_1 I_2}$ and $I_1 + I_2 + 2\sqrt{I_1 I_2}$.

The interference pattern, which results from the overlap of the two waves, is represented by the intensity variation across the region of overlap. This intensity pattern is plotted in the Fig. 3.15, assuming $I_1 = I_2 = 1$.

3.13.1 Interference of Two Plane Waves

Consider the interference of two planewaves propagating along the wave vectors $\bar{k}_1$ and $\bar{k}_2$ that which angles θ_1 and θ_2 with the z-axis as shown in Fig. 3.16. We will derive an expression for the field and intensity along the x-axis. The wave vectors are related to the wavelength of the light wave as follows:

$$|\bar{k}_1| = |\bar{k}_2| = k = \frac{2\pi}{\lambda}.$$

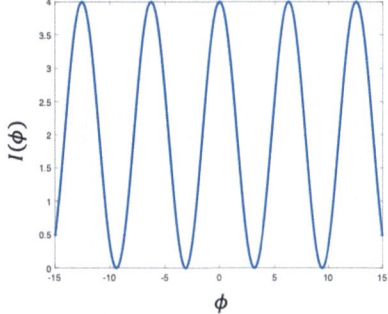

Fig. 3.15 The intensity pattern for the interference of two waves as given by Eq. (3.57)

The wave vectors are decomposed in two components: one along the x-axis and the other along z-axis, i.e.,

$$\bar{k}_1 = k(sin\theta_1 \bar{a}_x + cos\theta_1 \bar{a}_z) \quad (3.58)$$

and

$$\bar{k}_2 = k(-sin\theta_2 \bar{a}_x + cos\theta_2 \bar{a}_z) \quad (3.59)$$

Assume that the two planewaves have an initial phase given by δ_1 and δ_2, respectively. The planewaves are expressed in the following forms:

$$U_1(\bar{r}) = A_1 e^{-j(\bar{k}_1 \cdot \bar{r} + \delta_1)}$$

and

$$U_2(\bar{r}) = A_2 e^{-j(\bar{k}_2 \cdot \bar{r} + \delta_2)}$$

The addition of these two planewaves is

$$U(\bar{r}) = = U_1 + U_2$$

$$U(\bar{r}) = A_1 e^{-j(\bar{k}_1 \cdot \bar{r} + \delta_1)} + A_2 e^{-j(\bar{k}_2 \cdot \bar{r} + \delta_2)}$$

To observe the inference pattern on a screen placed at the $x - y$ plane, we derive an expression for the resultant intensity $I(r)$, assuming that the planewaves amplitudes A_1 and A_2 to be real quantities (Fig. 3.16). The resultant intensity will be based on the superposition of the two planewaves, taking into account their

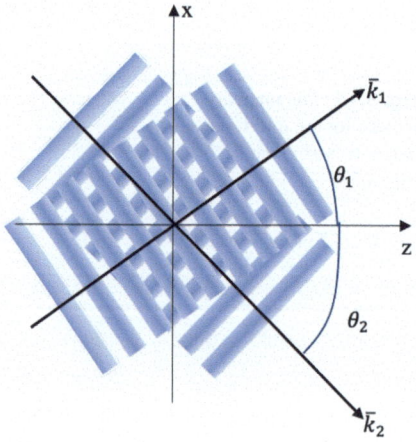

Fig. 3.16 Interference of two planewaves

3.13 Interference of Light Waves

amplitudes and phase differences. The resultant intensity is given by:

$$I = U \cdot U^*$$

Therefore,

$$I = \left[A_1 e^{-j(\bar{k}_1 \cdot \bar{r} + \delta_1)} + A_2 e^{-j(\bar{k}_2 \cdot \bar{r} + \delta_2)}\right]\left[A_1 e^{+j(\bar{k}_1 \cdot \bar{r} + \delta_1)} + A_2 e^{+j(\bar{k}_2 \cdot \bar{r} + \delta_2)}\right]$$
(3.60)

Multiplying complex conjugate terms in Eq. (3.60) results in

$$I = A_1^2 + A_2^2 + 2A_1 A_2 e^{j[(\bar{k}_2 - \bar{k}_1)\bar{r} + (\delta_2 - \delta_1)]}$$
$$I = A_1^2 + A_2^2 + 2A_1 A_2 \cos\left[(\bar{k}_2 - \bar{k}_1) \cdot \bar{r} + (\delta_2 - \delta_1)\right]$$
(3.61)

The resultant intensity forms a series of interference fringes that vary as a cosine function with the maximum intensity given by $(A_1 + A_2)^2$ and the minimum intensity by $(A_1 - A_2)^2$. The fringe spacing depends on the position vector $\bar{r}$, the wavelength, and the directions wave vectors direction, θ_1 and θ_2.

To analyze a specific case of the fringe pattern, let the phase difference be:

$$\phi = (\bar{k}_2 - \bar{k}_1) \cdot \bar{r} + (\delta_2 - \delta_1)$$

Consider the interference pattern along the x-axis, where $\bar{r} = x\bar{a}_x$. Let points **a** and **b** be located at consecutive points at along the x-axis, where the interference pattern has the same amplitude. At these points, the phase difference is 2π. The distance between two consecutive fringes, $x_b - x_a$ is the fringe spacing. Therefore, the fringe spacing is derived from this phase difference condition is given by:

$$\Delta\phi = \phi_b - \phi_a$$
$$= \left[= (\bar{k}_2 - \bar{k}_1) \cdot \bar{x}_b \bar{a}_x + (\delta_2 - \delta_1)\right] - \left[= (\bar{k}_2 - \bar{k}_1) \cdot \bar{x}_a \bar{a}_x + (\delta_2 - \delta_1)\right]$$
$$= \left[(\bar{k}_1 - \bar{k}_2) \cdot \bar{a}_x\right](x_b - x_a)$$
$$= 2\pi$$

Let the fringe spacing be $d = x_b - x_a$ then

$$\Delta\phi = \left[(\bar{k}_1 - \bar{k}_2) \cdot \bar{a}_x\right] d$$

Substituting from Eqs. (3.58 and 3.59) into $\Delta\phi = 2\pi$, we get

$$2\pi = [k \sin\theta_1 - (-k \sin\theta_2)] d.$$

Therefore,

$$d = \frac{2\pi}{k(\sin\theta_1 + \sin\theta_2)}$$

substituting for the value of k results in the following expression of the fringe spacing M as follows:

$$d = \frac{\lambda}{(\sin\theta_1 + \sin\theta_2)}. \tag{3.62}$$

The fringe spacing, d, is a function of the wavelength, which corresponds to the color of the light beams..

Thus, by measuring the distance between two adjacent fringes in the interference pattern of two planewaves, one can determine an unknown wavelength. Consequently, this principle forms the basis of interferometry, a technique used to measure wavelengths or phase difference $\delta_1 - \delta_2$. This concept is explored in more detail in later sections of this chapter.

Example If $\lambda = 0.5\ \mu$ m, the initial phase $\delta_1 = \delta_2$ and the wave vector directions are at $\theta_1 = \theta_2 = 30°$. Determine the fringe spacing.

Solution The fringe pattern is similar to the general two-wave interference given in Fig. 3.15 and the fringe spacing, d, is

$$d = \frac{0.5 \times 10^{-6}}{2sin(30)} = 0.5\ \mu\ \text{m}.$$

3.14 Interference by Division of Wavefronts

In this case, the inference is caused by the addition of two or multiple waves generated from the same wavefront emitted by a light source. Young's two-slit experiment is a typical example of this interference. This category includes multiple slits or pinholes illuminated by the same wavefront. Each of these pinholes can be considered as a source of a spherical wave.

3.14.1 Interference of Two Spherical Waves

Consider the interference of two spherical waves. Spherical waves can be generated from point sources that can be generated by a pinhole on a screen illuminated by

3.14 Interference by Division of Wavefronts

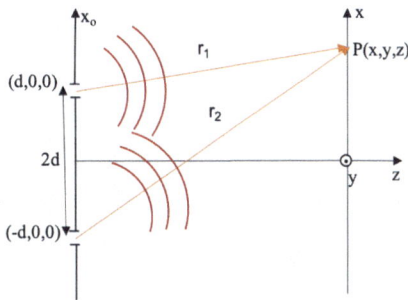

Fig. 3.17 Interference of two spherical waves emitted by two point sources located at $(d, 0, 0)$ and $(-d, 0, 0)$

a planewave. According to Huygens, any point on a wavefront emits a wavelet. Two pinholes located at a distance $2d$ apart at points $(d, 0, 0)$ and $(-d, 0, 0)$ are illuminated by a planewave, as shown in Fig. 3.17. At an observation point $P(x, y, z)$, is located a distance r_1 and r_2 from where the point sources are located. At P, the wavefronts of the spherical waves have radii r_1 and r_2 as shown.

The two spherical waves are denoted by ψ_1 and ψ_2 at point P are given by:

$$\psi_1(r) = \frac{a_1}{r_1} e^{-j(kr_1 - \delta_1)}$$

and

$$\psi_2(r) = \frac{a_2}{r_2} e^{-j(kr_2 - \delta_2)}$$

where δ_1 and δ_2 are the initial phases of the spherical waves. The resultant wave function at point P is given by the sum of the wave functions of the interfering spherical waves

$$\psi(r) = \psi_1(r) + \psi_2(r)$$

then

$$\psi(r) = \frac{a_1}{r_1} e^{-j(kr_1 - \delta_1)} + \frac{a_2}{r_2} e^{-j(kr_2 - \delta_2)}$$

The resultant intensity $I(r)$ at any point $P(x, y, z)$ in the x-y plane is given by

$$I = \psi \cdot \psi^* = \frac{a_1^2}{r_1^2} + \frac{a_2^2}{r_2^2} + 2\frac{a_1 a_2}{r_1 r_2} \cos[k(r_1 - r_2) + (\delta_1 - \delta_2)] \tag{3.63}$$

From the geometry of Fig. 3.17, we can determine the values of r_1 and r_2 as

$$r_1 = \left[(x - d)^2 + y^2 + z^2\right]^{\frac{1}{2}}$$

and

$$r_2 = \left[(x+d)^2 + y^2 + z^2\right]^{\frac{1}{2}}$$

Consider the case where $z \gg d, x$ and y, thus, $r_1 \approx r_2 \approx z$. This approximation is valid for the amplitudes but not for the phase because $(r_1 - r_2)$ is multiplied by $k = \frac{2\pi}{\lambda}$, which is very large. By taking z^2 outside of the square root, this leads to the following:

$$r_1 = z\left[1 + \left(\frac{x^2 + y^2 + d^2 - 2xd}{z^2}\right)\right]^{1/2}$$

and

$$r_2 = z\left[1 + \left(\frac{x^2 + y^2 + d^2 + 2xd}{z^2}\right)\right]^{1/2}$$

Using Taylor expansion as we have done in the previous section and keeping only the first two terms, we obtain the following approximations:

$$r_1 \approx z + \left(\frac{x^2 + y^2 + d^2 - 2xd}{2z}\right) \tag{3.64}$$

and

$$r_2 \approx z + \left(\frac{x^2 + y^2 + d^2 + 2xd}{2z}\right). \tag{3.65}$$

Therefore,

$$r_1 - r_2 = \frac{-2xd}{z} \tag{3.66}$$

Let

$$I_1 = |\psi_1|^2 = \frac{a_1^2}{z^2}, \text{ and } I_2 = \frac{a_2^2}{z^2}$$

Therefore, the resultant intensity is

$$I(x, z) = I_1 + I_2 + 2I_1 I_2 \cos\left[\frac{-2kxd}{z} + (\delta_1 - \delta_2)\right] \tag{3.67}$$

The fringe pattern is varying sinusoidally, as shown in Fig. 3.18, with a period $\frac{\lambda z}{2d}$. The phase difference $(\delta_1 - \delta_2)$ causes the fringes to shit along the $x - axis$

3.14 Interference by Division of Wavefronts

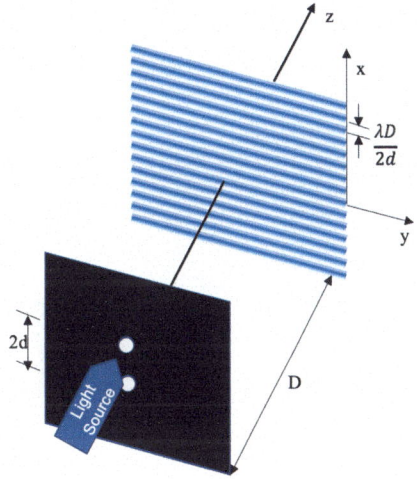

Fig. 3.18 The fringe pattern of the interference of two spherical waves emitted by two point sources

without affecting the fringe spacing. The amplitude of the intensity of the fringe pattern oscillates between $I_1 + I_2 + 2\sqrt{I_1 I_2}$ and $I_1 + I_2 - 2\sqrt{I_1 I_2}$, i.e. between $\left(\sqrt{I_1} + \sqrt{I_2}\right)^2$ and $\left(\sqrt{I_1} - \sqrt{I_2}\right)^2$.

Example Let the two spherical waves be in phase and have equal amplitudes, so $\delta_1 = \delta_2$ and $I_1 = I_2 = I_o$.

$$I(x,z) = I_o + I_o + \sqrt{I_o^2} 2 \cos\left[\frac{-2kxd}{z} + (\delta_1 - \delta_1)\right].$$

Derive an expression for the fringe spacing, and determine its value and plot the interference pattern for the following parameters: $d = 1$ mm, $z = 10$ m, $\lambda = 1000$ nm.

Solution The intensity of the interference pattern for the case where $\delta_1 = \delta_2$ and $I_1 = I_2 = I_o$ is given by

$$I(x) = 2I_o \left[1 + \cos\left(\frac{2kxd}{z}\right)\right] \quad (3.68)$$

Using trigonometric property, $\frac{1 + \cos\theta}{2} = \cos^2(\frac{\theta}{2})$, thus the intensity becomes

$$I(x) = 4I_o \cos^2\left(\frac{kxd}{z}\right) \quad (3.69)$$

(continued)

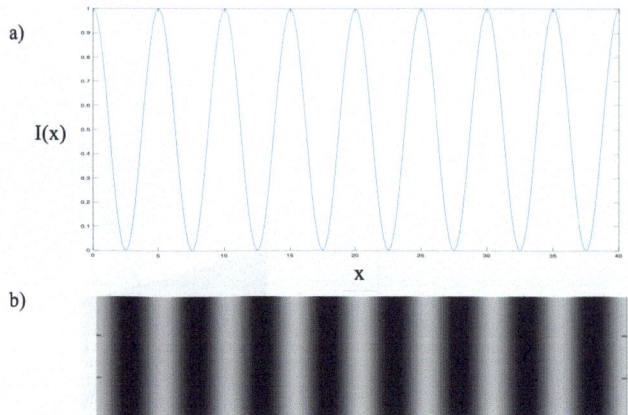

Fig. 3.19 MATLAB plots for (**a**) the intensity of the interference of two spherical waves for: $d = 1$ mm, $z = 10$ m, $\lambda = 1000$ nm, and (**b**) the fringe pattern as seen on an observation screen

The first zero of the fringe pattern is when $\frac{kxd}{z} = \frac{\pi}{2}$. Therefore, the first zero will be at $x = \frac{\lambda z}{4d}$ and the first order fringe, i.e., the first peak, is located at $x = \frac{\lambda z}{2d}$.

The fringe spacing is equal to

$$\frac{\lambda z}{2d} = \frac{10^{-6} \times 10}{2 \times 0.001} = 5 \text{ mm}.$$

The intensity pattern is plotted in Fig. 3.19 for $d = 1$ mm, $z = 10$ m, $\lambda = 1000$ nm.

Example In a Young's double-slit experiment, the slits are separated by 0.28 mm and the screen is placed 1.4 m away. The distance between the central bright fringe and the fourth bright fringe is 1.2 cm.

Determine the wavelength of the light source used in the experiment.

Solution As shown in Fig. 3.17, the spacing between the pinholes is $2d = 0.28$ mm or $d = 0.14$ mm and $D = 14$ m and $x = 1.2$ cm. The intensity pattern of the interference is given in 3.69

(continued)

3.14 Interference by Division of Wavefronts

$$I(x) = 4I_o \cos^2\left(\frac{kxd}{D}\right)$$

The Nth bright fringe will be when

$$\frac{kxd}{D} = 2N\pi$$

since $k = \frac{2\pi}{\lambda}$, then by substitution in the previous equation, we get

$$\lambda = \frac{xd}{ND} = \frac{1.2 \times 10^{-2} \times 0.14 \times 10^{-3}}{4 \times 1.4} = 0.6 \times 10^{-6} \text{ m}$$

Therefore, the wavelength of the light source is $0.6\ \mu$ m

3.14.1.1 Fringe Visibility

The quality of the fringes is measured by a term called the visibility, i.e., how easy to visualize the fringes. When $I_1 \neq I_2$ the fringes are of poor quality (i.e., lower contrast). When $I_1 = I_2$ the fringes have higher contrast (good quality). The definition of visibility that determines the quality of fringes is given by:

$$Visibility = \frac{I_{max} - I_{min}}{I_{max} + I_{min}} \tag{3.70}$$

Therefore,

$$0 \leq Visibility \leq 1 \tag{3.71}$$

Thus, the visibility is zero if $I_{max} = I_{min}$, and it is equal to 1 if $I_{min} = 0$.

3.14.2 Interference of Multiple Waves

Another case of interference by division of wavefronts is for a multiple pinholes in a screen that is illuminated by a planewave such that each pinhole has a phase shifter, it can be a thin piece of glass, such that the each pinhole has a phase shift difference from the neighboring pinhole by $j\phi$. The sum of m waves is

$$U = U_1 + U_2 + U_3 + \ldots\cdots + U_m$$
$$= U_o\left[e^{j\phi} + e^{j2\phi} + \ldots\cdots + e^{j(m-1)\phi}\right] \tag{3.72}$$

where $m = 1, 2, 3, \ldots, M$. Therefore,

$$U_m = \sqrt{I_o}e^{j(m-1)\phi}$$

Let $h = e^{j\phi}$. Then

$$U_m = \sqrt{I_o}h^{(m-1)}$$

Therefore, the total field is given by

$$U = \sqrt{I_o}\left(1 + h + h^2 + h^3 + \ldots + h^{M-1}\right)$$

The sum in the parenthesis is a geometric series. The sum of this finite geometric series is given by:

$$U = \sqrt{I_o}\frac{1 - h^M}{1 - h} = \sqrt{I_o}\frac{1 - e^{jM\phi}}{1 - e^{j\phi}}$$

Therefore, the intensity is the product of the field and its complex conjugate:

$$I = U \cdot U^* = I_o\frac{\left(1 - e^{jM\phi}\right)\left(1 - e^{-jM\phi}\right)}{\left(1 - e^{j\phi}\right)\left(1 - e^{-j\phi}\right)} \tag{3.73}$$

$$I = I_o\left|\frac{e^{jM\frac{\phi}{2}}}{e^{j\frac{\phi}{2}}}\frac{e^{-jM\frac{\phi}{2}} - e^{jM\frac{\phi}{2}}}{e^{-j\frac{\phi}{2}} - e^{j\frac{\phi}{2}}}\right|^2$$

Therefore, the resultant intensity of M waves with equal amplitude and equal phase difference is

$$I = I_o\frac{\sin^2\left(\frac{M\phi}{2}\right)}{\sin^2\left(\frac{\phi}{2}\right)} \tag{3.74}$$

This expression for the interference pattern is similar to that of phased array radar except for the difference in wavelength as phased array radars operate in the microwave range. The peak value intensity occurs at $\phi = 0, 2\pi, 4\pi, 6\pi, \ldots$, and at these points, the peak intensity will be approximately $I_{max} \approx I_o M^2$. This is because at these angles, the sine function can be approximated using the first term of its Taylor expansion. Figure 3.20 shows the intensity pattern for M=5 spherical waves. The number of the side lobes depends on M.

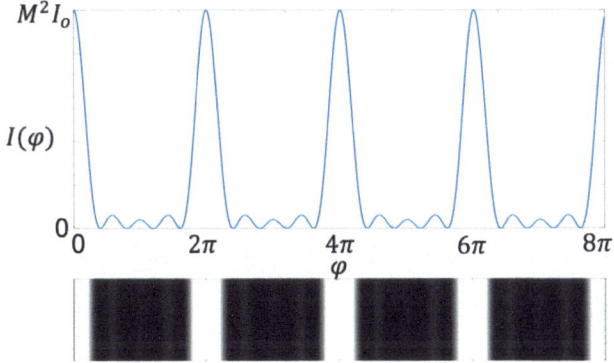

Fig. 3.20 MATLAB plots for the intensity of the interference of M=5 spherical waves

3.14.3 Interference by Division of Amplitudes

In this type of interference, the waves being added have equal differences in amplitude and phase. These waves are typically generated through multiple reflections and transmissions of the same initial wave. Assume that the waves are related as follows:

$$U_1 = \sqrt{I_0},\ U_2 = hU_1,\ U_3 = hU_2 = h^2 U_1,\ \ldots,\ U_m = h^{(m-1)}U_1,\ \ldots$$

where

$$h = |h|\,e^{j\phi}\ and\ |h| < 1$$

$$U = \sqrt{I_o}\left[1 + |h|\,e^{j\phi} + \left(|h|\,e^{j\phi}\right)^2 + \left(|h|\,e^{j\phi}\right)^3 + \ldots\right]$$

This is an infinite geometric series whose sum is given by

$$U = \frac{\sqrt{I_o}}{1 - |h|\,e^{j\phi}} \tag{3.75}$$

The intensity is given by $I = UU^*$

$$I = \frac{I_o}{\left(1 - |h|\,e^{j\phi}\right)\left(1 - |h|\,e^{-j\phi}\right)}$$

$$= \frac{I_o}{1 + |h|^2 - |h|\left(e^{j\phi} + e^{-j\phi}\right)}$$

$$= \frac{I_o}{1+|h|^2 - 2|h|\cos(\phi)}$$

$$= \frac{I_o}{1+|h|^2 - 2|h|\left[1 - 2sin^2\left(\frac{\phi}{2}\right)\right]}$$

$$= \frac{I_o}{1+|h|^2 - 2|h| + 4sin^2\left(\frac{\phi}{2}\right)}$$

Therefore, the intensity of the interference of infinite waves is given by

$$I = \frac{I_o}{(1-|h|)^2 + 4sin^2\left(\frac{\phi}{2}\right)} \qquad (3.76)$$

The maximum value of the intensity is when $4sin^2\left(\frac{\phi}{2}\right) = 0$. Hence, the maximum value of the intensity is when $4sin^2\left(\frac{\phi}{2}\right) = 0$, Hence,

$$I_{max} = \frac{I_0}{(1-|h|)^2}$$

and

$$I = \frac{I_{max}}{1 + \left(\frac{2F}{\pi}\right)^2 sin^2\left(\frac{\phi}{2}\right)} \qquad (3.77)$$

where the finesse F is defined as

$$F = \frac{\pi\sqrt{|h|}}{1-|h|}. \qquad (3.78)$$

The intensity pattern, as shown in Fig. 3.21, is a function of both the finesse and the phase difference ϕ.

3.14.4 Interference of Multiple Reflections from Two Parallel Partially Reflecting Surfaces

A practical example of interference by division of amplitude with equal phase difference, which has a wide range of applications, is the multi-reflections and transmissions between two parallel partially reflective plates. As shown in Fig. 3.22, a rectangular plate of thickness d, and a refractive index of n_2 is surrounded by a medium with refractive index n_1. The plate is illuminated by a planewave

3.14 Interference by Division of Wavefronts

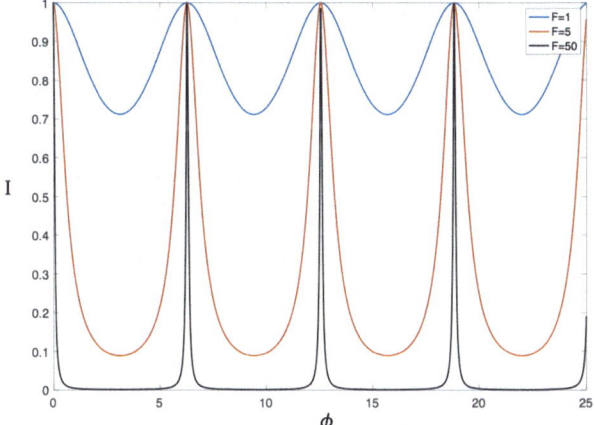

Fig. 3.21 The intensity pattern, Eq. (3.77) is plotted as a function of $\phi [radians]$ for the finesse, F=1, 5, and 50. The intensity peaks are sharper and narrower for larger finesse values

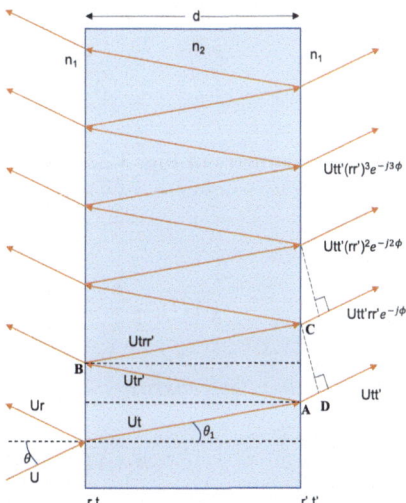

Fig. 3.22 General representation of incident, reflected, and transmitted beams

with amplitude U. The front surface of the plate has reflection coefficient r, and transmission coefficient t, while the second interface has corresponding reflection and transmission coefficients r' and t', respectively.

The amplitude of the incident beam on the first interface is U, and the angle of incidence is θ. At the first interface, the beam splits into a reflected beam with amplitude U_r, and a transmitted beam with amplitude U_t. The transmitted beam propagates through the plate and, at the second interface, splits again into two beams with amplitudes Utt' and Utr', as shown in Fig. 3.22. This process of propagation and splitting continues, with the beam undergoing multiple reflections and transmissions, accumulating phase changes at each step.

Now consider the first two transmitted beams. They are represented by Utt' and $Utt'rr'e^{-j\phi}$, where ϕ is the phase difference between them. This phase difference arises from the second beam traveling the distance $AB + BC$ and AD. The two beams are in phase after points C and D in their respective paths. The optical path length difference between the two beams is

$$\Lambda = n_2 \left(\overline{AB} + \overline{BC}\right) - n_1 \left(\overline{AD}\right)$$

Since $(\overline{AB}) = (\overline{BC}) = \frac{d}{cos\theta_1}$ then

$$\Lambda = \frac{2n_2 d}{cos\theta_1} - n_1 (\overline{AD})$$

From the geometry of the figure, $\overline{AD}$ is given by

$$(\overline{AD}) = (\overline{AC}) sin\theta$$

Using Snell's law, $n_1 \sin(\theta) = n_2 \sin(\theta_1)$, $\overline{AD}$ becomes

$$(\overline{AD}) = \frac{n_2}{n_1}(\overline{AC}) sin(\theta_1))$$

From the geometry of Fig. 3.22

$$(\overline{AC}) = 2d tan\theta_1$$

$$(\overline{AD}) = \frac{n_2}{n_1} 2d \cdot tan\theta_1 \cdot sin\theta_1 = \frac{n_2}{n_1} 2d \frac{sin^2\theta_1}{cos\theta_1}$$

Since $sin^2\theta_1 = 1 - cos^2\theta_1$

$$n_1(\overline{AD}) = \frac{2n_2 d}{cos\theta_1} - 2n_2 d\, cos\theta_1$$

Therefore, the optical path length difference between the two output beams is given by

$$\Lambda = \frac{2n_2 d}{cos\theta_1} - \left(\frac{2n_2 d}{cos\theta_1} - 2n_2 d\, cos\theta_1\right)$$

$$\Lambda = 2n_2 d cos\theta_1$$

The phase difference, the product of the wavenumber k and the optical path length difference, between the two beams is given by

3.14 Interference by Division of Wavefronts

$$\phi = k\Lambda = \frac{4\pi n_2 d\cos\theta_1}{\lambda} \tag{3.79}$$

This phase shift is the same for every two consecutive transmitted beams. The total transmitted beam, Ut, is the sum of all beams and is given by:

$$U_t = Utt'\left[1+(rr')e^{-j\phi}+(rr')^2 e^{-j2\phi}+(rr')^3 e^{-j3\phi}+...\right]. \tag{3.80}$$

This is the general expression for the total transmitted wave. Now, for the case where the input and output media have the same refractive index, then $r = r'$ and $t = t'$, and $R = |r|^2$ and $T = |t|^2 = 1 - R$ Let's change variables such that $x^2 = r^2 e^{-j\phi}$, the total transmitted wave is given by

$$U_t = UT\left[1 + x^2 + x^4 + x^6 + \ldots\right] \tag{3.81}$$

which is an infinite geometric series. Since $x2 < 1$, then the sum of the series is

$$1 + x^2 + x^4 + x^6 + \ldots = \frac{1}{1-x^2}$$

$$U_t = \frac{UT}{1 - r^2 e^{-j\phi}}$$

Because of conservation of energy, $t^2 = 1-r^2$. The intensity of the total transmitted beam is

$$I_t = U_t U_t^*$$

$$I_t = \left(\frac{UT}{1-r^2 e^{-j\phi}}\right)\left(\frac{U^*T}{1-r^2 e^{+j\phi}}\right)$$

$$I_t = \frac{|U|^2 T^2}{1 - 2r^2 \cos\phi + r^4}$$

Using trigonometric identity $\cos(\phi) = \cos^2\left(\frac{\phi}{2}\right) - \sin^2\left(\frac{\phi}{2}\right) = 1 - 2\sin^2\left(\frac{\phi}{2}\right)$, Hence,

$$I_t = \frac{|U|^2 (tt')}{1 + r^4 - 2r^2 + 4r^2 \sin^2\left(\frac{\phi}{2}\right)}$$

$$I_t = \frac{|U|^2 T^2}{1 + R^2 - 2R + 4R \sin^2\left(\frac{\phi}{2}\right)}$$

$$I_t = \frac{|U|^2 T^2}{(1-R)^2 + R\sin^2\left(\frac{\phi}{2}\right)}$$

$$I_t = \frac{T^2}{(1-R)^2} \frac{|U|^2}{1 + \frac{4R}{(1-R)^2} 4R\sin^2\left(\frac{\phi}{2}\right)}$$

The intensity of the input beam $I_i = |U|^2$

$$I_t = \frac{I_i}{1 + F\sin^2\left(\frac{\varphi}{2}\right)}$$

Therefore, when the transmitted beam is equal to the incident beam, $\frac{I_t}{I_i} = 1$, indicating total transmission, this occurs when $\sin^2\left(\frac{\phi}{2}\right) = 0$. In general, for a lossless medium between the reflecting plates, the intensity of the reflected beams is given by

$$I_r = I_i - I_t$$

therefore,

$$\frac{I_r}{I_i} = \frac{I_i - I_t}{I_i} = 1 - \frac{I_t}{I_i}.$$

The reflected intensity can also be expressed as:

$$\frac{I_r}{I_i} = \frac{F\sin^2\left(\frac{\phi}{2}\right)}{1 + F\sin^2\left(\frac{\phi}{2}\right)} \tag{3.82}$$

The intensity of both transmitted and reflected beams depends on the reflectance of the plate surfaces and the phase shift ϕ, which is a function of the plate refractive index, its thickens, the wavelength of light, and angle of incidence as shown in Eq.(3.79). The condition for total transmission is known as the condition of resonance, which will be discussed later in this book when discussing optical resonators.

As cam be observed in Fig. 3.23, for a high reflectance, R, the light waves pass through at specific values of ϕ, which is a function of wavelength, λ or frequency, ν of the light wave as shown in the following relationship:

$$\phi = \frac{4\pi n_2 d\cos\theta_1}{\lambda} = \frac{4\pi n_2 d\nu\cos\theta_1}{c}$$

These frequencies for which $\phi = 2m\pi$ where m is an integer are known as the resonant frequencies of the cavity.

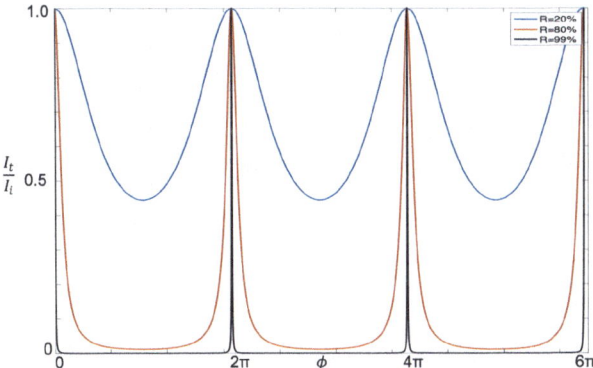

Fig. 3.23 The relative intensity of the transmitted beam as a function of surface reflectance $R = 20\%, 80\% \ and \ 99\%$. This plot is the same if plotted in terms of frequency or wavelength instead of the phase ϕ

3.15 Optical Interferometers

Interferometers are optical instruments that use the interference of two or more waves to perform precision measurements in a wide range of applications, including surface diagnostics, spectroscopy, gravitational wave detection, rotation sensing and optical communication exist, each designed according to its system architecture and application. While a comprehensive study of interferometers is beyond the scope of this book, we will highlight some of the well-known interferometers, focusing on their basic operating principles and practical applications through some examples.

3.15.1 Fabry–Perot Interferometer

One of the simplest configurations of an interferometer consists of two highly reflective mirrors separated with an air gap. When used interferometrically, the air gap typically ranges from several millimeters to several centimeters, but can be much larger when employed as a laser resonant cavity. If the gap can be mechanically adjusted by moving one of the mirrors, the device is referred to as a Fabry–Perot interferometer. When the mirrors are fixed and aligned for parallelism using a spacer, it is known as an etalon—though technically, it still functions as an interferometer.

The Fabry–Perot interferometer operates on the principle of multiple-beam interference. When a light beam strikes the first partially reflective surface, a portion of the beam is reflected while the remainder is transmitted into the medium between the mirrors. The transmitted beam is then reflected multiple times between the two surfaces, with a fraction of the light transmitted after each reflection, as illustrated in Fig. 3.24. This repeated reflection leads to interference patterns that can be analyzed for various measurement purposes.

Fig. 3.24 Typical Fabry–Perot interferometer configuration

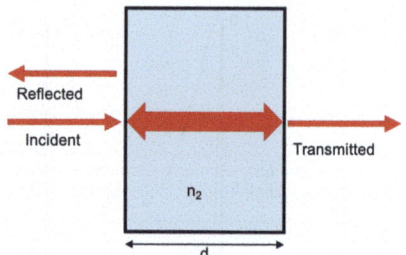

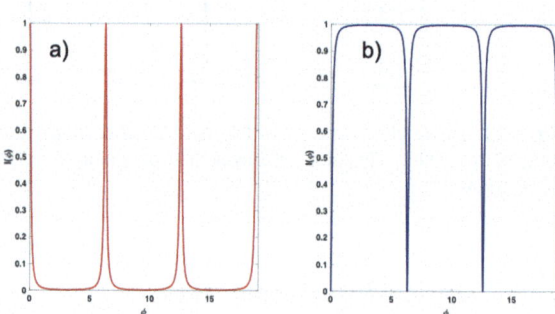

Fig. 3.25 Intensities of (**a**) transmitted and (**b**) reflected beams of a Fabry-Perot interferometer

The transmitted and reflected beams based on the analysis developed in Sect. 3.14 and the optical wave transmitted through the device are given by Eq. (3.82)

$$\frac{I_t}{I_i} = \frac{1}{1 + F sin^2\left(\frac{\phi}{2}\right)},$$

and Eq. (3.82)

$$\frac{I_r}{I_i} = \frac{F sin^2\left(\frac{\phi}{2}\right)}{1 + F sin^2\left(\frac{\phi}{2}\right)}.$$

The transmitted and reflected beams are plotted as a function of the phase ϕ in Fig. 3.25. It is clear that when the transmitted beam is at its peak, 100% transmission, the reflected beam amplitude is equal to zero. This takes place at all values of the phase $\phi = 2m\pi$, where m is an integer, m = 0, 1, 3, ... These are the conditions of resonance, where the reflected beam goes to zero. For a lossless cavity at resonance, all the light incident on the interferometer is transmitted through constructive interference, and the reflected beams are destructively interfere.

3.15.1.1 Free Spectral Range

The free-spectral (FSR) range of a Fabry–Perot Interferometer, λ_{fsr}, is the wavelength separation between two consecutive transmission peaks. It can also be

3.15 Optical Interferometers

Fig. 3.26 Fabry–Perot transmission spectrum labeled with the free spectral range, λ_{fsr} and linewidth, $\delta\lambda$

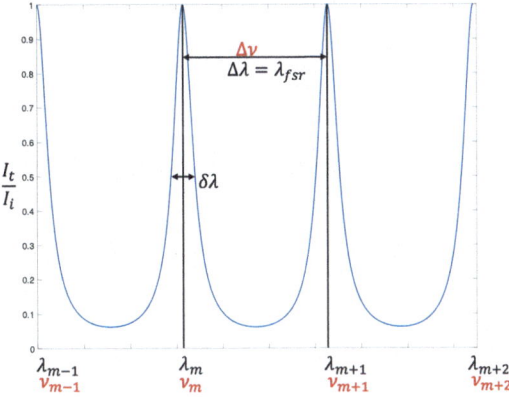

defined as the frequency range between two adjacent transmission peaks, $\Delta\nu$. The interferometer's transmission is at its maximum when the path difference between consecutive reflections is an integer multiple, m, of the wavelength. This occurs when the phase shift $\phi = 2m\pi$, where m is an integer, $m = 0, 1, 3, \ldots$ Figure 3.26 shows the interferometer spectrum as a function of wavelength. The wavelengths λ'_ms are the resonant wavelengths and $\delta\nu$ is the linewidth.

The Fabry–Perot transmission spectrum is labeled with the free spectral range, λfsr, and the linewidth $\Delta\nu$.

The phase shift condition for an interferometer peak is given by

$$\phi = \frac{4\pi n_2 d \cos\theta_1}{\lambda_m} = \frac{4\pi n_2 d \nu_m \cos\theta_1}{c} = 2m\pi$$

where

$$\nu_m = \frac{mc}{2n_2 d \cos\theta_1}$$

and

$$\nu_{m+1} = \frac{(m+1)c}{2n_2 d \cos\theta_1}.$$

Therefore, the free spectral range is

$$\Delta\nu = \nu_{m+1} - \nu_m = \frac{c}{2n_2 d \cos\theta_1} \tag{3.83}$$

In the case where the medium between the two reflecting surfaces is air, $n_2 = 1$ and $\theta = 0$, then the free spectral range is given by

$$\Delta\nu = \frac{c}{2d} \tag{3.84}$$

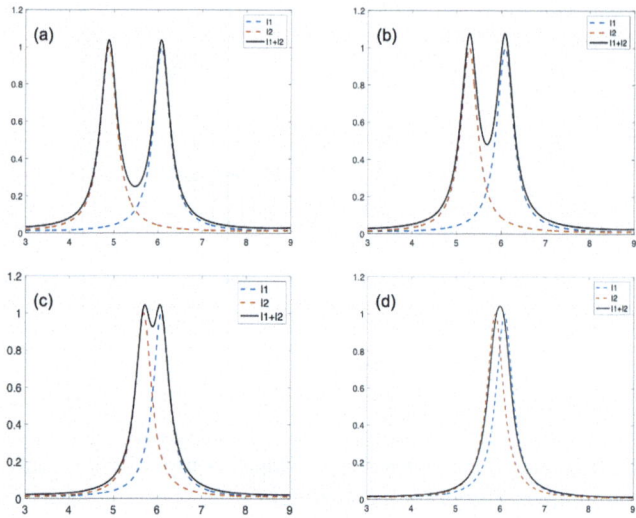

Fig. 3.27 Transmission plots for two adjacent wavelengths, plotted separately as I_1 and I_2, and the way it will show as I_3 for a polychromatic light source. In plots (**a**) and (**b**), the peaks are clearly resolved while in (**c**) are very close, and in (**d**), they are completely unresolved

This expression can be derived in terms of the wavelength λ, since $\nu = \frac{c}{\lambda}$ then $\frac{d\nu}{d\lambda} = -\frac{c}{\lambda^2}$; therefore, the free spectral range in wavelength terms is

$$\Delta\lambda = lambda_{fsr} = \frac{\lambda^2}{2d} \qquad (3.85)$$

We have dropped the negative sign, which means that the order of the peaks in wavelength and frequency scales are opposite to one another.

3.15.1.2 Resolving Power

One application of the Fabry–Perot interferometer is in spectroscopic analysis. When a light source is not monochromatic, its spectrum consists of several wavelengths. In this case, the transmitted beam will exhibit distinct peaks corresponding to the different wavelengths. To resolve adjacent peaks, they must be sufficiently separated. This ensures that the individual wavelengths can be distinguished, producing a discrete spectrum. The resolving power of the Fabry–Perot interferometer is defined as the minimum difference between two wavelengths that can be distinguished.

This concept is illustrated in Fig. 3.27. Plots I_1 and I_2 show the interference patterns for two monochromatic sources with different wavelengths. The difference between these wavelengths decreases progressively from subplots (a) to (d). Plot I_3 represents a polychromatic source emitting these wavelengths simultaneously.

The criterion that defines the resolving power of the interferometer is that the separation between adjacent peaks must be at least Full Width at Half Magnitude

3.15 Optical Interferometers

(FWHM). We begin by determining the FWHM, denoted as $\delta \lambda$ $\delta\lambda$, of the transmission spectrum, as shown in Fig. 3.26.

3.15.1.3 Rayleigh Criterion

The Rayleigh criterion is the widely accepted standard for defining the resolving power of an interferometer. It specifies the minimum separation between two peaks in the interferometer's spectrum required to distinguish between two adjacent wavelengths. To apply this criterion, we first need to determine the phase shift $\Delta\phi$ that reduces the transmission amplitude to 50% of its peak value. Using Eq. (3.83), we can proceed with this calculation

$$\frac{I_t}{I_i} = \frac{1}{2} = \frac{1}{1 + F \sin^2\left(\frac{\Delta\phi}{2}\right)} \tag{3.86}$$

This condition requires that

$$F \sin^2(\frac{\Delta\phi}{2}) = 1$$

$$\sin\left(\frac{\Delta\phi}{2}\right) = \frac{1}{\sqrt{F}}.$$

But $\phi = 0$ at the peak so $\Delta\phi$ is a very small angle. For sharp peaks, i.e., large finesse, the angle $\Delta\phi$ at half magnitude is very small, so the sine of the angle will be approximately equal to the angle in radians. Therefore, at the phase at half-magnitude is

$$\frac{\Delta\phi}{2} \approx \frac{1}{\sqrt{F}}$$

$$\Delta\phi \approx \frac{2}{\sqrt{F}}$$

The Rayleigh criterion for the minimum resolving power requires that the two peaks be separated by $2\Delta\phi$. So the phase difference between two peaks

$$\phi_{FWHM} = 2\Delta\phi \approx \frac{4}{\sqrt{F}}$$

But the phase of the transmitted wave for an interferometer with air gap and zero angle of incidence, Eq. (3.79), is $\phi = \frac{4\pi d}{\lambda}$. By taking the derivative of ϕ with respect to λ, we get the incremental change of ϕ as

$$\delta\phi = \frac{4\pi d \delta\lambda}{\lambda^2},$$

and

$$\delta\lambda = \frac{\lambda^2 \delta\phi}{4\pi d},$$

Therefore,

$$\delta\lambda_{min} = \frac{\lambda^2 \phi_{min}}{4\pi d}.$$

$$\delta\lambda_{min} = \frac{\lambda^2 \left(\frac{4}{\sqrt{F}}\right)}{4\pi d}.$$

Therefore, the resolving power of the Fabry–Perot interferometer is

$$\delta\lambda_{min} = \frac{\lambda^2}{\pi d\sqrt{F}}. \qquad (3.87)$$

Example A Fabry–Perot interferometer has a mirror reflectivity of 85% and a cavity length of 2 cm. The incident light has a wavelength of 500 nm and the refractive index of the medium inside the interferometer n=1.1. Calculate the finesse and the free spectral range (FSR) of the interferometer.

Solution Finesse (F) is a dimensionless parameter that characterizes the sharpness of the resonances in a Fabry–Perot interferometer. It can be calculated using the mirror reflectivity (R) as follows:

$$F = \pi\sqrt{R/(1-R)}$$

In this problem, the mirror reflectivity is 85%, so R = 0.85.

$$F = \pi\sqrt{0.85/(1-0.85)} F \approx 9.63$$

The finesse of the Fabry–Perot interferometer is approximately 9.63.

Free spectral range (FSR) is the separation between two adjacent resonance peaks and can be calculated using the cavity length (L) and the speed of light (c) as follows:

$$FSR = c/(2 \times L \times n)$$

(continued)

In this problem, the cavity length $L = 2$ cm, which is equal to 0.02 m. The incident light has a $\lambda = 500$ nm,, which is equal to 5×10^{-7} m. Now, we can calculate the FSR:

$$FSR \approx \frac{3 \times 10^8}{2 \times 1.1 \times 0.02} = 6.8 \times 10^9 \text{ Hz}$$

The free spectral range of the Fabry–Perot interferometer is 6.8×10^9 Hz.

3.15.2 Michelson Interferometer

The Michelson interferometer operates on the principle of amplitude-splitting two-wave interference. It is by far the best-known and historically the most important interferometer. A typical configuration of the Michelson interferometer is shown in Fig. 3.28. In this setup, an extended light source emits a beam that travels to the right. The beam splitter, BS, divides the beam into two parts, one segment travels to the right and the other travels upward. The two beams reflect off mirrors M1 and M2 and return to the beam splitter. A portion of the wave coming from M2 passes through the beam splitter and is transmitted downward, and part of the beam reflected from M1 is reflected by the beam splitter. The two beams are then recombined to form the output beam. The output beam falls on a screen or is directed toward a detector that is used for analyzing the interference of the two beams.

The two optical paths followed by the beams are known as the two arms of the interferometer. When the optical path lengths of the two arms are equal, constructive interference occurs between the two waves. If the optical path lengths are different by a fraction of a wavelength, a destructive interference occurs. Small changes in the optical path lengths cause a shift in the interference fringes. For different

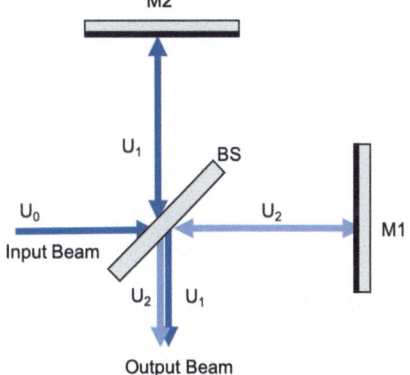

Fig. 3.28 A typical configuration for a Michelson Interferometer made of two mirror, M1 and M2, and a beam splitter, BS

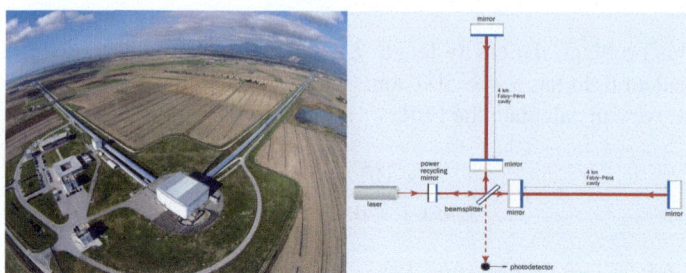

Fig. 3.29 Laser Interferometer Gravitational-Wave Observatory (LIGO) on the left is the facility in Florida and the right the LIGO Michelson interferometer configuration

applications of the interferometer, the two light paths can have different lengths, by moving one of the mirrors or tilting it or incorporating an optical element or even materials under test.

In 2015, Michelson interferometer was used in detecting gravitational waves for the first time in history. These waves were detected by two identical interferometers called, Laser Interferometer Gravitational-Wave Observatory (LIGO): one is located in Florida and the other in Washington State. Each arm of the interferometer is 4 km long. The LIGO system is shown in Fig. 3.29.

Example If a thin glass ($n_g = 1.520$) sheet 0.050 mm thick is inserted into one arm of a Michelson Interferometer illuminated by yellow helium light ($\lambda = 587.56$ nm), how many fringe pairs will thereupon be displaced?

Solution A shift in path of $\frac{\lambda_o}{2}$ corresponds, because the apparatus is in air, to a shift in OPL of $\frac{\lambda_o}{2}$, and a displacement of one fringe pair. By inserting glass plate of thickness D, thereby replacing a sheet of air, changes the OPL by an amount $Dn_g - Dn_{air} = D(n_g - 1)$. The light beam traverses twice through the glass plate, which corresponds to a distance of $N\lambda_o$ where N is the number of fringe pairs that shift due to the insertion of the glass plate in the path of the beam. Thus,

$$2D(n_g - 1) = N\lambda_o$$

and

$$N = \frac{2D(n_g - 1)}{\lambda_o} = \frac{2(0.050 \times 10^{-3})(0.520)}{587.56 \times 10^{-9}}$$

Therefore, $N = 88$

3.15 Optical Interferometers

Example A Michelson interferometer is used to measure the wavelength of monochromatic light. The movable mirror is shifted by a distance of 5.5 μm, and this results in 20 complete dark-to-bright fringe cycles. Calculate the wavelength of the light.

Solution In a Michelson interferometer, when the movable mirror is shifted by a certain distance, it causes a change in the optical path difference (OPD) that results in the formation of interference fringes. The relationship between the shift in the movable mirror (d), the number of fringe cycles (N), and the wavelength of the light (λ) is given by:

$$2 \times d = N \times \lambda$$

In this problem, the movable mirror is shifted by a distance of 5.5 μm, which is equal to 5.5×10^{-6} m, and the number of fringe cycles N is 20. We can now solve for the wavelength (λ):

$$\lambda = (2 * d)/N$$

$$\lambda = (2 \times 5.5 \times 10^{-6} \text{m})/20$$

$$\lambda = 11 \times 10^{-6} \text{m}/20$$

$$\lambda = 0.55 \times 10^{-6} \text{ m}$$

The wavelength of the monochromatic light is 0.55 μm or 550 nm.

Example A Michelson interferometer is used to measure the refractive index of a glass plate with a thickness of 2 mm. The glass plate is placed in the path of one of the interferometer's arms, and it results in a fringe shift of 300 fringes. The wavelength of the light source used is 600 nm. Calculate the refractive index of the glass plate.

Solution When a glass plate of thickness (t) and refractive index (n) is inserted in one of the arms of a Michelson interferometer, it results in a fringe shift (ΔN). The relationship between the fringe shift, the thickness of the glass plate, the refractive index, and the wavelength of the light (λ) is given by:

(continued)

$$\Delta N = (2 \times n \times t)/\lambda - (2 \times t)/\lambda$$

We need to solve for the refractive index (n):

$$n = (\Delta N \times \lambda + 2 \times t)/(2 \times t)$$

In this problem, the thickness of the glass plate is 2 mm, which is equal to 2×10^{-3} m, the fringe shift $\Delta N = 300$, and the wavelength of the light source is 600 nm, which is equal to 600×10^{-9} m. We can now calculate the refractive index (n):

$$n = (300 \times 600 \times 10^{-9} \text{ m} + 2 \times 2 \times 10^{-3} \text{ m})/(2 \times 2 \times 10^{-3} \text{ m})$$

$$n = (1.8 \times 10^{-4} \text{ m} + 4 \times 10^{-3} \text{ m})/4 \times 10^{-3} \text{ m}$$

$$n \approx 1.045$$

The refractive index of the glass plate is approximately 1.045.

3.15.3 Mach–Zehnder Interferometer

Mach–Zehnder Interferometer (MZI) is an amplitude-splitting two-wave interferometer. The input beam U0, as shown in Fig. 3.30, is split into two waves U1 and U2 by a beam splitter; then these two beams are reflected by separate mirrors and finally combined by another beam splitter forming the output beam U1+U2, which is in turn directed toward a detector. The paths of the two beams U1 and U2 form the two arms of the interferometer. When the two arms have the same optical path lengths, the two waves will constructively interfere. In applications of the apparatus, an optical component is placed in one of the arms and the interference fringes will be used to determine the properties of the object under test.

Probably the most important application of the Mach–Zehnder interferometer is an optical modulator in optical communication systems. MZI modulator takes the form of integrated photonic circuit made of Lithium Niobate as depicted Fig. 3.31. One of the arms of the modulator is electrically activated to change its index of refraction, which in turn changes the optical path length and alters the phase of the wave traveling through it. In digital communication application, when there is no voltage applied, both arms introduce the same phase and the two waves will add up, when a voltage applied that alters the phase by π, the two waves will cancel one another.

3.15 Optical Interferometers

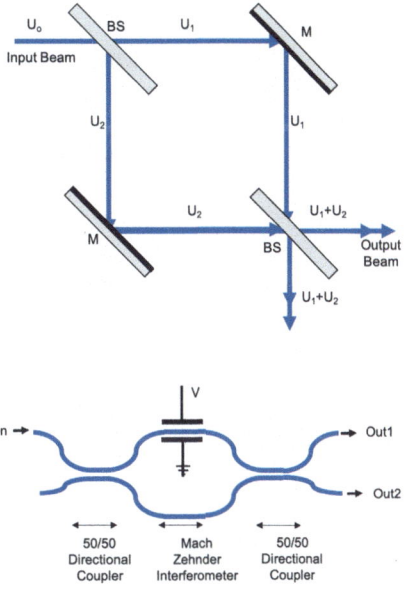

Fig. 3.30 Mach–Zehnder interferometer. The input beam is split into two beams U1 and U2 that follow different paths that combined at the output detector

Fig. 3.31 Mach–Zehnder integrated photonic circuit modulator

Example A Mach–Zehnder interferometer is used to measure the refractive index of a gas introduced into one of its arms. The length of each arm of the interferometer is 50 cm, and the wavelength of the light source used is 633 nm. When the gas is introduced, it results in a fringe shift of 150 fringes. Calculate the change in refractive index of the gas (Δn) compared to the initial condition (usually air).

Solution In a Mach–Zehnder interferometer, when a gas with a different refractive index is introduced into one of its arms, it results in a fringe shift (ΔN). The relationship between the fringe shift, the length of the arm (L), the change in refractive index (Δn), and the wavelength of the light (λ) is given by:

$$\Delta N = (2 \times L \times \Delta n)/\lambda$$

We need to solve for the change in refractive index (Δn):

$$\Delta n = (\Delta N \times \lambda)/(2 \times L)$$

In this problem, the length of each arm of the interferometer is 50 cm, which is equal to 0.5 m, the fringe shift ΔN is 150, and the wavelength of the

(continued)

light source is 633 nm, which is equal to 633×10^{-9} m. We can now calculate the change in refractive index (Δn):

$$\Delta n = (150 \times 633 \times 10^{-9}\text{m})/(2 \times 0.5\text{m})$$

$$\Delta n = (9.495 \times 10^{-5}\text{m})/1\text{m}$$

$$\Delta n \approx 9.495 \times 10^{-5}$$

The change in refractive index of the gas compared to the initial condition is approximately 9.495×10^{-5}.

3.15.4 Sagnac Interferometer

The Sagnac interferometer is an amplitude-splitting instrument that differs from the previous interferometers in many respects. It is very easy to align and quite stable. An interesting application of the device is its use as a gyroscope. The main feature of the device is that it has two identical but oppositely directed paths taken by the beams and that both form closed loops before they are united to produce interference. A deliberate slight shift in the orientation of one of the mirrors will produce a path length difference and a resulting fringe pattern. Since the beams are superimposed and therefore inseparable, the interferometer cannot be put to any of the conventional uses. A particular configuration of the instrument is shown in Fig. 3.32. The device is made of three mirrors and one beam splitter. The wave from a light source is split into two by the beam splitter. One of the waves will follow the path ABCDA, and the second wave will take the path ADCBA, then the beam splitter combines both waves at the output port. When the device is in stationary state, the two waves would have traveled the same optical path lengths and interfere constructively. When the device is rotated with radial speed ω, as shown, then the two waves will travel with different speeds resulting in different optical path lengths and the waves will be no longer constructively interfere. They will produce an output amplitude that is a function of the rotation speed. This is how they are used as gyroscopes. Sagnac interferometers are now used as gyroscopes for airplanes, guided missiles, and other rotation sensing applications.

If the interferometer is rotated clockwise with a linear speed $v = R\omega$, where R is the diagonal of the square as shown in Fig. 3.32, so the time travel of light between corners $A \, to \, B = t_{AB}$

$$t_{AB} = \frac{R\sqrt{2}}{c - \frac{v}{\sqrt{2}}}$$

3.15 Optical Interferometers

Fig. 3.32 Discrete Sagnac interferometer

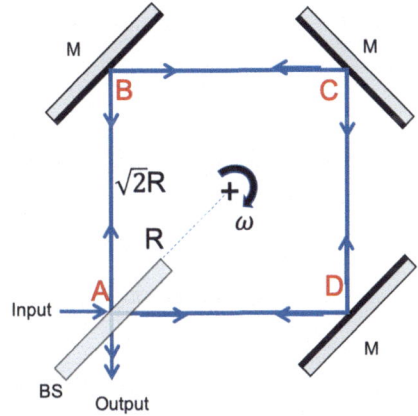

$$t_{AB} = \frac{2R}{\sqrt{2}c - \omega R}$$

$$t_{AD} = \frac{2R}{\sqrt{2}c - \omega R}$$

Hence, the time for the light to travel clockwise and anticlockwise:

$$t_{CW} = \frac{8R}{\sqrt{2}c + \omega R}$$

$$t_{CCW} = \frac{8R}{\sqrt{2}c - \omega R}$$

But $\omega R \ll c$. Therefore, the travel time difference between the clockwise and counterclockwise waves is

$$\Delta t = t_{CW} - t_{CCW}$$

$$\Delta t = \frac{8R}{\sqrt{2}c - \omega R} - \frac{8R}{\sqrt{2}c + \omega R} \tag{3.88}$$

$$\Delta t = \frac{16\omega R^2}{2c^2 - \omega^2 R^2} \approx \frac{8\omega R^2}{c^2} \tag{3.89}$$

This can be expressed in terms of area $A = 2R2$ of the square formed by the beam of light as

$$\Delta t = \frac{4A\omega R}{c^2}$$

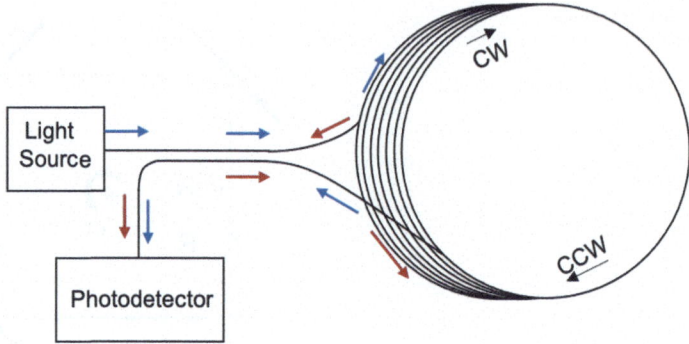

Fig. 3.33 Optical fiber gyroscope using several kilometers of fiber to increase the area that results in the increase of the gyroscope sensitivity

Let the period of the monochromatic light used be

$$period = \tau = \frac{\lambda}{c}$$

The fractional displacement of fringes

$$\Delta N = \frac{\Delta t}{\tau} = \frac{4A\omega R}{c\lambda} \tag{3.90}$$

It is obvious that the sensitivity of the gyroscope increases by increasing the area A. In order to increase the area gyroscopes, use optical fibers, Fig. 3.33 with very large length to increase the path length difference between the interfering waves and in turn increases its rotation sensitivity.

Example A Sagnac interferometer is used to measure the angular velocity of a rotating platform. The radius of the circular path of the light beams in the interferometer is 10 cm, and the wavelength of the light source used is 780 nm. When the platform is rotating, a fringe shift of 200 fringes is observed. Calculate the angular velocity (ω) of the rotating platform in rad/s.

Solution In a Sagnac interferometer, the fringe shift (ΔN) is related to the angular velocity (ω) of the rotating platform, the area (A) enclosed by the light beams, and the wavelength of the light (ω) as follows:

$$\Delta N = (4 \times A \times \omega)/\lambda$$

(continued)

We need to solve for the angular velocity (ω):

$$\omega = (\Delta N \times \lambda)/(4 \times A)$$

First, we need to calculate the area enclosed by the light beams in the interferometer. For a circular path with a radius (r), the area is given by:

$$A = \pi \times r^2$$

In this problem, the radius of the circular path is 10 cm, which is equal to 0.1 m:

$$A = \pi \times (0.1 \text{m})^2$$

$$A \approx 0.0314 \text{m}^2$$

Now, the fringe shift ΔN is 200, and the wavelength of the light source is 780 nm, which is equal to 780×10^{-9}m. We can now calculate the angular velocity (ω):

$$\omega = (200 \times 780 \times 10^{-9}\text{m})/(4 \times 0.0314 \text{m}^2)$$

$$\omega \approx 1.56 \times 10^{-5} \text{rad/s}$$

The angular velocity of the rotating platform is approximately 1.56×10^{-5} rad/s.

3.16 Summary

This chapter explored the principles of physical optics, focusing on how light behaves as an electromagnetic wave. Beginning with the electromagnetic nature of light, Maxwell's equations were introduced to describe how light propagates through space as coupled electric and magnetic fields. Key topics covered include the definitions of optical wave parameters such as polarization, magnitude, and phase, and their influence on wave propagation.

The chapter also discussed the behavior of light as it interacts with different media, including the laws of reflection and refraction derived from wave theory, and the transmission and reflection coefficients. Solutions to the wave equation, such as planewaves and spherical waves, were introduced as fundamental concepts in understanding wave propagation.

Interference of light waves, a major theme of the chapter, was examined in various contexts such as two-wave interference and interference caused by division of wavefronts and amplitudes. Important interferometric systems, including the Michelson, Mach–Zehnder, and Fabry–Perot interferometers, were also analyzed, along with their applications in optical measurement and communication systems.

3.17 Problems

1. Show that the spherical wave $u(r) = \frac{a}{r}e^{-jkr}$ is a solution of Helmholtz wave equation.
2. (a) Derive an expression for the interference of D planewave and a spherical wave emitted by a light source located a distance r from an observation screen.
   (b) Derive an expression for the fringe spacing on the screen.
   (c) Plot the interference pattern for $\lambda_o = 500$ nm and $D = 5m$
   (d) In a Young's double slit experiment, the light source used was emitting two wavelengths of 440 nm (blue) and 660 nm (red), the interference fringe pattern on a screen placed at a distance $D = 90$ cm shows two bright red fringes on either side of the central bright fringe, if the separation between the two slits is 0.3 mm, what is the separation between the bright red fringes?
3. In a Young's double-slit experiment using a monochromatic light of wavelength λ, the intensity of the light at a point on the screen when path difference is λ is K units. What is the intensity of light at a point when the path difference is $frac\lambda 3$.
4. In a Young's double-slit experiment with a monochromatic light source, the separation between the two slits is 0.5 mm and the screen is placed 1m away, when a thin transparent plastic with refractive index $n = 1.5$ sheet is placed over one slit the fringe pattern shifted 5 cm. What is the thickness of the sheet?
5. Two parallel narrow slits in an opaque screen are separated by 0.100 mm. They are illuminated by planewaves of wavelength 600 nm. A cosine-squared fringe pattern wherein consecutive maxima are 3.00 mm apart appears on a viewing screen. How far from the aperture screen is the viewing screen?
6. An expanded beam of red light from a HeNe laser ($\lambda_o = 632.8$ nm) is incident on a screen containing two very narrow horizontal slits separated by 0.200 mm. A fringe pattern appears on a white screen held 1.00 m away.
   (a) How far (in radians and millimeters) above and below the central axis are the first zeros of the intensity?
   (b) How far (in mm) from the axis is the fifth bright band?
7. Two pinholes in a thin sheet of aluminum are 1.00 mm apart and immersed in a large tank of water ($n = 1.33$). The holes are illuminated by $\lambda_o = 500$ nm planewaves, and the resulting fringe system is observed on a screen in the water, 3.00m from the holes. Determine the locations of the centers of the two maxima closest to the central axis of the apparatus.
8. A Mach–Zehnder interferometer is used to measure the coefficient of thermal expansion (α) of a glass rod. The length of the glass rod and each arm of the

interferometer is 30 cm. The wavelength of the light source used is 532 nm. When the temperature of the glass rod is increased by 5°C, it results in a fringe shift of 120 fringes. Calculate the coefficient of thermal expansion (α) of the glass rod.
9. A Sagnac interferometer is used as a gyroscope in a navigation system. The area enclosed by the light beams in the interferometer is $0.5 \, m^2$, and the wavelength of the light source used is 800 nm. When the navigation system undergoes a rotation, a fringe shift of 500 fringes is observed. Calculate the angular velocity (ω) of the rotation in degrees/s.

Bibliography

1. Born, M., & Wolf, E. (1999). Principles of Optics (7th ed.). Cambridge University Press.
2. Jackson, J. D. (1999). Classical Electrodynamics (3rd ed.). Wiley.
3. Hecht, E. (2017). Optics (5th ed.). Addison-Wesley.
4. Goodman, J. W. (2005). Introduction to Fourier Optics (3rd ed.). Roberts and Company Publishers.
5. Yariv, A., & Yeh, P. (2006). Photonics: Optical Electronics in Modern Communications (6th ed.). Oxford University Press.
6. Saleh, B. E. A., & Teich, M. C. (2019). Fundamentals of Photonics (3rd ed.). Wiley-Interscience.
7. Pedrotti, F. L., Pedrotti, L. S., & Pedrotti, L. M. (2017). Introduction to Optics (3rd ed.). Cambridge University Press.
8. Gaskill, J. D. (1978). Linear Systems, Fourier Transforms, and Optics. Wiley.
9. Lipson, S. G., Lipson, H., & Tannhauser, D. S. (2011). Optical Physics (4th ed.). Cambridge University Press.
10. Azzam, R. M. A., & Bashara, N. M. (1987). Ellipsometry and Polarized Light. North-Holland.
11. Born, M. (1980). The Mechanics of Light. Dover Publications.

Fourier Optics

4.1 Introduction

The branch of optical science known as "Fourier optics" originated in the 1940s through the 1960s with the application of communication theory and circuit design analysis techniques to optical diffraction theory. In 1968, Joseph W. Goodman published Introduction to Fourier Optics, a seminal textbook that consolidated fundamental concepts and applications, establishing Fourier optics as a significant and enduring discipline. Today, Fourier optics is applied to the analysis of diffraction and imaging, as well as specialized fields such as optical information processing, wavefront control, optical computing, and holography.

Optical systems, which are governed by Maxwell's equations for the electromagnetic field and the wave equation, behave as linear systems. This makes them well suited to mathematical techniques developed for electrical systems, such as Fourier transforms, convolutions, and correlations. In physical optics, important system characteristics such as the point-spread function and the optical transfer function can be described rigorously. This allows for the formulation of general transmission functions for lenses and the design of optical systems that are valid even in non-paraxial regimes. Fourier optics provides a rigorous theoretical framework for analyzing linear optical systems, enabling accurate calculations of resolution, imaging, and other interference phenomena.

The strength of the Fourier series and Fourier transform lies in their ability to decompose complex temporal or spatial functions into sets of fundamental harmonics. In the context of optics, this means that optical waves can be represented as a collection of planewaves, each propagating in a different direction or at different spatial frequencies. This chapter begins with a brief review of Fourier series and then extends the discussion to the Fourier transform, laying the groundwork for the deeper understanding and applications of Fourier optics.

4.2 Fourier Series

A Fourier series can represent periodic functions as sums of sine and cosine functions of different frequencies. Since optical functions are typically defined in the spatial domain, they are represented as two-dimensional functions. For a one-dimensional spatial periodic function $f_L(x)$ with a period L, the function can be represented as an infinite series, known as Fourier series, and is given by:

$$f_L(x) = a_0 + \sum_{n=1}^{\infty} \left(a_n \cos\left(\frac{2\pi n x}{L}\right) + b_n \sin\left(\frac{2\pi n x}{L}\right) \right) \quad (4.1)$$

where the constants a_n and b_n are the Fourier coefficients that are given by

$$a_0 = \frac{1}{L} \int_{-L/2}^{L/2} f_L(x)\,dx, \quad (4.2)$$

$$a_n = \frac{2}{L} \int_{-L/2}^{L/2} f_L(x) \cos\left(\frac{2\pi n x}{L}\right) dx, \quad (4.3)$$

and

$$b_n = \frac{2}{L} \int_{-L/2}^{L/2} f_L(x) \sin\left(\frac{2\pi n x}{L}\right) dx. \quad (4.4)$$

To obtain a complete and accurate representation of the function, all the infinite terms of the series must be summed. However, this is clearly impractical. In most real-world applications, only a finite number of terms of the series are used to approximate the function. The quality of the approximation improves as more terms are included. This concept will be illustrated through the following graphical examples.

Example Calculate and plot the Fourier series of a pulse train shown in Fig. 4.1, with a period L and pulse width $L/2$, for $N = 1, 3, 5$ *and* 50.

Fig. 4.1 A pulse with width $L/$

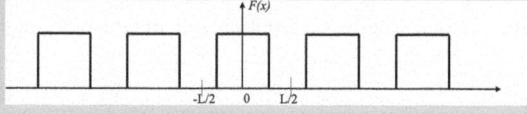

(continued)

4.2 Fourier Series

Solution The pulse can be represented by the function $f(x)$ given by

$$f(x) = \begin{cases} 1, \text{ for } -L/4 \leq x \leq L/4 \\ 0, \text{ elsewhere} \end{cases}$$

Using MATLAB to generate the series and plot it for the given number of terms of the series to be summed, we get the output in Fig. 4.2.

From the plots, it is clear that taking only the first term of the series does not accurately represent the input pulse. As the number of terms increases, the sum of the series more closely approximates the pulse shape. The vertical edges of the pulse correspond to very high frequencies, while the horizontal sections of the pulse represent low or zero frequencies.

Now let us consider a general representation of the Fourier series in a complex notation.

$$f_L(x) = \sum_{n=-\infty}^{\infty} \alpha_n e^{j\frac{2\pi x n}{L}} \qquad (4.5)$$

where

$$2\alpha_n = a_n - jb_n, \text{ for } n > 0,$$

$$2\alpha_n = a_n + jb_n, \text{ for } n < 0$$

and

$$\alpha_0 = a_0$$

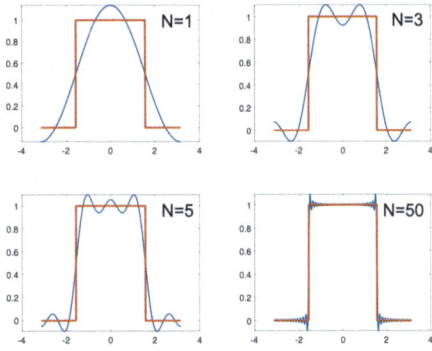

Fig. 4.2 Plots of the Fourier series for $f(x)$ for $N = 1, 3, 5$ and 50

The Fourier complex coefficient is given by

$$\alpha_n = \frac{1}{L} \int_{-L/2}^{L/2} f_L(x) e^{-j\frac{2\pi x n}{L}} \qquad (4.6)$$

4.3 Fourier Transform

The Fourier series decomposes periodic functions in an infinite series of its harmonics. Substitute Eq. (4.6) into Eq. (4.5) to get the Fourier series for $f_L(x)$ to become

$$f_L(x) = \sum_{n=-\infty}^{\infty} \left[\frac{1}{L} \int_{-L/2}^{L/2} f_L(x) e^{-j\frac{2\pi x n}{L}} \right] e^{j\frac{2\pi x n}{L}} \qquad (4.7)$$

Let $f_{x,n} = \frac{n}{L}$ and perform a change of variables in the Fourier coefficient by setting $u = x$. Thus, Eq. (4.7) becomes:

$$f_L(x) = \sum_{n=-\infty}^{\infty} \left[\frac{1}{L} \int_{-L/2}^{L/2} f_L(u) e^{-j2\pi x f_{x,n}} du \right] e^{j2\pi x f_{x,n}}$$

Define $\Delta f_x = f_{x,n+1} - f_{x,n} = \frac{n+1}{L} - \frac{n}{L} = \frac{1}{L}$, so the expression for $f_L(x)$ becomes:

$$f_L(x) = \sum_{n=-\infty}^{\infty} e^{j2\pi x f_{x,n}} \Delta f_x \left[\int_{-L/2}^{L/2} f_L(u) e^{-j2\pi x f_{x,n}} du \right] \qquad (4.8)$$

Now, let $L \to \infty$ and assume:

$$f(x) = \lim_{L \to \infty} f_L(x),$$

and $\int_{-\infty}^{\infty} |f(x)| dx$ exists. At this limit, the infinite series becomes an integral from $-\infty$ to ∞. Thus, we obtain:

$$f(x) = \int_{-\infty}^{\infty} df_x \int_{-\infty}^{\infty} f(u) e^{j2\pi f_x(x-u)} du \qquad (4.9)$$

This is, Eq. (4.9), the Fourier Integral Theorem. The Fourier transform of $f(x)$, denoted $F(f_x)$, is given by:

$$F(f_x) = \int_{-\infty}^{\infty} f(u) e^{-j2\pi f_x u} du \qquad (4.10)$$

4.3 Fourier Transform

where $F(f_x)$ is the Fourier transform of the function $f(x)$, and f_x is the spatial frequency in lines/mm. The inverse Fourier transform is given by:

$$f(x) = \int_{-\infty}^{\infty} F(f_x) e^{j2\pi x f_x} df_x \tag{4.11}$$

Example Determine the Fourier transform of the function $cos(2\pi x)$.

Solution To determine the Fourier transform of the function $\cos(2\pi x)$, we use the following Fourier transform definition as given by Eq. (4.10):

$$F(\omega) = \int_{-\infty}^{\infty} f(x) e^{-j\omega x} dx,$$

where $F(\omega)$ is the Fourier transform of the function $f(x)$, and ω is the angular frequency.

For the given function, $f(x) = cos(2\pi x)$. We substitute the function into the Fourier transform to get:

$$F(\omega) = \int_{-\infty}^{\infty} cos(2\pi x) * e^{-j\omega x} dx$$

To perform this integral, we use Euler's formula:

$$cos(\theta) = (e^{j\theta} + e^{-j\theta})/2.$$

Applying Euler's formula, the integral becomes:

$$F(\omega) = (1/2) \int_{-\infty}^{\infty} [(e^{j2\pi x} + e^{-j2\pi x}) * e^{-j\omega x}] dx.$$

Now, distribute the exponential term:

$$F(\omega) = (1/2) \int_{-\infty}^{\infty} \left(e^{j2\pi x} \times e^{-j\omega x} + e^{-j2\pi x} \times e^{-j\omega x} \right) dx.$$

Combining the exponentials leads to:

(continued)

$$F(\omega) = (1/2) \int_{-\infty}^{\infty} \left(e^{j(2\pi-\omega)x} + e^{-j(2\pi+\omega)x} \right) dx.$$

Now, we need to perform two separate integrations, i.e.,

$$F(\omega) = (1/2) \left(\int_{-\infty}^{\infty} e^{j(2\pi-\omega)x} dx + \int_{-\infty}^{\infty} e^{-j(2\pi+\omega)x} dx \right)$$

Using the integral property $\int e^{ax} dx = (1/a)e^{ax}$, we get:

$$F(\omega) = (1/2)[(1/j(2\pi-\omega))e^{j(2\pi-\omega)x} + (1/-j(2\pi+\omega))e^{-j(2\pi+\omega)x}]\Big|_{-\infty}^{+\infty}$$

Evaluating these terms at the limits of integration, we find that the integral converges only when $\omega = 2\pi$ and $\omega = -2\pi$. Using the definition of Dirac delta function, we obtain:

$$F(\omega) = \pi[\delta(\omega-2\pi) + \delta(\omega+2\pi)]$$

where $\delta(x)$ is the Dirac delta function, which is equal to zero everywhere except at $x = 0$, where it is infinite (with an integral equal to 1 over the entire range of the variable x).

Example Determine the Fourier transform of the rectangular function $f(x) = rect(x/L)$ defined as

$$f(x) = rect(x/L) = \begin{cases} 1 & -L/2 \leq x \leq L/2 \\ 0 & elsewhere \end{cases} \quad (4.12)$$

Solution Substitute in the Fourier transform integral to get

$$F(f_x) = \int_{-\infty}^{\infty} f(x) e^{-j2\pi x f_x} dx$$

$$= \int_{-L/2}^{L/2} e^{-j2\pi x f_x} dx$$

(continued)

4.3 Fourier Transform

Fig. 4.3 Plot of the Fourier transform of the rectangular function

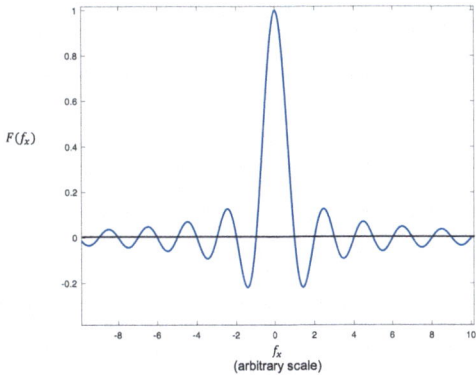

$$= \frac{-1}{j2\pi f_x} \left[e^{-j\pi L f_x} - e^{j\pi L f_x} \, dx \right]$$

$$= L \frac{sin(\pi L f_x)}{\pi L f_x}.$$

Therefore, the Fourier transform of the rectangular function is

$$F(f_x) = L sinc(L f_x). \tag{4.13}$$

The plot of $F(f_x)$ is shown in Fig. 4.3 for an arbitrary value of L.

Example Derive the Fourier transform of the Gaussian function $g(x)$

$$g(x) = e^{\frac{j\pi x^2}{\lambda F}}. \tag{4.14}$$

Solution The Fourier transform of the function $g(x)$ is given by:

$$\mathcal{F}\{g(x)\} = G(f_x)$$

where $\mathcal{F}\{\ldots\}$ is the Fourier transform operator. Therefore, the Fourier transform of the function $g(x)$ denoted by $G(f_x)$ is given by:

(continued)

$$G(f_x) = \int_{-\infty}^{\infty} dx\, e^{\frac{j\pi x^2}{\lambda F}} e^{-j2\pi x f_x}$$

Using the typical technique of completing the squares in the exponents results in:

$$G(f_x) = \int_{-\infty}^{\infty} dx\, e^{j\frac{\pi}{\lambda F}[x^2 - 2\lambda F x f_x + (\lambda F f_x)^2]} e^{-j\pi(\lambda F) f_x^2}$$

$$e^{j\pi(\lambda F) f_x^2} \int_{-\infty}^{\infty} dx\, e^{\frac{j\pi}{\lambda F}(x - \lambda F f_x)^2}$$

Perform the following change of variables:

$$u = \sqrt{\frac{2}{\lambda F}}(x - \lambda F f_x)$$

$$\frac{u^2}{2} = \frac{1}{\lambda F}(x - \lambda F f_x)^2$$

Hence,

$$du = \sqrt{\frac{2}{\lambda F}}\, dx$$

$$G(f_x) = \sqrt{\frac{2}{\lambda F}}\, e^{-j\pi(\lambda F) f_x^2} \int_{-\infty}^{\infty} du\, e^{\frac{j\pi}{2} u^2}$$

Since the integral $\int_{-\infty}^{\infty} du\, e^{\frac{j\pi}{2} u^2}$ is known as the Fresnel integral, and it is equal to

$$\int_{-\infty}^{\infty} du\, e^{\frac{j\pi}{2} u^2} = 1 + j, \tag{4.15}$$

Therefore,

$$G(f_x) = \sqrt{\frac{2}{\lambda F}}\, e^{-j\pi(\lambda F) f_x^2} (1 + j)$$

Therefore, the Fourier transform of the Gaussian function $g(x)$ is given by

(continued)

$$G(f_x) = \sqrt{j\lambda F} e^{-j\pi(\lambda F)f_x^2}$$

which is also a Gaussian function in the spatial frequency domain.

4.3.1 Dirac Delta Function

The Dirac delta function, $\delta(x - x_0)$, is defined as a function that has a non-zero value at only one specific point, x_o, and is equal to zero everywhere else. The total area under the function is equal to 1. Hence, $\delta(x - x_0)$ is a function whose amplitude is zero at all points except at x_o, where the area under the function is unity. In other words, the delta function "picks out" the value at x_o while ensuring that the integral over the entire domain is equal to 1, despite the function being infinitely narrow at that point:

$$\int_{-\infty}^{\infty} \delta(x - x_0)\, dx = 1 \tag{4.16}$$

Thus, the delta function allows us to express a function $f(x)$ as:

$$f(x) = \int_{-\infty}^{\infty} f(s)\delta(s - x)\, ds. \tag{4.17}$$

The Fourier transform of the Dirac delta function is particularly simple. For $\delta(x)$, its Fourier transform is:

$$\mathcal{F}[\delta(x)] = \int_{-\infty}^{\infty} dx\, \delta(x) e^{-j2\pi f_x x} = 1 \tag{4.18}$$

This result shows that the Fourier transform of the delta function is a constant, reflecting its property of being localized in space.

Taking the inverse transform of Eq. (4.18) leads to

$$f(x) = \int_{-\infty}^{\infty} df_x [\mathcal{F}[\delta(x)]] e^{j2\pi f_x x} \tag{4.19}$$

Therefore, we can define the delta function by the following integral:

$$\delta(x) = \int_{-\infty}^{\infty} df_x\, e^{j2\pi f_x x} \tag{4.20}$$

which is another definition of the delta function.

Example Derive the Fourier transform of the cosine function

$$f(x) = cos(2\pi a x)$$

Solution

$$F(f_x) = \int_{-\infty}^{\infty} dx \, cos(2\pi a x) e^{-j2\pi x f_x}$$

$$= \int_{-\infty}^{\infty} dx \left[\frac{e^{j2\pi a x} + e^{-j2\pi a x}}{2} \right] e^{-j2\pi x f_x}$$

Therefore,

$$F(f_x) = \frac{1}{2} [\delta(f_x - a) + \delta(f_x + a)]$$

which is a set of two delta functions located at the spatial frequency of the function $f(x)$.

4.3.2 Fourier Analysis in Three Dimensions

Consider a scalar function $g(x, y, t)$ with two transverse spatial coordinates and one temporal coordinate. Now define the Fourier transform $G(f_x, f_y, \nu)$ by the following equation:

$$G(f_x, f_y, \nu) = \iiint_{-\infty}^{\infty} g(x, y, t)) e y^{-j2\pi(f_x x + f_y + \nu t)} \, dx dy dt \qquad (4.21)$$

The inverse transform in three dimensions is given by

$$g(x, y, t) = \iiint_{-\infty}^{\infty} G(f_x, f_y, \nu) e y^{+j2\pi(f_x x + f_y + \nu t)} \, df_x df_y d\nu \qquad (4.22)$$

This notation and sign convention, where the exponents in the transform kernel are negative and those in the inversion are positive, are used throughout this work. It provides a consistent and convenient framework for performing and interpreting Fourier transforms and their inversions in multiple dimensions.

4.3.3 Fourier Analysis in Optics

Optical systems, being linear, are well suited for the use of Fourier analysis. This method describes the propagation of light based on harmonic analysis and linear systems theory. Harmonic analysis involves the expansion of arbitrary functions into sums of harmonic functions with different frequencies. A function $f(t)$ in the time is expressed as a sum of harmonic functions:

$$F(\nu)e^{j2\pi \nu t}, \tag{4.23}$$

where ν is the temporal frequency.

In optical systems, however, functions are represented in spatial domain. For example, an optical function can generally be represented in four dimensions, Fourier Analysis in Optics.

Optical systems are linear and can benefit from the Fourier analysis method. It describes the propagation of light based on harmonic analysis and linear systems. Harmonic analysis is based on the expansion of arbitrary functions in terms of harmonic functions of different frequencies. The function $f(t)$ in time is expressed as the sum of the harmonic functions:

$$F(\nu)e^{j2\pi \nu t}, \tag{4.24}$$

where ν is the temporal frequency.

In optical systems, however, functions represented in the spatial domain. For example, an optical function can generally be represented in four dimensions, such as $f(x, y, z, t)$. An image, e.g., is a two-dimensional function in space. So an arbitrary function $f(x, y)$ of variables x and y, representing the spatial coordinates in a plane, can be expressed as the sum of the harmonic functions:

$$F(f_x, f_y)e^{-j2\pi(f_x x + f_y y)} \tag{4.25}$$

where f_x and f_y are spatial frequencies (typically in cycles per millimeter). Figure 4.4 shows examples of low and high spatial frequency patterns in one dimension. An image can thus be decomposed into a series of planewaves, each propagating in different directions with different spatial frequencies, as depicted in Fig. 4.5.

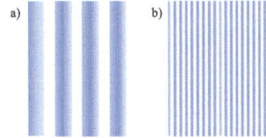

Fig. 4.4 Examples of the wavefronts of planewaves with different spatial frequencies: (**a**) low spatial frequency and (**b**) high spatial frequency

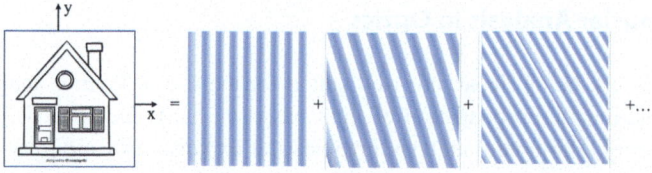

Fig. 4.5 An image in two dimensions can be decomposed using Fourier analysis into a series of harmonics shown here as plane waves propagating in different direction and frequencies

Fig. 4.6 Lines of zero phase of the function $e^{j2\pi(xf_x+yf_y)}$ where the angle $\theta = atan\left(f_x/f_y\right)$

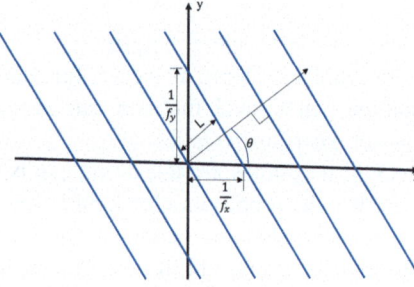

The two-dimensional Fourier transform, $F(f_x, f_y)$, can be viewed as a decomposition of a function $f(x, y)$ into a linear combination of elementary functions of the form $e^{j2\pi(xf_x+yf_y)}$. For any particular spatial frequency pair (f_x, f_y), the corresponding elementary function has a phase that is zero or an integer multiple of 2π radians along lines described by the equation

$$xf_x + yf_y = m, \qquad (4.26)$$

where m is an integer, $m = 0, 1, 2, 3, \ldots$. This equation can be rewritten as:

$$y = -\frac{f_x}{f_y}x + \frac{m}{f_y}, \qquad (4.27)$$

which an equation of a line in the $x - y$ plane with a slope f_x/f_y, as shown in Fig. 4.6. This is a function of a line making an angle θ with the x-$axis$, where θ is given by:

$$\theta = atan\left(\frac{f_y}{f_x}\right), \qquad (4.28)$$

The spatial period of this function is:

$$L = \frac{1}{\sqrt{f_x^2 + f_y^2}}. \qquad (4.29)$$

4.4 Scalar Diffraction Theory

Thus, the Fourier transform can be interpreted as a decomposition of the function $f(x, y)$ into its frequency components. The function can be represented by planewaves propagating at angles determined by the spatial frequencies. For instance, zero-frequency content corresponds to a planewave propagating at an angle $\theta = 0°$.

4.4 Scalar Diffraction Theory

The diffraction theory describes how optical fields evolve as they propagate, from one location to another in space, as it it encounters obstacles on its path. It is based on solving the scalar wave equation. It is also known as the Helmholtz-wave equation introduced in Chap. 3. The Helmholtz-wave equation for a monochromatic optical field $U(x, y, z)$ is given by:

$$\left(\nabla^2 + k^2\right) U(x, y, z) = 0.$$

This wave equation was developed under the assumption that there are no sources of radiation present in the medium. However, to account for the presence of a localized radiation source $f(x, y, z)$, the wave equation can be generalized as:

$$\left(\nabla^2 + k^2\right) U(x, y, z) = -f(x, y, z), \tag{4.30}$$

or equivalently,

$$\left(\nabla^2 + k^2\right) U(\bar{r}) = -f(\bar{r}) \tag{4.31}$$

Solving this partial differential equation can be approached using infinite series techniques. However, these series often converge slowly, making it difficult to extract general insight into the solution overall behavior. A more efficient approach is to obtain a closed-form solution, possibly it is expressed as an integral involving known functions. One such method is the Green's function technique.

The physical reasoning behind this approach is straightforward. To determine the optical field produced by a distributed source, we calculate the contribution of each elementary portion of the source and some all of these contributions. If $G(\underline{r}, \underline{r}')$ represents the field at the observation point $\underline{r}$ caused by a unit point source at the source point $\underline{r}'$, then the field at $\bar{r}$ produced by the source distribution $f(\bar{r})$ is obtained by integrating the function $G(\underline{r}, \underline{r}')$ over the entire region of $\bar{r}'$ occupied by the source. The function G is called the Green's function, and it satisfies the following wave equation:

$$\left(\nabla^2 + k^2\right) G(\bar{r}, \bar{r}') = -\delta(\bar{r} - \bar{r}'). \tag{4.32}$$

The Green's function method is valuable because it provides a general, closed-form solution to the wave equation, enabling the analysis of diffraction and wave propagation in optical systems.

Let the operator $\mathcal{L}$ be defined as:

$$\mathcal{L} = \nabla^2 + k^2, \tag{4.33}$$

The inhomogeneous wave equation, Eq. (4.32), then becomes:

$$\mathcal{L}\{G\} = -\delta(\vec{r} - \vec{r}'). \tag{4.34}$$

There are an infinite number of Green's functions associated with this operator because the solution is not unique unless we specify certain domain & boundary conditions, as shown in Fig. 4.7.

To solve this class of wave equation, we will use a standard technique. First, multiply Eq. (4.29) with G and Eq. (4.32) with U to get:

$$G\left(\nabla^2 + k^2\right)U = 0$$

$$U\left(\nabla^2 + k^2\right)G = -U\delta(\vec{r} - \vec{r}')$$

and subtract these two equations to get

$$G\nabla^2 U - U\nabla^2 G = U\delta(\vec{r} - \vec{r}'). \tag{4.35}$$

Using a vector identity, Eq. (4.35) can be reduced to the following:

$$\nabla \cdot (G\nabla U - U\nabla G) = U\delta(\vec{r} - \vec{r}'). \tag{4.36}$$

Next we specify the region of concern. In diffraction problem, we need to find the field in the right semi-infinite half space as illustrated in Fig. 4.8.

Now integrate Eq. (4.36) over the volume V bounded by the surface S.

Fig. 4.7 Green's boundary conditions

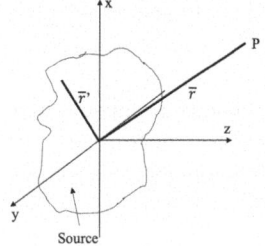

Fig. 4.8 The semi-infinite half-space where diffracted field is to be determined, $S = S_1 + S_2$

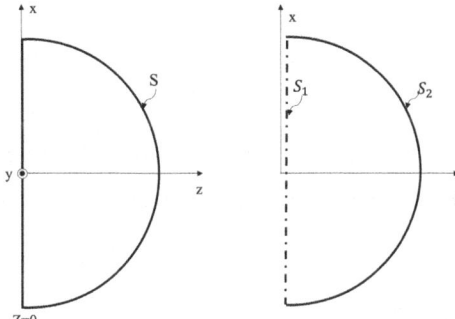

$$\int_V \nabla \cdot (G\nabla U - U\nabla G) \, d\sigma = \int_V U(\bar{r}')\delta(\bar{r} - \bar{r}') \, d\sigma = U(\bar{r}). \qquad (4.37)$$

Using the vector identity of the divergence, we obtain:

$$\int_V \nabla \cdot \underline{A} \, d\sigma = \oint_S \underline{A} \cdot \underline{n} \, ds, \qquad (4.38)$$

Eq. (4.37) can be rewritten as

$$\int_S (G\nabla U - U\nabla G) \cdot \bar{n} \, ds = U(\bar{r}), \qquad (4.39)$$

where the vector $\bar{n}$ is the normal to the surface S. Equation (4.39) states that the field at any point inside the region S is determined by the field on the surface of the bounding region. Thus, knowing the field on the surface is sufficient. The field on the surface will be divided into two parts, S_1 and S_2, as shown in Fig. 4.8.

$$\int_S (G\nabla U - U\nabla G) \cdot \bar{n} \, ds = \int_{S_1} (G\nabla U - U\nabla G) \cdot \bar{n} \, ds$$
$$+ \int_{S_2} (G\nabla U - U\nabla G) \cdot \bar{n} \, ds. \qquad (4.40)$$

Now we specify the boundary conditions, There are three possible boundary conditions for G:

(1) $G = 0$ on S, $\qquad Dirichlet \qquad (4.41)$

(2) $\dfrac{\partial G}{\partial n} = 0$ on S, $\qquad Neuman \qquad (4.42)$

$$(3)\ G + \frac{\partial G}{\partial n} = 0 \quad \text{on } S. \qquad Cauchy \qquad (4.43)$$

4.4.1 Solving the Green's Wave Equation

According to Eq. (4.33), the operator $\mathcal{L}$ is a differential operator such that

$$\mathcal{L}\{U\} = -\rho, \qquad (4.44)$$

so that

$$\therefore U = +\mathcal{L}^{-1}\{\rho,\} \qquad (4.45)$$

where $\mathcal{L}^{-1}$ is an integral operator. Consider

$$U(r) = + \int_{v'} G(\underline{r},\underline{r}')\rho(\underline{r}')\,d\sigma' \qquad (4.46)$$

to check the validity of this assumption

$$\mathcal{L}\{U\} = +\mathcal{L}\left\{\int G(\underline{r},\underline{r}')\rho(\underline{r}')\,d\sigma'\right\},$$

Exchanging the order of operations, we get

$$\mathcal{L}\{U\} = \int \mathcal{L}\{G(\underline{r},\underline{r}')\}\rho(\underline{r}')\,d\sigma',$$

using Eq. (4.34) we obtain:

$$\mathcal{L}\{U\} = -\int \delta(\underline{r}-\underline{r}')\rho(\underline{r}')\,d\sigma'.$$

Therefore,

$$\mathcal{L}\{U\} - -\rho.$$

Considering spherical coordinate system, for spherical symmetry G is not a function of θ, and for $\bar{\bar{r}} \neq \bar{r}'$, hence

$$\left(\nabla^2 + k^2\right) G = 0,$$

4.4 Scalar Diffraction Theory

Substituting for the Laplacian using the spherical coordinates, we get

$$\frac{1}{r}\frac{d^2}{dr^2}(rG) + k^2 G = 0. \tag{4.47}$$

Let us assume a solution for Eq. (4.47)

$$rG = A_1 e^{jkr} + A_2 e^{-jkr}. \tag{4.48}$$

But from the boundary condition, we know that the wave is an outgoing wave so $A_2 = 0$. Therefore,

$$G = A_1 \frac{e^{jkr}}{r}, \text{ where } r = |\bar{r} - \bar{r}'|. \tag{4.49}$$

Now we need to determine the value of A.

$$\nabla^2 G + k^2 G = -\delta(\underline{r} - \underline{r}'),$$

$$\int \nabla^2 G \, d\sigma + \int k^2 G \, d\sigma = -\int \delta(\bar{r} - \bar{r}') \, d\sigma,$$

$$\int \nabla^2 \left(\frac{A_1}{r} e^{jkr}\right) r^2 \sin\theta \, dr \, d\theta \, d\phi + \int k^2 \left(\frac{A}{r} e^{jkr}\right) \underbrace{r^2 \sin\theta \, dr \, d\theta \, d\phi}_{\to 0 \text{ as } r = 0} = -1,$$

$$\int \nabla^2 G \, r^2 \sin\theta \, dr \, d\theta \, d\phi = -1,$$

$$\int_v \nabla \cdot \nabla G \, d\sigma = -1,$$

$$\oint_s \nabla G \cdot d\underline{s} = -1$$

$$\nabla G = \frac{\partial}{\partial r}\left(\frac{A}{r} e^{jkr}\right) = A_1 \left(jk + \frac{1}{r}\right) \frac{e^{jkr}}{r},$$

$$\oint_s A_1 \left(jk - \frac{1}{r}\right) \frac{e^{jkr}}{r} r^2 \sin\theta \, d\theta \, d\phi = -1,$$

at $r = 0$

$$-A_1 \oint \sin\theta \, d\theta \, d\phi = -1,$$

therefore,

$$A_1 = \frac{1}{4\pi},$$

The green function is given by:

$$G = \frac{e^{-jkr}}{4\pi r}. \tag{4.50}$$

$$U = \oint_{s_1} (G\nabla U - U\nabla G) \cdot \underline{n}\, ds + \oint_{s_2} (G\nabla U - U\nabla G) \cdot \underline{n}\, ds, \tag{4.51}$$

$$= \int_{s_1+s_2} \left(G\frac{\partial U}{\partial n} - U\frac{\partial G}{\partial n}\right) ds, \tag{4.52}$$

on the surface S_2, $R \to \infty$., as shown in Fig. 4.9.

As S_2 approaches the hemisphere ($R \longrightarrow \infty$) and considering that both U and G have their values falling off as $\frac{1}{R}$, it is tempting to drop the integral on S_2, but at the surface, the integration increases as R^2, so this argument is incomplete. Also, it is tempting to assume that since the wave is propagating with a finite speed & S_2 is at infinity, the waves at that location do not exist. However, this assumption contradicts our premise of harmonic fields, which are assumed to exist at all times.

4.4.2 Sommerfeld Condition

Sommerfeld broke this integration in Eq. (4.51) into parts asserting that the integral over the hemisphere in the far zone decreases to zero for a finite aperture source. On the hemisphere S_2 the Green's function is given by:

$$G = \frac{e^{-jkR}}{4\pi R}, \tag{4.53}$$

Fig. 4.9 Notation for the calculation of the Green's function in the right-half-sphere

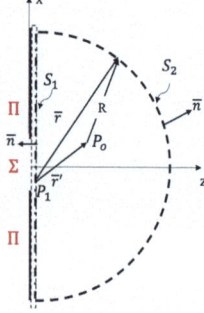

4.4 Scalar Diffraction Theory

Let the integral on S_2 in Eq. (4.51) be given by I_2

$$I_2 = \int_{S_2} (G\nabla U - U\nabla G) \cdot \underline{n}\, d\underline{s} = \int_{S_2} \left(G\frac{\partial U}{\partial n} - U\frac{\partial G}{\partial n} \right) ds. \tag{4.54}$$

Using G as given by Eq. (4.53) to obtain

$$\therefore \frac{\partial G}{\partial n} = \left(jk - \frac{1}{R} \right) \frac{e^{jkR}}{4\pi R} \simeq jkG, \quad \text{as } R \longrightarrow \infty,$$

$$I_2 = \int_{S_2} \left[G\frac{\partial U}{\partial n} - U(jkG) \right] ds$$

$$= \int_{S_2} G\left(\frac{\partial U}{\partial n} - jkU \right) ds$$

$$I_2 = \int_{S_2} \frac{e^{jkR}}{4\pi R} \left(\frac{\partial U}{\partial n} - jkU \right) R^2\, d\Omega \tag{4.55}$$

where Ω is the solid angle that subtends by S_2 at P_o. Hence,

$$I_2 = \frac{1}{4\pi} \int R\left(\frac{\partial U}{\partial n} - jkU \right) e^{jkR}\, d\Omega, \tag{4.56}$$

Thus, for I_2 to vanish, the following condition must be satisfied:

$$\lim_{R\to\infty} R\left(\frac{\partial U}{\partial n} - jkU \right) = 0. \tag{4.57}$$

The Sommerfeld radiation condition addresses this issue. This condition can be satisfied if U vanishes at least as quickly as the spherical wave. Given that the field illuminating the aperture consists of a spherical wave or a linear combination of spherical waves, the Sommerfeld radiation condition will be satisfied. As a result, Eq. (4.51) simplifies to:

$$U(P_o) = \int_{S_1} \left(G\frac{\partial U}{\partial n} - U\frac{\partial G}{\partial n} \right) ds. \tag{4.58}$$

4.4.3 Kirchhoff Diffraction

Since we did dispose off the integration over S_2, thus, the field at point P_o can be expressed in terms of the field and its normal derivative over the infinite plane S_1, as well as the first term of the expression in Eq. (4.51)

$$U(P_o) = \iint_{s_1} \left(G \frac{\partial U}{\partial n} - U \frac{\partial G}{\partial n} \right) ds.$$

Kirchhoff adopted the following boundary conditions assumptions:

1. Across aperture Σ, U and $\partial U/\partial n$ are exactly the same as they would be in the absence of the screen.
2. On surface Π, which lies in the geometrical shadow of the screen, $U = 0$ & $\partial U/\partial n = 0$.

Therefore,

$$U(P_o) = \iint_{\Sigma} \left(G \frac{\partial U}{\partial n} - U \frac{\partial G}{\partial n} \right) ds. \qquad (4.59)$$

Of course, physically, neither of Kirchhoff's conditions can be true. Except for dimensions of the aperture $\Sigma \gg \lambda$, then the results of Kirchhoff boundary conditions are in agreement with experiments. Based on these boundary conditions, the field $U(P_o)$ will be given by

$$U(P_o) = \frac{1}{4\pi} \iint_{\Sigma} \left(\frac{e^{-jkr}}{r} \frac{\partial}{\partial n} U(P_1) - U(P_1) \frac{\partial}{\partial n} \frac{e^{-jkr}}{r} \right) ds, \qquad (4.60)$$

This is the Kirchhoff diffraction integral.

Let us consider the case of a point source illuminating the aperture, Fig. 4.10. Point P is the source of a diverging spherical wave so the field at a point P_1 at the aperture is given by

$$U(P_1) = A \frac{e^{-jkr_1}}{r_1}, \qquad (4.61)$$

where r_1 is the distance the spherical wave travels to reach P_1. Therefore,

Fig. 4.10 Point source at P illuminating an aperture

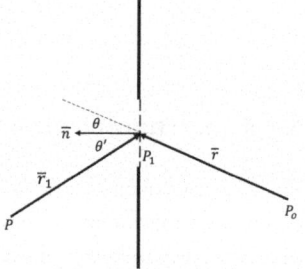

4.4 Scalar Diffraction Theory

$$\frac{\partial}{\partial n} U(P_1) = A \frac{\partial}{\partial n}\left(\frac{e^{-jkr_1}}{r_1}\right) = A \frac{\partial r_1}{\partial n} \frac{\partial}{\partial r_1}\left(\frac{e^{-jkr_1}}{r_1}\right), \quad (4.62)$$

$$\frac{\partial}{\partial n} U(P_1) = A \frac{\partial r_1}{\partial n}\left(-jk - \frac{1}{r_1}\right) \frac{e^{-jkr_1}}{r_1}, \quad (4.63)$$

where n is the normal to the aperture plane and

$$r_1 = \sqrt{x^2 + y^2 + z^2}, \quad (4.64)$$

Therefore,

$$\frac{\partial r_1}{\partial n} = \frac{\partial r_1}{\partial z} = \frac{z}{r_1} = \cos(\underline{s}, \underline{n}) = \cos\theta', \quad (4.65)$$

and

$$\frac{\partial}{\partial n} U(P_1) = A\left(-jk - \frac{1}{r_1}\right) \frac{e^{-jkr_1}}{r_1} \cos\theta'. \quad (4.66)$$

According to Huygen's principle, every point on the wavefront across the aperture Σ emits a wavelet. So point P_1 is a source of a spherical wave. Thus, following the steps from Eqs. (4.61) to (4.66), we get

$$\frac{\partial}{\partial n} A \frac{e^{-jkr}}{r} = A\left(-jk - \frac{1}{r}\right) \frac{e^{-jkr}}{r} \cos\theta. \quad (4.67)$$

For the case $k \gg \frac{1}{r}$ & $k \gg \frac{1}{s}$, Eqs. (4.66) and (4.67) becomes

$$\frac{\partial}{\partial n} U(P_1) = -jkA \frac{e^{-jkr_1}}{r_1} \cos\theta', \quad (4.68)$$

and

$$\frac{\partial}{\partial n} A \frac{e^{-jkr}}{r} = -jkA \frac{e^{-jkr}}{r} \cos\theta. \quad (4.69)$$

Substituting Eqs. (4.68) to (4.69) into Eq. (4.60) and using $k = 2\pi/\lambda$, we obtain:

$$U(P_o) = \frac{-j}{2\lambda} \iint_\Sigma A \frac{e^{jkr_1}}{r_1} \frac{e^{jkr}}{r} (\cos\theta + \cos\theta') \, ds. \quad (4.70)$$

This is the Fresnel–Kirchhoff diffraction integral. Rewriting Eq. (4.70) results in the following equation:

$$U(P_o) = \iint_\Sigma U'(P_1) \frac{e^{jkr}}{r} ds,$$

where

$$U'(P_1) = \frac{1}{j\lambda} \left(A \frac{e^{jks}}{s} \right) \left(\frac{\cos\theta + \cos\theta'}{2} \right),$$

j introduces a phase shit and $(\cos\theta + \cos\theta')/2$ is a directionality factor ≤ 1.

4.5 The Rayleigh–Sommerfeld Diffraction Theory

Kirchhoff's diffraction theory aligns well with experiments, but it contains an internal inconsistency. First, a well-known potential theorem states that if $U = 0$ and $\partial U/\partial n = 0$ along any finite curve, the potential function must vanish over the entire plane. Second, for a three-dimensional wave equation if the solution vanishes on any finite surface element, it must vanish through the entire space. Therefore, if the field is zero on surface Π, as shown in Fig. 4.9, it should be zero on Σ as well. Third, the Fresnel–Kirchhoff diffraction formula fails to reproduce the assumed boundary conditions as the observation point approaches the screen or the aperture.

Sommerfeld removed these inconsistencies. By not imposing boundary values on both the field & its normal derivative, consider again Eq. (4.58)

$$U(P_o) = \int_{S_1} \left(G \frac{\partial U}{\partial n} - U \frac{\partial G}{\partial n} \right) ds.$$

To remove Kirchhoff's theory inconsistencies, we can choose either G or $\partial G/\partial n$ to vanish by selecting an appropriate Green's function G. This eliminates the need to impose simultaneous boundary conditions on both U and $\partial U/\partial n$.

Sommerfeld identified the existence of such Green's function. Suppose that G is generated not only by a point source located at P_1, but also by a second point source located at P_2, which is the mirror image of P_1 on the opposite side of the screen, Fig. 4.11. The Sommerfeld Green function, G_s, is given by

$$G_s(P_o) = \frac{e^{-jkR_1}}{4\pi R_1} - \frac{e^{-jkR_2}}{4\pi R_2}. \tag{4.71}$$

Thus, the field $U(x, y, z)$ can be expressed as:

$$U(x, y, z) = \int_{z=0} U \frac{\partial G_s}{\partial n} dx\, dy, \tag{4.72}$$

4.5 The Rayleigh–Sommerfeld Diffraction Theory

Fig. 4.11 Two sources setting to derive Sommerfeld Green function

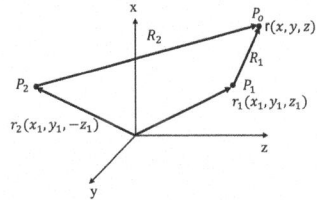

where the distances R_1 and R_2 are:

$$R_1 = \sqrt{(x-x_1)^2 + (y-y_1)^2 + (z-z_1)^2}, \tag{4.73}$$

$$R_2 = \sqrt{(x-x_2)^2 + (y-y_2)^2 + (z+z_2)^2}. \tag{4.74}$$

The derivative of the Green function with respect to the normal ($\partial G/\partial n$) to the aperture is given by

$$\frac{\partial G}{\partial z} = \frac{1}{4\pi}\left[\left(-jk-\frac{1}{R_1}\right)\frac{e^{-jkR_1}}{R_1}\frac{\partial R_1}{\partial z} - \left(-jk-\frac{1}{R_2}\right)\frac{e^{-jkR_2}}{R_2}\frac{\partial R_2}{\partial z}\right], \tag{4.75}$$

where

$$\frac{\partial R_1}{\partial z} = -\frac{2(z-z_1)}{2R_1} = -\frac{z-z_1}{R_1}, \tag{4.76}$$

At the aperture plane $z_1 = 0$

$$\left.\frac{\partial R_1}{\partial z}\right|_{z=0} = -\frac{z_1}{R_1}. \tag{4.77}$$

Similarly,

$$\left.\frac{\partial R_2}{\partial z}\right|_{z=0} = \frac{z_1}{R_2}, \tag{4.78}$$

At $z_1 = 0$, $R_1 = R_2$. Therefore by substituting in Eq. (4.72) and then in Eq. (4.75), we get:

$$\left.\frac{\partial G}{\partial z}\right|_{z=0} = \frac{e^{-jkR_1}}{2\pi R_1}\frac{z_1}{R_1}\left(\frac{1}{R_1} + jk\right), \tag{4.79}$$

$$U(x, y, z > 0) = \frac{1}{2\pi} \iint_\Sigma U(x_o, y_o, 0) \frac{e^{-jkR_1}}{R_1} \frac{z}{R_1} \left(\frac{1}{R_1} + jk\right) dx_o\, dy_o, \tag{4.80}$$

In Eq. (4.80) P_1 in Fig. 4.11 has been moved to the aperture so that the coordinates for the aperture plane as $(x_o, y_o, 0)$ and (x, y, z) are the coordinates for the output field. In this case, R_1 can be expressed as:

$$R_1 = \sqrt{(x - x_o)^2 + (y - y_o)^2 + z^2}. \tag{4.81}$$

4.6 Propagation Into the Right-Half-Space

Let us consider a scenario where a monochromatic electromagnetic wave is traveling along the optical z-axis and is incident on an aperture, as shown in Fig. 4.12. We assume that radiation sources are only present in the region $z < 0$, and we have prior knowledge of the "field" at the aperture. The radiation then propagates into the right-half-space (RHS), a simple source-free region. The diffraction integral, denoted as Eq. (4.80), provides an exact solution to Maxwell's equations in the right-half-space.

The field radiating through the aperture is obstructed by the screen. As a result, we can modify the diffraction integral as follows:

$$U(x, y, z > 0) = \frac{1}{2\pi} \iint_{-\infty}^{\infty} U(x_o, y_o, 0) \frac{e^{-jkR}}{R} \frac{z}{R} \left(\frac{1}{R} + jk\right) dx_o\, dy_o. \tag{4.82}$$

In optics applications, the wavelength of light $\lambda \cong 10^{-6} m$. The wavenumber $k \cong 10^6 \, m^{-1}$. Consequently, $jk \gg 1/R$, which holds true for all practical cases. If the output field $U(x, y, z)$ is derived near the optic axis (z-axis), then $R \approx z$, so the fraction $z/R \approx 1$. Utilizing these approximations, Eq. (4.82) simplifies to:

$$U(x, y, z) \approx \frac{jk}{2\pi z} \iint_{-\infty}^{\infty} U(x_o, y_o) e^{-jkR} dx_o dy_o. \tag{4.83}$$

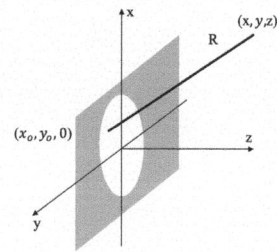

Fig. 4.12 Radiation in the right-half-space through an aperture

4.6 Propagation Into the Right-Half-Space

Now, let us consider the expression for the distance R in Eq. (4.81) rewritten in a different form here.

$$R = z\sqrt{1 + \frac{(x-x_o)^2 + (y-y_o)^2}{z^2}}, \qquad (4.84)$$

using binomial expansion, if $(x-x_o)^2+(y-y_o)^2/z^2 < 1$, then R can be expressed as

$$R = z + \frac{(x-x_o)^2 + (y-y_o)^2}{2z} - \frac{[(x-x_o)^2 + (y-y_o)^2]^2}{8z^3}$$
$$+ \frac{[(x-x_o)^2 + (y-y_o)^2]^3}{16z^5} - + \dots \qquad (4.85)$$

The number of terms in the expansion of R used in Eq. (4.84) depends on the location of the output plane (z), as well as the size of the aperture and the area of the output field. For most problems of interest, only the first two terms of the expansion are needed. This approximation is known as the Fresnel approximation, and the integral in Eq. (4.84) is referred to as the Fresnel diffraction integral. Therefore, the Fresnel approximation can be expressed as:

$$R \approx z + \frac{(x-x_o)^2 + (y-y_o)^2}{2z}, \qquad (4.86)$$

or

$$R \approx z + \frac{(x^2+y^2) + (x_o^2+y_o^2) - 2(xx_o + yy_o)}{2z}. \qquad (4.87)$$

Substituting Eq. (4.86) in Eq. (4.83) leads to:

$$U(x, y, z) = \frac{je^{-jkz}}{\lambda z} \iint_{-\infty}^{\infty} U(x_o, y_o) e^{-j\frac{k}{2z}[(x-x_o)^2+(y-y_o)^2]} dx_o \, dy_o, \qquad (4.88)$$

where we have incorporated the finite limits of the aperture in the definition of $U(x_o, y_o)$, in accord with the usual assumed boundary conditions. Equation (4.88), is known as theFresnel diffraction integral, can be easily identified as a convolution, which can be expressed in the following form:

$$U(x, y) = \iint_{-\infty}^{\infty} U(x_o, y_o) h(x - x_o, y - y_o) \, dx_o \, dy_o, \qquad (4.89)$$

where the convolution kernel is:

$$h(x, y) = \frac{je^{-jkz}}{\lambda z} e^{-jk\frac{(x^2+y^2)}{2z}}, \tag{4.90}$$

Accordance of the linear systems theory, $h(x, y)$ is the impulse response of the system, and accordingly the transfer function is

$$H(f_x, f_y) = e^{-jkz} e^{-j\pi\lambda z\left(f_x^2 + f_y^2\right)}. \tag{4.91}$$

Therefore, the propagation of the lightwave into the right-half-space is a convolution between the field at the aperture and the impulse response of the system.

4.7 Fresnel Diffraction

The optical field in the right half-space can be represented by the general diffraction integral given in Eq. (4.91). Upon closer examination of this integral, the terms outside the integral represent two key components: a planewave propagating in the positive z-direction, e^{-jkz}, and a chirp function, $e^{-jk\frac{(x^2+y^2)}{2z}}$. Meanwhile, the integral is the Fourier transform integral of the product of the input field and the chirp function $e^{-jk\frac{(x_o^2+y_o^2)}{2z}}$. We can rewrite the Fresnel diffraction integral in the typical Fourier transform form as follows:

$$U(f_x, f_y, z)$$
$$\approx \frac{je^{-jkz}}{\lambda z} e^{-jk\frac{(x^2+y^2)}{2z}} \iint\limits_{-\infty}^{\infty} \left\{ U(x_o, y_o) e^{-jk\frac{(x_o^2+y_o^2)}{2z}} \right\} e^{-j2\pi(x_o f_x + y_o f_y)} \, dx_o \, dy_o. \tag{4.92}$$

where $f_x = \frac{x}{\lambda z}$, and $f_y = \frac{y}{\lambda z}$ are the spatial frequencies.

For the Fresnel approximation to be valid, the third term in the binomial expansion of Eq. (4.83) must be significantly small in order to satisfy the following condition:

$$z^3 >> \left(\frac{k}{8}\left[(x-x_o)^2 + (y-y_o)^2\right]^2\right)_{maximum}, \tag{4.93}$$

or equivalently,

$$z^3 >> \left(\frac{\pi}{4\lambda}\left[(x-x_o)^2 + (y-y_o)^2\right]^2\right)_{maximum}. \tag{4.94}$$

4.7 Fresnel Diffraction

The maximum value on the right-hand-side of this equation is determined by the geometry of both the source and observation area. The Fresnel approximation condition is more valid when we consider the case when $x_0 = 0$ and $y_o = 0$. Under these circumstances, Eq. (4.94) becomes

$$z^3 \gg \left(\frac{\pi}{4\lambda}\left[x^2 + y^2\right]^2\right)_{maximum},$$

$$z^3 \gg \left(\frac{\pi}{4\lambda}\left[r^2\right]^2\right)_{maximum},$$

$$z^3 \gg \left(\frac{\pi r^4}{4\lambda}\right)_{maximum},$$

$$z^3 \gg \frac{\pi}{4\lambda}\left(r^4\right)_{maximum}, \qquad (4.95)$$

or equivalently,

$$z \gg \left(\frac{\pi}{4\lambda}r_{max}^4\right)^{1/3}, \qquad (4.96)$$

where r_{max} is the maximum radius of the observation area, or half the width of the square, of the observation area where the Fresnel approximation holds. A similar formula can be derived to define the size of the input aperture. In this case, z will determine the beginning of the Fresnel region. For example, for an aperture with 1 mm radius illuminated by a planewave with a wavelength of $1\,\mu m$, $z \gg 10\,mm$. Clearly for small apertures, the Fresnel zone can start at distances very close to the aperture.

A quantity known as the Fresnel number given by

$$N_F = \frac{r^2}{\lambda z}. \qquad (4.97)$$

If N_F is less than 1 for a given scenario, it is generally accepted that the observation plane is in the Fresnel region, where the Fresnel approximations often lead to useful results. However, for relatively "small" fields over the source aperture, the Fresnel expression can be applicable up to Fresnel numbers on the order of 10. In the previous example where $r = 1\,mm$ and $\lambda = 1\,\mu m$, for $N_F = 1$, $z = 1\,m$. Therefore, the Fresnel number can be much greater than 1 while still satisfying the Fresnel zone condition given by Eq. (4.96).

4.7.1 Fresnel Diffraction Examples

In this section, we will examine how the light propagates in the right half as a monochromatic light illuminates an aperture using the Fresnel approximation. The Fresnel diffraction integral given in Eq. (4.92) does not have a simple analytic solution. To determine the field distribution in the Fresnel zone, we use computer simulation employing the Fast Fourier Transform (FFT) function in MATLAB software to perform these calculations. We can use the convolution properties of the linear systems theory, where the output of the system is the convolution of input function with the impulse response of the system. Let $u_1(x_o, y_o)$ be the input function that represents the optical field at the aperture and $u_2(x, y)$ be the output field at the observation plane, then

$$u_2(x, y) = u_1(x_o, y_o) \bigotimes h(x_o, y_o), \quad (4.98)$$

where the symbol $\bigotimes$ denotes the convolution operation, and $h(x_o, y_o)$ is the impulse response given by Eq. (4.90). Equation (4.98) can be expressed in terms of the transfer function of the system $H(f_x, t_y)$ as

$$u_2(x, y) = \mathcal{F}^{-1}\left\{\mathcal{F}\{u_1(x, y)\} \cdot H(f_x, f_y)\right\}, \quad (4.99)$$

where the symbol $\mathcal{F}$ denotes the Fourier transform operation and $\mathcal{F}^{-1}$ demotes the inverse Fourier transform operation. To evaluate the Fresnel diffraction pattern, we can use either technique outlined in Eqs. (4.98) and (4.99). The following are computed Fresnel diffraction pattern from a square aperture 50 mm in width, light wavelength $\lambda = 0.5\,\mu\text{m}$ for an observation screen placed at several distances, z, from the aperture. According to Fresnel approximation validity condition given by Eq. (4.96), for the given example $z >> 0.85$.

In Fig. 4.13, the intensity pattern across the diffracted was computed at 0.1 m from the aperture. the intensity replicates that across aperture with some ringing at the edges, which is expected in the edge of the shadow region. In the following figure, we plot the intensity pattern across the diffracted beam at $z = 10$ and 100 m. As z increases, the intensity pattern deviates from a replication of the aperture images and becomes an interference pattern between the waves passing through

Fig. 4.13 The intensity distribution of the diffracted beam at 1 m from the square aperture

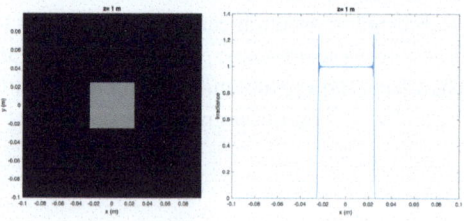

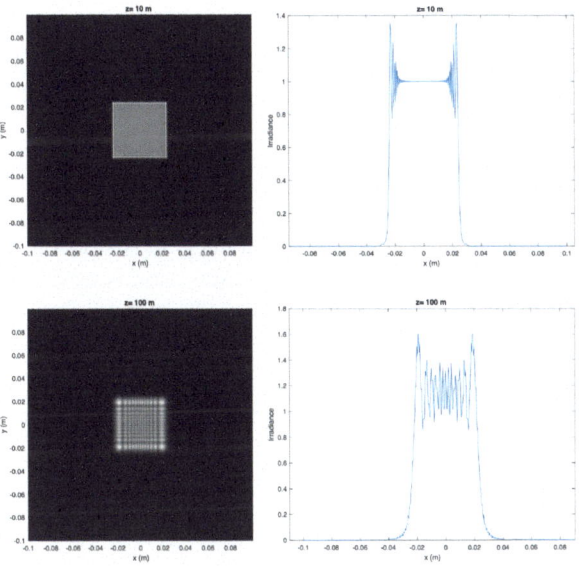

Fig. 4.14 The intensity of the diffracted field at $z = 10$ and 100 m

the aperture. Additionally, there is a spreading of the light beyond the width of the aperture as z increases (Fig. 4.14).

4.8 Fraunhofer Diffraction

In the previous section, we demonstrated that the Fresnel approximation becomes valid at a distance z that satisfies equation (4.96). As the observation screen moves further away, z can become so large that the phase exponent of the chirp function $e^{-jk\frac{(x_o^2 + y_o^2)}{2z}}$ becomes so small that its value approaches unity. The distance, z, that satisfies such a condition is:

$$z \gg \frac{\pi}{\lambda}\left(x_o^2 + y_o^2\right). \tag{4.100}$$

For example, at a wavelength of $0.5\,\mu$m and an aperture width of 50 mm, the observation distance z must satisfy

$$z \gg 4000\,\text{m}.$$

Consequently, using this approximation, the value of the chirp function under the integral is approximately unity. As a result, the Fresnel diffraction integral given by

Eq. (4.92) becomes

$$U(f_x, f_y, z) \approx \frac{je^{-jkz}}{\lambda z} e^{-jk\frac{(x^2+y^2)}{2z}} \iint_{-\infty}^{\infty} U(x_o, y_o) e^{-j2\pi(x_o f_x + y_o f_y)} \, dx_o \, dy_o$$

(4.101)

This is known as the Fraunhofer diffraction integral. Apart from the multiplicative phase factors preceding the integral, this expression is essentially the Fourier transform of the input field distribution across the aperture, evaluated at spatial frequencies:

$$f_x = \frac{x}{\lambda z}$$

and

$$f_y = \frac{y}{\lambda z}.$$

The Fraunhofer diffraction is also known as the far-field, while the Fresnel diffraction is known as the near-field. The Fraunhofer approximation imposes a stringent condition for the distance z. However, the approximation condition is satisfied in many important problems. Notably, the Fraunhofer diffraction pattern can be observed at distances much closer than those required by Eq. (4.100), provided that the aperture is illuminated by a converging spherical wave or by a positive lens placed between the aperture and the observation screen. In the latter case, the Fraunhofer diffraction pattern appears at the focal plane of the lens.

Unlike Fresnel diffraction, Fraunhofer diffraction does not involve a transfer function. In the Fresnel case, eliminating the chirp function from the Fresnel integral disrupts the space invariance of the diffraction integral.

Next, we consider several examples of Fraunhofer diffraction patterns produced by different aperture geometries. In these examples, we use the Fraunhofer diffraction integral, Eq. (4.101), to derive an expression for the complex field distribution at the observation plane. In each case, we derive an expression for the intensity distribution, which is more relevant and can be related to in many applications.

4.8.1 Fraunhofer Diffraction for a Rectangular Aperture

To explore a significant problem that yields insightful results through theoretical analysis, let us examine a perfectly conducting, infinitesimally thin plane sheet with a rectangular aperture located in the (x_o, y_o) plane at $z = 0$. The amplitude of the

4.8 Fraunhofer Diffraction

transmittance function of the aperture is given by

$$t(x_o, y_o) = rect\left(\frac{x}{2p}\right) rect\left(\frac{y}{2q}\right), \quad (4.102)$$

where p is half the aperture width along the x-axis, and q is half the aperture width along the y-axis. If the aperture is illuminated by a unit-amplitude monochromatic planewave, incident normally on the aperture plane, the amplitude of the incident field distribution matches the aperture transmittance function $t(x_o, y_o)$. The Fraunhofer diffraction pattern can be obtained by substituting Eq. (4.102) in the Fraunhofer diffraction integral, Eq. (4.101) as

$$U(f_x, f_y, z) = \frac{je^{-jkz}}{\lambda z} e^{-jk\frac{(x^2+y^2)}{2z}} \iint_{-\infty}^{\infty} t(x_o, y_o) e^{-j2\pi(x_o f_x + y_o f_y)} dx_o \, dy_o,$$

$$U(f_x, f_y, z) = K \iint_{-\infty}^{\infty} rect\left(\frac{x_o}{2p}\right) rect\left(\frac{y_o}{2q}\right) e^{-j2\pi(x_o f_x + y_o f_y)} dx_o \, dy_o, \quad (4.103)$$

where

$$K = \frac{je^{-jkz}}{\lambda z} e^{-jk\frac{(x^2+y^2)}{2z}}.$$

The rect-functions in Eq. (4.103) change the limits of the integral; hence, the Fraunhofer diffraction integral becomes

$$U(f_x, f_y, z) = K \int_{-p}^{+p} dx_o \int_{-q}^{+q} dy_o \, e^{-j2\pi(x_o f_x + y_o f_y)},$$

$$U(f_x, f_y, z) = K \frac{e^{-j2\pi x_o f_x}}{-j2\pi f_x} \Big|_{-p}^{+p} \frac{e^{-j2\pi y_o f_y}}{-j2\pi f_y} \Big|_{-q}^{+q},$$

$$U(f_x, f_y, z) = K \left\{ \frac{e^{-j2\pi p f_x} - e^{+j2\pi p f_x}}{-j2\pi f_x} \right\} \left\{ \frac{e^{-j2\pi q f_y} - e^{+j2\pi q f_y}}{-j2\pi f_y} \right\},$$

$$U(f_x, f_y, z) = K \left\{ \frac{sin(2\pi p f_x)}{\pi f_x} \right\} \left\{ \frac{sin(2\pi q f_y)}{\pi f_y} \right\},$$

$$U(f_x, f_y, z) = K p \, sinc(2p f_x) q \, sinc(2q f_y), \quad (4.104)$$

$$U(f_x, f_y, z) = K A \, sinc(2p f_x) sinc(2q f_y), \quad (4.105)$$

where $A = 4pq$, which is the area of the aperture. We can express the diffraction pattern in terms of observation field axes using $f_x = x/\lambda z$ and $f_y = y/\lambda z$ as

$$U(x, y, z) = KA \, sinc\left(\frac{2px}{\lambda z}\right) sinc\left(\frac{2qy}{\lambda z}\right), \quad (4.106)$$

therefore, the intensity is expressed as:

$$I(x, y, z) = |U(x, y, z)|^2,$$

$$I(x, y, z) = \frac{A^2}{\lambda^2 z^2} sinc^2\left(\frac{2px}{\lambda z}\right) sinc^2\left(\frac{2qy}{\lambda z}\right). \quad (4.107)$$

Figure 4.15 shows a plot of the normalized Fraunhofer diffraction intensity pattern along the x-axis and y-axis using for the following example $\lambda = 0.5\,\mu m$, $p = 20\,mm$ and $z = 2000\,m$. It is a typical sinc function plot with the zeros of the function depending on p, λ and z. The first zero of the pattern is at $x = \lambda z/2p$ and the width of the main loop is given by

$$\Delta x = \frac{\lambda z}{p} \text{ and } .\Delta y = \frac{\lambda z}{q}, \quad (4.108)$$

for this example $\Delta x = 0.05\,m$ and $\Delta y = 0.1\,m$.

Figure 4.16 shows the two dimensional intensity diffraction pattern of a rectangular aperture, based on the parameters specified in the given example.

4.8.2 Fraunhofer Diffraction for a Circular Aperture

Let us consider first the diffraction pattern for a rectangular aperture illuminated by a planewave. The amplitude of the transmittance function of the aperture is given by

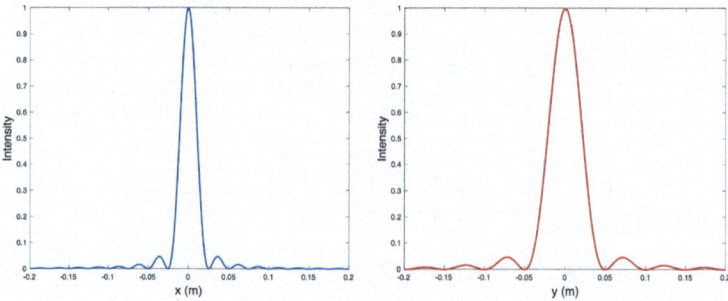

Fig. 4.15 Fraunhofer diffraction pattern along the x-axis for a rectangular aperture

4.8 Fraunhofer Diffraction

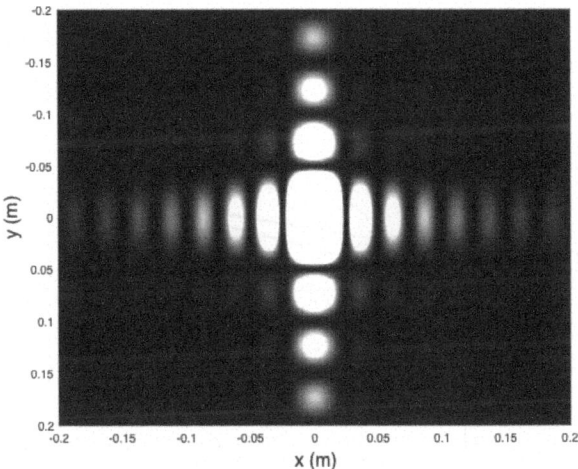

Fig. 4.16 Two-dimensional image of the Fraunhofer diffraction intensity pattern for a rectangular aperture

$$t(x_o, y_o) = circ\left(\frac{\rho}{a}\right) \begin{cases} 1, & \rho \leq a \\ 0, & \rho > a \end{cases}, \quad (4.109)$$

where a is the radius of the aperture. If the aperture is illuminated by a unit-amplitude monochromatic planewave, incident normally to the aperture plane, then the amplitude of the incident field distribution is equal to the aperture transmittance function $t(x_o, y_o)$. The aperture has circular symmetry so we need to spherical coordinate system (r, θ, φ). On the plane of the aperture, as shown in Fig. 4.17:

$$x_o = \rho cos\varphi, \quad y_o = sin\varphi, \quad (4.110)$$

and on the plane of the screen

$$f_x = q cos\phi, \quad f_y = sin\phi, \quad (4.111)$$

and deferential surface element in the aperture is

$$dx_o dy_o = \rho d\rho d\varphi. \quad (4.112)$$

The Fraunhofer diffraction pattern is given by substituting Eq. (4.109) in the Fraunhofer diffraction integral, Eq. (4.101) as

$$U(f_x, f_y, z) = K \iint circ\left(\frac{\rho}{a}\right) e^{-j2\pi(\rho q cos\varphi cos\phi + \rho q sin\varphi sin\phi)} \rho d\rho d\varphi, \quad (4.113)$$

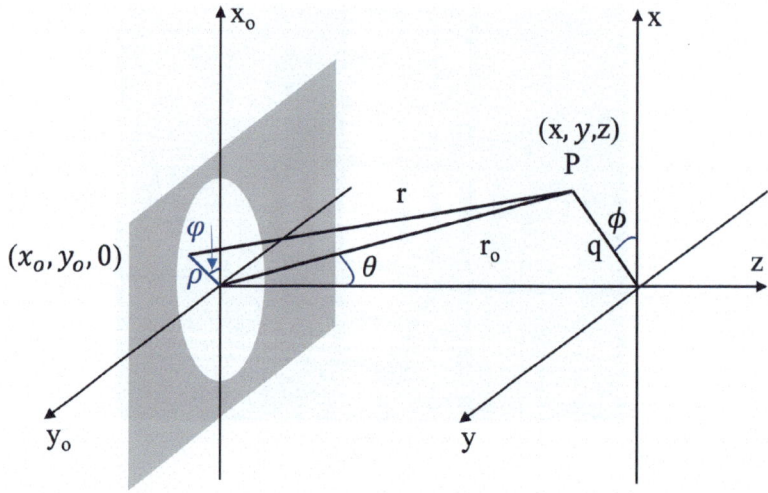

Fig. 4.17 C circular aperture in a spherical coordinate system

where K is as in Eq. (4.103). We use the following identity:

$$cos\varphi cos\phi + sin\varphi sin\phi = cos(\varphi - \phi),$$

in Eq. (4.113), and rewrite the integral by using the limits defined by the circular aperture

$$U(q, \phi) = K \int_{\rho=0}^{a} \rho d\rho \int_{\varphi=0}^{2\pi} e^{-j2\tau\rho q \cos(\varphi - \phi)} d\varphi, \quad (4.114)$$

and since point P is an arbitrary point on the screen we can take $\phi = 0$, then the diffraction integral becomes

$$U_P = K \int_{\rho=0}^{a} \rho d\rho \int_{\varphi=0}^{2\pi} e^{-j2\tau\rho q \cos\varphi} d\varphi. \quad (4.115)$$

The second integral in Eq. (4.115) has a solution in terms of Bessel function. Bessel function of the first kind and order m is given by

$$J_m(u) = \frac{j^{-m}}{2\pi} \int_0^{2\pi} e^{j(mv + u\cos v)} dv, \quad (4.116)$$

4.8 Fraunhofer Diffraction

for $m = 0$

$$J_0(u) = \frac{1}{2\pi} \int_0^{2\pi} e^{ju\cos v} dv, \tag{4.117}$$

using this result, we can rewrite Eq. (4.115) and since J_0 is an even function, therefore,

$$U_P = \frac{K}{2\pi} \int_{\rho=0}^{a} \rho J_0(2\pi\rho q) d\rho. \tag{4.118}$$

Using the following property of Bessel functions of the first kind

$$\int_{u=0}^{v} u J_0(u) du = v J_1(v) \tag{4.119}$$

$$U_P(q) = K' \frac{J_1(2\pi a q)}{2\pi a q}, \tag{4.120}$$

and the intensity of the Fraunhofer diffraction pattern for the circular aperture with radius a is

$$I(q) = I_o \left[\frac{J_1(2\pi a q)}{2\pi a q} \right]^2, \tag{4.121}$$

if we substitute $q = f_x = x/\lambda z$ as given by Eq. (4.111), the diffraction intensity becomes

$$I(x) = I_o \left[\frac{J_1(2\pi a z/\lambda z)}{2\pi a x/\lambda z} \right]^2. \tag{4.122}$$

Figure 4.18 shows the diffraction pattern of Eq. (4.122). This is referred to as the Airy pattern, and the main loop is called the Airy disc. There is no exact analytical expression of the zeros of the Bessel function. The approximate location of the first zero of $I(x)$ is at

$$\frac{2\pi a x}{\lambda z} = 1.22\pi,$$

or

$$x = 1.22 \frac{\lambda z}{2a},$$

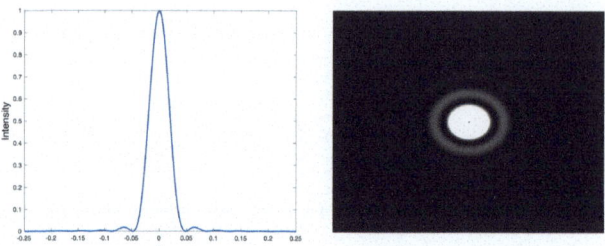

Fig. 4.18 Diffraction pattern for a circular aperture at the Fraunhofer zone

Fig. 4.19 Fraunhofer diffraction pattern of two square apertures

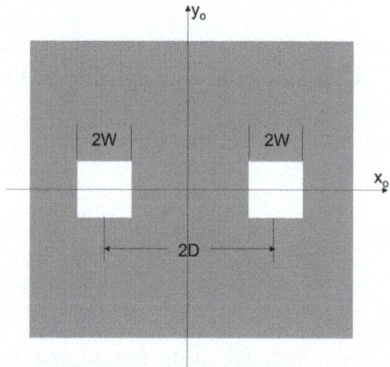

therefore, the width of the main lobe of the Airy disc is

$$d = 1.22 \frac{\lambda z}{a}. \tag{4.123}$$

4.8.3 Fraunhofer Diffraction for Two Square Apertures

Consider the diffraction from two square apertures, each with a width 2w centered on the x-axis and separated by a distance 2D, as shown in Fig. 4.19.

The transmittance function for these aperture is defined as:

$$t(x_o, y_o) = rect\left(\frac{x_o - D}{W}, \frac{y_o}{W}\right) + rect\left(\frac{x_o + D}{W}, \frac{y_o}{W}\right), \tag{4.124}$$

4.8 Fraunhofer Diffraction

substituting the transmittance function in the Fraunhofer diffraction integral results in the following integral:

$$U(f_x, f_y, z) = K \iint\limits_{-\infty}^{\infty} \left\{ rect\left(\frac{x_o - D}{W}, \frac{y_o}{W}\right) + rect\left(\frac{x_o + D}{W}, \frac{y_o}{W}\right) \right\}$$

$$e^{-j2\pi(x_o f_x + y_o f_y)} \, dx_o \, dy_o, \qquad (4.125)$$

the rect-function rectifies the integration limits, hence

$$U(f_x, f_y, z) = K \left\{ \int_{-D-W}^{-D+W} e^{-j2\pi x_o f_x} \, dx_o \int_{-W}^{+W} e^{-j2\pi y_o f_y} \, dy_o \right.$$

$$\left. + \int_{D-W}^{D+W} e^{-j2\pi x_o f_x} \, dx_o \int_{-W}^{+W} e^{-j2\pi y_o f_y} \, dy_o \right\},$$

these four integrals are carried out to get

$$U(f_x, f_y, z)$$

$$= K \left\{ \frac{e^{-j2\pi x_o f_x}}{-j2\pi f_x} \Big|_{-D-W}^{-D+W} \frac{e^{-j2\pi y_o f_y}}{-j2\pi f_y} \Big|_{-W}^{+W} + \frac{e^{-j2\pi x_o f_x}}{-j2\pi f_x} \Big|_{D-W}^{D+W} \frac{e^{-j2\pi y_o f_y}}{-j2\pi f_y} \Big|_{-W}^{+W} \right\},$$

$$U(f_x, f_y, z) = K \left\{ e^{j2\pi D f_x} \left[\frac{sin(2\pi W f_x)}{2\pi f_x} \cdot \frac{sin(2\pi W f_y)}{2\pi f_y} \right] \right.$$

$$\left. + e^{-j2\pi D f_x} \left[\frac{sin(2\pi W f_x)}{2\pi f_x} \cdot \frac{sin(2\pi W f_y)}{2\pi f_y} \right] \right\},$$

$$U(f_x, f_y, z) = 2KW^2 \cdot sinc(2\pi W f_x) \cdot sinc(2\pi W f_y) \cdot cos(2\pi D f_x). \qquad (4.126)$$

The diffraction pattern of the two square apertures is a sinc function as the case for a square aperture, but it is modulated by a cosine term representing the interference between two point sources located at the centers of the apertures. As a result, overall the diffraction pattern displays both the characteristic diffraction of the square apertures and the effect of their separation. The term $2W^2$ represents the area of each square aperture.

Since $D > W$, the interference pattern represented by the cosine function is modulated by the sinc function is illustrated in Fig. 4.20. Furthermore, for multiple square apertures, the overall diffraction pattern can be described as the product of

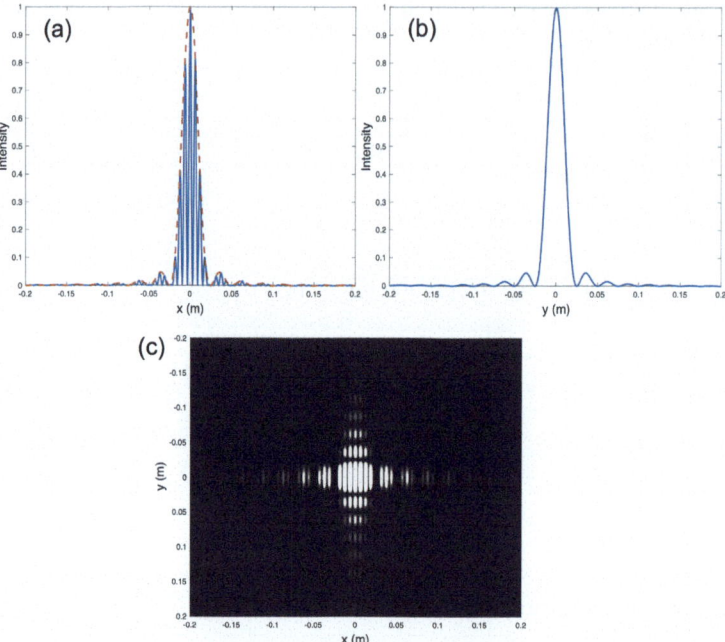

Fig. 4.20 Diffraction by two square apertures: (**a**) cross section along the x-axis of the pattern with a dashed red lines showing the diffraction pattern by a singe square aperture, (**b**) cross section along the y-axis, and (**c**) the two-dimensional diffraction intensity pattern

the diffraction pattern of a single square aperture and the interference pattern of multiple point sources, as given by Eq. (3.74).

4.8.4 Diffraction by a Thin Sinusoidal Amplitude Grating

In practical applications, the transmittance function can be more complex than the clear apertures considered in previous examples, where light transmission only affected amplitude or phase uniformly. In this section, we examine a transmittance function that modulates the amplitude of incident light across the aperture. Consider the transmittance function of a sinusoidal amplitude grating:

$$t_A(x_o, y_o) = \left[\frac{1}{2} + \frac{m}{2}cos(2\pi f_o x_o)\right] rect\left(\frac{x_o}{2W}\right) rect\left(\frac{y_o}{2W}\right). \quad (4.127)$$

This transmittance function describes a square aperture of width $2W$ where the amplitude varies sinusoidally along the x-axis with a frequency f_o and modulation index m. Figure 4.21a shows the cross section of the transmittance function along

4.8 Fraunhofer Diffraction

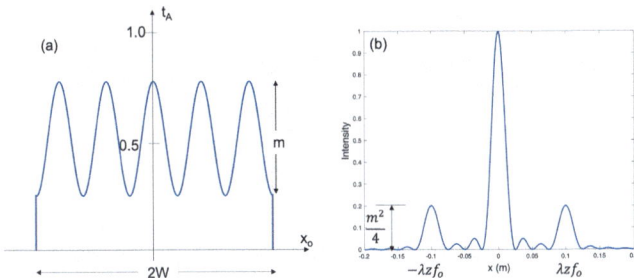

Fig. 4.21 (a) Cross section of the transmittance function of a thin sinusoidal amplitude grating and (b) cross section of the Fraunhofer intensity pattern across the x-axis

the x-axis, so along the y-axis, the amplitude of the beam does not experience any changes.

If the screen is normally illuminated by a unit-amplitude monochromatic plane, the field distribution across the screen emerges as $t_A)(x, y_o)$. The resulting Fraunhofer diffraction pattern at an observation screen located at a distance z is given by:

$$U(f_x, f_y, z) = K \iint\limits_{-\infty}^{\infty} \left\{ \left[\frac{1}{2} + \frac{m}{2} cos(2\pi f_o x_o) \right] rect\left(\frac{x_o}{2W}\right) rect\left(\frac{y_o}{2W}\right) \right\}$$
$$e^{-j2\pi(x_o f_x + y_o f_y)} dx_o dy_o, \qquad (4.128)$$

$$U(f_x, f_y, z) = \frac{K}{2} \iint\limits_{-\infty}^{\infty} \left\{ [1 + mcos(2\pi f_o x_o)] rect\left(\frac{x_o}{2W}\right) rect\left(\frac{y_o}{2W}\right) \right\}$$
$$e^{-j2\pi(x_o f_x + y_o f_y)} dx_o dy_o. \qquad (4.129)$$

We will treat this integral as the addition of two integrals and express the $cos(2\pi f_o x_o)$ term in its exponential form, therefore,

$$U(f_x, f_y) = U_1(f_x, f_y) + U_2(f_x, f_y),$$

$$U_1(f_x, f_y) = \frac{K}{2} \iint\limits_{-\infty}^{\infty} rect\left(\frac{x_o}{2W}\right) rect\left(\frac{y_o}{2W}\right) e^{-j2\pi(x_o f_x + y_o f_y)} dx_o dy_o,$$

$$= 2KW^2 sinc(2W f_x) sinc(2W f_y), \qquad (4.130)$$

and

$$U_2(f_x, f_y) = \frac{mK}{4} \iint\limits_{-\infty}^{\infty} \left(e^{j2\pi f_o x_o} + e^{-j2\pi f_o x_o}\right) rect\left(\frac{x_o}{2W}\right) rect\left(\frac{y_o}{2W}\right)$$

$$e^{-j2\pi(x_o f_x + y_o f_y)} \, dx_o \, dy_o,$$

$$= \frac{mK}{4} \int_{-W}^{+W} \left(e^{j2\pi f_o x_o} + e^{-j2\pi f_o x_o}\right) e^{-j2\pi x_o f_x} dx_o \int_{-W}^{+W} e^{-j2\pi y_o f_o} dy_o,$$

the second integral on y_o is $sinc(2\pi W f_y)$, so

$$U_2(f_x, f_y) = \frac{mK}{4} W sinc(2\pi W f_y) \int_{-W}^{+W} \left(e^{-j2\pi x_o (f_x - f_o)} + e^{-j2\pi x_o (f_x + f_o)}\right) dx_o$$

$$= \frac{mK}{4} W sinc(2\pi W f_y) \left\{ \frac{e^{-j2\pi x_o (f_x - f_o)}}{-j2\pi (f_x - f_o)} \Big|_{-W}^{+W} + \frac{e^{-j2\pi x_o (f_x + f_o)}}{-j2\pi (f_x + f_o)} \Big|_{-W}^{+W} \right\}$$

$$= \frac{mK}{2} sinc(2\pi W f_y) \left\{ \frac{sin[2\pi W(f_x - f_o)]}{2\pi W(f_x - f_o)} + \frac{sin[2\pi W(f_x + f_o)]}{2\pi W(f_x + f_o)} \right\},$$

$$== mKW^2 sinc(2\pi W f_y) \{sinc(2W(f_x - f_o)) + sinc(f_x + f_o)\} \qquad (4.131)$$

Adding Eqs. (4.130) and (4.131), we get Fraunhofer diffraction for thin sinusoidal amplitude grading

$$U(f_x, f_y) = KW^2 sinc(2W f_y) \left\{ sinc(2W f_x) + \frac{m}{2} sinc[2W(f_x - f_o)] \right.$$
$$\left. + \frac{m}{2} sinc[2W(f_x + f_o)] \right\}, \qquad (4.132)$$

and the normalized intensity of the Fraunhofer diffraction is given by

$$I(f_x, f_y) = sinc^2(2W f_y) \left\{ sinc^2(2W f_x) + \frac{m^2}{4} sinc^2[2W(f_x - f_o)] \right.$$
$$\left. + \frac{m^2}{4} sinc^2[2W(f_x + f_o)] \right\}. \qquad (4.133)$$

If we substitute for the values of $fx = x/\lambda z$ and $f_y = y/\lambda z$ in previous equation, we get the intensity pattern to become

$$I(x, y) = sinc^2\left(\frac{2Wy}{\lambda z}\right)\left\{sinc^2\left(\frac{2Wx}{\lambda z}\right) + \frac{m^2}{4}sinc^2\left[\frac{2W}{\lambda z}(x - \lambda z f_o)\right]\right.$$
$$\left. + \frac{m^2}{4}sinc^2\left[\frac{2W}{\lambda z}(x + \lambda z f_o)\right]\right\}. \quad (4.134)$$

Figure 4.21b illustrates the cross-sectional normalized intensity pattern along the x-axis for a thin sinusoidal amplitude grating with parameters $m = 0.9, W = 20$ mm, $f_o = 100$ and $z = 2000$ m. The plot shows some of the incident light is absorbed by the grating, while the sinusoidal transmittance variation across the aperture has deflected some of the energy out of the central diffraction pattern into two additional side patterns. The central diffraction pattern is know as the zero order, while the side patterns referred to as the first orders. All the lobes have the same width $\lambda z/W$ and the first-order lobes are located at $\pm \lambda z f_o$.

From the preceding examples, it is evident that the Fraunhofer diffraction zone is at a very large distance from the aperture, which can make it very impractical in many real-life applications. In the next section will introduce a way to bring the Fraunhofer zone to very close distances from the aperture.

4.9 Babinet's Principle

The Fresnel–Kirchhoff diffraction formula reveals an interesting feature known as Babinet's principle. According to this principle, the Fraunhofer diffraction integral given by Eq. (4.101) is calculated over the clear portion of an aperture. Notably, the field at the observation point remains unchanged if the integral is expressed as the sum of any number of integrals provided that the combined areas of these integrals cover the original aperture area. For example, if the aperture is divided into several segments, the field at the observation point equals to the sum of the integrals over these segments. In Fig. 4.22, the diffraction integral over aperture A is equal to the sum of the integrals over aperture A_1, A_2, A_3 and A_4, i.e.,

$$\int_A \ldots = \int_{A_1} \ldots + \int_{A_2} \ldots + \int_{A_3} \ldots + \int_{A_4} \ldots,$$

where each of these integrals corresponds to the Fraunhofer diffraction integral.

Fig. 4.22 Aperture A is segmented into four apertures A_1, A_2, A_3 and A_4

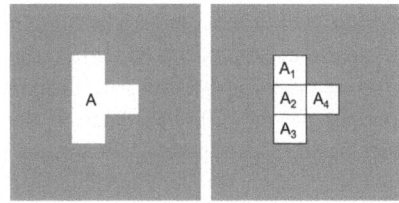

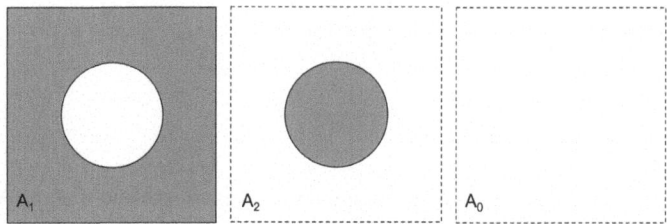

Fig. 4.23 Demonstration of Babinet's principle for complementary aperture A_1 and A_2

Now, consider the apertures depicted in Fig. 4.23. The clear regions of apertures A_1 and A_2 together cover the entire xy plane. as represented by aperture A_0. Apertures A_1 and A_2 are complementary to each other since the clear portions of aperture A_1 exactly correspond to the opaque portions of aperture A_2 and vice versa. In the case of the uninterrupted field, the field at the observation point, P, is $E_0(P)$. If all three cases are illuminated by the same unit amplitude monochromatic wave and the diffracted field at the observation point, P, are E_1, E_2 and E_0, then the following relation should hold

$$E_0(P) = E_1(P) + E_2(P), \tag{4.135}$$

This formula is known as the Babinet's principle. Let us examine this formula further by rearranging its terms as follows:

$$E_2(P) = E_0(P) - E_1(P). \tag{4.136}$$

Now, consider a special case where $E_2(P) = 0$ at some observation point in the absence of apertures. In this scenario, the diffracted fields for the circular aperture and the disc are related by:

$$E_2(P) = -E_1(P), \tag{4.137}$$

and the intensity of the diffracted fields of the complementary apertures is

$$I_1(P) = I_2(P). \tag{4.138}$$

We can get to the same formula when considering that the field illuminating these aperture is collimated with a laser beam with circular cross section much larger than the apertures. As discussed earlier, the diffraction from the circular aperture as we have seen previously is an Airy disc. Further, if the same beam illuminates the disc, aperture A_2, the diffracted beam is the combined diffraction from the disc and the laser beam. Since the laser beam has larger circular cross section, it results in an Airy disc much smaller than that of the circular aperture. Therefore, Eq. (4.138) holds except at one point where $x = 0$ and $y = 0$. Hence, in general, we can apply

4.10 Image Formation and Optical Filtering Using a Lens

the Babinet' principle as expressed in Eqs. (4.137) and (4.138) for complimentary apertures.

4.10 Image Formation and Optical Filtering Using a Lens

The theory presented in this chapter focuses on the propagation of wavefronts in the right-half-space. However, one of the core problems in Fourier optics is to analyze systems composed of a series of lenses arranged along an optical axis, either for image formation or as part of a filtering system. As discussed in Chap. 3, a planewave passing through a thin lens emerges as a spherical wave converging into the focal point of the lens, Fig. 3.14.

For a double-convex lens, where two surfaces have different radii of curvatures R_1 and R_2, and refractive index n, the focal length of the lens is given by

$$\frac{1}{f} = (n-1)\left(\frac{1}{R_1} - \frac{1}{R_2}\right), \tag{4.139}$$

and its transmittance function is given by

$$t(x, y) = e^{jk\left(\frac{x^2+y^2}{2f}\right)}. \tag{4.140}$$

We have discussed in the previous section while discussing ways of meeting the Fraunhofer diffraction condition in short distances, z, from the aperture, either by illuminating the aperture with a converging spherical wave or placing a positive lens behind the aperture. In the following, we will explore using a positive lens to achieve such goal. We consider three possible configurations of where to place the lens with respect to the input. In all these cases, the output plane is the back-focal-plane of the lens. In all cases, the illumination is assumed to be monochromatic. Under this condition, the systems studied are "coherent" systems, which means that they are linear in complex amplitude. Figure 4.24 shows the three possible configurations to perform Fourier transformation by a lens. In all cases shown, the illumination is a monochromatic collimated planewave, which is incident either on the input object transparency or on the lens.

In case (a), the input transparency is placed in front and directly against the lens; in case (b), the input transparency is placed a distance d in front of the lens; and in case (c), the input transparency is placed behind the lens a distance d from the focal plane. In the following subsections, we examine each one of these cases.

4.10.1 Input is Placed in-Front and Directly Against the Lens

Consider an input with transmittance function $t_A(x_o, y_o)$ placed immediately against a positive lens with focal length f. If unit amplitude monochromatic planewave

Fig. 4.24 System configuration for performing Fourier transform by a positive lens. In all the cases, a collimated planewave (blue) is illuminating the system

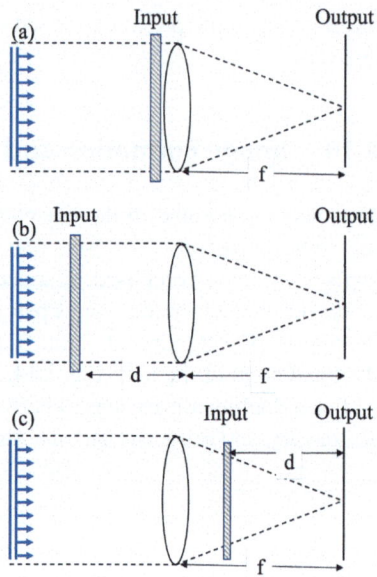

illuminates the input, the field distribution $U_1(x_o, y_o)$ emerging from the input is given by

$$U_1(X_o, y_o) = t_A(x_o, y_o). \tag{4.141}$$

Since the thin lens has a finite diameter, it has an associated pupil function defined as

$$P(x_o, y_o) = \begin{cases} 1 & \text{within the aperture of the lens} \\ 0 & \text{otherwise} \end{cases} \tag{4.142}$$

Using the transmittance function of the lens, Eq. (4.140), the field distribution immediately after the lens becomes

$$U_2(x_o, y_o) = U_1(x_o, y_o) P(x_o, y_o) e^{jk\left(\frac{x_o^2 + y_o^2}{2f}\right)}. \tag{4.143}$$

To determine the field distribution at the focal plane, $U_f(x, y)$, we use the Fresnel diffraction integral given by Eq. (4.92), with $z = f$. Thus, the field distribution at the focal plane is given by:

$$U_f(x, y_y) = \frac{je^{-jkf}}{\lambda f} e^{-jk\frac{(x^2+y^2)}{2f}} \iint_{-\infty}^{\infty} \left\{ U_2(x_o, y_o) e^{-jk\frac{(x_o^2+y_o^2)}{2zf}} \right\}$$

$$e^{-j2\pi(x_o f_x + y_o f_y)} \, dx_o \, dy_o, \tag{4.144}$$

4.10 Image Formation and Optical Filtering Using a Lens

where $f_x = x/\lambda f$ and $f_y = y/\lambda f$ and

$$U(x,y) = \frac{je^{-jkf}}{\lambda f} e^{-jk\frac{(x^2+y^2)}{2f}} \iint\limits_{-\infty}^{\infty} \left\{ U_1(x_o, y_o) P(x_o, y_o) e^{jk\left(\frac{x_o^2+y_o^2}{2f}\right)} e^{-jk\frac{(x_o^2+y_o^2)}{2f}} \right\}$$
$$e^{-j2\pi\left(\frac{x_o y}{\lambda f} + \frac{y_o x}{\lambda f}\right)} dx_o\, dy_o. \tag{4.145}$$

The two chirp functions inside the integral cancel out, and the field at the focal plane becomes

$$U_f(x,y) = \frac{je^{-jkf}}{\lambda f} e^{-jk\frac{(x^2+y^2)}{2f}} \iint\limits_{-\infty}^{\infty} \{U_1(x_o, y_o) P(x_o, y_o)\}$$
$$e^{-j2\pi\left(\frac{x_o y}{\lambda f} + \frac{y_o x}{\lambda f}\right)} dx_o\, dy_o. \tag{4.146}$$

Thus, the field distribution at the focal plane is promotional to the Fraunhofer diffraction integral of the input field, modified by the lens's pupil function, even though the distance f does not satisfy the Fraunhofer condition. If the lens aperture is larger than the extent of the input transparency, the pupil function can be neglected. Therefore, Eq. (4.146) simplifies to:

$$U_f(x,y) = \frac{je^{-jkf}}{\lambda f} e^{-jk\frac{(x^2+y^2)}{2f}} \iint\limits_{-\infty}^{\infty} t_A(x_o, y_o) e^{-j2\pi\left(\frac{x_o y}{\lambda f} + \frac{y_o x}{\lambda f}\right)} dx_o dy_o, \tag{4.147}$$

This can be written in terms of Fourier transform notation as

$$U_f(f_x, f_y) = \frac{je^{-jkf}}{\lambda f} e^{-jk\frac{(x^2+y^2)}{2f}} \{\mathcal{F}\}\{t_A\}, \tag{4.148}$$

where $f_x = x/\lambda f$ and $f_y = y/\lambda f$.

The Fourier transform of the input at the focal plane is altered by the phase term preceding the integral, which introduces a phase curvature. As a result, the output is not exact Fourier transform of the input. However, in many practical applications, the intensity distribution, rather than the phase, is of primary importance.

However, in many practical applications, the intensity distribution, rather than the phase, is of primary importance.

The intensity at the focal plane, denoted as $I_f(x,y)$, is given by the modulus squared of the field distribution:

$$I_f(x,y) = |U_f(f_x, f_y)|^2,$$

$$I_f(x, y) = |\frac{1}{\lambda^2 f^2}| \int\!\!\!\int_{-\infty}^{\infty} t_A(x_o, y_o) \, e^{-j2\pi\left(\frac{x_o y}{\lambda f} + \frac{y_o x}{\lambda f}\right)} \, dx_o \, dy_o|^2, \tag{4.149}$$

$$I_f(x, y) = \frac{1}{\lambda^2 f^2} |\mathcal{F}\{t_A\}|^2. \tag{4.150}$$

Since the phase curvature affects only the phase of the field, it does not influence the intensity. Thus, the intensity at the focal plane corresponds to the modulus squared of the Fourier transform of the input transmittance function, making it a more relevant measure in most optical applications.

4.10.2 Input is Placed at a Distance d in Front of the Lens

In Fig. 4.24b, the input positioned at a distance d in front of the lens. To derive an expression for the field at the focal plane, we need to perform two steps.

1. Propagate the input field at a distance d to reach the lens. This requires using the Fresnel diffraction integral over the distance d.
2. Transform the field by the lens: As in the previous subsection, this step involves applying the Fourier transform.

The propagation of the input field over a distance d can performed using the Fresnel diffraction integral, or transfer function as given in Eq. (4.92), which is rewritten here for the current case

$$H(\xi, \eta) = e^{-jkd} e^{-j\pi\lambda d\left(\xi_x^2 + \eta_y^2\right)}, \tag{4.151}$$

where (ξ, η) are the spatial frequencies corresponding to the plane of the lens.

Let the Fourier transform of the input object be denoted as:

$$T_A(\xi, \eta) = \mathcal{F}\{t_A\}. \tag{4.152}$$

with the spatial frequencies defined as:

$$\xi = \frac{x_1}{\lambda d}, \text{ and } \eta = \frac{y_1}{\lambda d}, \tag{4.153}$$

where (x_1, y_1) are the coordinates at the lens plane.

Thus, the field distribution that illuminates the lens is denoted by:

$$U_1(x_1, y_1) = \mathcal{F}^{-1}\{T_A \cdot H\}. \tag{4.154}$$

4.11 The Canonical 4-F Optical Processor

This field will then undergo a Fourier transform by the lens, yielding the field distribution at the focal plane as

$$U_f(f_x, f_y) = \frac{je^{-jkf}}{\lambda f} e^{-jk\frac{(x^2+y^2)}{2f}} \mathcal{F}\left\{\mathcal{F}^{-1}\{T_A \cdot H\}\right\}, \quad (4.155)$$

Using the proper coordinate transformation, the field distribution at the focal plane will be given by

$$U_f(f_x, f_y) = \frac{je^{-jkf}}{\lambda f} e^{-jk\frac{(x^2+y^2)}{2f}} T_A \cdot H, \quad (4.156)$$

$$U_f(f_x, f_y) = \frac{je^{-jkf}}{\lambda f} e^{-jk\frac{(x^2+y^2)}{2f}} e^{-jkd} e^{-j\pi\lambda d\left(\xi_x^2+\eta_y^2\right)} T_A,$$

$$U_f(f_x, f_y) = \frac{je^{-jk(f-d)}}{\lambda f} e^{-jk\frac{(x^2+y^2)}{2}\left(\frac{1}{f}\frac{1}{d}\right)} e^{-jkd} T_A, \quad (4.157)$$

therefore, the field distribution at the focal plane, $U_f(f_x, f_y)$, is given by:

$$U_f(f_x, f_y) = \frac{K_f}{\lambda f} \mathcal{F}\{t_A\}, \quad (4.158)$$

where K_f is a phase term that does not affect the intensity. The intensity at the focal plane is expressed as:

$$I_f(x, y) = \frac{1}{\lambda^2 f^2} |\mathcal{F}\{t_A\}|^2. \quad (4.159)$$

Once again, the intensity at the focal pane of the lens is the modulus square of the Fourier transform of the input as seen previously. In the special case when $d = f$, the phase term $K_f = 0$, as evident from Eq. (4.157).

This is a significant result, when the object is placed at a distance f in front of the lens, the field distribution at the back focal plane is the exact Fourier transform of the input. This configuration is known as $f - to - f$ system, which serves as a fundamental building block for many optical filtering systems.

4.11 The Canonical 4-F Optical Processor

In Fourier optics, the Fourier optical transform configuration is a fundamental concept, both theoretically and practically. A critical extension of this is the cascade of two Fourier transform configurations, commonly known as the $4f$ optical system.

Fig. 4.25 The 4-F optical processor

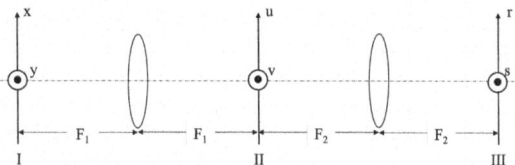

Figure 4.25 illustrates this setup. It consists of an initial lens with a focal length F_1, followed by a stop and filter transparency $P(u, v)$ at plane II, then a second lens of focal length F_2, and finally, an output plane $III(r, s)$. To maintain clarity, distinct labels are used for the transverse coordinates at each plane.

In a two-lens cascade system, the output is analyzed at plane III. However, when a circular stop is present in plane II, we must first recognize that the output plane III is an image plane of I. The resulting image is inverted and magnified by factor of F_2/F_1. This system provides insights in imaging and diffraction limitations, particularly when using rectangular or circular function apertures for the transmission function $P(u, v)$.

Various filtering operations are possible by selecting appropriate choices of the mask $P(u, v)$ offering a deeper understanding of the fundamental principles of optical information processing.

Using results from the single Fourier presented earlier, we can express the scalar field component in plane III, $U_3(r, s)$, in terms of the input at plane I, $U_1(x, y)$, and the transmission mask $P(u, v)$. The transverse plane coordinates are denoted by $I(x, y)$, $II(u, v)$, and $III(r, s)$ with different symbols to avoid confusion when integrating over planes I and II, as shown.

Under monochromatic illumination and for an arbitrary input (both incorporated in the function $U_1(x, y)$), the output $U_3(r, s)$ is obtained by completing a convolution-like integration of $U_1(x, y)$ with an impulse response kernel. To clarify this linear system behavior, we define the impulse response

$$p\left(r + \frac{F_2}{F_1}x, s + \frac{F_2}{F_1}y\right) = \frac{e^{-j2k(F_1+F_2)}}{F_1/F_2} \iint\limits_{-\infty}^{\infty} df_r df_s \, e^{-j2\pi\left[f_r\left(r+\frac{F_2}{F_1}x\right)+f_r\left(s+\frac{F_2}{F_1}y\right)\right]}$$

(4.160)

where the spatial frequencies f_r and f_s are

$$f_r = \frac{u}{F_2\lambda} \text{ and } f_s = \frac{v}{F_2\lambda}. \tag{4.161}$$

The output field $U_3(r, s)$ is given in terms of the impulse response as

$$U_3(r, s) = \iint\limits_{-\infty}^{\infty} dx dy \, U_1(x, y) p\left(r + \frac{F_2}{F_1}x, s + \frac{F_2}{F_1}y\right). \tag{4.162}$$

Consider the simple case if the input is a point source located at $x = x_o$ and $y = y_o$, then the input field is given by

$$U_1(x, y) = \delta(x - x_o, y - y_o). \tag{4.163}$$

Substituting for $U_1(x, y)$ in Eq. (4.162) the output field becomes

$$U_3(r, s) = p\left(r + \frac{F_2}{F_1}x_o, s + \frac{F_2}{F_1}y_o\right) \tag{4.164}$$

which is the impulse response of the system.

The 4-F optical processor has numerous applications, including the implementation of optical spatial filters. The transfer function of the filter is represented by $P(u, v)$, determining the specific filtering operation. For instance, if $P(u, v)$ is a transparent circular aperture, it functions as a low-pass filter, with the spatial frequency cutoff determined by the radius of the circular function. Conversely, if $P(u, v)$ is an opaque disc, the filter is a high-pass filter.

These filtering capabilities make the 4-F system a versatile tool in optical information processing, allowing for the manipulation of spatial frequency components based on specific requirements.

4.12 Problems

1. Derive the Fourier transforms of the following functions:
   (a) $rect(x)rect(y)$
   (b) $Triangular\ function \wedge (x)$
   (c) $f(x - x_o, y - y_o)$
   (d) $f(x)e^{-jkx}$
   (e) $e^{-a(x^2+y^2)}$
2. Prove the following Fourier transform theorems:
   (a) $\mathcal{F}\mathcal{F}\{g(x, y)\} = g(-x, -y)$ at all points of continuity of g.
   (b) $\mathcal{F}^{-1}\mathcal{F}^{-1}\{g(x, y)\} = g(-x, -y)$ at all points of continuity of g.
   (c) $\mathcal{F}\{g(x, y)h(x, y)\} = \mathcal{F}\{g(x, y)\} \otimes \mathcal{F}\{h(x, y)\}$.
3. Derive and plot the 2D Fourier transform of a rectangular aperture of width 2 a 2a and height 2 b 2b. Assume that the aperture is centered at the origin.
4. Derive and plot the 2D Fourier transform of a two rectangular apertures of width 2a and height 2b. Assume the apertures is centered at $x = -D$ and $x = +D$, where $D = 4a$.
5. Derive and plot the 2D Fourier transform of two circular apertures of radius 2a. Assume that the apertures are centered at $x = -D$ and $x = +D$, where $D = 4a$.
6. Prove that that the Fresnel propagation over a sequence of successive distances $z_1, z_{23}, ..., Z_5$ must be equivalent to Fresnel propagation over the single distance $z = z_1 + z_2 + \cdots \bullet + Z_n$.

7. A thin square-wave phase grating has a thickness that varies periodically (period L) such that the phase of the transmitted light jumps between $0\ radians$ and $\text{cf}\phi\ radians$
   (a) Find the diffraction efficiency of this grating for the first diffraction orders.
   (b) (b) What value of ϕ yields the maximum diffraction efficiency, and what is the value of that maximum efficiency?
8. Derive Eq. (4.28).

Bibliography

1. Goodman, J. W. (2005). Introduction to Fourier Optics (3rd ed.). Roberts and Company Publishers.
2. Gaskill, J. D. (1978). Linear Systems, Fourier Transforms, and Optics. Wiley.
3. Bracewell, R. N. (2000). The Fourier Transform and Its Applications (3rd ed.). McGraw-Hill.
4. VanderLugt, A. (1992). Optical Signal Processing. Wiley.
5. O'Shea, D. C., Suleski, T. J., Kathman, A. D., & Prather, D. W. (2004). Diffraction, Fourier Optics and Imaging. SPIE Press.
6. Yu, F. T. S. (2001). Fourier Optics and Image Processing. In R. G. Driggers (Ed.), Handbook of Optical Engineering (pp. 497–528). CRC Press.
7. Yu, F. T. S., & Gregory, D. A. (Eds.). (1998). Optical Pattern Recognition. Cambridge University Press.
8. Saleh, B. E. A., & Teich, M. C. (2019). Fundamentals of Photonics (3rd ed.). Wiley.
9. Stark, H. (Ed.). (1982). Applications of Optical Fourier Transforms. Academic Press.
10. Abushagur, M. A. G., & Caulfield, H. J. (Eds.). (1995). Selected Papers on Fourier Optics. SPIE Milestone Series, Vol. MS105. SPIE Press.
11. James, J. F. (2011). A Student's Guide to Fourier Transforms: With Applications in Physics and Engineering (3rd ed.). Cambridge University Press.
12. Voelz, D. G. (2011). Computational Fourier Optics: A MATLAB Tutorial. SPIE Press.
13. Banerjee, P. P., & Banerjee, T. (2008). Computational Fourier Optics: The Mathematica Perspective. CRC Press.

Light Polarization

5.1 Introduction

Polarization is a property that is inherent to all vector waves. Light, as a transverse electromagnetic wave, has an electric field that oscillates perpendicular to the direction of propagation. Light is considered unpolarized when the direction of electric field fluctuates randomly over time. Common sources of unpolarized light include sunlight, halogen lamps, LED spotlights, and incandescent bulbs. In contrast, when the electric field fluctuation is well-defined, light is referred to as polarized light. The most common source of polarized light is a laser beam.

5.2 States of Polarization

Polarized light can be classified into three states based on the orientation of its electric field:

- **Linear Polarization**. The electric field of light is confined to a single plane along the direction of propagation, as shown in Fig. 5.1.
- **Circular Polarization**. The electric field comprises two perpendicular linear components of equal in amplitude, with a phase difference of $\pi/2$. This causes electric field to rotate in a circle pattern around the direction of propagation, depending on the rotation direction, it is classified as left- or right-hand circularly polarized light, as shown in Fig. 5.2.
- **Elliptical Polarization**. The electric field traces an ellipse, resulting from the combination of two linear components with differing amplitudes or a phase difference that is not $\pi/2$. This is the most general form of polarized light, as shown in Fig. 5.3.

Fig. 5.1 Linear polarization

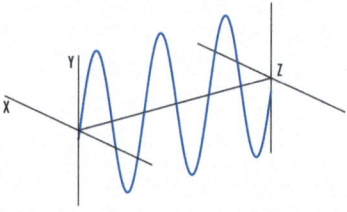

Fig. 5.2 Circular polarization

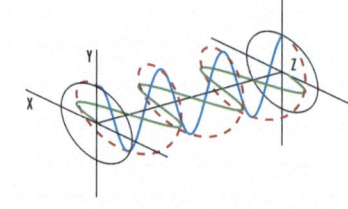

Fig. 5.3 Elliptical polarization

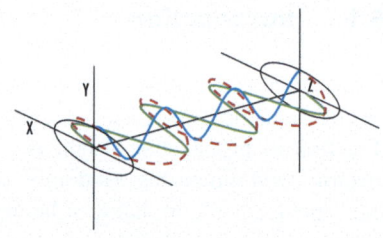

5.2.1 Linear Polarization State

Linearly polarized, or plane-polarized light has a constant orientation of the electric field, though, its magnitude and sign vary in time. The electric field remains confined to a fixed plane known as the plane-of-vibration, which contains both the electric field vector $\overline{E}$ and the propagation vector $\overline{k}$ aligned along the direction of motion.

Consider two harmonics, linearly polarized lightwaves with the same frequency, propagating in the same region of space and direction. If their electric field vectors are collinear, their superposition results in a new linearly polarized wave.

The harmonic electromagnetic wave propagating along the z z-axis can be represented as:

$$\overline{E}(z,t) = \left(E_{ox}\overline{a}_x + E_{oy}\overline{a}_y\right) e^{-j(kz-\omega t)} = E_{ox} e^{j(\omega t - kz)} \overline{a}_x + E_{oy} e^{j(\omega t - kz + \phi)} \overline{a}_y \tag{5.1}$$

and the real part of the electric field is given by

$$\overline{E}(z,t) = E_{ox} \cos(\omega t - kz)\overline{a}_x + E_{oy} \cos(\omega t - kz + \phi)\overline{a}_y \tag{5.2}$$

where ϕ is the initial phase shift between the two orthogonal components of the electric field.

5.3 Circular Polarization

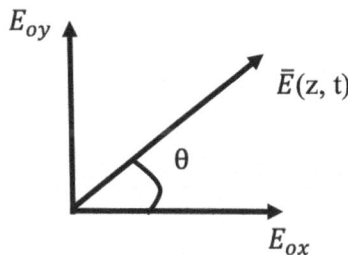

Fig. 5.4 The polarization is the vectorial sum of the $E_{ox}\bar{a}_x + E_{oy}\bar{a}_y$ and is oriented with angle θ with respect to the x-axis

- For positive phase difference, ($\phi > 0$), the y-component of the electric field, E_y lags behind the x-component E_x.
- For negative phase difference, ($\phi < 0$), the y-component of the electric field E_y leads E_x.

The resultant optical field $\bar{E}(z, t)$ is the vector sum of the two perpendicular fields, as described by Eq. (5.2). The resultant wave also exhibits linear polarization, with a fixed amplitude of $(E_{ox}\bar{a}_x + E_{oy}\bar{a}_y)$. As the waves propagate toward a plane of observation, the resultant field $\bar{E}$ oscillating, along a tilted line, cosinusoidally in time, as shown in Fig. 5.4. The amplitude of the electric field $\bar{E}$ is determined by the amplitudes of the original orthogonal waves.

The amplitude of the resultant field is given by

$$|\bar{E}| = \sqrt{E_{ox}^2 + E_{oy}^2}, \tag{5.3}$$

and the angle θ is given by

$$\theta = tan^{-1}\left(\frac{E_{oy}}{E_{ox}}\right). \tag{5.4}$$

If $E_{oy} = E_{ox}$, the electric field oscillates at an angle $\theta = 45°$. If the phase ϕ is equal to multiple integers of $\pm \pi$, the resultant field remains linearly polarized.

As time progresses, the electric field continues to oscillate within the same plane, but its magnitude varies as a cosine function.

5.3 Circular Polarization

Circular polarization arises when both constituent waves have equal amplitudes, i.e., $E_{ox} = E_{oy} = E_0$, and the phase difference $\phi = -\frac{\pi}{2} + 2m\pi$, where $m = 0, \pm 1, \pm 2, \ldots$ Substituting $\phi = -\frac{\pi}{2}$ in Eq. (5.2) we obtain:

$$\bar{E}(z, t) = E_{ox}cos(\omega t - kz)\bar{a}_x + E_{oy}cos\left(\omega t - kz - \frac{\pi}{2}\right)\bar{a}_y,$$

Simplifying further, this becomes:

$$\bar{E}(z,t) = E_o \left[\cos(\omega t - kz)\bar{a}_x + \sin(\omega t - kz)\bar{a}_y \right] \quad (5.5)$$

Noting that the scalar amplitude of the electric field, E_o remains constant, while the direction of field $\bar{E}$ is time-dependent, rotating continuously rather than remaining confined to a single plane as in linear polarization.

Examine the direction of the electric field as the time progresses.

- At time $t = 0$:

$$\bar{E}_x = E_o \cos(kz_0)\bar{a}_x, \quad \bar{E}_y = E_o \sin(kz_0)\bar{a}_y.$$

Thus, the electric field $\bar{E}$ lies in the first quadrant.

- At a later time, $t = kz_0/\omega$:

$$\bar{E}_x = E_o \bar{a}_x, \quad \bar{E}_y \bar{a}_y = 0.$$

The electric field now aligns along the x x-axis.

Right-Circular Polarization As time progressed, the electric field rotates clockwise with an angular frequency ω, as observed by an individual facing the incoming wave, i.e., looking back at the source,.This wave is called right-circularly polarized or simply right-circular polarized light. The $\bar{E}$−vector completes one full rotation as the wave propagates through one wavelength. Figure 5.5a–e illustrate the orientation

Fig. 5.5 The orientation of the electric field for a right-circular polarization as the time progresses from (**a**) to (**e**)

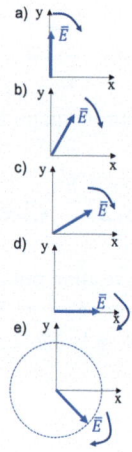

5.4 Elliptical Polarization

of the electric field $\bar{E}$ at five successive moments in time for a **right-circular polarization**.

Left-Circular Polarization In contrast, if the phase difference between the x and y components of the electric field is $\phi = +\pi/2 + 2m\pi$, where $m = 0, \pm 1 \pm 2 \pm 3 \pm \ldots$ the resultant electric field vector becomes:

$$\bar{E}(z,t) = E_o \left[cos(\omega t - kz)\bar{a}_x - sin(\omega t - kz)\bar{a}_y \right] \tag{5.6}$$

Here, the amplitude remains the same, but the electric field now rotates counter-clockwise over time, indicating left-circular polarization.

Synthesis of Linear Polarization from Circular Waves Adding the expressions for right-circular and left-circular polarization Eqs. (5.5) and (5.6) results in the resultant electric field to be given by

$$\bar{E}(z,t) = 2E_o cos(\omega t - kz)\bar{a}_x$$

This expression corresponds to a linearly polarized wave. Thus, it demonstrates that a linearly polarized wave can be synthesized by superimposing two oppositely polarized circular waves of equal amplitude. The combined effect results in a constant-direction electric field that oscillates in magnitude, characteristic of linear polarization.

5.4 Elliptical Polarization

Linear and circular polarizations are special cases of light wave polarizations. Linear polarization occurs when the phase difference $\phi = \pm m\pi$, while circular polarization occurs when $\phi = \pm \frac{\pi}{2} \pm 2m\pi$, where $m = 0, 1, 2, 3, \cdots$. When the phase difference ϕ does not meet these these conditions, the resulting polarization is elliptical, which is the most general case of polarization. Therefore, linear and circular polarizations are special cases of elliptical polarization. Using Eq. (5.2), we can express the x and y-components of the electric fields as

$$\bar{E}_x = E_o cos(kz - \omega t)\bar{a}_x,$$

and

$$\bar{E}_y = E_{0y} cos(kz - \omega t + \phi)\bar{a}_y.$$

Applying the trigonometric expansion to the y y-component:

$$\bar{E}_y = E_{0y}[cos(kz - \omega t)cos\phi - sin(kz - \omega t)sin\phi]\bar{a}_y$$

Therefore,

$$\frac{E_x}{E_{ox}} = cos(kz - \omega t)$$

and

$$\frac{E_y}{E_{oy}} = \cos(kz - \omega t)\cos\phi - \sin(kz - \omega t)\sin\phi.$$

Rearranging terms, we get:

$$\frac{E_y}{E_{0y}} - \frac{E_x}{E_{0x}}\cos\phi = -\sin(kz - \omega t)\sin\phi$$

Since,

$$1 - \left(\frac{E_x}{E_{0x}}\right)^2 = 1 - \cos^2(kz - \omega t) = \sin^2(kz - \omega t),$$

we can write:

$$sin(kz - \omega t) = \left[1 - \left(\frac{E_x}{E_{ox}}\right)^2\right]^{1/2}.$$

Substituting we can write:

$$\left(\frac{E_x}{E_{0y}} - \frac{E_x}{E_{0x}}cos\phi\right)^2 = \left[1 - \left(\frac{E_x}{E_{0x}}\right)^2\right]\sin^2\phi.$$

Expanding further:

$$\left(\frac{E_y}{E_{0y}}\right)^2 + \left(\frac{E_x}{E_{0x}}\right)^2\cos^2\phi - 2\left(\frac{E_y}{E_{0y}}\right)\left(\frac{E_x}{E_{0x}}\right)cos\phi = \sin^2\phi - \left(\frac{E_x}{E_{0x}}\right)^2\sin^2\phi.$$

Simplifying:

$$\left(\frac{E_y}{E_{0y}}\right)^2 + \left(\frac{E_x}{E_{0x}}\right)^2 - 2\left(\frac{E_y}{E_{0y}}\right)\left(\frac{E_x}{E_{0x}}\right)cos\phi = \sin^2\phi. \tag{5.7}$$

5.4 Elliptical Polarization

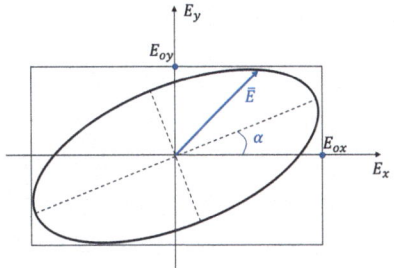

Fig. 5.6 Elliptical polarization; the endpoint of the electric field vector sweeps out an ellipse as it rotates once around

This is the equation of an ellipse making an angle α with the (E_x, E_y)-coordinate system Fig. 5.6 such that

$$\tan(2\alpha) = \frac{2E_{0x}E_{0y}\cos\phi}{E_{0x}^2 - E_{0y}^2} \tag{5.8}$$

Let us examine Eq. (5.7) for special values the phase difference ϕ.

1. If $\phi = \pm\pi/2, \pm 3\pi/2, \pm 5\pi/2, \ldots$, Eq. (5.7) reduces to:

$$\left(\frac{E_x}{E_{0x}}\right)^2 + \left(\frac{E_y}{E_{0y}}\right)^2 = 1 \tag{5.9}$$

which is an equation for an ellipse with semi-major axes $E0x$ E_{ox} and E_{oy}. When $E_{ox} = E_{oy} = E_o$ Eq. (5.9) becomes

$$E_x^2 + E_y^2 = E_o^2$$

which is the equation of a circle with radius E_o. Therefore, for these phase differences, the wave is circularly polarized.

2. If $\phi = 0, \pm 2\pi, \pm 4\pi, \pm 6\pi, \ldots$, then Eq. (5.7) becomes

$$\left(\frac{E_x}{E_{ox}} - \frac{E_y}{E_{oy}}\right)^2 = 0$$

which can be rewritten as

$$E_y = E_{oy}\frac{E_x}{E_{ox}}$$

representing a straight line, indicating linear polarization.

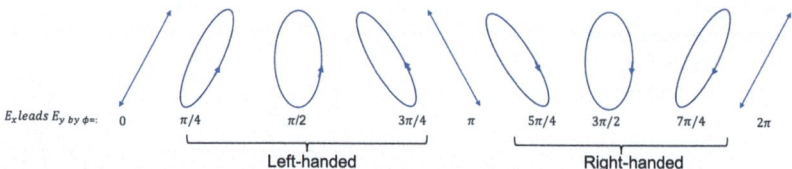

Fig. 5.7 The state of polarization for a number of values of the phase difference ϕ for the case when E_x leads E_y

Therefore, for all values of ϕ, the wave is elliptically polarized except when ϕ is equal a multiples of 2π or $\pi/2$, at which point the polarization becomes either linear or circular.

Figure 5.7 symbolically demonstrates the state of polarization for the values of $\phi = 0,\ \pi/4, \pi/2, 3\pi/4,\ \pi,\ 5\pi/4, 3\pi/2, 7\pi/4,\ and\ \pi$. The polarization changes from linear to elliptical to circular to elliptical and back to linear. Additionally:

- For $\phi = \pi/4, \pi/2, 3\pi/4$, the polarization is left-handed.
- For $\phi = 5\pi/4, 3\pi/27\pi/4$ the polarization is right-handed.

In elliptical polarization, the electric field vector traces out an elliptical path as time progresses, with the shape and orientation of the ellipse determined by the relative amplitudes E_{ox} and E_{oy}, and the phase difference ϕ. The resultant electric field rotates continuously, either clockwise or counterclockwise, depending on the sign of the phase difference ϕ, creating an elliptical trajectory.

5.5 Mathematical Representation of Polarization

The polarization state of the light wave is determined by the transverse components of its electric field. Using Eq. (5.1), the electric field can be expressed as: y

$$\bar{E}(z,t) = E_{ox} e^{j(\omega t - kz)} \bar{a}_x + E_{oy} e^{j(\omega t - kz + \phi)} \bar{a}_y.$$

This expression be represented in vector form:

$$\bar{E} = \begin{bmatrix} E_{ox} e^{j(\omega t - kz)} \\ E_{oy} e^{j(\omega t - kz + \phi)} \end{bmatrix} = e^{j(\omega t - kz)} \begin{bmatrix} E_{ox} \\ E_{oy} e^{j\phi} \end{bmatrix}$$

The complex amplitudes of the electric field can be written in a vector form known as the Jones vector, which is commonly used to represent the state of polarization:

$$\bar{J} = \begin{bmatrix} E_{ox} \\ E_{oy} e^{j\phi} \end{bmatrix}. \tag{5.10}$$

5.5 Mathematical Representation of Polarization

In many applications, the exact values of the amplitudes of the electric field are not critical, and Jones vectors are typically normalized to represent the intensity su that:

$$\bar{J} = \frac{1}{\sqrt{E_{ox}^2 + E_{oy}^2}} \begin{bmatrix} E_{ox} \\ E_{oy} e^{j\phi} \end{bmatrix}. \quad (5.11)$$

The Jones vector provides a mathematical representation of different states of polarization, describing the orientation and phase of the electric field components.

Examples of Jones Vectors for Common Polarization States

1. Horizontal Linearly Polarized Light (along the x-$axis$):

$$\bar{J}_h = \begin{bmatrix} 1 \\ 0 \end{bmatrix}. \quad (5.12)$$

2. Vertical Linearly Polarized Light (along the y-$axis$):

$$\bar{J}_v = \begin{bmatrix} 0 \\ 1 \end{bmatrix}. \quad (5.13)$$

3. 45° Linearly Polarized Light (making a 45° angle with the x-$axis$):

$$\bar{J} = \frac{1}{\sqrt{2}} \begin{bmatrix} 1 \\ 1 \end{bmatrix}. \quad (5.14)$$

4. Right-Handed Circularly Polarized Light:

$$\bar{J} = \frac{1}{\sqrt{2}} \begin{bmatrix} 1 \\ -j \end{bmatrix}. \quad (5.15)$$

5. Left-Handed Circularly Polarized Light:

$$\bar{J} = \frac{1}{\sqrt{2}} \begin{bmatrix} 1 \\ j \end{bmatrix} \quad (5.16)$$

6. Right-Handed Elliptically Polarized Light:

$$\bar{J} = \frac{1}{\sqrt{a^2 + b^2 + c^2}} \begin{bmatrix} a \\ b - jc \end{bmatrix} \quad (5.17)$$

7. Left-Handed Elliptically Polarized Light:

$$\bar{J} = \frac{1}{\sqrt{a^2 + b^2 + c^2}} \begin{bmatrix} a \\ b + jc \end{bmatrix} \quad (5.18)$$

Table 5.1 Jones vectors for some polarization states

Polarization state	Jones vector
Horizontal linear polarization	$\begin{bmatrix} 1 \\ 0 \end{bmatrix}$
Vertical linear polarization	$\begin{bmatrix} 0 \\ 1 \end{bmatrix}$
Linear polarization at $+45^o$	$\frac{1}{\sqrt{2}} \begin{bmatrix} 1 \\ 1 \end{bmatrix}$
Linear polarization at -45°	$\frac{1}{\sqrt{2}} \begin{bmatrix} 1 \\ -1 \end{bmatrix}$
Linear polarization at angle θ	$\begin{bmatrix} cos\theta \\ sin\theta \end{bmatrix}$
Right circular polarization	$\frac{1}{\sqrt{2}} \begin{bmatrix} 1 \\ -j \end{bmatrix}$
Left circular polarization	$\frac{1}{\sqrt{2}} \begin{bmatrix} 1 \\ j \end{bmatrix}$
Right elliptical polarization	$\frac{1}{\sqrt{a^2+b^2+c^2}} \begin{bmatrix} a \\ b-jc \end{bmatrix}$
Left elliptical polarization	$\frac{1}{\sqrt{a^2+b^2+c^2}} \begin{bmatrix} a \\ b+jc \end{bmatrix}$

The angle α α, which represents the tilt of the ellipse with respect to the x x-axis, is given by:

$$\alpha = \mp tan^{-1}\left(\frac{c}{b}\right) \tag{5.19}$$

Table 5.1 provides a summary of Jones vectors for several polarization states, illustrating how different states of polarization can be represented and analyzed mathematically using this vector formalism.

5.6 Natural Light

An ordinary light source such as incandescent light bulb emits light because of the thermal excitations of electrons transition resulting in the mission of photons. The tungsten filament consists of a very large number of randomly oriented atomic emitters. Each excited atom radiates a polarized wave-train for a short period of time in the order of 10^{-8} seconds.

Emissions with the same frequency combine to form a single resultant polarized wave, but this persists for no longer than 10^{-8} seconds. New wave-trains are constantly emitted, causing the overall polarization to change randomly. If these changes occur rapidly, no single resultant polarization state can be distinguished, the wave is referred to as natural light. It is also known as unpolarized or randomly polarized light, characterized by a rapidly varying succession of different polarization state.

Mathematically, natural light can be represented as two arbitrary, incoherent, orthogonal, linearly polarized waves of equal amplitude, with a relative phase difference ϕ, varies rapidly and randomly.

5.7 Polarizers

Polarizers are optical elements that change the state of polarization as the light wave passes through them. Polarizers are widely used in various applications, most commonly in displays, 3D movie glasses, and sunglasses. Polarizing devices include:

Polarizing devices include:

- Linear polarizers.
- Phase retarders.
- Polarization rotators.

Polarizers come in many different configurations that are based on one of four fundamental physical mechanisms:

1. **Dichroism (Polarization by Absorption)**: Certain materials selectively absorb one component of the electric field while transmitting the other, producing linearly polarized light.
2. **Reflection**: When light is incident at a specific angle (the Brewster angle), the reflected light becomes partially or completely polarized parallel to the surface.
3. **Scattering**: Scattered light, such as that in the atmosphere, becomes polarized perpendicular to the direction of propagation.
4. **Birefringence (Double Refraction)**: Certain materials have different refractive indices along different axes, splitting the incoming light into two orthogonal, linearly polarized components.

Polarizers play a significant role in controlling and analyzing the state of light's polarization in optical systems.

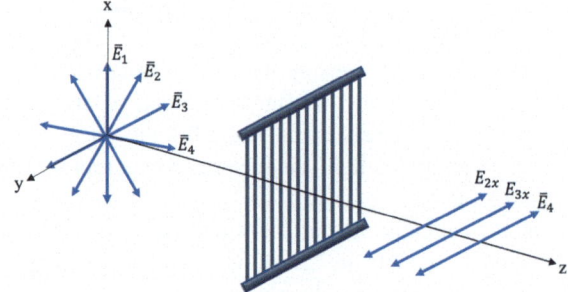

Fig. 5.8 Wire-grid polarizer passes only the x-components of the electric fields of the light wave impinging on it

5.7.1 Dichroism

Dichroism is based on the phenomena where one of the orthogonal components of the electric field is absorbed while the other component passes through. This selective absorption of one of the two orthogonal polarization components of an incident beam creates linearly polarized light. Dichroism can be realized through wire-grid polarizers or dichroic crystals.

5.7.1.1 Wire-Grid Polarizer

The wire-grid polarizer is a simple type of polarizer, consisting of a grid of parallel conducting wires, spaced in the order of the wavelength of light. These polarizers are mainly used in infrared (IR) and longer wavelengths region.

When light impinges on the polarizer, the electric field component parallel to the conducting wires induces electric current along the wires, resulting in absorption, meanwhile, the electric field component perpendicular to the wires passes unattenuated. Thus, the wire-grid polarizer passes the linear polarized light perpendicular to the wires, as shown in Fig. 5.8.

For an unpolarized light wave propagating along the z-axis, only the y-component of the electric field passes through the polarizer, while x-component of the electric field is absorbed. As a result, the wire-grid polarizer acts as a linear polarizer.

5.7.1.2 Dichroic Crystals

Certain materials exhibit inherent dichroic due to an anisotropy in their crystalline structures. Probably the best known example is the naturally occurring mineral tourmaline. Such crystals have a specific direction known as the principal or optic axis, depending on their atomic structure. The component of the electric field perpendicular to the optic axis is strongly absorbed by the crystal. The degree of absorption increases with the thickness of the crystal, as shown in Fig. 5.9.

A plate cut from a tourmaline crystal parallel to its optic axis and several millimeters thick can serve as a linear polarizer. In this configuration, the crystal's optic axis becomes transmission axis of the polarizer. Other materials with similar anisotropic properties can be used to construct linear polarizers.

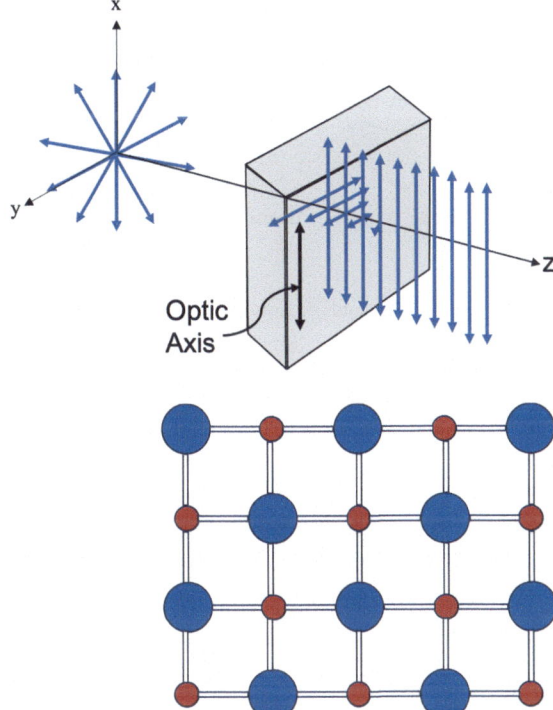

Fig. 5.9 A dichroic crystal allows the E-field component parallel to its optic axis to pass through without any absorption, while the E-field component perpendicular to the optic axis is absorbed as the light propagates through the crystal

Fig. 5.10 One layer of a cubic crystalline structure of sodium chloride. Each chloride ion (red spheres) is surrounded by six individual sodium ions (blue spheres) and vice versa for the sodium ions

5.7.2 Birefringence

Birefringence is defined as the double refraction of light in a transparent, molecularly ordered material, characterized by the orientation-dependent differences in refractive index. optically isotropic transparent solids, refractive index is the same in all directions throughout the crystalline lattice. Examples of isotropic solids include glass, table salt, many polymers, and a wide variety of organic and inorganic compounds.

In contrast, birefringent materials have a refractive index that varies based on the direction of the electric field polarization. The simplest crystalline lattice structure is cubic, as illustrated by the molecular model of sodium chloride, shown in Fig. 5.10, an arrangement where all of the sodium and chloride ions are ordered with uniform spacing along three mutually perpendicular axes. Each chloride ion is surrounded by (and electrostatically bonded to) six individual sodium ions and vice versa for the sodium ions.

When light enters an isotropic crystal like sodium chloride, it is refracted at a constant angle propagating through the crystal at a single velocity without being polarized by interaction with the electronic structure of the crystalline lattice. In contrast, anisotropic crystals, such as quartz, calcite, and tourmaline, have distinct crystallographical axes, which interact with light by a mechanism that is dependent

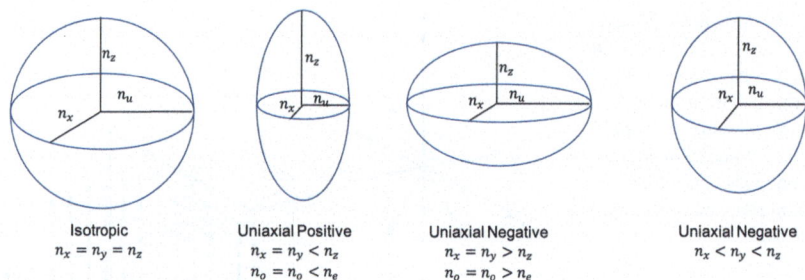

Fig. 5.11 Index ellipsoids of Isotropic and anisotropic crystals classifications

upon the orientation of the crystalline lattice with respect to the incident light differently depending on the orientation of the crystal relative to the incident light angle.

- When light enters along the optic axis of anisotropic crystals, it behaves in a manner similar to the interaction with isotropic crystals, propagating at a single velocity.
- When light enters a non-equivalent axis, it splits into two rays, each polarized with vibration directions oriented at right angles (mutually perpendicular) to one another and traveling at different velocities. This phenomenon is known as double refraction or birefringence, and it is exhibited to a greater or lesser degree in all anisotropic crystals.

Figure 5.11 illustrates the difference between isotropic and anisotropic crystals.

In uniaxial crystals, the refractive properties can be described by an index ellipsoid with the following representation:

$$\frac{x^2}{n_o^2} + \frac{y^2}{n_o^2} + \frac{z^2}{n_e^2} = 1, \tag{5.20}$$

where n_o is the ordinary refractive index, and n_e is the extraordinary refractive index.

5.7.2.1 Uniaxial Crystals

Uniaxial crystals are a special type of anisotropic material structure. These crystals exhibit two distinct refractive indices, one for light polarized along the optic axis (n_e) and another for light polarized in either of the two directions perpendicular to it (n_o). Light polarized along the optic axis is called the extraordinary ray, while, light polarized perpendicular to it is called the ordinary ray. These polarization directions are know as the crystal "principal axes." The Birefringence (B) of an uniaxial crystal is defined as

$$B = |n_e - n_o| \tag{5.21}$$

5.7 Polarizers

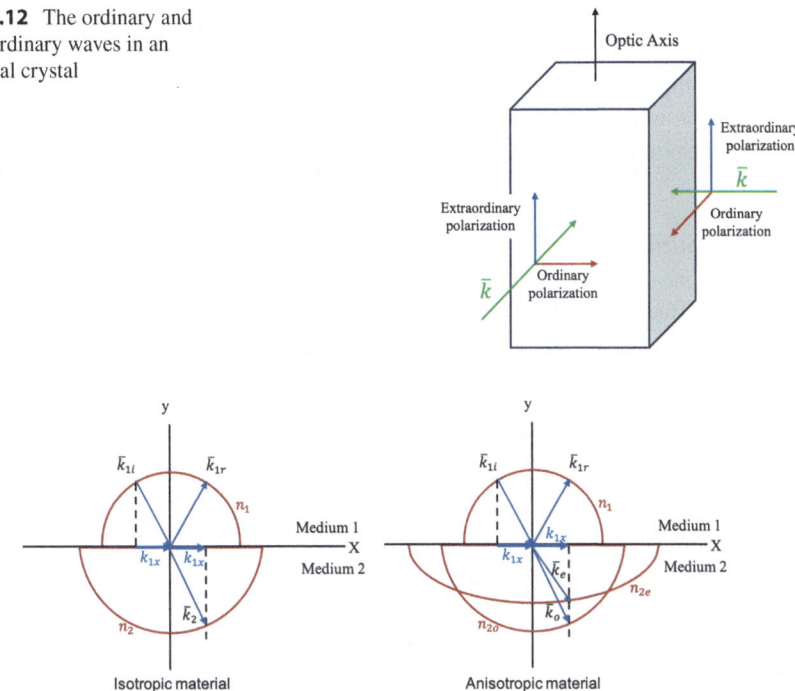

Fig. 5.12 The ordinary and extraordinary waves in an uniaxial crystal

Fig. 5.13 Incident and refracted waves' relations for isotropic and anisotropic materials

Figure 5.12 illustrates the behavior of the extraordinary and ordinary waves as they pass through a uniaxial crystal. The difference of refractive indices (n_o and n_e) causes the electric fields traveling along these axes propagate with different speeds, resulting in a phase shift that leads to a change in the polarization-state change.

Birefringence can be classified in two types:
Positive birefringence when $n_e > n_o$, and
negative birefringence when $n_e < n_o$.

Effect of Birefringence on Refraction To understand the birefringence effect on refraction of light waves, refer to Fig. 5.13.

- In the case of isotropic material, the refractive indices are the same in all directions, represented by the red-colored half semicircle. The incident, reflected, and refracted waves represented by their wave vectors $\bar{k}_{1i}$, $\bar{k}_{1r}$ and $\bar{k}_2$, respectively. Based on the boundary conditions at the interface, the tangential components of the electric field are equal on both sides of the interface. Thus, k_{1x} remains the same for the incident and refracted fields. In this case, Snell's law, applies to the relation between the incident and refracted angles.

- In the case for an anisotropic material, the second medium exhibits two indices of refraction, represented in Fig. 5.12 by the red circle for the ordinary wave, and the red ellipse for the extraordinary wave. This leads to the occurrence of double refraction, where the light splits into two refracted fields represented by $\bar{k}_0$ and $\bar{k}_e$, corresponding to the ordinary and extraordinary rays, respectively.

5.7.2.2 Double Refraction

When light passes through a calcite crystal, it splits into two rays. This phenomenon is double refraction. These two rays are each plane polarized by the crystal such that their planes of polarization are mutually perpendicular. For normal incidence (a Snell's law angle of 0°), the two planes of polarization are also perpendicular to the plane of incidence. For normal incidence (a 0° angle of incidence), Snell's law predicts that the angle of refraction will be 0°. In the case of double refraction of a normally incident ray of light, at least one of the two rays must violate Snell's law as we know it. For calcite, one of the two rays does indeed obey Snell's law; this ray is called the ordinary ray (or o-ray). The other ray (and any ray that does not obey Snell's law) is an extraordinary ray (or e-ray).

For **ordinary** rays, the vibration direction indicated by the electric vectors in Fig. 5.14 is perpendicular to the ray path. For **extraordinary** rays, the vibration direction is not perpendicular to the ray path. The direction that is perpendicular to the vibration direction is known as the wave normal. While the ray path of extraordinary rays does not satisfy Snell's law, the wave normal does, meaning that the wave normal direction for the refracted ray aligns with Snell's law relative to the wave normal of the incident ray.

Uniaxial crystals such as calcite can be used as a polarizers exploiting the double refraction resulting from the difference between the two indices of refraction associ-

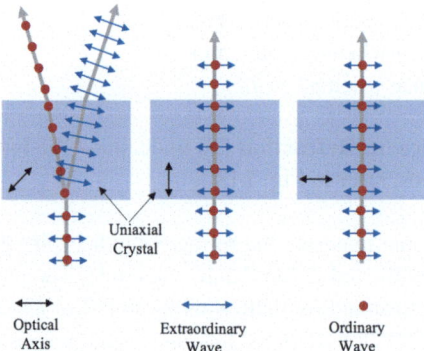

Fig. 5.14 Propagation of a light wave through a calcite crystal with various optical axis orientations. Ordinary waves, which are polarized perpendicular to the plane of the paper, and extraordinary waves, which are polarized parallel to the plane of the paper, are depicted for three different orientations of the optical axis. Double refraction occurs when the optical axis is neither parallel nor perpendicular to the wave's propagation direction

5.7 Polarizers

ated with the index ellipsoid. Figure 5.14 demonstrates how unpolarized light interacts with a uniaxial crystal, depending on the orientation of the optic axis of the crystal relative to the light wave direction of propagation. The refractive index of extraordinary ray, $n(\theta)$, depends on the direction of propagation relative to optic axis.

$$\frac{1}{n^2(\theta)} = \frac{cos^2(\theta)}{n_o^2} + \frac{sin^2(\theta)}{n_e^2} \tag{5.22}$$

where θ is the angle between the propagation vector and the optic axis of the crystal. The refractive index varies as follows:

- For $\theta = 0°$, $n(\theta) = n_o$.
- For $\theta = 90°$, $n(\theta) = n_e$.
- For $0° < \theta < 90°$ the two polarizations are separated by the crystal.

5.7.2.3 Polarizing Prisms

Birefringence can be utilized to construct polarizing prisms that separate specific polarization components from incoming light waves. Two common examples are the Glan–Foucault prism and the Glan–Taylor prism.

Glan–Foucault Polarizer The Glan–Foucault polarizer, shown in Fig. 5.15a, is made of a calcite crystal. The incoming light ray strikes the crystal surface normally, and the electric field E E can be resolved into components that are either parallel or perpendicular to the optic axis. Both components traverse the first calcite section without deviation.

To achieve total internal reflection of the ordinary ray (o-ray) and transmission of the extraordinary ray (e-ray), the angle of incidence u u at the calcite–air interface must satisfy the condition:

$$n_e < 1/sin\theta < n_o$$

As a result, the ordinary ray is totally internally reflected, while the transmitted light is 100% linearly polarized. The reflected beam, however, is not fully polarized. When two such prisms are cemented together with an adjusted interface angle, the device becomes a Glan–Thompson polarizer.

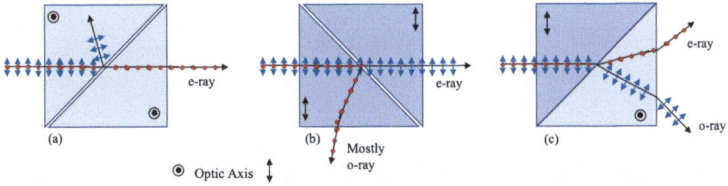

Fig. 5.15 Polarizing prisms: (**a**) Glan–Foucault, (**b**) Glan-Taylot, and (**c**) Wollaston prism

Glan–Taylor Prism The Glan–Taylor prism, illustrated in Fig. 5.15b, has improved transmission compared to the Glan–Foucault polarizer. This design allows for a more highly polarized reflected beam, making it suitable for use as a polarizing beamsplitter.

Wollaston Prism

The Wollaston prism, depicted in Fig. 5.15c, functions as a polarizing beamsplitter, transmitting both orthogonal polarization components. It can be made from calcite or quartz. At the diagonal interface, the two component rays separate:

- The *e-ray* becomes an o-ray, changing its refractive index accordingly. In calcite, where $n_e < n_o$, the emerging o-ray bends toward the normal.
- The *o-ray*, initially perpendicular to the optic axis, becomes an *e-ray* in the right-hand section.

These prisms are widely used in optical systems to manipulate the polarization state of light and achieve efficient polarization separation.

5.8 Polarization By Reflection

When an unpolarized light reflects off a non-metallic surface, it can become completely polarized, partially polarized or remain unpolarized depending on the angle of incidence. Complete polarization occurs for an angle known as Brewster's angle. As discussed in Chap. 3, the reflection coefficient $r_{||} = 0$ at the polarization angle for internal reflection, i.e., $n_i < n_t$. According to Fresnel equations, the reflection coefficient for parallel polarization is:

$$r_{||} = +\frac{tan(\theta_p - \theta_t)}{tan(\theta_p + \theta_t)} = 0 \tag{5.23}$$

For internal reflection, where $\theta_p > \theta_t$, the numerator is non-zero, making the denominator approach infinity; therefore,

$$tan(\theta_p + \theta_t) = \infty.$$

Thus,

$$\theta_p + \theta_t = 90° \tag{5.24}$$

and

$$\theta_t = 90° - \theta_p.$$

5.9 Retarders

Fig. 5.16 Polarization by reflection

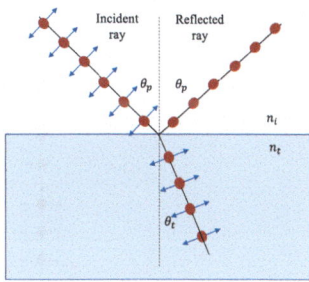

Applying Snell's law and substituting the angle of incidence with the polarization angle, we get:

$$n_i sin(\theta_p) = n_t sim(\theta_t) = n_t sin(90° - \theta_p) = n_t cos(\theta_p). \quad (5.25)$$

Therefore,

$$tan(\theta_p) = \frac{n_t}{n_i} \quad (5.26)$$

or

$$\theta_p = atan\left(\frac{n_t}{n_i}\right) \quad (5.27)$$

When the angle of incidence equals Brewster's angle θ_p, only the perpendicular polarization is reflected as illustrated in Fig. 5.16.

5.9 Retarders

In anisotropic crystals, double refraction causes two rays of light to travel at different velocities through the crystal due to the difference between the refractive indices, n_e and n_o. As a result, the slow ray takes longer to traverse the crystal than the fast ray. By the time the slow ray reaches the surface, the fast ray has already traversed the crystal and traveled an additional distance, Δ beyond it. This distance Δ is called **retardation**, as graphically illustrated in Fig. 5.17 The retardation causes one of the polarization states to lag behind the other.

In cases of positive birefringence $n_e > n_o$, the extraordinary ray (e-ray) with parallel polarization, lags behind the ordinary ray (o-ray) with perpendicular polarization. This lag corresponds a phase shift, φ, which depends on Δ and is given by

$$\varphi = k\Delta$$

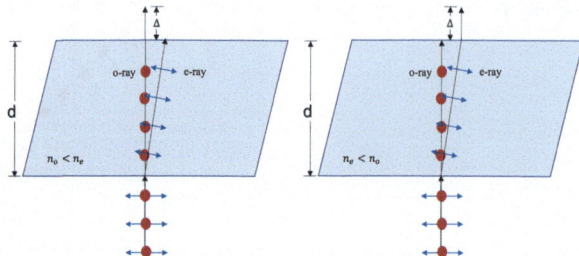

Fig. 5.17 Wave retardation for positive birefringence $n_e > n_o$ and negative birefringence, $n_e < n_o$

Fig. 5.18 The state of polarization as E_x leads E_y by φ in increments of $\pi/4$

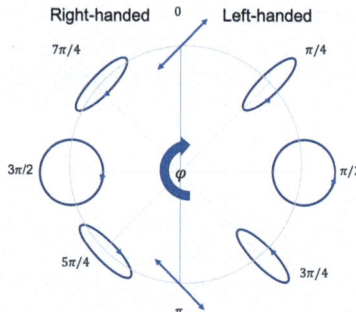

where

$$\Delta = d(n_e - n_o) \tag{5.28}$$

therefore,

$$\varphi = kd(n_e - n_o) \tag{5.29}$$

In Eq. (5.29), the phase shift, φ, is positive when the e-ray lags behind the o-ray, and negative when it leads. Figure 5.18 demonstrates the state of polarization as E_x leads E_y by φ in increments of $\pi/4$.

Retarders are used to transform a plane polarization into any of the states of polarization shown in Fig. 5.18 by adjusting the crystal's thickness to produce the desired phase shift φ defined by Eq. 5.29. Two common types of retarders are **half-wave plates** and **quarter-waveplates**.

5.9.1 The Half-Wave Plates

A half-wave plate introduces phase difference of π radians (180°), between the ordinary and extraordinary waves, making it a half-wave retarder. If an incoming

Fig. 5.19 A half-wave plate rotates light initially linearly polarized at an angle θ through a total angle of 2θ. Here light was incident oscillating in the first and third quadrants, and it emerged oscillating in the second and fourth quadrants

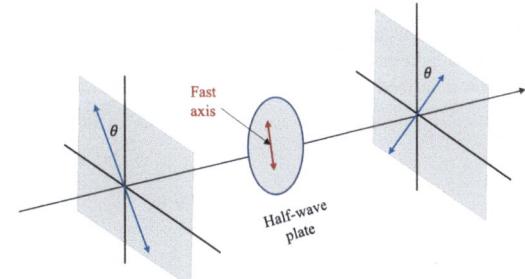

beam of linearly polarized light makes angle θ with the fast axis (as shown in Fig. 5.19), the half-waveplates rotate the plane of polarization by twice this angle. For this reason, half-wave plates are also known as **polarization rotators**.

A half-wave plate also affects elliptical light, flipping its orientation, and inverts the handedness of circular or elliptical polarization (e.g., converting right-handed to left-handed circular polarization and vice versa). It shifts the polarization state halfway in Fig. 5.18. As the e- and o-waves propagate through a retarder, their relative phase difference φ, Eq. (5.29), increases continuously, causing the polarization state to change gradually. Therefore, when the thickness of the plate satisfies:

$$d(|n_e - n_o|) = (2m + 1)\frac{\lambda_o}{2} \tag{5.30}$$

where $m = 0, 1, 2, 3, \ldots$, it functions as a half-wave pate with phase shifts as $\varphi = \pi, 2\pi, 3\pi, \ldots$.

5.9.2 The Quarter-Wave Plate

A quarter wave-plate is an optical element that introduces a relative phase shift of $\varphi = \pi/2$ between the orthogonal ordinary and extraordinary components of a wave. A shown Fig. 5.18, a 90° phase shift transforms linear polarization into elliptical polarization, or into circular polarization if $E_{ox} = E_{oy}$, and vice versa.

As the ordinary and extraordinary propagate through a retarder, their relative phase difference φ increases, leading to a gradual change in the wave's polarization state. If the quarter-wave plate's thickness meets the condition:

$$d(|n_e - n_o|) = (4m + 1)\frac{\lambda_o}{4} \tag{5.31}$$

where $m = 0, 1, 2, 3, \ldots$ the quarter-wave plate effectively converts linear polarization at $-45°$ through a quarter-wave plate that results in a left-handed circular polarization state as depicted in Fig. 5.20, into left-handed circular polarization, as

Fig. 5.20 A quarter-wave plate transforms a linearly polarized light at 45^0 into a circular polarized state

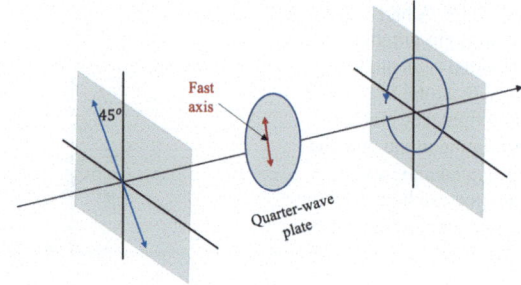

5.10 Orthogonality of Jones Vectors

Two vectors, $\bar{A}$ and $\bar{B}$, are orthogonal if they satisfy the inner product condition:

$$\bar{A} \cdot \bar{B} = 0 \text{ or } \bar{A} \cdot \bar{B}^* = 0 \tag{5.32}$$

Here, $\bar{B}^*$ represents the complex conjugate of vector $\bar{B}$. Jones vectors representing the horizontal and vertical linear polarization states satisfy this condition and therefore constitute a pair of orthogonal polarization states.

$$\bar{J}_h^T \cdot \bar{J}_v = \begin{bmatrix} 1 & 0 \end{bmatrix} \cdot \begin{bmatrix} 0 \\ 1 \end{bmatrix} = 0 \tag{5.33}$$

Similarly, right and left circular polarization states are also orthogonal to each other:

$$\bar{J}_{CR}^T \cdot \bar{J}_{CL}^* = \frac{1}{\sqrt{2}} \begin{bmatrix} 1 & -j \end{bmatrix} \cdot \frac{1}{\sqrt{2}} \begin{bmatrix} 1 \\ j \end{bmatrix}^* = \frac{1}{2} \left[(1)(1)^* + (-j)(j)^* \right] = \frac{1}{2}[1 - 1] = 0 \tag{5.34}$$

Orthogonal elliptical polarization states also exist. For example, the Jones vectors

$$\frac{1}{5} \begin{bmatrix} 4 \\ 3j \end{bmatrix} \text{ and } \frac{1}{5} \begin{bmatrix} 3 \\ -4j \end{bmatrix}$$

represent an orthogonal pair of elliptically polarized states. Therefore, any polarization has a corresponding orthogonal state.

Furthermore, the sum of any set of Jones vectors is itself a Jones vector. This means that the superposition of different polarization states in the Jones vectors representation always results in a state with definite polarization. For example, adding a right and left circular polarization states in a linear horizontal polarization

5.11 Matrix Representation For Optical Polarizing Devices

state with twice the amplitude of either of the circular polarization states, i.e.,

$$\frac{1}{\sqrt{2}}\begin{bmatrix}1\\-j\end{bmatrix} + \frac{1}{\sqrt{2}}\begin{bmatrix}1\\j\end{bmatrix} = \frac{1}{\sqrt{2}}\begin{bmatrix}2\\j\end{bmatrix} = \sqrt{2}\begin{bmatrix}1\\0\end{bmatrix}. \tag{5.35}$$

This confirms that the superposition principle holds for Jones vectors, leading to a well-defined polarization state.

5.11 Matrix Representation For Optical Polarizing Devices

Optical polarizing devices include linear polarizers, phase retarders, and polarization rotators. In Sect. 5.5, we introduced Jones vectors to represent the state of polarization. In this section, we are introducing Jones matrices to mathematically represent polarizers and other optical elements. This approach enables us to determine the stare of polarization of a light beam as it passes through polarizing optical elements. Similar to the ray matrix concept introduced in Chap. 2 in representing optical components and systems. This process is depicted in a form of a block diagram in Fig. 5.21.

The input–output relationship for an optical polarizing device can be represented by

$$\bar{J}_1 = M\bar{J}_0 \tag{5.36}$$

where M is 2×2 is Jones matrix of the optical polarizing device.

5.11.1 Jones Matrix For Linear Polarizers

A linear polarizer has a preferred direction, called the transmission axis of the polarizer. It transmits plane polarized light with minimum (maximum) loss when its transmission axis is oriented along (orthogonal to) the direction with the oscillation direction of the electric field of the incident light. An ideal linear polarizer transmits light polarized parallel to its transmission axis with no loss at all and blocks completely light polarized perpendicular to it.

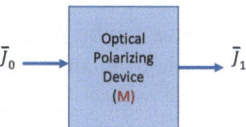

Fig. 5.21 Representation of the relation between the input and output of an optical polarizing device in terms of Jones calculus

5.11.1.1 Derivation of the Jones Matrix for a Horizontal Linear Polarizer

For a horizontal linear polarizer, light polarized horizontally passes through and emerges horizontally polarized. In contrast, a vertical polarized light is blocked by it. Based on these two conditions, the linear polarizer leads to the following:

$$M \begin{bmatrix} 1 \\ 0 \end{bmatrix} = \begin{bmatrix} A & B \\ C & D \end{bmatrix} \begin{bmatrix} 1 \\ 0 \end{bmatrix} = \begin{bmatrix} A \\ C \end{bmatrix} = \begin{bmatrix} 1 \\ 0 \end{bmatrix}$$

and

$$M \begin{bmatrix} 0 \\ 1 \end{bmatrix} = \begin{bmatrix} A & B \\ C & D \end{bmatrix} \begin{bmatrix} 0 \\ 1 \end{bmatrix} = \begin{bmatrix} A \\ C \end{bmatrix} = \begin{bmatrix} 0 \\ 0 \end{bmatrix}.$$

Hence, $A = 1$ and $B = C = D = 0$. Therefore, the Jones matrix for a horizontal linear polarizer is:

$$M_h = \begin{bmatrix} 1 & 0 \\ 0 & 0 \end{bmatrix}. \tag{5.37}$$

Similarly, we can show that Jones matrix for a vertical linear polarizer is:

$$M_v = \begin{bmatrix} 0 & 0 \\ 0 & 1 \end{bmatrix}. \tag{5.38}$$

When the transmission axis of the polarizer makes an angle θ with the polarization direction of the incident light, the transmitted electric field and intensity are given by:

$$E = E_o \cos\theta, \text{ and } I = I_o \cos^2\theta, \tag{5.39}$$

This relationship is known as **Malus Law**.

5.11.1.2 Extinction Ratio of Polarizers

Practical polarizers cannot meet the stringent requirement of passing one polarization state with no loss and completely blocks the orthogonal polarization state. The **extinction ratio** of a polarizer is defined as the ratio of the maximum ($\theta = 0°$) to the minimum ($\theta = 90°$) transmitted intensities. The extinction ratio is always greater than 1. Typical extinction ratios can be between 10^3 and 10^7, depending of the polarizer material. Polarizers made from anisotropic crystals exhibiting double refraction usually have high extinction ratios.

5.11.1.3 Jones Matrices for Linear Polarizers at Specific Angles

For linear polarizer with transmission axis oriented at 45° to the horizontal, the Jones matrix is:

5.11 Matrix Representation For Optical Polarizing Devices

$$M_{45°} = \frac{1}{2}\begin{bmatrix} 1 & 1 \\ 1 & 1 \end{bmatrix}, \quad (5.40)$$

For a linear polarizer oriented at $-45°$ to the horizontal, the Jones matrix becomes:

$$M_{-45°} = \frac{1}{2}\begin{bmatrix} 1 & -1 \\ -1 & 1 \end{bmatrix}, \quad (5.41)$$

For a linear polarizer with a transmission axis oriented at an arbitrary angle θ to the horizontal direction, the following Jones matrix is:

$$M_\theta = \begin{bmatrix} cos^2\theta & sin\theta\, cos\theta \\ sin\theta\, cos\theta & sin^2\theta \end{bmatrix}. \quad (5.42)$$

5.11.2 Jones Matrix for Phase Retarders

Phase retarders, as discussed previously, make use of the birefringence property of anisotropic crystals. The refractive index of an appropriately cut uniaxial crystal has maximum and minimum values along orthogonal directions called the **slow axis** (SA) and the **fast axis** (FA), respectively. Figure 5.19 illustrates an incident wave polarized at an angle θ with respect to the fast axis, which is oriented along vertically. Inside the crystal, the wave decomposes into two components: one polarized along the fast axis and the other along the slow axis.

The component polarized along the fast axis (y-$axis$) moves faster (inside the retarder) than the one polarized along the slow axis (x-$axis$). Both waves, however, propagate along the same direction but with different wavenumbers $k = 2\pi/\lambda$, where λ is the wavelength along the polarization axis. This difference results in a phase shift given by:

$$\varphi = \varphi_x - \varphi_y = \frac{2\pi}{\lambda_o}(|n_e - n_o|)d \quad (5.43)$$

after traversing a thickness d inside the crystal. Here, n_e and n_o are the refractive indices, for light polarized along the o-wave and e-waves, axes respectively, and λ_o is the wavelength of light in vacuum. The phase shift φ between the x- and y-components of the electric field determines the polarization of the light that emerging from the phase retarder. The retarder transforms a plane polarized wave into an elliptically polarized wave, with circular and linear polarizations as special cases, depending on the retarder thickness d and the angle between the polarization direction of the incident light. and the fast axis. The ideal phase retarder changes only the phase, not the amplitude.

When the slow and fast axes of the phase retarder are aligned horizontally and vertically, respectively, the retarder matrix has non-zero diagonal elements only. Thus, the Jones matrix for a phase retarder is:

$$M_r = \begin{bmatrix} e^{j\varphi_x} & 0 \\ 0 & e^{j\varphi_y} \end{bmatrix} \quad (5.44)$$

In general, phase retarders delay one component of the field by a phase Γ. If the fast axis is along the $x\text{-}axis$, the y-component of the field is delayed by Γ. The corresponding Jones matrix is:

$$M_r = \begin{bmatrix} 1 & 0 \\ 0 & e^{-j\Gamma} \end{bmatrix}. \quad (5.45)$$

As discussed previously for quarter-wave and half-wave pates, they transform the polarization of an incident linearly polarized light making an angle $-\theta$ with the fast axis. In the case of the half-wave plate (HWP), the phase shift between the polarization components $\varphi = (2m + 1)\lambda_o/2$. The Jones matrix for a half-wave plate is:

$$M_{HWP} = \begin{bmatrix} 1 & 0 \\ 0 & e^{j\pi} \end{bmatrix} = \begin{bmatrix} 1 & 0 \\ 0 & -1 \end{bmatrix} \quad (5.46)$$

For a quarter-wave plate (QWP), the phase shift is $\varphi = (4m + 1)\lambda_o/4$. Thus, the Jones matrices for the half-wave plate is given by:

$$M_{QWP} = \begin{bmatrix} 1 & 0 \\ 0 & e^{\pm j\frac{\pi}{2}} \end{bmatrix} = \begin{bmatrix} 1 & 0 \\ 0 & \pm j \end{bmatrix}. \quad (5.47)$$

The sign of matrix element $\pm j$ depends on the orientation of the fast axis of the crystal:

- Positive j j when the fast axis is horizontal,
- Negative j j when the fast axis is vertical.

Table 5.2 lists standard polarizers and their Jones matrices:

This table summarizes the Jones matrices for different types of polarizers, illustrating their effects on light polarization.

5.11 Matrix Representation For Optical Polarizing Devices

Table 5.2 Jones matrices for polarizing optical elements

Optical element	Jones matrix
Horizontal linear polarizer	$\begin{bmatrix} 1 & 0 \\ 0 & 0 \end{bmatrix}$
Vertical linear polarizer	$\begin{bmatrix} 0 & 0 \\ 0 & 1 \end{bmatrix}$
Linear polarizer at $+45°$	$\frac{1}{2}\begin{bmatrix} 1 & 1 \\ 1 & 1 \end{bmatrix}$
Linear polarizer at $-45°$	$\frac{1}{2}\begin{bmatrix} 1 & -1 \\ -1 & 1 \end{bmatrix}$
Linear polarizer at θ	$\frac{1}{2}\begin{bmatrix} \cos^2\theta & \sin\theta\cos\theta \\ \sin\theta\cos\theta & \sin^2\theta \end{bmatrix}$
Linear polarization rotator by an angle θ	$\begin{bmatrix} \cos\theta & -\sin\theta \\ \sin\theta & \cos\theta \end{bmatrix}$
Quarter-wave plate with fast-axis vertical	$e^{j\pi/4}\begin{bmatrix} 1 & 0 \\ 0 & -j \end{bmatrix}$
Quarter-wave plate wit fast-axis horizontal	$e^{j\pi/4}\begin{bmatrix} 1 & 0 \\ 0 & j \end{bmatrix}$
Half-wave plate	$\frac{1}{2}\begin{bmatrix} 1 & 0 \\ 0 & -1 \end{bmatrix}$
Right-handed circular polarizer	$\frac{1}{2}\begin{bmatrix} 1 & j \\ -j & 1 \end{bmatrix}$
Left-handed circular polarizer	$\frac{1}{2}\begin{bmatrix} 1 & -j \\ j & 1 \end{bmatrix}$

Example What does a quarter-wave plate do if the input polarization is linear but at an arbitrary angle α?

Solution

$$\begin{bmatrix} 1 & 0 \\ 0 & j \end{bmatrix} \begin{bmatrix} 1 \\ \tan\alpha \end{bmatrix} = \begin{bmatrix} 1 \\ j\tan\alpha \end{bmatrix}$$

(continued)

Fig. 5.22 Polarization states for the output of a quarter-wave plate for an incident light wave linearly polarized at angle α

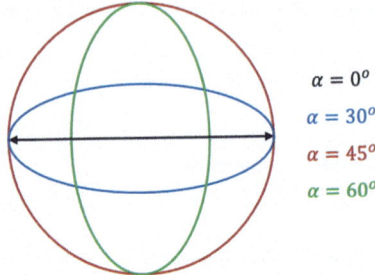

Let us assume that the angle $\alpha = 0°, 30°, 45°$ and 60. So, the Jones vectors for the incident light wave are given by

$$\begin{bmatrix} 1 \\ 0 \end{bmatrix}, \begin{bmatrix} 1 \\ 0.58 \end{bmatrix}, \begin{bmatrix} 1 \\ 1 \end{bmatrix} \text{ and } \begin{bmatrix} 1 \\ 1.7 \end{bmatrix}.$$

By multiplying the QWP matrix by each of these, we get the following Jones vectors for the output beam polarizations:

$$\begin{bmatrix} 1 \\ 0 \end{bmatrix}, \begin{bmatrix} 1 \\ j0.58 \end{bmatrix}, \begin{bmatrix} 1 \\ j \end{bmatrix} \text{ and } \begin{bmatrix} 1 \\ j1.7 \end{bmatrix},$$

these vectors represent: Linear, elliptical, circular, and elliptical polarizations, respectively, which are shown in Fig. 5.22.

Example A yellow light of wavelength 589 nm incident on a thin mica sheet with $n_o = 1.5997$ and $n_e = 1.5941$. What is the minimum thickness of a mica sheet that would serve as a quarter-wave plate?

Solution For a quarter-wave plate, the optical path difference (OPD) between the o-wave and e-wave must be an odd whole number of $\lambda_0/4$. The OPD for a quarter-wave plate of thickness d is given by

$$OPD = d() = (4m+1)\lambda_0/4 \qquad (5.48)$$

where $m = 0, 1, 2, 3, \ldots$. Therefore, the thickness of the mica sheet is

(continued)

$$d = \frac{(4m+1)\lambda_0/4}{|n_o - n_e|} \quad (5.49)$$

Therefore, the minimum thickness is when $m = 0$ and is given by

$$d = \frac{\lambda_0/4}{|n_o - n_e|} = \frac{589}{4(1.5997 - 1.5941)} = 26.3 \, \mu m \quad (5.50)$$

Hence, the minimum thickness of the mica sheet to make a quarter-wave plate is 26.3 μm.

5.11.3 Jones Matrices for Cascaded Elements

To model the effect of multiple polarizing optical components on polarization state of a light wave, we use a series of Jones matrices. The overall effect of a set of optical elements is represented by the product of the individual Jones matrices. Figure 5.23 illustrates the input–output relation for a system of cascaded optical polarizing elements on the polarization of an incident light.

The relationship between the output, E_o, and output, E_i, is given by

$$E_o = M_3 M_2 M_1 E_i, \quad (5.51)$$

where $M_1, M_2,$ and M_3 are the Jones matrices of the individual components, arranged from right to left, similar to the order used in ray tracing matrices.

Consider the case when a linear polarized light at $-45°$ passes through a QWP with its fast axis is along the vertical axis. The output is given by:

$$\begin{bmatrix} 1 & 0 \\ 0 & -j \end{bmatrix} \begin{bmatrix} 1 \\ -1 \end{bmatrix} = \begin{bmatrix} 1 \\ j \end{bmatrix}$$

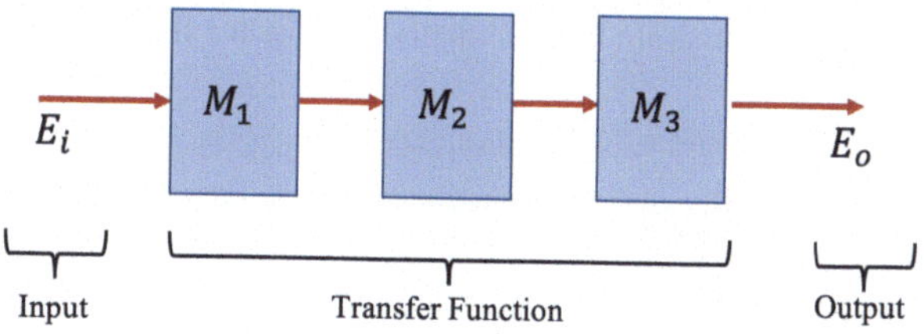

Fig. 5.23 A system of cascaded Jones matrices for polarizing optical components

Thus, the output is a left-handed circular polarization, which is shown in Fig. 5.20. Let us examine what happens if we followed the QWP with a linear polarizer at $+45°$. The output if this cascaded system is

$$\begin{bmatrix} 1 & 1 \\ 1 & 1 \end{bmatrix} \begin{bmatrix} 1 & 0 \\ 0 & -j \end{bmatrix} \begin{bmatrix} 1 \\ -1 \end{bmatrix} = \begin{bmatrix} 1 \\ 1 \end{bmatrix}$$

which is a linearly polarized light at $+45°$, which is the same effect of passing this polarized light through a HWP as shown here

$$\begin{bmatrix} 1 & 0 \\ 0 & -1 \end{bmatrix} \begin{bmatrix} 1 \\ -1 \end{bmatrix} = \begin{bmatrix} 1 \\ 1 \end{bmatrix}.$$

Notice that we have dropped some of the normalizing factors to simply show the effect on the polarization state.

5.11.4 Intensity Propagation Through Polarizers

As light traverses a series of polarizers, its intensity changes according to the transmittance of the polarizer. In special cases, a linearly polarized light passing through a linear polarizer oriented in the same polarization plane transmitted without attenuation, while, a polarizer oriented perpendicular to the polarization plane blocks the light completely.

In general, we assume ideal polarizers, meaning there is no attenuation when the polarization of the light matches the polarizer's transmission axis. If the state of polarization of the light beam makes an angle θ ϑ with a linear polarizer, Malus's law applies.

$$I = I_o cos^2\theta,$$

where I is the transmitted intensity, I_o is the initial intensity of the incident light. This relationship reflects the cosine-squared dependence of transmitted intensity on the angle between the light's polarization direction and the polarizer's axis. The following examples demonstrate how intensity propagates through polarizers:

> **Example** Determine the output intensity and unpolarized light of intensity, I_o, is incident on a vertical linear polarizer and a second linear polarizer oriented at angle $\theta = 45°$. as shown in the Fig. 5.24

(continued)

Fig. 5.24 Unpolarized light passing through two linear polarizers

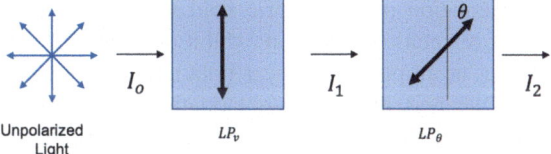

Solution

$$I_1 = 1/2 I_o,$$

$$I_2 = I_1 cos^2(\theta) = I_2 = 1/2 I_0 cos^2(\theta) = 1/2 I_0 cos^2(45^o)$$

Therefore,

$$I_2 = \frac{I_o}{8}$$

Example Unpolarized light with $I_o = 72$ W/m^2 passes through two polarizing filters, as shown in Fig. 5.24. If the light that emerges from the second filter has $I_2 = 9$ W/m2^2. What is the angle θ between the two filters?

Solution

$$I_2 = I_1 cos^2(\theta) = 1/2 I_o cos^2(\theta),$$

hence,

$$9 = 1/2 \times 72 \times cos^2\theta,$$

$$cos^2\theta = \frac{1}{4} \text{ and } cos\theta = \frac{1}{2},$$

therefore, $\theta = 60°$.

5.12 Summary

In this chapter, we explored the fundamental concepts of light polarization, an intrinsic property of all vector waves such as light. Polarization describes how the orientation of the electric field vector varies as light propagates. Depending on this orientation, light can exhibit different states of polarization, such as linear, circular, or elliptical. In linear polarization, the electric field oscillates in a single plane, whereas in circular polarization, it has two perpendicular components of equal magnitude with a phase difference of $\pi/2$, resulting in a rotating electric field. Elliptical polarization is the most general form, occurring when the electric field components have unequal magnitudes or a phase difference that is not $\pi/2$, tracing an elliptical path.

The mathematical representation of polarization is important for understanding its behavior in various optical systems. Jones vectors were introduced to describe the state of polarization of a light wave. To model the effect of optical devices on light polarization, we used Jones matrices, which represent polarizing components such as linear polarizers, phase retarders, and polarization rotators. These matrices allow us to predict how the state of polarization changes as light interacts with different optical elements.

Different optical devices influence light's polarization in characteristic ways. Linear polarizers transmit the component of light aligned with their transmission axis and block the orthogonal component. Phase retarders introduce a phase shift between the components of the electric field, transforming the polarization state, while polarization rotators alter the orientation of the electric field vector. To understand the combined effect of multiple polarizing elements, we used cascaded Jones matrices, where the overall effect is represented by the product of the individual matrices.

Intensity variations in polarized light were analyzed using Malus's law, which describes how the intensity of transmitted light depends on the angle between the light's polarization and the polarizer's transmission axis. Additionally, the concept of birefringence was explored, showing how anisotropic materials can split light into ordinary and extraordinary rays that propagate at different velocities, resulting in phase differences that alter the polarization state. This phenomenon leads to retardation, where devices such as half-wave and quarter-wave plates are used to convert between different polarization states.

Polarization control plays a crucial role in various optical applications, ranging from controlling laser beams to enhancing display technologies, optical communications, and imaging systems. This chapter establishes a foundational understanding of polarization, its mathematical representation, and how to analyze and manipulate it using optical devices, preparing readers to design and analyze optical systems that effectively manage light's polarization state.

5.13 Problems

1. Two light waves $E_x = E_o cos(kz - \omega t)$ and $E_y = -E_o cos(kz - \omega t)$ overlap in space. Show that the resultant is linear light and determine its amplitude and tilt angle θ.

2. Two waves $E_x = 4sin(ky - \omega t)$ and $E_y = 3sin(ky - \omega t)$ overlap in space. Describe completely the state of polarization of the resultant wave.
3. Two linearly polarized waves are described by $E_x = E_0 cos(kz - \omega t)$ and $Ey = -E_o cos(kz - \omega t)$. Show that the resultant wave is linearly polarized, and determine its amplitude and tilt angle θ relative to the x-axis.
4. Two linearly polarized waves, $E_x = 4sin(ky - \omega t)$ and $E_y = 3sin(ky - \omega t)$, overlap in space. Determine the state of polarization of the resultant wave, indicating whether it is elliptical, and determine the orientation of the major axis of the ellipse.
5. A beam of light with Jones vector

$$\bar{\ } = \frac{\bar{1}}{\sqrt{2}} \begin{bmatrix} 1 \\ 1 \end{bmatrix}$$

representing 45° linearly polarized light) passes through a quarter-wave plate with its fast axis oriented along the horizontal axis. Find the resulting Jones vector and describe the polarization state of the output wave.
6. Unpolarized light with initial intensity $I_o = 80 \text{ W/m}^2$ passes through two polarizing filters. The first polarizer is aligned vertically, and the second is oriented at 60° to the vertical axis. Calculate the final intensity of the transmitted light.
7. A linearly polarized light at −45° passes sequentially through a quarter-wave plate (fast axis vertical) and a linear polarizer oriented at +45°. Use Jones matrices to determine the final polarization state and the intensity of the light.
8. A linearly polarized beam at 60° passes through a half-wave plate. Determine the new polarization angle after passing through the half-wave plate, and explain the transformation in polarization state.
9. A calcite crystal has refractive indices $n_o = 1.658$ and $n_e = 1.486$. Determine the minimum thickness of the crystal needed to create a quarter-wave plate for light of wavelength 600 nm.
10. LCD screens for laptops use liquid crystals to influence the polarization according to the figure below. A certain color was obtained by applying a voltage to give transmission maxima for that certain color. However, this produces very pale colors since also surrounding wavelengths are partly transmitted. Assume that one wants to produce green (546 nm) color. The liquid crystal has a thickness d that will act as a retarder with phase shift $\Delta \varphi$, which is controlled by the voltage applied. The input light is unpolarized white light.

Bibliography

1. Hecht, E. (2017). Optics (5th ed.). Pearson Education.
2. Born, M., & Wolf, E. (1999). Principles of Optics: Electromagnetic Theory of Propagation, Interference, and Diffraction of Light (7th ed.). Cambridge University Press.
3. Saleh, B. E. A., & Teich, M. C. (2019). Fundamentals of Photonics (3rd ed.). Wiley.

4. Yariv, A., & Yeh, P. (2007). Photonics: Optical Electronics in Modern Communications (6th ed.). Oxford University Press.
5. Collett, E. (1992). Polarized Light: Fundamentals and Applications. CRC Press.
6. Azzam, R. M. A., & Bashara, N. M. (1987). Ellipsometry and Polarized Light. Elsevier.
7. Chipman, R. A., Lam, W. S. T., & Young, G. (2018). Polarized Light and Optical Systems. CRC Press.

Optical Beams and Resonators 6

6.1 Introduction

Plane and spherical waves are simple solutions of the Helmholtz wave equation, as discussed in Chap. 3. However, planewaves have infinite extent and do accurately represent beams emitted by light sources like lasers. Laser beams are typically well collimated and have a finite extent. The wavefront normals of these beams form a small angle with the direction of propagation giving them the characteristics of paraxial rays described in Chap. 3. Therefore, a more precise description of optical beams is needed. This is achieved through an approximate solution of the Helmholtz wave equation known as the paraxial approximation, and the corresponding paraxial wave equation. This approximate analytic solution of the paraxial wave equation is known as the **Gaussian Beam**, characterized by a Gaussian intensity distribution across the beam.

Laser sources generate beams using optical cavities that rely on optical feedback. Optical cavities are also known as optical resonators that typically use two or more reflectors to achieve feedback. Optical resonators can be constructed from discrete components such as mirrors in gas lasers or from waveguides found in semiconductor lasers..

This chapter will discuss the propagation and properties of Gaussian beams, as well as the theory of operation, and design of discrete and integrated optical resonators.

6.2 Gaussian Beams

In general, Both the amplitude and shape of the electric field distribution change during propagation. However, there exists a special class of fields in which the shape of the distribution remains constant, while the amplitude and lateral extent of the field vary. Gaussian beams belong to this class, and they are solutions of an approximate wave equation.

6.2.1 Paraxial Wave Equation

The Helmholtz wave equation describes the propagation of electromagnetic waves. For monochromatic electromagnetic waves, it is expressed as:

$$\nabla^2 U(x, y, z) + k^2 U(x, y, z) = 0, \tag{6.1}$$

where ∇^2 is the Laplacian operator and $k = 2\pi/\lambda$ is the wavenumber. Plane and spherical waves are both simple solutions of this equation. However, to accurately describe waves with a finite extent, such as laser beams, we need a modified wave equation. Consider the planewave solution given by the complex amplitude $U(x, y, z)$, where

$$U(\bar{r}) = A(x, y, z))e^{-jkz}, \tag{6.2}$$

which represents a planewave propagating along the z-axis, modulated transversely by the complex amplitude $A(x, y, z))$. We assume that the variation of the envelope, $A(x, y, z)$, with z is slow within a wavelength, λ, allowing the wave approximately maintains its planewave nature. Such a wave is considered as a paraxial wave meaning its wavefront normals satisfy the conditions of paraxial rays. For a paraxial wave to satisfy the Helmholtz equation, the complex envelope $A(r)$ must satisfy another partial differential equation. Assuming that $A(r)$ varies slowly with respect to z implying that within a distance $\Delta Z = \lambda$, the change in amplitude, ΔA, is much smaller than A itself. Thus,

$$\Delta A = \left(\frac{\partial A}{\partial z}\right)\Delta z = \left(\frac{\partial A}{\partial z}\right)\lambda, \tag{6.3}$$

and since $\Delta A \ll A$, we have

$$\left(\frac{\partial A}{\partial z}\right)\lambda \ll A,$$

or:

$$\left(\frac{\partial A}{\partial z}\right) \ll \frac{A}{\lambda} = \frac{Ak}{2\pi}, \tag{6.4}$$

Differentiating both sides with respect to z, we obtain:

$$\frac{\partial^2 A}{\partial z^2} \ll k\frac{\partial A}{\partial z}. \tag{6.5}$$

6.2 Gaussian Beams

By substituting from Eq. (6.4) into Eq. (6.5) we get

$$\frac{\partial^2 A}{\partial z^2} \ll k^2 A. \tag{6.6}$$

We now define the transverse Laplacian operator as

$$\nabla_t^2 = \frac{\partial^2}{\partial x^2} + \frac{\partial^2}{\partial y^2}. \tag{6.7}$$

This allows us to rewrite the Helmholtz wave equation as:

$$\left(\nabla_t^2 + \frac{\partial^2}{\partial z^2}\right) U + k^2 U = 0. \tag{6.8}$$

Assuming the paraxial wave solution $U(r) = A(x, y, z)e^{-jkz}$, we substitute it into the wave equation:

$$\left(\nabla_t^2 + \frac{\partial^2}{\partial z^2}\right) Ae^{-jkz} + k^2 Ae^{-jkz} = 0,$$

which expands to:

$$\nabla_t^2 Ae^{-jkz} + \frac{\partial^2}{\partial z^2}\left(Ae^{-jkz}\right) + k^2 Ae^{-jkz} = 0. \tag{6.9}$$

Taking the second derivative with respect to z results in:

$$\nabla_t^2 Ae^{-jkz} + e^{-jkz}\frac{\partial^2 A}{\partial z^2} - j2k\frac{\delta A}{\delta z}e^{-jkz} - k^2 Ae^{-jkz} + k^2 Ae^{-jkz} = 0, \tag{6.10}$$

Simplifying further, we get:

$$\nabla_t^2 A + \frac{\partial^2 A}{\partial z^2} - j2k\frac{\delta A}{\delta z} = 0. \tag{6.11}$$

According to our earlier assumption, the second term is negligible compared to the third term, so the equation reduces to:

$$\nabla_t^2 A - j2k\frac{\delta A}{\delta z} = 0. \tag{6.12}$$

which is known as the paraxial wave equation. This equation governs waves with slowly varying amplitudes, such as laser beams.

6.2.2 The Fundamental Gaussian Beam Solution

Laser beams as described earlier have finite extent and exhibit cylindrical symmetry. To proceed, let us rewrite the paraxial wave equation in cylindrical coordinate system with $r_t = \sqrt{x^2 + y^2}$. Equation (6.12) is expressed in the cylindrical coordinate system as:

$$\frac{1}{r_t}\frac{\delta}{\delta r_t}\left(r_t \frac{\delta A}{\delta r_t}\right) - j2k\frac{\delta A}{\delta z} = 0. \tag{6.13}$$

As shown in Eq. (6.2), the complex function A represents the envelope of the planewave propagating along the z-axis. To solve the second-order differential equation (6.13), we assume a solution that includes phase terms to introduce the curvature as the wave propagates. Consider the solution

$$A(r_t, z) = A_o e^{-j\left[P(z) + \frac{kr_t^2}{2q(z)}\right]}, \tag{6.14}$$

where A_o is a constant, and $p(z)$ and $q(z)$ are functions to be determined to satisfy the paraxial wave equation. This solution resembles the paraboloidal waves discussed in Chap. 3. Substituting the assumed solution in Eq. (6.13) yields

$$\frac{1}{r_t}\frac{\delta}{\delta r_t}\left(r_t \frac{\delta}{\delta r_t}\left[A_o e^{-j\left[P(z) + \frac{kr_t^2}{2q(z)}\right]}\right]\right) - j2k\frac{\delta}{\delta z}\left[A_o e^{-j\left[P(z) + \frac{kr_t^2}{2q(z)}\right]}\right] = 0, \tag{6.15}$$

Now, let us evaluate these derivatives one step at a time.

1. First derivative with respect to r_t:

$$r_t \frac{\delta}{\delta r_t}\left[A_o e^{-j\left[P(z) + \frac{rk_t^2}{2q(z)}\right]}\right] = -jkA_o \frac{r_t^2}{q(z)} e^{-j\left[P(z) + \frac{kr_t^2}{2q(z)}\right]}$$

2. Second derivative with respect to r_t:

$$\frac{1}{r_t}\frac{\delta}{\delta r_t}\left\{-jkA_o \frac{r_t^2}{q(z)} e^{-j\left[P(z) + \frac{kr_t^2}{2q(z)}\right]}\right\}$$

$$= \frac{1}{r_t}\left\{-jkA_o \left[\frac{2r_t}{q(z)} + \frac{r_t^2}{q(z)}\left(\frac{-jkr_t}{q(z)}\right)\right] e^{-j\left[P(z) + \frac{kr_t^2}{2q(z)}\right]}\right\}$$

6.2 Gaussian Beams

$$= \left[-j\frac{2k}{q(z)} - \frac{k^2 r_t^2}{q^2(z)}\right] A_o e^{-jk\left[P(z) + \frac{r_t^2}{2q(z)}\right]} = \left[-j\frac{2k}{q(z)} - \frac{k^2 r_t^2}{q^2(z)}\right] A,$$

The second term of the wave equation becomes

$$-j2k\frac{\delta}{\delta z}\left[A_o e^{-j\left[P(z) + \frac{kr_t^2}{2q(z)}\right]}\right] = -j2k\left\{-j\left[p'(z) - \frac{rk_t^2 q'(z)}{q^2(z)}\right]\right\} A$$

$$= -\left[2kp'(z) + 2k^2\frac{r_t^2 q'(z)}{q^2(z)}\right] A.$$

Substituting these derivatives back into the paraxial wave equation (6.15) and rearranging terms to get:

$$\left[-j\frac{2k}{q(z)} - \frac{k^2 r_t^2}{q^2(z)}\right] A - 2k\left[p'(z) - \frac{kr_t^2 q'(z)}{q^2(z)}\right] A = -2k\left[p'(z) + \frac{j}{q(z)}\right] r_t^0 A$$

$$+ \frac{k^2}{q^2(z)}\left[q'(z) - 1\right] r_t^2 A = 0,$$

Simplifying,

$$-2k\left[p'(z) + \frac{j}{q(z)}\right] r_t^0 + \frac{k^2}{q^2(z)}\left[q'(z) - 1\right] r_t^2 = 0, \qquad (6.16)$$

For Eq. (6.16) to be equal to zero, the coefficients of r_t^0 and r_t^2 must be each equal to zero. Hence,

$$q'(z) - 1 = 0, \qquad (6.17)$$

and

$$p'(z) + \frac{j}{q(z)} = 0. \qquad (6.18)$$

Equation (6.17) is a simple first-order differential equation that has a solution

$$q(z) = z + q_o, \qquad (6.19)$$

where q_o is a constant. To determine q_o, let us find the solution for $A(r_t, z = 0)$ such that the amplitude of the wave U is Gaussian. By substituting $z = 0$ in Eq. (6.14) and ignoring the factor $P(z=0)$, we get:

$$A(r_t, z=0) = A_o e^{-j\left[\frac{kr_t^2}{2q_o}\right]},$$

For the envelope $A(r_t, 0)$ to be Gaussian, then q_o must be an imaginary number. So let $q_o = jz_o$, making

$$A(r_t, z=0) = A_o e^{-\frac{kr_t^2}{2z_o}}, \qquad (6.20)$$

which is a Gaussian distribution. Therefore,

$$q(z) = z + jz_o. \qquad (6.21)$$

The Gaussian distribution of Eq. (6.20) is shown in Fig. 6.1.
Therefore, we can find a relationship between z_o and the beam radius ω_o as

$$\frac{kr_t^2}{2z_o} = \frac{r_t^2}{w_o^2},$$

rearranging the equation results in:

$$z_o = \frac{\pi w_o^2}{\lambda}, \text{ and } w_o = \sqrt{\frac{\lambda z_o}{\pi}}. \qquad (6.22)$$

z_o is known as the Rayleigh range and the confocal parameter. Using the value of $q(z)$, we can get

$$\frac{1}{q(z)} = \frac{1}{z + jz_o} = \frac{z}{z^2 + z_o^2} - j\frac{z_o}{z^2 + z_o^2} \qquad (6.23)$$

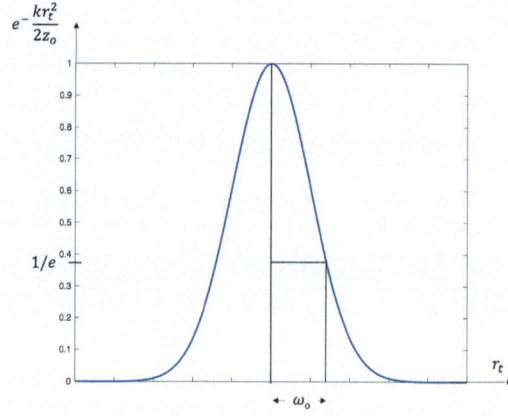

Fig. 6.1 The cross section of the Gaussian profile of Eq. (6.20) is plotted along the radial distance r_t showing the the radius ω_o at amplitude $1/e$

6.2 Gaussian Beams

Rewriting Eq. (6.14) in terms of the q parameter and leave p for now leads to:

$$A(r_t, z) = A_o e^{-\frac{kz_o r_t^2}{2(z^2+z_o^2)}} e^{-j\frac{kz r_t^2}{2(z^2+z_o^2)}}, \qquad (6.24)$$

The first term is also a Gaussian function at any point z. Thus, the width of the beam at that point $\omega(z)$ is given by:

$$w^2(z) = \frac{2(z^2+z_o^2)}{kz_o} = \frac{\lambda(z^2+z_o^2)}{\pi z_o} = \frac{\lambda z_o}{\pi}\left[1+\left(\frac{z}{z_o}\right)^2\right]. \qquad (6.25)$$

By substituting Eq. (6.22) into Eq. (6.25), we get a value for the beam waist at any point z along the direction of propagation to be:

$$w^2(z) = w_o^2\left[1+\left(\frac{z}{z_o}\right)^2\right]. \qquad (6.26)$$

Therefore, the beam waist increases as z^2/z_o^2 as shown in Fig. 6.2.

The parameter z_o is equal to z when the beam radius $w(z) = \sqrt{2}w_o$. As the beam propagates away from the beam waist, it diverges. We can find the angle of divergence of the beam by letting $z \gg z_o$ so

$$w(z) \cong w_o \frac{z}{z_o},$$

where ω_o is the beam waist. The divergence angle of the beam is:

$$\theta_{div} = \frac{dw(z)}{dz} = \frac{w_o}{z_o} = \frac{w_o}{\pi w_o^2/\lambda},$$

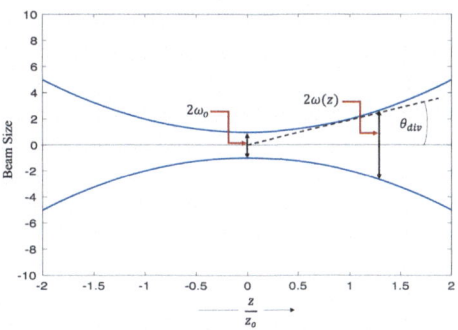

Fig. 6.2 The size of the Gaussian beam along the direction of propagation

therefore,

$$\theta_{div} = \frac{\lambda}{\pi w_o}. \tag{6.27}$$

Now, let us look at the second exponential term in Eq. (6.24), $e^{-j\frac{kzr_t^2}{2(z^2+z_o^2)}}$, it is a phase term, when combined with e^{-jkz} of the wave function $U(r_t, z)$ as given by Eq. (6.2), we obtain:

$$U(r_t, z) = A_o e^{-\frac{kz_o r_t^2}{2(z^2+z_o^2)}} e^{-j\frac{kzr_t^2}{2(z^2+z_o^2)}} e^{-jkz}. \tag{6.28}$$

The phase of the wave function is given by:

$$\phi = kz + \frac{kzr_t^2}{2(z^2 + z_o^2)}, \tag{6.29}$$

illustrates the curvature added to the wavefronts as it propagates along the z z-axis, making the beam diverge. This what makes the beam spread as it propagates along z-axis. Figure 6.3 illustrates the wavefronts of the Gaussian beam at distance z from the beam waist. These wavefronts are parabolic, but for small r_t they can be approximated by spherical wavefronts. If $R(z)$ is the radius of curvature of a spherical wave, then

$$R = \sqrt{x^2 + y^2 + z^2} = \sqrt{r_t^2 + z^2}$$

$$R = z\sqrt{1 + \frac{r_t^2}{z^2}}, \tag{6.30}$$

for large values of z, $z^2 \gg r_t^2$, which is the case since Gaussian beams divergence is very small. Therefore,

Fig. 6.3 The Gaussian beam wavefront at a distance z from the beam waist

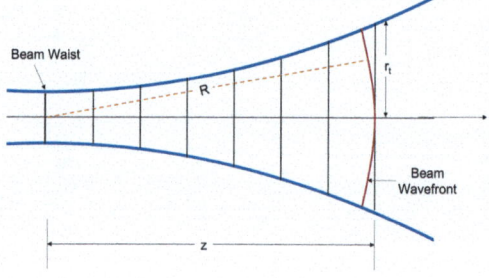

6.2 Gaussian Beams

$$R \approx z + \frac{r_t^2}{2z} \approx z + \frac{r_t^2}{2R(z)}. \tag{6.31}$$

Thus, we can represent the approximate spherical wave with the following expression:

$$U_{spherical} = \frac{e^{-jkz}}{R} e^{-j\frac{kr_t^2}{2R(z)}}. \tag{6.32}$$

By comparing Eq. (6.28) of the Gaussian beam with Eq. (6.32), we can rewrite the Gaussian beam equation as

$$U(r_t, z) = A_o e^{-\frac{kz_o r_t^2}{2(z^2+z_o^2)}} e^{-j\frac{kr_t^2}{2R(z)}} e^{-jkz}, \tag{6.33}$$

where R(z) is the radius of curvature of the Gaussian beam wavefronts

$$R(z) = \frac{z^2 + z_o^2}{z}, $$

or

$$R(z) = z\left(1 + \left(\frac{z_o}{z}\right)^2\right). \tag{6.34}$$

Now, let us take another look at the expression for $1/q$ in Eq. (6.23) and use the expressions we derived for $\omega(z)$, Eq. (6.25), and $R(z)$, Eq. (6.34), we get

$$\frac{1}{q} = \frac{1}{R(z)} - j\frac{\lambda}{\pi w^2(z)}. \tag{6.35}$$

Thus, knowing q (z) q(z) provides the properties of the Gaussian beam.

Example A Gaussian beam at $\lambda = 1$ μm and $q = 100 + j20$ determine the beam size, location, radius of curvature, beam waist, and angle of divergence.

Solution From the expression of $q = z + jz_o$ cm, we get

$$z_0 = 20,$$

$$R = 100\left(1 + \left(\frac{20}{100}\right)^2\right) = 104 \, \text{cm},$$

(continued)

$$w_o = \sqrt{\frac{\lambda z_o}{\pi}} = \sqrt{\frac{1 \times 10^{-4} \times 20}{\pi}} = 0.0252\,\text{cm}$$

$$w(z) = w_o\left[1 + \left(\frac{z}{z_o}\right)^2\right]^{1/2} = .0250\sqrt{6} = 0.062$$

and

$$\theta_{div} = \frac{\lambda}{\pi w_o} = 0.0013\,radians.$$

To complete the derivation of the optical beam wave function, we need to determine the term $p(z)$ in Eq. (6.18) is rewritten here as

$$p'(z) + \frac{j}{q(z)} = 0,$$

rearranging the terms results in:

$$p'(z) = \frac{-j}{q(z)} = \frac{-j}{z + jz_o} = -\frac{z_o}{z^2 + z_o^2} - j\frac{z}{z^2 + z_o^2}. \tag{6.36}$$

Therefore, by integrating both sides, we get:

$$P(z) = -j\ln\sqrt{[z^2 + z_o^2]} + tan^{-1}\left(\frac{z}{z_o}\right). \tag{6.37}$$

Recall the assumed solution for the paraxial wave equation, Eq. (6.14):

$$A(r_t, z) = A_o e^{-j\left[P(z) + \frac{kr_t^2}{2q(z)}\right]}.$$

Substituting for $p(z)$ from Eq. (6.37)

$$e^{-jP(z)} = e^{-j\left\{-j\ln\sqrt{[z^2+z_o^2]} + tan^{-1}\left(\frac{z}{z_o}\right)\right\}}, \tag{6.38}$$

$$e^{-jP(z)} = e^{-\ln\sqrt{[z^2+z_o^2]}} \times e^{-j tan^{-1}\left(\frac{z}{z_o}\right)}$$

6.2 Gaussian Beams

$$= \frac{1}{\sqrt{[z^2 + z_o^2]}} e^{-j\tan^{-1}\left(\frac{z}{z_o}\right)}$$

$$e^{-jP(z)} = \frac{1}{z_o\sqrt{1 + \left(\frac{z}{z_o}\right)^2}} e^{-j\tan^{-1}\left(\frac{z}{z_o}\right)},$$

using Eq. (6.34)

$$e^{-jP(z)} = \frac{w_o/z_o}{w(z)} e^{-j\tan^{-1}\left(\frac{z}{z_o}\right)}. \tag{6.39}$$

By using Eqs. (6.29) and 6.39, we obtain the full Gaussian beam wave equation:

$$U(r_t, z) = A_o \frac{w_o/z_o}{w(z)} e^{-j\tan^{-1}\left(\frac{z}{z_o}\right)} e^{-\frac{k z_o r_t^2}{2(z^2+z_o^2)}} e^{-j\frac{k r_t^2}{2R(z)}} e^{-jkz}.$$

Simplifying this expression:

$$U(r_t, z) = A_1 \frac{w_o}{w(z)} e^{-\frac{r_t^2}{2w^2(z)}} e^{-jkz} e^{-j\tan^{-1}\left(\frac{z}{z_o}\right)} e^{-j\frac{k r_t^2}{2R(z)}}. \tag{6.40}$$

This is the fundamental Gaussian beam solution of the paraxial wave equation. The first term $w_o/w(z)$, indicates that the amplitude decreases as z increases, since the beam width $w(z)$ is proportional to z. The second term $exp[\frac{r_t^2}{2w^2(z)}]$, represents the Gaussian profile of the amplitude the plane perpendicular to z. The third term $exp(-jkz)$, is the planewave, while $exp(-jtan^{-1}(z/z_o))$ is the phase correction, and the last exponential $e^{-j\frac{k r_t^2}{2R(z)}}$, accounts for the phase curvature discussed earlier. Thus, Eq. (6.40) provides the full description of the Gaussian beam as it propagates along the z z-axis.

At $z = 0$, the radius of curvature of the Gaussian beam is:

$$R(z) = z\left(1 + \left(\frac{z_o}{z}\right)^2\right),$$

yields $R(0) = \infty$, which represents the curvature of the planewave at the beam waist. As $z \gg z_o$ $R(z)$ becomes extremely large, and the Gaussian beam wavefront approaches that of a planewave. To find the minimum $R(z)$, we take the derivative:

$$\frac{dR(z)}{dz} = 1 - \frac{z_o^2}{z^2} = 0,$$

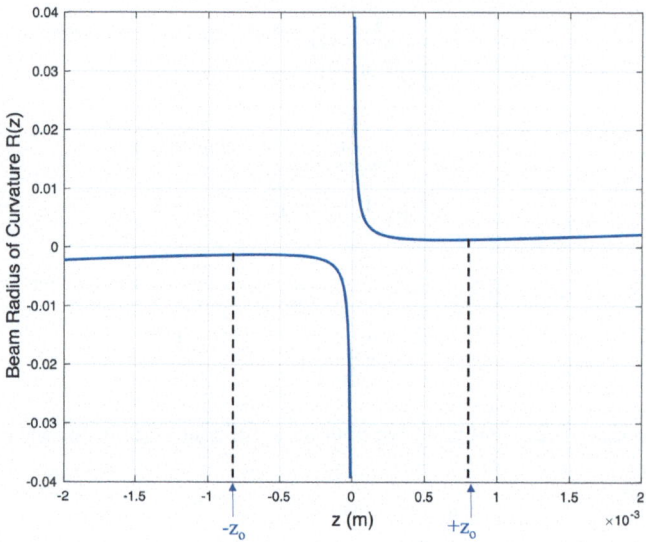

Fig. 6.4 The Gaussian beam radius of curvature $R(z)$

implying $z = \pm z_o$. Figure 6.4 shows a plot for the Gaussian beam radius of curvature on both sides of the beam waist along the z-axis. It is clear that $R(0) = \infty$ is located at the beam waist at $z = 0$, indicating a planewave As z increases, the radius of curvature decreases reaching a minimum at $z = z_o$ then increases again.

The intensity of the Gaussian beam, $I(r_t, z)$ is given by

$$I(r_t, z) = |U(r_t, z)|^2 = I_o \left(\frac{w_o}{w(z)}\right)^2 e^{-\frac{2r_t^2}{w^2(z)}}. \tag{6.41}$$

Figure 6.5 shows an intensity plot for a Gaussian beam with $w_0 = 10\,\mu$ m, and $\lambda = 0.5\,\mu$m, as the beam starting from the beam waist and propagating along the z-axis.

If we rewrite Eq. (6.41) for $r_t = 0$, we get

$$I(0, z) = I_o \left(\frac{w_o}{w(z)}\right)^2 = \frac{I_o}{1 + (z/z_o)^2}, \tag{6.42}$$

Figure 6.6 shows a plot of the intensity $I(0, z)$ of the same Gaussian beam in Fig. 6.5 labeling the positions of z_o and $-z_o$ corresponding to $I(0, z) = 0.5 I_o$. In this case for the given parameters $z_o = 0.63$ mm.

The phase, $\varphi(z)$, of the Gaussian beam, Eq. (6.40), as it propagates along z is given by:

$$\varphi(r_t, z) = kz + \tan^{-1}\left(\frac{z}{z_0}\right) + \frac{kr_t^2}{2R(z)}, \tag{6.43}$$

Fig. 6.5 Gaussian beam intensity plot

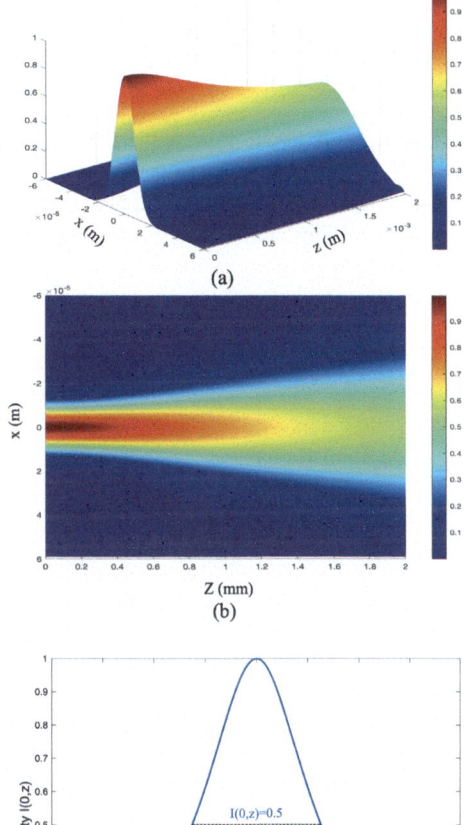

Fig. 6.6 The Gaussian beam intensity, $I(0, z)$ along the z-axis for $r_t = 0$

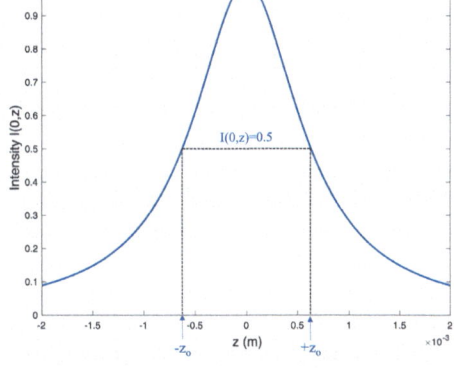

where kz represents the accumulated planewave phase, $\zeta(z) = tan^{-1}(z/z_o)$ is called the Guoy phase shift (varying between $-\pi/2$ to $\pi/2$), and the last term $kr_t^2/2R(z)$ accounts for the spherical wave distortion. The phase at the center of the beam, i.e., at $r_t = 0$, is

$$\varphi(0, z) = kz + \zeta(z). \tag{6.44}$$

The Gaussian beam solutions derived here assume cylindrical symmetry. Solving the paraxial wave equation in Cartesian coordinates would yield higher-order solutions, known as Hermite–Gaussian beams, which are beyond the scope of this discussion but can be found in other references.

Example A HeNe laser with $\lambda = 633$ nm produces a Gaussian beam with spot size $2w_o = 0.1$ mm is pointed at the moon. Determine the following: The angular of divergence, depth of focus, beam diameter at the moon, $z = 3.844 \times 10^8$ m, and radius of curvature at the moon.

Solution Angular divergence

$$2\theta_{div} = \frac{2\lambda}{\pi w_o} = \frac{2 * 633 \times 10^{-9}}{3.14 \times 0.05 \times 10^{-3}} = 0.008 \, radians$$

Depth of focus

$$2z_o = \frac{2\pi w_o^2}{\lambda} = \frac{2 \times 3.14 \times (0.05 \times 10^{-3})^2}{633 \times 10^{-9}} = 0.0248 \text{ m}$$

Beam diameter at the moon

$$w(z) = w_o \sqrt{1 + \left(\frac{z}{z_o}\right)^2} = 0.05 \times 10^{-3} \sqrt{1 + \left(\frac{3.844 x 10^8}{0.0124}\right)^2} = 1.55 \times 10^6 \text{m}$$

Beam radius of curvature at the moon

$$R(z) = z \left(1 + \left(\frac{z_o}{z}\right)^2\right) = 3.844 x 10^8 \left(1 + \left(\frac{0.0124}{3.844 x 10^8}\right)^2\right) = 3.844 x 10^8 \text{m}.$$

6.2.3 Gaussian Beam Propagation

The parameter $q(z)$ known the complex radius of the Gaussian beam allows us to determine all the properties of the beam as demonstrated in the previous subsection. In practice, Gaussian beams such as laser beams can propagate through free space, passive or active media as well as through optical components. It is important to develop a simple technique to define the beam parameters as the beam propagates through optical systems.

Consider a Gaussian beam propagating along $+z$, as shown in Fig. 6.7. Let q_1 is the beam complex radius at z_1 from the beam waist and q_2 the beam complex radius at z_2. If the beam propagates in free space a distance d, therefore we can represent q_1 and q_2 as

6.2 Gaussian Beams

Fig. 6.7 A Gaussian beam propagating along the z-axis of a complex radius of curvature q_1 at distance z_1 from the beam waist and q_2 at distance z_2 from the beam waist

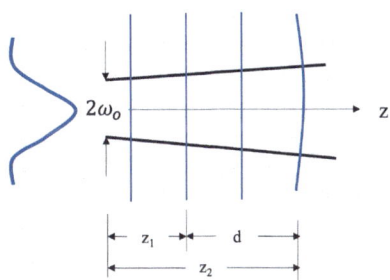

$$q_1 = z_1 + jz_o, \quad (6.45)$$

$$q_2 = z_2 + jz_o = z_1 + d + jz_o. \quad (6.46)$$

Thus, we can write

$$q_2 = q_1 + d. \quad (6.47)$$

In Chap. 2, we derived the ray matrix for the propagation of optical rays through optical systems based on Snell's law. The ray matrix for a ray propagation a distance d in the same medium is:

$$\begin{bmatrix} A & B \\ C & D \end{bmatrix} = \begin{bmatrix} 1 & d \\ 0 & 1 \end{bmatrix}. \quad (6.48)$$

Rewrite Eq. (6.47) in this scenario:

$$q_2 = \frac{q_1 + d}{0 + 1},$$

which can expressed in terms of the ray matrix coefficients as:

$$q_2 = \frac{Aq_1 + B}{Cq_1 + D}. \quad (6.49)$$

This is known as the ABCD law and clearly holds for this specific case. We can show the same for a Gaussian beam propagation through other optical elements.

6.2.3.1 Gaussian Beam Passing Through a Thin Spherical Lens

Consider a Gaussian beam passing through a thin spherical lens with focal length f, as shown in Fig. 6.8. Using Eq. (6.14), the Gaussian beam at the input of the lens is represented by:

$$U(r_t, z) \propto e^{-j\left[\frac{kr_t^2}{2q_1(z)}\right]}. \quad (6.50)$$

Fig. 6.8 Gaussian beam propagation through a thin lens. The complex radius of curvature, q_1, represents the beam at the input (left surface) of the lens, while q_2 represents the beam at the output of the lens

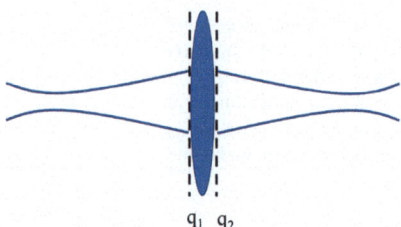

Since the lens is thin, we expect the beam waist to remain unchanged as it passes through the lens. The lens transmittance function is given by:

$$t_{lens}(r_t) = e^{j\frac{kr_t^2}{2f}}, \quad (6.51)$$

The wave at the output of the lens becomes:

$$U(r_t, z) \propto e^{-j\left[\frac{kr_t^2}{2q_1(z)}\right]} \times e^{j\frac{kr_t^2}{2f}} = e^{-j\frac{kr_t^2}{2}\left(\frac{1}{q_1} - \frac{1}{f}\right)} = e^{j\frac{kr_t^2}{2q_2(z)}}. \quad (6.52)$$

Thus,

$$\frac{1}{q_2} = \frac{1}{q_1} - \frac{1}{f}, \quad q_2 = \frac{q_1}{-q_1/f + 1}, \quad (6.53)$$

The ray matrix for a thin lens is:

$$\begin{bmatrix} A & B \\ C & D \end{bmatrix} = \begin{bmatrix} 1 & 0 \\ -1/f & 1 \end{bmatrix}. \quad (6.54)$$

Using this matrix and Eq. (6.53), we can express the beam propagation through the lens as:

$$q_2 = \frac{Aq_1 + B}{Cq_1 + D}. \quad (6.55)$$

This confirms that the ABCD law is valid for Gaussian beam propagation through a lens.

6.2.3.2 Generalization of the ABCD Law

The ABCD matrices derived for paraxial rays (rays at small angles with respect to the propagation axis) in geometrical optics also applied to paraxial Gaussian beams. This is because geometrical optics rays are perpendicular to spherical wavefronts, which align with the spherical wavefronts found in Gaussian beam solutions derived from the paraxial wave equation. Thus, the ABCD law links ray and wave optics for paraxial approximations.

6.2 Gaussian Beams

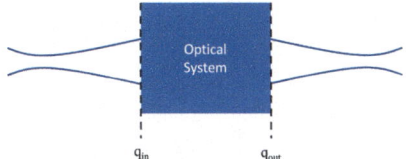

Fig. 6.9 A Gaussian beam with complex radius of curvature of q_{in} at the input plane of an optical system that can be described with the paraxial ABCD matrix. The Gaussian beam exiting the optical system has an output complex radius of curvature of q_{out}

The ABCD law can be generalized for any optical system consisting of several optical elements. Let q_{in} be the complex radius of the input Gaussian beam, and q_{out} is the complex radius of an output beam after passing through the optical system, as shown in Fig. 6.9. The relationship between the input and output beams is given by:

$$q_{out} = \frac{A q_{in} + B}{C q_{in} + D}, \tag{6.56}$$

or alternatively:

$$\frac{1}{q_{out}} = \frac{C + D\,(1/q_{in})}{A + B\,(1/q_{in})}. \tag{6.57}$$

These two equations are used very frequently in laser design to describe beam propagation through complex optical systems..

Example A Gaussian beam with beam waist $w_o = 5\,\text{mm}$ is incident on a lens of focal length $f = 10\,\text{cm}$, such that its waist is located at the lens's front surface. Determine the beam waist at the focal plane of the lens if the wavelength $\lambda = 0.5\,\mu\text{m}$.

Solution Since the beam waist is at the lens, thus, its $R = \infty$ at the front surface of the lens.

$$q_{in} = z + j z_o = 0 + j \frac{\pi w_o^2}{\lambda} = j \frac{3.14 \times (0.5 \times 10^{-3})}{0.5 \times 10^{-6}} = j1.57.$$

The ray matrix for the optical system, a lens followed by a transition distance f is:

$$\begin{bmatrix} A & B \\ C & D \end{bmatrix} = \begin{bmatrix} 1 & 0.01 \\ 0 & 1 \end{bmatrix} \cdot \begin{bmatrix} 1 & 0 \\ -1/0.01 & 1 \end{bmatrix} = \begin{bmatrix} 1 & 0.01 \\ -100 & 1 \end{bmatrix},$$

(continued)

applying the ABCD yields:

$$q_{out} = \frac{0.01}{-j157+1} = \frac{0.01 + j1.57}{24650} = 0 + j63.7 \times 10^{-6}.$$

The output beam waist $w_{o\text{-}out}$ is:

$$w_{o\text{-}out} = \sqrt{\frac{63.7 \times 10^{-6} \times 0.5 \times 10^{-6}}{3.14}} = 3.18\,\mu m.$$

Therefore, the lens reduced the beam waist from 0.5 mm to 3.18 μm. The reduction in size of the beam depends on the focal length of the lens, the wavelength, and the diffraction limit of the lens size.

6.3 Optical Resonators

Generating and amplifying light require feedback at specific frequency ranges, much like electronic oscillators that rely on an amplifier coupled with a resonant circuit. Optical resonators, also known as optical cavities, enhance the resonance of light waves at particular wavelengths. Typically, these devices consist of two or more reflective surfaces, such as mirrors, that reflect light back and forth within the cavity, resulting in constructive interference of light waves at specific resonant frequencies. Resonators, as optical cavities, store electromagnetic energy and are critical components of laser sources. In a laser, the resonator provides necessary feedback to the gain medium, enabling the stimulated emission of coherent light.

The simplest form of an optical resonator is the Fabry–Perot cavity, which consists of two parallel plane mirrors that reflect light back and forth, creating standing waves at resonant wavelengths. This type of resonator was previously discussed as a class of interferometers in Chap. 3. Other configurations include resonators with spherical mirrors, and two or three-dimensional ring arrangements, available in both discrete and integrated forms. Additionally, integrated optical micro-ring resonators, microdisks, and microspheres are examples of optical resonators.

6.3.1 Fabry–Perot Resonators

The Fabry–Perot is a simple 1-D optical cavity consisting of two parallel mirrors, as illustrated in Fig. 6.10a. As the light wave bounces back and forth, they accumulates

6.3 Optical Resonators

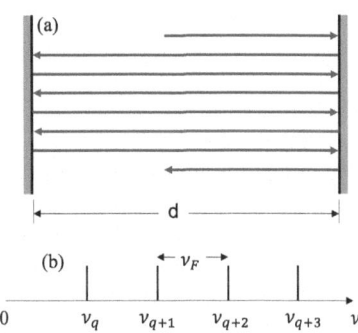

Fig. 6.10 Fabry–Perot resonator comprised two parallel planar mirrors separated by a distance d. (**a**) Light rays perpendicular to the mirrors reflect back and forth without escaping the mirrors. (**b**) Adjacent resonant frequencies, v_i, separated by frequency v_F

a phase shift, φ, for a complete round trip through the cavity, given by:

$$\varphi = k2d, \tag{6.58}$$

where k is the wavenumber, and d is the separation between the mirrors.

For constructive interference to occur within the cavity, the total phase shift for a round trip must be an integer multiple of 2π, i.e.,

$$\varphi = k2d = q(2\pi), \quad q = 1, 2, 3, \ldots \tag{6.59}$$

where q represents the resonant mode, corresponding to discrete wavenumbers, k_q, that satisfy this condition. Therefore:

$$2k_p d = 2q\pi, \tag{6.60}$$

or equivalently

$$\frac{2\pi}{\lambda_p} d = q\pi, \tag{6.61}$$

where λ_q is the resonant wavelength of the resonator. The resonant wavelengths are given by:

$$\lambda_q = \frac{2d}{q}, \tag{6.62}$$

and the corresponding resonant frequencies are:

$$v_q = q \frac{c}{2d}. \tag{6.63}$$

The separation between these resonant frequencies, known as the **free spectral range** (v_F), is:

$$v_F = \frac{c}{2d}, \text{ so that } v_q = qv_F \qquad (6.64)$$

where c is the speed of light in free space.

If the medium between the mirrors has a refractive index n, the expressions for the resonant frequencies and free spectral range become:

$$v_q = q\frac{c}{2nd}, \text{ and } v_F = \frac{c}{2nd}. \qquad (6.65)$$

Example A HeNe laser operating at $\lambda = 633$ nm, with a cavity of length $d = 30$ cm. What is the mode number q, the resonant frequency, and the free spectral range of this cavity?

Solution From Eq. (6.62) we get:

$$d = q\frac{\lambda_q}{2}, \text{ and } q = \frac{2d}{\lambda}$$

Therefore, d must be equal to an integer multiple of half the resonant wavelengths. Therefore,

$$q = \frac{2 \times 0.30}{633 \times 10^{-9}} = 9.4787 \times 10^5.$$

Thus, the resonant frequency is

$$v_q = q\frac{c}{2d} = 9.4787 \times 10^5 \times \frac{3 \times 10^8}{2 \times 0.3} = 4.7393 \times 10^{14} = 474 \text{ THz},$$

and the free spectral range is

$$v_F = \frac{c}{2d} = \frac{3 \times 10^8}{2 \times 0.3} = 500 \text{ MHz}.$$

6.3.1.1 The Density of Modes

The density of modes $M(v)$, defined as the number of modes per unit frequency per unit length of the resonator. The number of modes per frequency, N_v, is the inverse of the free spectral range, v_F, between modes, i.e.,

$$N_v = \frac{1}{v_F} = \frac{2d}{c}, \qquad (6.66)$$

where d d is the length of the resonator and c is the speed of light in free space.

6.3 Optical Resonators

Since there are two orthogonal modes, corresponding to two orthogonal polarizations that can exist for each frequency, the density of modes becomes:

$$M(\nu) = \frac{2N_\nu}{d} = \frac{4}{c}. \qquad (6.67)$$

Using this expression, the density of modes is:

$$M(\nu) = \frac{4}{c} = 1.33 \times 10^{-8} \text{ m}^{-1}.$$

6.3.2 Resonator Losses

As light waves reflect back and forth in the resonator experience losses due to the mirrors' reflection coefficient, r, as well as absorption and scattering in the medium between the mirrors. The wave undergoes a phase shift φ after completing a round trip around the cavity. Let the reflection and attenuation facto for a round trip be denoted by

$$h = |r|e^{-j\varphi}, \qquad (6.68)$$

where $|r|$ is the round-trip attenuation factor, and the phase shift is given by:

$$\varphi = 2kd = \frac{4\pi \nu d}{c} = \frac{2\pi \nu}{\nu_F}. \qquad (6.69)$$

Assume that the original wave has amplitude U_0, so after one round trip, the amplitude becomes:

$$U_1 = hU_0. \qquad (6.70)$$

Thus, the total sum of all waves, including multiple reflection, is:

$$U = U_0 + U_1 + U_2 + \ldots == U_0 \left(1 + h + h^2 + h^3 + \ldots\right). \qquad (6.71)$$

This is the sum of an infinite geometric series in h, where $|h| < 1$:

$$U = \frac{U_0}{1-h}. \qquad (6.72)$$

Hence, the intensity of the light is given by:

$$I = |U|^2 = \frac{I_0}{|1 - |r|e^{-j\varphi}|^2}.$$

Expanding this expression:

$$I = \frac{I_0}{1 + |r|^2 - 2|r|\cos(\varphi)}. \tag{6.73}$$

The intensity of the beam, I, is maximized when $\cos(\varphi) = 1$, leading to:

$$I_{max} = \frac{I_0}{(1 - |r|)^2}. \tag{6.74}$$

By substituting I_{max} in the previous equations, we get:

$$I = \frac{I_{max}}{1 + (2F/\pi)^2 \sin^2(\pi \nu/\nu_F)}, \tag{6.75}$$

where finesse, F, is defined as:

$$F = \frac{\pi \sqrt{|r|}}{1 - |r|}. \tag{6.76}$$

The minimum value of the intensity is

$$I_{min} = \frac{I_{max}}{1 + (2F/\pi)^2}. \tag{6.77}$$

At resonance, $\nu_q = q\nu_F$, so $\pi \nu/\nu_F = q\pi$. Since q is an integer then $\sin^2(\pi \nu/\nu_F) = \sin^2(q\pi) = 0$.

Therefore at resonance, the output intensity of the resonator becomes:

$$I = I_{max}.$$

Indicating perfect transmission at resonant frequency.

Figure 6.11 illustrates the intensity plot for a Fabry–Perot around a resonant frequency, assuming that both mirrors have reflectance R. As R increases, the curve becomes sharper, indicating higher selectivity and reduced bandwidth. The broadening of the intensity curve is directly related to the losses within the cavity.

Example Derive an expression for the spectral width, $\delta\nu$, of the intensity for a Fabry–Perot resonator.

Solution The spectral with is the Full-Width at Half-Magnitude (FWHM) of the intensity at resonance. At FWHM, the intensity is equal to 0.5. Thus,

(continued)

6.3 Optical Resonators

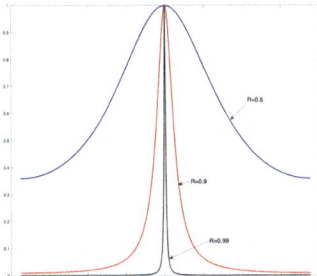

Fig. 6.11 Intensity of the output beam for a Fabry–Perot resonator as a function of mirror reflectance R around a resonant frequency. As the reflectance R increases, the peak intensity becomes sharper and more pronounced, indicating higher finesse and narrower bandwidth of the resonator

at frequency $v = v_q + \delta v/2$, the intensity $I = 0.5 I_{max}$. Therefore, at this frequency the denominator of Eq. (6.75) is equal 2. Hence,

$$(2F/\pi)^2 \sin^2\left(\pi(v_q + \delta v/2)/v_F\right) = 1,$$

thus,

$$\sin^2\left(\pi(v_q + \delta v/2)/v_F\right) = (\pi/2F)^2,$$

by taking the square root of both sides, we get:
using the trigonometric expansion for $sin(x+y)$ in the previous equation we get

$$\sin(\pi \delta v/2 v_F) = \pi/2F.$$

For large finesse δv is very small, thus, we can use Taylor expansion for the sine function to get

$$\delta v \approx \frac{v_F}{F}. \tag{6.78}$$

The spectral width of a resonator δv represents the range of frequencies over which the resonator can effectively store and transfer energy. Resonators with high losses tend to have a larger spectral width compared to those with low losses. This is because high losses result in a broader resonance peak and thus a larger range of frequencies over which the resonator can operate.

In Fig. 6.12a, a monochromatic beam with 1 Watt of power is directed at a mirror with a reflectivity of $R = 0.999$, allowing only 1 mW of light to pass through.

Fig. 6.12 (a) A light beam reflected from a mirror of $R = 0.999$. (b) The same beam as it passes through a resonator

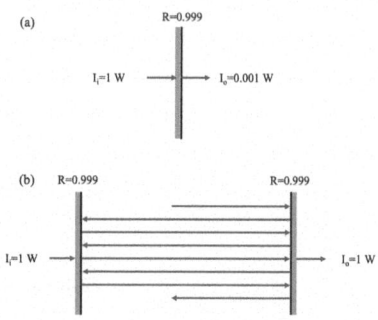

However, when a second mirror is added, as shown in Fig. 6.12b, and the separation between the mirrors is adjusted so that the resonant frequency of the resonator matches the input beam frequency, the light will bounce back and forth between the mirrors, building up the energy within the cavity. Eventually, the output power reaches 1 $Watt$ and remains at that level until the input beam is switched off. The time it takes for the beam to reach this steady state is known as the **photon lifetime.**

6.3.2.1 Resonator Total Losses
To determine the total losses in a resonator, let the reflection coefficients of the two mirrors be r_1 and r_2, with reflectances:

$$R_1 = |r_1|^2, \text{ and } R_2 = |r_2|^2. \tag{6.79}$$

Let α_s be the loss coefficient of the medium between the mirrors, accounting for absorption and scattering. The round-trip intensity attenuation factor is given by

$$|r|^2 = R_1 R_2 e^{-2\alpha_s d} = e^{-2\alpha_r d}, \tag{6.80}$$

where α_r is the total loss coefficient. By taking the natural logarithm of both sides of Eq. (6.80) and rearranging terms we get

$$\alpha_r = \alpha_s + \frac{1}{2d} \ln\left(\frac{1}{R_1 R_2}\right). \tag{6.81}$$

Substituting from Eq. (6.80) for $|r|$ in Eq. (6.76) we get the following expression for the finesse:

$$F = \frac{\pi e^{-\alpha_r d/2}}{1 - e^{-\alpha_r d}}. \tag{6.82}$$

If $\alpha_r d \ll 1$, then using the Taylor expansion we get

$$e^{\alpha_r d} \approx 1 - \alpha_r d,$$

the expression for finesse simplifies to:

$$F \approx \frac{\pi}{\alpha_r d}. \tag{6.83}$$

Example Determine the finesse and the spectral width for a Fabry–Perot resonator with the following parameters: $R_1 = R_2 = 0.99$, and $d = 10$ cm.

Solution The finesse is

$$F = \frac{\pi \sqrt{|r|}}{1 - |r|} = \frac{\pi \sqrt{0.995}}{0.005} = 626.$$

The spectral width is

$$\delta \nu \approx \frac{\nu_F}{F} = \frac{c/2d}{F} = \frac{3 \times 10^8}{2 \times 0.1 \times 626} = 2.4 \text{ MHz}.$$

6.3.2.2 Quality Factor and Photon Lifetime

The resonator quality factor, also known as Q-factor, is a measure of the quality of an optical resonator or cavity. It quantifies the energy stored in the cavity per cycle of oscillation, relative to the energy lost per cycle due to various sources of loss, such as absorption and scattering. A higher Q-factor indicates that the energy stored in the resonator lasts longer before being lost, resulting in a narrower linewidth and a longer photon lifetime. The Q-factor is an important parameter in determining the performance of a resonator, as it affects the efficiency, stability, and spectral purity of the resonant cavity.

The energy decay lifetime, i.e., the photon lifetime τ_p of a cavity mode is defined by

$$\frac{d\mathcal{E}}{dt} = -\frac{\mathcal{E}}{\tau_p}, \tag{6.84}$$

where $\mathcal{E}$ is the energy stored in a cavity mode. So the energy in a passive resonator, i.e., no active material exists in the resonator, the stored energy is given by

$$\mathcal{E}(t) = \mathcal{E}(0))e^{-t/\tau_p}. \tag{6.85}$$

The resonator loss per unit length is α_r as given by Eq. (6.81). Hence, the loss per unit time is $c\alpha_r$, and the photon lifetime is given by

$$\tau_p = \frac{1}{c\alpha_r}. \tag{6.86}$$

Consider Eq. (6.78) for the spectral width and the Finesse equation (6.83), we get

$$\delta\nu \approx \frac{\nu_F}{F} = \frac{\nu_F}{\pi/\alpha_r d} = \frac{c/2d}{\pi/\alpha_r d} = \frac{c\alpha_r}{2\pi}$$

$$\delta\nu = \frac{1}{2\pi\tau_p}. \tag{6.87}$$

The quality factor or the Q-factor is defined as

$$Q = 2\pi \frac{Energy\ stored\ in\ the\ cavity}{Wnergy\ loss\ per\ cycle}. \tag{6.88}$$

At a resonant frequency ν_o, the loss per cycle is $c\alpha_r/\nu_o$, therefore

$$Q = \frac{2\pi\nu_o}{c\alpha_r} = \frac{\nu_o}{\delta\nu}, \tag{6.89}$$

so the Q-factor is equal to the resonant or central frequency divided by the spectral width. Therefore, the narrower the spectral width of the resonator, the higher the Q-factor. Also Q can be expressed in terms of the photon lifetime as

$$Q = \omega\tau_p. \tag{6.90}$$

Based on the previous set of results, the spectral width of the resonator and energy storing capacity of the cavity depend on the resonator losses. Resonators with low losses have a higher energy storage capacity because they can retain energy for a longer period of time before it dissipates. Conversely, resonators with high losses have a lower energy storage capacity because the energy is dissipated more quickly.

Therefore, reducing resonator losses can help increase the energy storage capacity and improve the spectral width of a resonator. This is an important consideration in the design and optimization of resonators for various applications, such as in microwave and optical communication systems, lasers, and sensors.

6.3.3 Spherical Mirror Resonators

Optical resonators consist of two or more reflective components, such as mirrors, gratings, or other reflective surfaces. One of the most common types of optical resonators is the laser resonator, which typically consists of two spherical mirrors. The beams inside the resonator reflect back and forth between the mirrors, remaining confined between them to form a stable resonator.

A typical spherical resonator consists of two mirrors with radii R_1 and R_2 separated by a distance d, as illustrated in Fig. 6.13. The centers of curvature of both

6.3 Optical Resonators

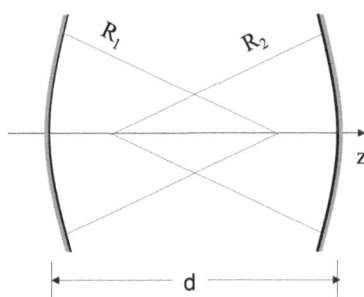

Fig. 6.13 Geometry for spherical mirror resonator constructed with two concave mirrors

mirrors lie on the optical axis of the system. The mirrors can be concave ($R < 0$) or convex ($R > 0$), depending on the desired properties of the resonator.

The **Fabry–Perot** resonator is a special case of a spherical mirror resonator where both of the mirrors have an infinite radius of curvature ($R = \infty$). This resonator exhibit a cylindrical symmetry around the z-axis.

In Chap. 2, Sect. 2.8, we discussed periodic optical systems and derived the stability condition for these systems in terms of their ray **ABCD** matrix. For the resonator to be stable, the following condition must hold:

$$\frac{|A + D|}{2} \leq 1. \tag{6.91}$$

This stability condition ensures that the rays remain confined within the resonator as they reflect between the mirrors. The ABCD ray matrix for the spherical-mirror resonator in Fig. 6.13 is given by

$$\begin{bmatrix} A & B \\ C & D \end{bmatrix} = \begin{bmatrix} 1 & 0 \\ -2/R_1 & 1 \end{bmatrix} \begin{bmatrix} 1 & d \\ 0 & 1 \end{bmatrix} \begin{bmatrix} 1 & 0 \\ -2/R_2 & 1 \end{bmatrix} \begin{bmatrix} 1 & d \\ 0 & 1 \end{bmatrix}. \tag{6.92}$$

Expanding this matrix gives:

$$\begin{bmatrix} A & B \\ C & D \end{bmatrix} = \begin{bmatrix} \frac{R_2-2d}{R_2} & \frac{-2d^2+2dp}{p} \\ \frac{-2R_2-2(R_1-2d)}{R_1 R_2} & \frac{4d^2-2R_1 d-4d R_2+R_1 R_2}{R_1 R_2} \end{bmatrix}.$$

Thus,

$$A + D = \frac{R_2 - 2d}{R_2} + \frac{4d^2 - 2R_1 d - 4d R_2 + R_1 R_2}{R_1 R_2}.$$

Rewriting in terms of simplified parameters:

$$\frac{A+D}{2} = 1 - \frac{2d}{R_2} - \frac{2d}{R_1} + \frac{2d}{R_1 R_2} = 2\left(1 - \frac{d}{R_1}\right)\left(1 - \frac{d}{R_2}\right) - 1.$$

Therefore, the stability condition becomes:

$$0 \le \left(1 - \frac{d}{R_1}\right)\left(1 - \frac{d}{R_2}\right) \le 1, \tag{6.93}$$

This condition can be expressed in terms of the parameters $g_1 = 1 - d/R_1$ and $g_2 = 1 - d/R_2$, known as the **g-parameters**:

$$0 \le g_1 g_2 \le 1. \tag{6.94}$$

For a symmetric resonator, where both mirrors have the same radius R, i.e., $R_1 = R_2 = R$, the stability condition is expressed as:

$$0 \le \left(1 - \frac{d}{R}\right) \le 1 \text{ or } 0 \le g \le 1.$$

6.3.4 Gaussian Beams in Stable Resonators

Consider a Gaussian beam reflected back and forth in a stable spherical-mirror resonator, as shown in Fig. 6.14. for the resonator to be stable, the beam must maintain its constant shape and size at all points within the cavity. This implies that the beam's q-parameter remains constant and repeats at the same point in the cavity as the beam reflects back and forth.

Additionally, the curvature of the phase fronts of the Gaussian beam at each mirror matches the curvature of the mirrors themselves. This ensures that the Gaussian mode reproducible and stable within the resonator, a critical requirement for applications in lasers and optical amplifiers.

Overall, a stable resonator is designed to maintain a stable mode shape and size, which is critical for many applications such as lasers and optical amplifiers.

As the Gaussian beam completes a round trip within the resonator, the q-parameter remains unchanged at corresponding locations. Using the Gaussian beam propagation formula given in Eq. (6.56), the q-parameter inside the cavity where

Fig. 6.14 A spherical-mirror resonator with a stable Gaussian mode

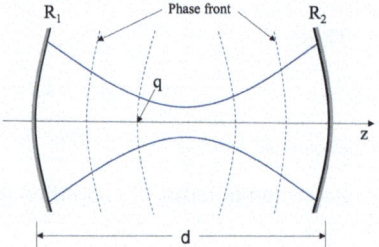

6.3 Optical Resonators

$q_1 = q_2 = q$ is given by:

$$q = \frac{Aq + B}{Cq + D}. \tag{6.95}$$

The constancy of the q-parameter as the beam propagates a round trip in the resonator is a key property of a stable resonator, which ensures a reproducible mode. Such that the mode shape and size are maintained throughout the resonator.

Expanding Eq. (6.95) results in the following quadratic equation:

$$Cq^2 + (D - A)q - B = 0, \tag{6.96}$$

The ABCD coefficients in the Gaussian beam propagation formula, also known as the beam transfer matrix, can be selected to realize a particular resonator design that produces a specific Gaussian beam. If a cavity to be designed to produce a beam waist w_0, the beam propagation formula can be expressed in terms of $1/q$ as given in Eq (6.63):

$$\frac{1}{q} = \frac{1}{R(z)} - \frac{j\lambda}{\pi \omega^2(z)}.$$

The Gaussian beam propagation in terms of $1/q$ $1/q$ is given by Eq. (6.57):

$$\frac{1}{q} = \frac{C + D/q}{A + B/q},$$

which, when expanded, results in the following expression:

$$\frac{B}{q^2} + \frac{A - D}{q} - C = 0. \tag{6.97}$$

This is a quadratic equation in $1/q$. Solving this equation yields:

$$\frac{1}{q} = -\frac{(A - D)}{2B} \pm \sqrt{\frac{(A - D)^2 + 4BC}{4B^2}}. \tag{6.98}$$

Since the ABCD matrix is a unitary matrix, (i.e., $AD - BC = 1$), we have:

$$4BC = 4AD - 4,$$

which simplifies the term under the square root to:

$$\frac{A^2 + D^2 - 2AD + 4AD - 4}{4B^2} = \frac{\left(\frac{A+D}{2}\right)^2 - 1}{B^2}.$$

Fig. 6.15 An example of a simple optical resonator

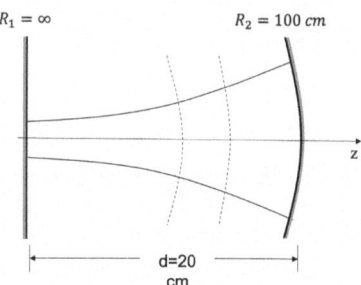

Therefore,

$$\frac{1}{q} = -\frac{(A-D)}{2B} \pm \frac{j}{B}\sqrt{1 - \left(\frac{A+D}{2}\right)^2}, \quad (6.99)$$

where the stability condition ensures that:

$$\left(\frac{A+D}{2}\right) \leq 1. \quad (6.100)$$

Within the cavity, the beam radius of curvature is given by:

$$R(z) = \frac{2B}{A-D}, \quad (6.101)$$

and the beam waist is:

$$w^2(z) = \frac{\lambda B}{\pi} \frac{1}{\sqrt{1 - \left(\frac{A+D}{2}\right)^2}}. \quad (6.102)$$

Example Consider the resonator shown in Fig. 6.15, which consists of a plane mirror and a spherical mirror with radius $R_2 = 100$ cm, and a cavity length $d = 20$ cm. Determine the beam waists at the plane mirror and at the spherical mirror, as well as the radius of curvature of the beam at the spherical mirror, given a wavelength $\lambda = 1$ μm.

Solution The ABCD matrix of the resonator, following the beam from the plane mirror for a round trip, is

(continued)

6.3 Optical Resonators

$$\begin{bmatrix} A & B \\ C & D \end{bmatrix} = \begin{bmatrix} 1 & 0 \\ 0 & 1 \end{bmatrix} \begin{bmatrix} 1 & 20 \\ 0 & 1 \end{bmatrix} \begin{bmatrix} 1 & 0 \\ -2/100 & 0 \end{bmatrix} \begin{bmatrix} 1 & 20 \\ 0 & 1 \end{bmatrix} = \begin{bmatrix} \frac{3}{5} & 32 \\ -\frac{1}{50} & \frac{3}{5} \end{bmatrix}.$$

The radius of curvature R is

$$R = \frac{2B}{A-D} = \frac{64}{0.6 - 0.6} = \infty.$$

This infinite radius of curvature corresponds to the plane mirror, where we expect the wavefront to be flat, matching the surface of the mirror.

Now, let us calculate the beam waist at the plane mirror using the formula:

$$w^2 = \frac{\lambda B}{\pi} \frac{1}{\sqrt{1 - \left(\frac{A+D}{2}\right)^2}} = \frac{32 \times 10^{-4}}{3.14} \frac{1}{\sqrt{1 - \left(\frac{0.6+0.6}{2}\right)}} = 12.25 \times 10^{-4} \, \text{cm}^2.$$

Therefore, the beam waist is

$$w_0 = \sqrt{16.1 \times 10^{-4}} = 0.035 \, \text{cm}.$$

To determine the beam waist and radius of curvature at the spherical mirror, we need to find z_o, which is given by:

$$z_o = \frac{\pi w_o^2}{\lambda} = \frac{3.14 \times 12.25 \times 10^{-4}}{1 \times 10^{-4}} = 40$$

Therefore, the radius of curvature at the spherical mirror is:

$$R(20) = 20\left[1 + \left(\left[1 + \left(\frac{40}{20}\right)^2\right]\right)\right] = 100 \, \text{cm}$$

which exactly matches the radius of curvature of the spherical mirror. Finally, the beam size at the spherical mirror is:

$$w(20) = 0.035\left[1 + \left(\frac{20}{40}\right)^2\right]^{1/2} = 0.039 \, \text{cm}.$$

6.4 Micro-Ring Resonators

Micro-ring resonators are a type of optical device that are used to filter and manipulate light at the microscale. They are based on the principle of resonant cavities, which are structures that can trap and amplify light by reflecting it back and forth between two or more mirrors.

In micro-ring resonators, light is guided around a circular or annular waveguide, creating a resonant cavity. When the circumference of the ring is an integer multiple of the wavelength of the light, constructive interference occurs and a resonance is formed. This resonance creates a sharp spectral peak in the transmission spectrum of the device, which can be used to filter or modulate the light.

Micro-ring resonators have a number of advantages over other optical devices. They are small, typically on the order of micrometers in size, which makes them compatible with microfabrication techniques and allows them to be integrated with other optical and electronic components. They are also highly tunable, with their resonant wavelength and bandwidth dependent on the geometry of the device and the refractive index of the surrounding material.

6.4.1 Single Ring Resonator

Consider the directional coupling between two single-mode waveguides as shown in Fig. 6.16. Let E_{i1} arid E_{i2} be the field amplitudes of the input, and E_{t1} and E_{t2} be the field amplitude of the output. The linear input–output relation can be expressed as:

$$E_{t1} = E_{i1}\kappa_{11} + E_{i2}\kappa_{21} \tag{6.103}$$

$$E_{t2} = E_{i1}\kappa_{12} + E_{i2}\kappa_{22} \tag{6.104}$$

where κ_{11} and κ_{22} are the straight-through coupling coefficients, and κ_{12} and κ_{21} are the cross-coupling coefficients. These equations can be represented in the following matrix form:

$$\begin{bmatrix} E_{t1} \\ E_{t2} \end{bmatrix} = \begin{bmatrix} \kappa_{11} & \kappa_{21} \\ \kappa_{12} & \kappa_{22} \end{bmatrix} \begin{bmatrix} E_{i1} \\ E_{i2} \end{bmatrix}. \tag{6.105}$$

The coupling coefficient is a function of the coupling length, waveguide separation, and wavelength of the light propagating into the waveguide.

The coupler can be modified to for a ring-resonator, as shown in Fig. 6.17, by connecting the waveguides labeled E_{i2} and E_{o2}.

In order to simplify the analysis, let the input modal field $E_{i1} = 1$. We can express the following relationship between the different field components shown in the figure:

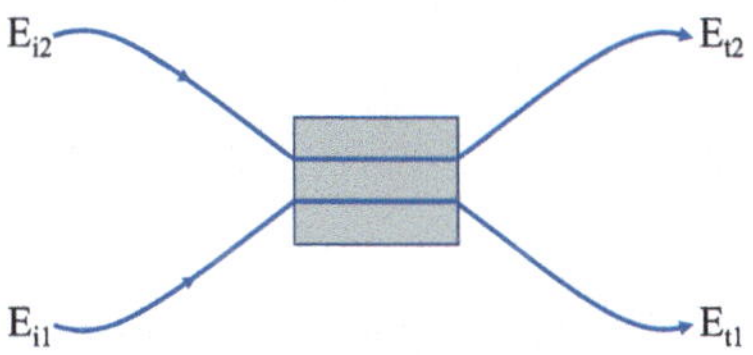

6.4 Micro-Ring Resonators

Fig. 6.16 Optical directional coupler

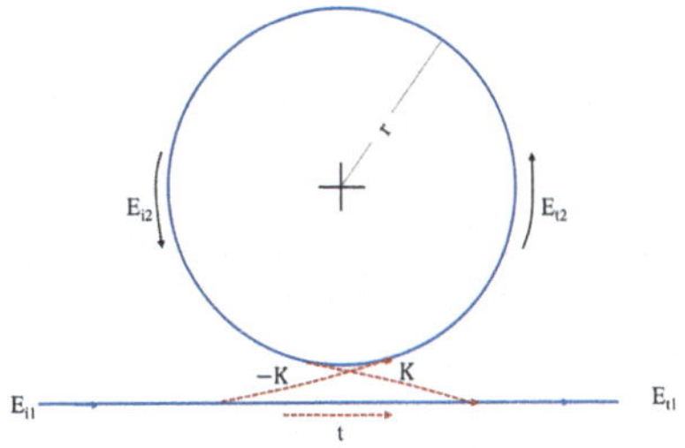

Fig. 6.17 A typical single ring resonator

$$\begin{bmatrix} E_{t1} \\ E_{t2} \end{bmatrix} = \begin{bmatrix} t & \kappa \\ -\kappa^* & t^* \end{bmatrix} \begin{bmatrix} E_{i1} \\ E_{i2} \end{bmatrix} \quad (6.106)$$

The field E_{t2} as it coupled in the ring waveguide it travels around the ring and creates resonant modes for the round-trip phase φ that satisfies the following:

$$\varphi = k(2\pi r)n_{eff} = 4\pi^2 n_{eff} \frac{r}{\lambda} = 2q\pi, \quad q = 1, 2, 3, \ldots \quad (6.107)$$

where $k = 2\pi/\lambda$ is the wavenumber, r is the radius and n_{eff}, is the effective refractive index of the waveguide.

$$E_{i2} = \alpha E_{t2} e^{-j\varphi}. \quad (6.108)$$

where α is the attenuation in the waveguide, for a lossless waveguide $\alpha = 1$. Using Eqs. (6.106) and (6.108) assuming $\alpha = 1$, we get

$$E_{t1} = t E_{i1} + \kappa E_{i2} = t E_{i1} + \kappa E_{t2} e^{-j\varphi}$$

$$E_{t2} = -\kappa^* E_{i1} + t^* E_{i2} = -\kappa^* E_{i1} + t^* E_{t2} e^{-j\varphi},$$

rearranging these equations, we get

$$\frac{E_{t1}}{E_{i1}} = \frac{(t + \kappa^2) - t^2 e^{-j\varphi}}{1 - t^* e^{-j\varphi}}. \quad (6.109)$$

A typical characteristic of ring resonators is given in Fig. 6.18.

Micro-ring resonators are an essential component in photonic circuits, which can be configured to serve as an optical add-drop wavelength filter. Combined

Fig. 6.18 Typical characteristic of a ring resonator

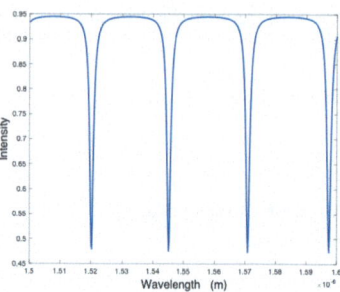

with Mach–Zehnder interferometers, they can form the basis for highly functional photonic integrated circuits.

Their small size, high tunability, and compatibility with microfabrication techniques make them an ideal platform for the development of new optical devices and technologies.

6.5 Summary

This chapter explored the fundamental concepts of optical beams and resonators, focusing on the theory, behavior, and practical applications of Gaussian beams and various resonator configurations. Gaussian beams were introduced as a solution to the paraxial wave equation, which closely approximates laser beam behavior. Unlike idealized planewaves, Gaussian beams have a finite extent and exhibit a Gaussian intensity profile. Key parameters such as beam waist, Rayleigh range, and divergence angle were discussed, highlighting their importance in controlling beam propagation and focusing characteristics.

The propagation and stability of Gaussian beams were analyzed through the complex beam parameter, $q(z)$, which enables a systematic evaluation of beam properties as it propagates through free space or optical elements. We examined how the beam shape and amplitude evolve along the propagation axis and applied the ABCD matrix method to model beam behavior through various optical components, such as lenses.

Optical resonators, central for laser operation, were then discussed as devices that provide feedback and enable light amplification. The Fabry–Perot resonator and spherical mirror resonators were analyzed, with a focus on the stability conditions required to maintain confined oscillations within the cavity. These resonators allow constructive interference for specific resonant frequencies, promoting energy build-up and selective frequency filtering.

This chapter also covered resonator losses and quality factors, discussing how loss mechanisms, including mirror reflectance and medium absorption, impact energy storage and spectral width. Key parameters such as finesse and quality factor (Q) were introduced to quantify a resonator's spectral sharpness and energy storage efficiency.

6.6 Problems

1. An optical resonator is made of two spherical mirrors with radii R_1 and R_2 and separated with a distance d as shown in the figure. The paraxial rays are reflected repeatedly between the two mirrors. Determine the condition of stability of the resonator.
2. A Gaussian beam has a wavelength of 633 nm and a beam waist $w_o = 0.5$ mm. Calculate the Rayleigh range, divergence angle, and beam radius at a distance of 1 m from the beam waist.
3. Consider a spherical mirror resonator with two mirrors of radii $R_1 = 10$ cm and $R_2 = 20$ cm, separated by a distance $d = 15$ cm. Determine whether this resonator is stable.
4. A Gaussian beam with waist $w_o = 0.3$ mm is focused by a lens with a focal length of 20 cm. Assuming that the beam waist is located at the lens, calculate the new beam waist size and location after the beam passes through the lens.
5. A Gaussian beam with $q = 10 + j5$, determine the beam waist, and depth of focus.
6. Derive the ABCD matrix for a resonator consisting of a plane mirror and a spherical mirror with radius $R = 50$ cm, separated by a distance $d = 10$ cm. Use this matrix to determine the radius of curvature of the beam at the spherical mirror.
7. A Fabry–Perot resonator consists of two mirrors with reflectance $R = 0.99$ and spacing $d = 1$ cm. Calculate the free spectral range (FSR) and finesse of the resonator, as well as the transmission of the resonator at resonance.
8. A micro-ring resonator has a radius of $r = 5\,\mu$m and an effective refractive index $n_{eff} = 1.5$. Calculate the free spectral range (FSR) in terms of frequency and wavelength for light at a wavelength of 1550 nm.
9. Design a micro-ring resonator with a free spectral range of 20 GHz for light at 1550 nm. Determine the required radius of the ring resonator, assuming an effective refractive index of $n_{eff} = 1.45$.
10. A micro-ring resonator has a coupling coefficient $\kappa = 0.1$ and an internal field attenuation factor $\alpha = 0.9$. Calculate the resonance condition and plot the transmission spectrum over a range of wavelengths near the resonance. Indicate the FSR on your plot.

Bibliography

1. Siegman, A. E. (1986). Lasers. University Science Books.
2. Yariv, A., & Yeh, P. (2007). Photonics: Optical Electronics in Modern Communications (6th ed.). Oxford University Press.
3. Kogelnik, H., & Li, T. (1966). "Laser Beams and Resonators." Applied Optics, 5(10), 1550-1567.
4. Svelto, O. (2010). Principles of Lasers (5th ed.). Springer.
5. Saleh, B. E. A., & Teich, M. C. (2019). Fundamentals of Photonics (3rd ed.). Wiley.

6. Haus, H. A. (1984). Waves and Fields in Optoelectronics. Prentice Hall.
7. Lipson, M., Gleskova, H., & Lipson, S. G. (2017). Optical Physics (4th ed.). Cambridge University Press.
8. Huang, Y. J., & Yariv, A. (1993). "Theory of Microdisk Lasers as Geometrically Defined Resonant Cavities." IEEE Journal of Quantum Electronics, 29(4), 1339-1350.

Optical Waveguides

7.1 Introduction

Waveguides serve as conduits to transport light in an optical medium such as an optical fiber or photonic integrated circuit, in the same way metallic wires transport electrons in an integrated circuit. However, unlike traditional wires, waveguides can function as passive elements. Optical waveguides can be few micrometers long, as in the case in photonic circuits, or thousands of kilometers, as in the case of optical fibers. Waveguides confine the light as it propagates within a high-refractive-index medium, known as the core, surrounded by a lower-refractive-index medium, called the cladding. Waveguides confine light within a high-refractive-index region, known as the core, which is surrounded by a lower-refractive-index region called the cladding. This confinement is achieved through total internal reflection (TIR), a fundamental optical phenomenon detailed in Chap. 3.

This chapter introduces the core principles, structures, and applications of optical waveguides. We will discuss the basics of waveguide theory, covering essential concepts such as modes, dispersion, and coupling.

7.2 Waveguide Structures

Optical waveguides come in a variety of two-dimensional (2D) and three-dimensional (3D) structured, designed to suit specific applications. Regardless of the shape of the waveguide, light travels within the core, which is encased by a cladding layer that has a lower refractive index, to ensure confinement. Waveguides can have either planar or nonplanar configurations. The most basic form of a planar waveguide is the optical slab waveguide, while a common example of a nonplanar waveguide is an optical fiber.

Figure 7.1 provides a cross-sectional view along the light propagation direction (z-axis) of standard waveguide structures including their respective designations.

Fig. 7.1 Cross section of several standard waveguide structures

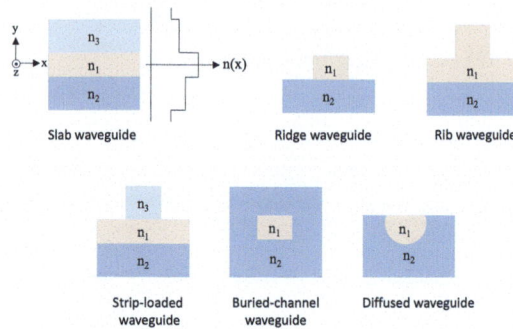

The following is a brief description of there waveguides:

1. **Slab Waveguide**: A slab waveguide, which provides optical confinement solely in one transverse direction, the core is nestled between cladding layers specifically in one direction, for instance, the y-direction, and is associated with an index profile $n(y)$. The core of a planar waveguide, also referred to as the film, is bounded by the upper and lower cladding layers, termed the cover and the substrate, respectively.
2. **Ridge Waveguide**: A ridge waveguide resembles a strip waveguide in structure, but with a crucial distinction: the elevated strip or "ridge" that tops its planar structure possesses a high refractive index and serves as the actual core. Due to its envelopment on three sides by low-index air (or cladding material), a ridge waveguide offers robust optical confinement.
3. **Rib Waveguide**: A rib waveguide exhibits a structure akin to that of strip or ridge waveguides, but the key difference lies in the strip. The strip shares the same index as the high-index planar layer below it, making it an integral part of the core.
4. **Buried Waveguides:** A buried channel waveguide is characterized by a high-index waveguiding core is embedded within a low-index surrounding medium. While the cross-sectional geometry of the core can vary, it often a rectangular shape.
5. **Strip-Loaded Waveguide**: A strip-loaded waveguide is created by augmenting a planar waveguide, which inherently ensures optical confinement in the y-direction, with a dielectric strip of index $n_3 < n_1$ or a metal strip. This modification fosters optical confinement in the x-direction. The core of a strip waveguide is the n_1 region located beneath the loading strip. The thickness of this core is determined by the thickness of the n_1 layer, and its width is defined by the width of the loading strip.
6. **Diffused Waveguide**: A diffused waveguide is created by introducing a high-index region within a substrate via diffusion of dopants. An example of this is a $LiNbO_3$ waveguide.

7.3 Optical Slab Waveguides

In the following sections, we will explore selected waveguide types in more detail, examining their properties in terms of light guidance and their performance in photonic systems.

7.3 Optical Slab Waveguides

The slab waveguide is the most fundamental and simplistic waveguide structure. Its analysis provides insights into the essential principles of light guidance in optical media. This type of waveguide is viewed as a planar structure. Figure 7.2 depicts a 3D slab waveguide structure. The typical dimensions of the waveguide are such that

$$l >> w >> d, \tag{7.1}$$

and the light is confined to the *y-axis* and propagates along the *z-axis*.

There are two types of waveguides based on the refractive index profile between the core and the cladding. A waveguide where the core refractive index, denoted as n_1, presents an abrupt change with respect to the cladding index, n_2, is referred to as a **step-index** waveguide. Conversely, a waveguide where the refractive index profile of the core varies gradually is known as a **graded-index** waveguide. In this chapter, we consider only step-index waveguides.

7.3.1 Qualitative Approach to Optical Beam Confinement

We begin by exploring optical beam confinement in a symmetric planar dielectric waveguide using a qualitative approach, illustrated in in Fig. 7.3a. Here, we consider a planewave traveling along a wave vector denoted $\bar{k}_1$. The planewave has wavefronts, represented by the blue dashed lines, which are perpendicular to the wave vector k_1. As the wave propagates along the z-axis, it reflects back and forth between the core and cladding interfaces, as depicted in Fig. 7.3b. Within the core, these reflections result in interference.

Fig. 7.2 Basic structure of an optical slab waveguide

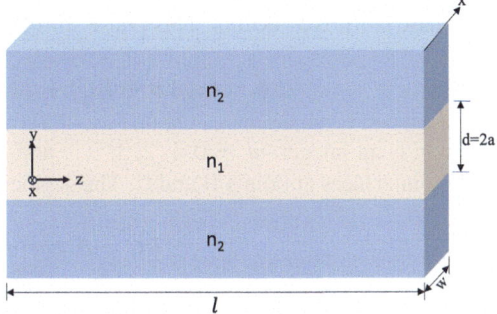

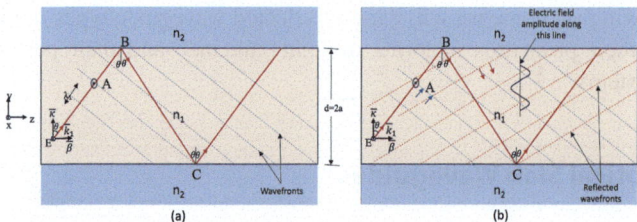

Fig. 7.3 Propagation of a wave within the core of a waveguide

A guided mode can exist only if a transverse resonance condition along the *y-axis* is satisfied, ensuring that the wave, upon multiple reflections, interferes constructively with itself. This resonance condition creates a stable pattern of standing waves within the core, allowing the optical wave to be confined.

To maintain propagation within the waveguide core without significant refraction into the cladding, the wave must satisfy the condition for total internal reflection (*TIR*) condition:

$$\theta \geq \theta_c = sin^{-1}\left(\frac{n_2}{n_1}\right). \tag{7.2}$$

where m is the angle of incidence at the core–cladding boundary, n_1 is the refractive index of the core, and n_2 is the refractive index of the cladding. Waves that satisfy this condition are referred to as waveguide modes, whereas those that do not are classified as substrate radiation modes.

In the core region, the y component of the wave vector is $\kappa = k_1 cos\theta$, for a ray with an angle of incidence ϑ, while the z component is $\beta = k_1 sin\theta$. Here, the wavenumber k_1 in the core of the waveguide is:

$$k_1 = \frac{2\pi}{\lambda_1} = \frac{2n_1\pi}{\lambda}, \tag{7.3}$$

where λ is the wavelength in free space, and λ_1 is the wavelength in the core of the waveguide. As light propagates from point A to point C, reflecting at points B and C, a phase shift, $\Delta\phi$, occurs. This phase shift can be expressed as:

$$\Delta\phi = k_1(AB + BC) + 2\phi(\theta) = m(2\pi),$$

where m is an integer, $m = 0, 1, 2, 3, \ldots$, and $\phi(\theta)$ is the phase shift at the core–blading interfaces at points B and C. The segment BC is calculated as:

$$BC = d/cos(\theta),$$

7.3 Optical Slab Waveguides

where d d is the core thickness. Since triangle ABC is a right triangle, the length AB can be expressed as:

$$AB = BC\cos(2\theta).$$

Thus, the total path length $AB + BC$ is:

$$AB + BC = BC\cos(2\theta) + BC = BC[(2\cos 2\theta - 1) + 1] = 2d\cos\theta.$$

Substituting into the phase shift equation, we get:

$$2dk_1 \cos\theta + 2\phi(\theta) = m(2\pi). \tag{7.4}$$

Since m is an integer, only specific, discrete angle ϑ will satisfy this resonance condition, Eq. (7.4), resulting in discrete values of the propagation constant β_m associated with guided modes labeled by the mode number m. The guided mode with $m = 0$ is known as the fundamental mode, while those with $m \neq 0$ are referred to as high-order modes.

Since guided modes are discrete, only specific values of θ can fulfill the trainee's resonance condition. This guiding mode condition can be rewritten as:

$$dk_1 \cos(\theta_m) + \phi(\theta_m) = m\pi. \tag{7.5}$$

Substituting $k_1 = 2\pi n_1/\lambda$ into this equation, we obtain:

$$\left[\frac{2dn_1}{\lambda}\right]\cos\theta_m + \frac{\phi_m}{\pi} = m. \tag{7.6}$$

Example Derive Eq. (7.4) using the transverse resonance condition for constructive interference.

Solution Consider the transverse component of the wave vector $\overline{\kappa}$, the roundtrip phase shift is:

$$\delta\phi = 2d\kappa + 2\phi(\theta),$$

$$\delta\phi = 2dk_1 \cos(\theta) + 2\phi(\theta).$$

The condition for constructive interference is that the round-trip phase shift must be equal to an integer number of 2π. Therefore,

$$dk_1 \cos(\theta) + \phi(\theta) = m\pi.$$

> **Example** Determine the angle θ_m for the modes $m = 1, 5, 50$ assuming the following: $d = 10\,\mu m$, $\lambda = 1.5\,\mu m$, $n_1 = 1.45$ and $\phi_m = \pi$ for all m.
>
> **Solution** Rearranging Eq. (7.6) to get
>
> $$\theta_m = \cos^{-1}\left[\frac{\lambda(m-1)}{2dn_1}\right].$$
>
> Substituting for the the values of for $m = 1, 5, 20$, the values of the angle θ_m are listed in the Table 7.1:

The phase shift ϕ_m experienced by the electric field as it reflects from the core–cladding interface within a waveguide depends on factors such as the refractive indices of the core and cladding, the angle of incidence, and the polarization of the electric field. The exact phase shift can be determined using the Fresnel equations, detailed in Chap. 3, which dictate how the phase of an electromagnetic wave changes as it reflects off or transmits through a boundary between two different media.

Waves with the electric field pointing in the x-direction correspond to perpendicular or s-polarization. These waves are referred to as transverse electric (TE) because $E_z = 0$, as shown in Fig. 7.4a. In contrast, parallel, or p-polarization no longer involves an exclusively transverse electric field since it includes a component along the z direction. This arrangement is known as transverse magnetic (TM) because $H_z = 0$, as illustrated in Fig. 7.4b.

Table 7.1 Guided mode angle of incidence

Mode number	Angle of incidence
1	90
5	84.0622
20	60.5687

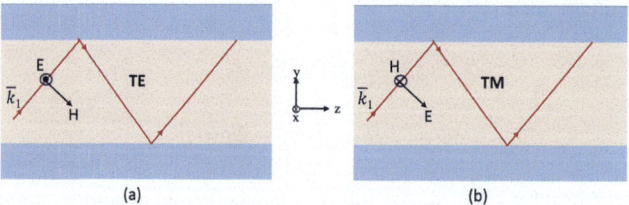

Fig. 7.4 Transverse electric (TE) and transverse magnetic polarization (TM)

7.3 Optical Slab Waveguides

The wave propagation constant k_1 has a transverse component given by:

$$\kappa_m = k_1 \cos\theta_m = \left(\frac{2\pi n_1}{\lambda}\right) \cos\theta_m, \tag{7.7}$$

and a longitudinal component given by:

$$\beta_m = k_1 \sin\theta_m = \left(\frac{2\pi n_1}{\lambda}\right) \sin\theta\theta_m. \tag{7.8}$$

The wave vectors $\overline{k}_1$, $\overline{\kappa}$ are related through the so-called k-vector triangle, shown in Fig. 7.5. The relationship between these propagation vectors is given by:

$$\beta^2 + \kappa^2 = \left(\frac{2\pi n_1}{\lambda}\right)^2. \tag{7.9}$$

The boundary conditions governing the transition of the propagation vectors across the core–cladding interface are also illustrated in Fig. 7.5.

At the boundary, the spatial phase change rate (projection of wavefront propagation) on the waveguide core side must equal that on the waveguide cladding side. This **phase-matching** condition requires:

$$\beta_1 = \beta_2 = \beta, \tag{7.10}$$

which ensures the coupling of the electromagnetic field between two media.

Substituting Eq. (7.8) into Eq. (7.9), we get

$$n_1 k \sin\theta_1 = n_2 k \sin\theta_2,$$

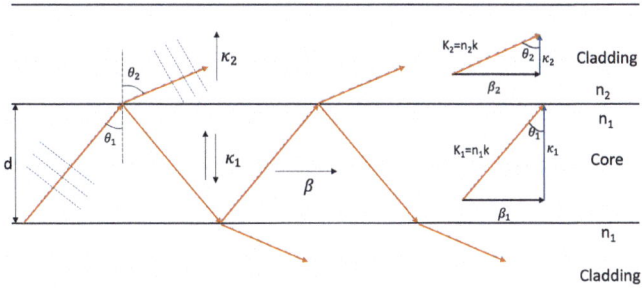

Fig. 7.5 K-vector triangles at the core–cladding interface and the boundary conditions that govern them

which simplifies to Snell's law once the common factor k k is removed from both sides. For total internal reflection (TIR)$\theta_1 > \theta_c$, therefore $sin\theta_1 > n_2/n_1$. Using the k-vector triangle, we obtain:

$$\kappa_2 = \sqrt{(n_2k)^2 - \beta^2} = \sqrt{(n_2k)^2 - (n_1ksin\theta_1)^2},$$

Since $n_1 sin\theta_1 > n_2$, this expression simplifies to:

$$\kappa_2 = \pm jk\sqrt{(n_1 sin\theta_1)^2 - n_2^2} = \pm j\alpha, \tag{7.11}$$

where is defined as:

$$\alpha = k\sqrt{(n_1 sin\theta_1)^2 - n_2^2},$$

and is a real, positive quantity. This α α represents the rate of exponential decay of the evanescent field in the cladding.

7.3.2 Mode Field Profile

A planewave propagating within the waveguide core along a wave vector $\bar{k}_1$, at an angle $\theta_1 > \theta_c$, can be considered to be composed two primary components. The first component is a wave propagating along the z-axis with a wavenumber β, and the second component is wave reflecting back and forth between the two core–cladding interfaces with a transverse wavenumber κ_1, as shown in Fig. 7.5.

Similarly, within the cladding, the planewave consists of two components. The first is a wave propagating along the $z\text{-}axis$ with a wavenumber β, and the second is a transverse wave with a wavenumber κ_2.

The guided modes that reflect within the core create oscillating standing waves with a spatial period along the y-axis given by:

$$\Lambda_y = \frac{2\pi}{\kappa_m} = \frac{\lambda}{n_1 cos\theta_m}. \tag{7.12}$$

The discrete guiding modes propagating along the z-axis in the core with a wavenumber β_m have a period along z given by:

$$\Lambda_z = \frac{2\pi}{\beta_m} = \frac{\lambda}{n_1 sin\theta_m}. \tag{7.13}$$

For TE modes, which involve electric fields oscillating perpendicular to the direction of propagation, the planewaves traveling in the core can be represented by two distinct fields: one directed upward (positive y-direction) and the other downward (negative y-direction). Both fields propagate simultaneously along the

7.3 Optical Slab Waveguides

positive z-direction. Using the notation for planewaves introduced in Chap. 3:

$$E(\bar{r}, t) = E_o e^{j(\omega t - \bar{k} \cdot \bar{r})}, \tag{7.14}$$

we omit the ωt term for simplicity. The vectors $\bar{k}$ and $\bar{r}$ are given by:

$$\bar{k} = +\kappa \bar{a}_y + \beta \bar{a}_z, \tag{7.15}$$

for upward moving field, and

$$\bar{k} = -\kappa \bar{a}_y + \beta \bar{a}_z \tag{7.16}$$

for downward moving field, and

$$\bar{r} = y\bar{a}_y + z\bar{a}_z. \tag{7.17}$$

The total electric field within the core can be written as the sum of these two components:

$$E = E_o \left[e^{-j\kappa y - j\beta z} + e^{j\kappa y - j\beta z} \right]. \tag{7.18}$$

Simplifying, we get:

$$E = 2E_o \cos(\kappa y) e^{-j\beta z}. \tag{7.19}$$

Converting Eq. (7.9) to its real instantaneous form gives:

$$E(y, z, t) = 2E_o \cos(\kappa y) \cos(\omega t - \beta z), \tag{7.20}$$

which represents a planewave propagating in the positive z-direction. For discrete guided modes, this expression becomes:

$$E(y, z, t) = 2E_o \cos(\kappa_m y) \cos(\omega t - \beta_m z). \tag{7.21}$$

The amplitude profile depends on the mode number m. For even modes (i.e., $m = 0, 2, 4, \ldots$), the profile takes the form of $cos(\kappa_m y)$, whereas, for the odd modes (i.e., $m = 1, 3, 5, \ldots$), the profile is represented by $sin(\kappa_m y)$. This pattern becomes clearer when we analyze the waveguide's optical confinement using the rigorous electromagnetic method, which we'll delve into in the subsequent section.

Next, we examine the waves that propagate from the core into the cladding. Although we are operating in the total-internal-reflection condition, light doesn't abruptly vanish at the core–cladding interface. Instead, a small amount of the light, known as then **"evanescent wave,"** penetrates slightly into the cladding. The electric field component that propagates into the cladding can be represented by the following planewave:

$$E(y, z, t) = E_o e^{j(\omega t - \bar{k}_2 \cdot \bar{r})}, \qquad (7.22)$$

where

$$\bar{k}_2 \cdot \bar{r} = \kappa_2 y + \beta z. \qquad (7.23)$$

Applying Eq. (7.11) for the value of κ_2, we have:

$$\bar{k}_2 \cdot \bar{r} = \pm j\alpha y + \beta z, \qquad (7.24)$$

where α is a real quantity defined by:

$$\alpha = k\sqrt{(n_1 \sin\theta_1)^2 - n_2^2}. \qquad (7.25)$$

Substituting Eq. (7.24) into Eq. (7.22), we obtain:

$$E(y, z, t) = E_o e^{j[\omega t - (\pm j\alpha y + \beta z)]},$$

which simplifies to:

$$E(y, z, t) = E_o e^{\pm \alpha y} e^{j(\omega t - \beta z)}. \qquad (7.26)$$

The first exponential represents either a gain, if the sign is positive, or a decay if the sign is negative. As it is physically implausible to have a gain, Eq. (7.26) can be refined to:

$$E(y, z, t) = E_o e^{-\alpha y} e^{j(\omega t - \beta z)}, \qquad (7.27)$$

indicating a planewave that decays exponentially along the y-direction, away from the core–cladding interface. This decaying wave in the cladding is known as the evanescent wave. Consequently, guided modes are confined primarily to the core of the waveguide.

A sketch of the mode profiles for $m = 0, 1, 2$ is shown in Fig. 7.6. It should be noted that within the core, the field amplitudes of these modes intersect the zero value m times.

The waveguide's design confines light primarily within the core due to its higher refractive index compared to the cladding. This leads to total internal reflection at the core–cladding boundary. Each mode experiences an "effective" refractive index, a value between the refractive indices of the core and cladding, which depends on the spatial distribution of the mode field.

7.3 Optical Slab Waveguides

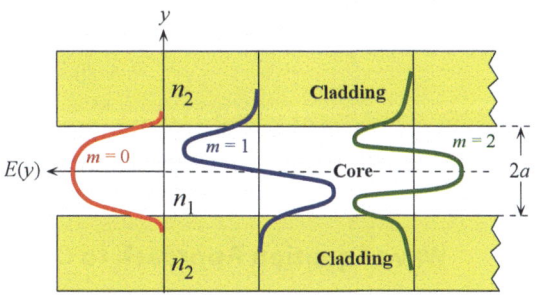

Fig. 7.6 The electric field profile for the first three guided modes for $m = 1, 2$ and 3

7.3.3 Waveguide Effective Refractive Index

The effective refractive index (n_{eff}) of a waveguide is a parameter used to characterize the propagation of light within a waveguide. It reflects the influence of the waveguide's structure and the resulting confinement of light, which together affect the speed at which light propagates along the waveguide.

As discussed in the previous section, light propagating within a waveguide does not only travel in the core but also extends into the cladding, creating an "evanescent" tail. This causes the light to effectively experience a refractive index that is lies between the refractive index of the core and the cladding.

The effective refractive index is calculated from the mode propagation constant (β) by:

$$n_{eff} = \frac{\beta}{k}, \quad (7.28)$$

where k is the free space wavenumber. Substituting the value for β, we get:

$$n_{eff} = \frac{k n_1 sin\theta_1}{k} = n_1 sin\theta_1. \quad (7.29)$$

Since θ_1 determines the specific guided mode, each mode has different effective index. This in turn determines the mode's phase velocity given by:

$$v_{mode} = \frac{c}{(n_{eff})_{mode}}. \quad (7.30)$$

When $\theta_1 = \theta_c$, we have $sin\theta_c = n_2/n_1$, resulting in an effective index $n_{eff} = n_2$, which corresponds the index of the cladding. This is then the effective index of the highest-order guided mode. For the fundamental mode $\theta_1 = 90°$, the effective index $n_{eff} = n_1$. Thus, for other modes, the effective index will satisfy $n_1 > n_{eff} > n_2$. This range signifies that the effective index decreases with increasing mode order. Consequently, the wavelength within the core of the waveguide for modes propagating along the z-direction is:

$$\lambda_z = \lambda/n_{eff}. \tag{7.31}$$

where λ is the free-space wavelength. This variation in effective index among modes allows for a distinct propagation behavior for each mode within the waveguide.

7.4 Wave Equation Approach to Optical Beam Confinement

The wave equation approach to optical beam confinement describes how light waves propagate within optical waveguides, applying fundamental principles of wave optics and leveraging the wave equation to model light behavior.

The propagation of light waves is governed by the electromagnetic wave equation, derived from Maxwell's equations. In an isotropic and homogeneous medium, this wave equation is is in the form of a second-order partial differential equation, as introduced in Chap. 3. The solutions of this equation within a waveguide represent distinct modes, each characterized by a unique spatial pattern and propagation constant. These modes serve as eigenfunctions of the wave equation, with the propagation constants functions as the corresponding eigenvalues.

Consider the waveguide structure shown in Fig. 7.7. In this configuration, The three refractive indices satisfy $n_1 > n_2 > n_3$, and the guiding layer has a thickness d. Two possible electric field polarizations, transverse electric (TE) and transverse magnetic (TM) must be taken into account. The waveguide is oriented along the z-$axis$, with the wave vector $\bar{k}$ of the guided wave making repeated reflections at the interfaces at angles exceeding the critical angle, thus remaining confined within the waveguide. The electric field may be oriented either transversely (TE) or magnetically (TM) to the direction of propagation, as depicted in Fig. 7.4.

For the TE case, the electric field E is polarized along the x-$axis$ (perpendicular of the page of Fig. 7.4.) Let the waveguide be excited by a source with frequency ω_o, and a vacuum wave vector of magnitude $k_o = \omega_o/c$. To determine the allowed modes of the waveguide, we first solve the Helmholtz wave equation in each dielectric region, then apply the boundary conditions to connect these solutions

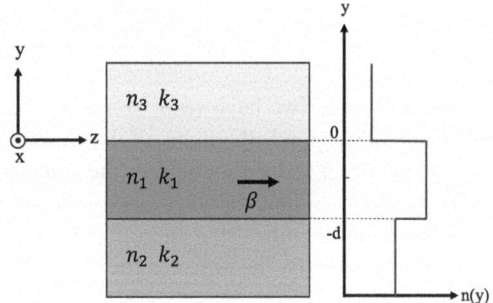

Fig. 7.7 A cross section of an asymmetric slab waveguide

7.4 Wave Equation Approach to Optical Beam Confinement

across the interfaces. For a sinusoidal wave with angular frequency ω_o, the wave equation for the electric field components in each region simplifies in the scalar form:

$$\nabla^2 E_x + k_o^2 n_i^2 = 0 \tag{7.32}$$

where $n_i = n_1, n_2 \text{ or } n_3$, depending on the region of the waveguide. The electric field in the waveguide $E_x(y, z)$ is a function of both y and z, but since the slab extends infinitely in the x direction, E_x is independent of x. Due to the translational invariance of the structure in the z-$axis$ n, we do not expect the amplitude to vary along the z-axis, but we do expect that the phase varies since the wave is propagating along the z-axis. We assume a solution to Eq. (7.32) in the form

$$E_x(y, z) = E_x(y) e^{-j\beta z} \tag{7.33}$$

where β is a propagation coefficient along the z-$axis$. Substituting this solution into Eq. (7.32), and noting that the electric field along the x-$axis$ is constant, i.e. $dE_x/dx = 0$ we get

$$\frac{d^2 E_x(y)}{dy^2} + \left(k_o^2 n_i^2 - \beta^2\right) E_x(y) = 0 \tag{7.34}$$

This is an ordinary second-order differential equation with a solution in the form of

$$E_x(y) = E_o e^{-j\left(k_o^2 n_i^2 - \beta^2\right)^{1/2} y}, \tag{7.35}$$

where E_o is the field amplitude at $y = 0$. The solution depends on the factor $\left(k_o^2 n_i^2 - \beta^2\right)$. If it is a positive quantity, then the solution will be harmonic, sinusoidal, and if it is negative, then the solution will be decaying. The general solution depends on the relative magnitude of β with respect to $k_o n_i$. Consider the case where $\beta > k_o n_i$. The solution, Eq. (7.35), will have a real exponential form

$$E_x(y) = E_o e^{\pm\sqrt{\beta^2 - k_o^2 n_i^2}\, y} \quad for\ \beta > k_o n_i \tag{7.36}$$

Since the waveguide is not a gain medium, the exponent in the solution is chosen to be negative, resulting in a decaying electric field. This decay characterizes an evanescent field, which occurs the condition for total internal reflection (TIR). The attenuation coefficient, defining this decay rate, is given by:

$$\alpha_i = \sqrt{\beta^2 - k_o^2 n_i^2}. \tag{7.37}$$

Therefore, the wave equation solution in the decaying region becomes:

$$E_x(y) = E_o e^{-\alpha y} \text{ for } \beta > k_o n_i, \text{ and } y > 0, \tag{7.38}$$

and

$$E_x(y) = E_o e^{+\alpha y} \text{ for } \beta > k_o n_i, \text{ and } y < -d. \tag{7.39}$$

Now, consider the case where $\beta < k_o n_i$ and by using the *k-vector* triangle shown in Fig. 7.5, the transverse wave vector is defined as:

$$\kappa_i = \sqrt{k_o^2 n_i^2 - \beta^2}. \tag{7.40}$$

The solution, Eq. (7.35), has a real oscillatory form:

$$E_x(y) = E_o e^{-j\kappa y}, \tag{7.41}$$

and

$$E_x(y) = E_o e^{\pm j\kappa_i y} \text{ for } \beta < k_o n_i \text{ and } -d < y < 0. \tag{7.42}$$

Here, β and κ represent the longitudinal and transverse wavevectors, respectively, within the guiding core, and these terms characterize different guiding modes. Rearranging Eq. (7.28) for the effective refractive index (n_{eff}), we obtain

$$\beta = k_o n_{eff} = k_o n_1 sin\theta_1 \tag{7.43}$$

Wave Equation Solution for the Waveguide-Guided Modes Let us consider the solution conditions in the core of the waveguide by substituting Eq. (7.43) into Eq. (7.40) to get

$$\kappa_1 = \sqrt{k_o^2 n_1^2 - k_o^2 n_1^2 sin^2\theta_1} = k_o n_1 cos\theta_1 \tag{7.44}$$

Therefore, the transverse wavevector (κ_1) is real and positive value. Hence, the solution to the wave equation in the core of the waveguide is given by Eq. (7.42). Therefore, the general solution within the waveguide core is expressed as:

$$E_x(y) = Bcos(\kappa_1 y) + Csin(\kappa_1 y), \quad -d < y < 0 \tag{7.45}$$

where B and C are constants to be determined using the boundary conditions at the interfaces between the core and cladding of the waveguide.

Wave Equation Solution in the Cladding of the Waveguide In the cladding regions, the solution condition satisfies Eq. (7.37). Therefore, the wave equation

7.4 Wave Equation Approach to Optical Beam Confinement

solution in the upper and lower cladding regions then takes the form:

$$E_x(y) = Ae^{-\alpha_3 y} \qquad y > 0, \tag{7.46}$$

and

$$E_x(y) = De^{+\alpha_2(y+d)} \qquad y < -d \tag{7.47}$$

Similarly, the constant coefficients A and D are determined using the boundary conditions. Inside the core, the electric field exhibits an oscillatory solution given by Eq. (7.45) for all values of the incidence angle θ_1. As the refractive indices satisfy $n_1 > n_2 > n_3$ increasing angle θ_1 will eventually reach the critical angle, resulting in total internal reflection (TIR) at the core–upper cladding interface. At this point, the field becomes evanescent in the upper cladding, while remaining oscillatory in the core and substrate regions. This configuration is known as a substrate mode. When θ_1 reaches the critical angle for both interfaces, the field is oscillatory within the core and decays in both cladding regions. These solutions correspond to the guided modes of the waveguide core, as illustrated in Fig. 7.8.

In summary, for a confined guided wave, β must satisfy the condition:

$$k_o n_2 < \beta < k_o n_1. \tag{7.48}$$

7.4.1 Eigenvalues of the Slab Waveguide

To find the values of β that lead to allowed solutions to the wave equation, we apply the boundary conditions to the general solutions developed in Eqs. (7.45) to (7.47). Since $\beta = k_o n_1 \sin\theta_1$ in the waveguide core we expect β to vary between $k_o n_2$ and $k_o n_1$ to satisfy the condition in Eq. (7.48). The remainder of this section details how to find the allowed values of β by applying the boundary conditions at the core–cladding interfaces. The boundary conditions that connect the solutions at the

Fig. 7.8 Sketch of the electric field distribution as a function of β for a slab waveguide

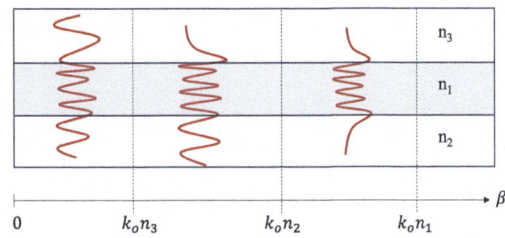

interfaces are:

1. Continuity of the tangential component of the electric field (E).
2. Continuity of the tangential component of the electric field (H).

The continuity of the normal components of D and B is inherently satisfied when the tangential conditions are met. Since E_x, is transverse to the interface, the first boundary condition is straightforward to apply. If we assume that the fields are harmonic, then we can express the magnetic intensity in terms of the electric intensity and derive a simple boundary condition for the magnetic field. From Maxwell equations, we have:

$$\nabla \times \overline{E} = -\frac{\partial \overline{B}}{\partial t}. \tag{7.49}$$

For a harmonic magnetic field:

$$\overline{B} = \mu \overline{H} = \mu \overline{H}_o e^{j\omega t}. \tag{7.50}$$

Thus,

$$\nabla \times \overline{E}(t) = -j\mu\omega \overline{H}(t) \tag{7.51}$$

Focusing on the tangential component of the magnetic field (H_z) we expand the curl in terms of its z-component:

$$\left(\frac{\partial E_y}{\partial x} - \frac{\partial E_x}{\partial y}\right) = -j\mu\omega H_z \tag{7.52}$$

Since there is no E_y component to the electric field, then Eq. (7.52) is reduced to

$$\frac{\partial E_x}{\partial y} = j\mu\omega H_z. \tag{7.53}$$

Thus,

$$H_z = -\frac{j}{\mu\omega}\frac{\partial E_x}{\partial y} \tag{7.54}$$

Applying the first boundary condition at the interface at $y = 0$, we get:

$$Ae^{-\alpha_3 0} = B\cos(\kappa_1 0) + C\sin(\kappa_1 0)$$

This can be satisfied only if:

$$A = B. \tag{7.55}$$

7.4 Wave Equation Approach to Optical Beam Confinement

Ensuring continuity of the magnetic field at $y = 0$ requires the first derivative of the electric field to be continuous at $y = 0$. Taking the derivative and applying the continuity condition for the magnetic field, we obtain:

$$-\alpha_3 A e^{-\alpha_3 0} = -B\kappa_1 sin(\kappa_1 0) + C\kappa_1 cos(\kappa_1 0)$$

which simplifies to:

$$-\alpha_3 A = C\kappa_1,$$

yielding

$$C = -\frac{\alpha_3}{\kappa_1} A. \tag{7.56}$$

Applying boundary condition to the interface at $y = -d$, we get:

$$D e^{-\alpha_2(-d+d)} = A\left[cos(-\kappa_1 d) - \frac{\alpha_3}{\kappa_1} sin(-\kappa_1 d)\right]. \tag{7.57}$$

Noting that $cos(-\theta) = cos(\theta)$ and $sin(-\theta) = -sin(\theta)$, Eq. (7.57) is reduced to:

$$D = A\left[cos(\kappa_1 d) + \frac{\alpha_3}{\kappa_1} sin(\kappa_1 d)\right]. \tag{7.58}$$

Thus, we we were able to express the coefficients C and D in terms of the coefficient A. We can now rewrite the solutions for E_x, in the core and cladding regions, using these coefficients:

$$E_x(y) = \begin{cases} A e^{-\alpha_3 y} & y > 0 \\ A cos(\kappa_1 y) - \frac{\alpha_3}{\kappa_1} A sin(\kappa_1 y) & -d < y < 0 \\ A\left[cos(\kappa_1 d) + \frac{\alpha_3}{\kappa_1} sin(\kappa_1 d)\right] e^{-\alpha_2(y+d)} & y < -d \end{cases} \tag{7.59}$$

Finally, applying the boundary condition of $\partial E_x/\partial x$ across the core–substrate interface at $y = -d$, we obtain:

$$A\left[\kappa_1 sin(\kappa_1 d) - \alpha_3 cos(\kappa_1 yd)\right] = A\alpha_2 \left[cos(\kappa_1 d) + \frac{\alpha_3}{\kappa_1} sin(\kappa_1 d)\right] \tag{7.60}$$

The term on the left of Eq. (7.60) is the core term, while the term on the right is the substrate term. Dividing both sides of the equation by $A cos(\kappa_1 d)$ gives:

$$\kappa_1 tan(\kappa_1 d) - \alpha_3 = \alpha_2 + \frac{\alpha_2 \alpha_3}{\kappa_1} tan(\kappa_1 d) \tag{7.61}$$

Therefore,

$$tan(\kappa_1 d) = \frac{\alpha_2 + \alpha_3}{\kappa_1 \left(1 - \frac{\alpha_2 \alpha_3}{\kappa_1^2}\right)}. \tag{7.62}$$

This equation is a transcendental equation. It is known as the characteristic equation for TE modes in the slab waveguide. This equation does not have a closed-form solution and must be solved either numerically or graphically.

Example Solution for a Specific Waveguide Structure Consider the solution of the transcendental equation for a specific waveguide structure, such that $n_1 = 1.5, n_2 = 1.45, n_3 = 1.4, d = 5$ µm, and $\lambda = 1$ µm. We can use κ_1 as the variable for plotting the transcendental equation. The remaining of the parameters can be written in terms of κ as follows:

$$\beta = \sqrt{k_o^2(1.5)^2 - \kappa_1^2}, \tag{7.63}$$

$$\alpha_2 = \sqrt{\beta^2 - k_o^2(1.45)^2}, \tag{7.64}$$

$$\alpha_3 = \sqrt{\beta^2 - k_o^2(1.4)^2}. \tag{7.65}$$

The graphical solution is shown in Fig. 7.9. The intersections between the right-hand-side (RHS) and left-hand-side (LHS) of the transcendental equation, Eq. (7.62), on the figure represent the allowed solutions for the waveguide parameters. These are the allowed guided modes within the range of κ_1. From the k-vector triangle, we have:

$$\kappa_1 = n_1 k_o cos\theta_1,$$

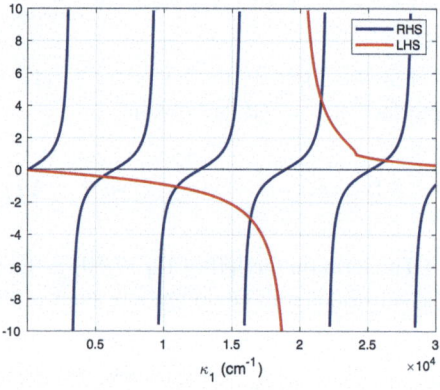

Fig. 7.9 Graphical solution for the transcendental equation of asymmetric slab waveguide

7.4 Wave Equation Approach to Optical Beam Confinement

so the allowed values of κ_1 corresponding to indicates to the angles of incidence for the guided mode wavevectors.

The approximate values of κ_1 for the guided modes, along with the corresponding β values and angles of incidence, are presented in the following table:

κ_1	β	θ_1
5541	94,085	86.63
10,893	93,616	83.36
16,375	92,814	80
21,567	91,747	76.77
26,248	90,519	76.83

These are only the lowest-order modes of the waveguide.

This example shows that the values of β are discrete, meaning there are only a limited number of guided modes in the waveguide.

7.4.2 The Symmetric Waveguide

In the symmetric waveguide where both the top and bottom claddings have the same refractive index n_2, while the core has a higher refractive index n_1. By setting the center of the y-axis at the middle of the core, the core extends between $-d/2 \leq y \leq d/2$. Following a similar approach to that in the previous section, we derive the electric field $E_x(y)$ for TE modes. Due to the symmetry, the field distribution is centered around $y = 0$. The electric field for TE guided modes in a symmetric waveguide is given by:

$$E_x(y) = \begin{cases} Ae^{-\alpha(y-d/2)} & for \ y \geq d/2 \\ A\frac{cos(\kappa y)}{cos(\kappa d/2)} \ or \ A\frac{sin(\kappa y)}{sin(\kappa d/2)} & for \ -d/2 \leq y \leq d/2 \\ \pm Ae^{-\alpha(y+d/2)} & for \ y \pm -d/2 \end{cases} \quad (7.66)$$

There are two choices for the description of the field in the guiding layer, depending on whether a symmetric (cosine) for even, or antisymmetric (sine) for odd modes, depending on the mode excitation. This distinction into even and odd modes arises naturally from the waveguide's symmetrical refractive index profile

To find the characteristic equation for the symmetric waveguide, we apply the boundary condition for $\partial E_x(y)/\partial y$ at $y = d/2$, which yields:

1. For even modes (symmetric), we get:

$$-\alpha = -\kappa tan(\kappa d/2),$$

results in:
$$tan(\kappa d/2) = \frac{\alpha}{\kappa}. \tag{7.67}$$

2. For odd modes (symmetric), we get:
$$-\alpha = \kappa \frac{1}{tan(kd/2)},$$

results in:
$$tan(\kappa d/2) = -\frac{\kappa}{\alpha}. \tag{7.68}$$

Thus, the characteristic equations for the TE modes in a symmetric slab waveguide are given by:

$$tan(\kappa d/2) = \begin{cases} \frac{\alpha}{\kappa} & for\ even\ modes \\ -\frac{\kappa}{\alpha} & for\ odd\ modes \end{cases} \tag{7.69}$$

Similarly, the characteristic equations for the TM modes, which account for the difference in polarization, are given by:

$$tan(\kappa d/2) = \begin{cases} \left(\frac{n_1}{n_2}\right)^2 \frac{\alpha}{\kappa} & for\ even\ modes \\ -\left(\frac{n_1}{n_2}\right)^2 \frac{\kappa}{\alpha} & for\ odd\ modes \end{cases} \tag{7.70}$$

Example Determine the number of TE guided modes for a symmetric slab waveguide with the following parameters:
$\lambda = 1550\,\text{nm}, n_1 = 1.5, n_2 = 1.45\ and\ d = 2,\ and\ 10\,\mu\text{m}$.

Solution To find the number of TE guided modes, we analyze the characteristic equation Eq. (7.65) by plotting the curves for the LHS and RHS in terms of κ, the transverse wave vector. The necessary expressions to plot these curves are:
$$\beta = \sqrt{k_o^2 n_1^2 - \kappa^2}, \tag{7.71}$$

and
$$\alpha = \sqrt{\beta^2 - k_o^2 n_2^2} = \sqrt{k_o^2(n_1^2 - n_2^2) - \kappa^2}. \tag{7.72}$$

(continued)

7.4 Wave Equation Approach to Optical Beam Confinement

To identify the eigenvalues of the TE modes, we graphically solve Eq. (7.69). In this graphical approach, we plot the functions $tan(\kappa d/2)$, α/κ, and $-\kappa/\alpha$ as functions of κ. These are plotted as functions of κ for Fig. 7.10a shows these curves for $d = 2\,\mu m$:

- The top curve, α/κ α/χ, corresponds to even modes and starts at $+\infty+\infty$, eventually decreasing to zero.
- The function $tan(\kappa d/2)$ starts at zero and increases; it intersects with α/κ, ensuring at least one mode is present, regardless of waveguide thickness.
- The lower curve $-\kappa/\alpha$, associated with odd modes, starts at zero and decreases rapidly with increasing κ.

For $d = 2\,\mu m$, we observe only one intersection between α/κ and $tan(\kappa d/2)$, indicating that only one even mode is allowed in this waveguide.

When the waveguide thickness increases, more modes are supported. Figure 7.10b shows the plot for $d = 10\,\mu m$:

- As κ increases, the modes alternate between even and odd structures.
- For $d = 10\,\mu m$, there are three even modes and two odd modes, totaling five guided modes in the waveguide.

7.4.2.1 Summary of Guided Mode Properties

The analysis and previous example reveal several properties of guided modes:

1. Only a limited number of guided modes exist, each with district and discrete value of the traverse propagation vector β representing the eigenvalues of the electric field solutions across the waveguides.

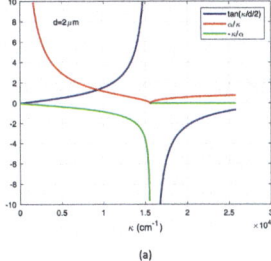
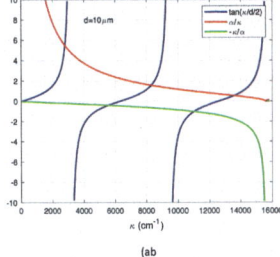

Fig. 7.10 Solution for Eq. (7.69) for a systemic slab waveguide where the core thickness (**a**) $d = 2\,\mu m$ and (**b**) $d = 10\,\mu m$

2. Modes with β values different from those of guided modes are not confined within the waveguide and are known as radiation modes, which generally are not guided.
3. Each guided mode has a specific electric field distribution in both the core and cladding regions of the waveguide, reflecting the mode's unique structure.

7.4.3 Normalized Waveguide Parameters

The properties of waveguide modes are often characterized by dimensionless parameters that simplify analysis and allow comparisons between different waveguide structures. Three essential normalized parameters for a step-index planar waveguide are the **normalized frequency** (also called the **V-number**), the **numerical aperture,** and the **normalized guide index**.

1. Normalized Frequency (V-Number)

    The V-number for a waveguide, which relates the wavelength, core thickness, and refractive indices, is defined as:

$$V = \frac{2\pi}{\lambda} d \sqrt{n_1^2 - n_2^2}, \tag{7.73}$$

   where d is the thickness of the waveguide core, λ is the wavelength of light, n_1 and n_2 are the refractive indices of the core and cladding, respectively.

    The *V-number* is a key parameter that determines the number of modes that can be supported by the waveguide. When $V < \pi$, the waveguide typically supports only a single mode.
2. The Numerical Aperture

    The numerical aperture of the waveguide (NA) is given by:

$$NA = \sqrt{n_1^2 - n_2^2}$$

    The NA quantifies the light acceptance angle of the waveguide and is proportional to the V-number.
3. Normalized Guide-Index (b)

    The propagation constant β can be represented by the **normalized guide index** b, defined as:

$$b = \frac{\beta^2 - k_1^2}{k_1^2 - k_2^2} = \frac{n_{eff}^2 - n_2^2}{n_1^2 - n_2^2} \tag{7.74}$$

   where n_{eff} is the effective refractive index of the waveguide associated with the propagation constant β, and k_1 and k_2 are the wavenumbers in the core and cladding, respectively.

7.4 Wave Equation Approach to Optical Beam Confinement

The parameter b ranges from 0 to 1 and indicates the confinement of the mode within the core: $b = 1$ corresponds to a strongly guided mode, while $b = 0$ represents a mode with minimal confinement.

4. **The Asymmetry Factor (a)**

The asymmetry factor a a quantifies the effect of waveguide asymmetry and varies depending on the polarization of the mode. For an asymmetric waveguide, the asymmetry factor is defined as:

$$a = \begin{cases} a_E = \frac{n_2^2 - n_3^2}{n_1^2 - n_2^2} & for\ TE\ modes \\ a_M = \frac{n_1^4}{n_3^4} \frac{n_2^2 - n_3^2}{n_1^2 - n_2^2} & for\ TM\ modes \end{cases} \quad (7.75)$$

where n_3 is the refractive index of the bottom cladding. Notably, $a_M > a_E$ for a given asymmetric waveguide, reflecting a stronger asymmetry effect on TM modes. For a symmetric waveguide (where $n3 = n2$), $a_E = a_M = 0$, as the asymmetry factor becomes negligible.

7.4.4 Modal Dispersion

Different modes generally have different propagation constants and hence different propagation speeds. This variation in mode speed causes **modal dispersion**, which can result in the spreading of light pulses as they travel along the waveguide.

The transverse resonance condition states that the waveguide acts like a standing wave cavity in the transverse direction. In order to be resonant, the round-trip phase of a transverse component of k must add up to an integer number of 2π. The resonance condition given in Eq. (7.5) is modified here for an asymmetric waveguide to become

$$2kn_1 d \cos\theta - \Phi_1 - \Phi_2 = 2\pi m \quad (7.76)$$

where $m = 0, 1, 2, \ldots$ is the mode order, Φ_1 is the reflection phase shift between the core and substrate, and Φ_2 is the reflection phase shift between the core and cover of the waveguide. Substituting the normalized parameters into this equation yields the normalized dispersion relation

$$V\sqrt{1-b} = m\pi + atan\left(\sqrt{b/(1-b)}\right) + atan\left(\sqrt{(b+a)/(1-b)}\right) \quad (7.77)$$

This dispersion equation can be numerically calculated for $m = 0, 1, 2$ and $a = 0, 5$. Figure 7.11 illustrates these calculations, showing how the V-number determines the cutoff values for each mode. The cutoff values correspond to the minimum V-number required for a particular mode to propagate, with higher-order modes requiring larger V-numbers for guidance.

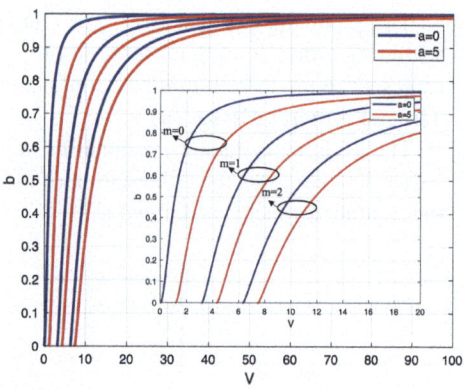

Fig. 7.11 The normalized index b as a function of the normalized frequency V for three mode orders m and two asymmetry coefficients $a = 0$ and $a = 5$. The inset provides a detailed view of the plots in the range $0 \leq V \leq 20$

7.4.5 Waveguide Cutoff Conditions

Each mode in a waveguide has a minimum frequency (or equivalently, a maximum wavelength) above which it can be guided. This limit is known as the **cutoff condition**. Below this frequency, the mode becomes a radiation mode and is not confined to the core.

The cutoff frequency for each mode occurs when $b = 0$. The value of V at this point represents the longest wavelength that the mode can support. In other words, the intersection of each line for a mode number m with a specific asymmetry factor a a indicates the normalized frequency of the longest wavelength that will propagate in that mode. From Fig. 7.11, we observe that the lowest-order symmetric mode ($m = 0$) reaches $b = 0$ at $V = 0$, implying that this mode is always guided and does not have a cutoff.

To generalize, setting $b = 0$ in the normalized dispersion equation (Eq. (7.77)) and solving for V gives the cutoff condition for all modes in a slab waveguide:

$$V_c = m\pi + atna(\sqrt{a}) \tag{7.78}$$

where V_c is the critical V-number for each mode to be guided. In symmetric slab waveguides, where the asymmetry parameter $a = 0$, this reduces to:

$$V_c = m\pi.$$

Thus, the cutoff condition in terms of the V-number becomes:

$$V_c = \frac{2\pi d}{\lambda_c}\sqrt{n_1^2 - n_2^2} = m\pi. \tag{7.79}$$

Rearranging to find the cutoff wavelength λ_c, we get:

$$\lambda_c = \frac{2d}{m}\sqrt{n_1^2 - n_2^2}. \tag{7.80}$$

7.4 Wave Equation Approach to Optical Beam Confinement

Example Determine the cutoff wavelength for the lowest-order modes in a symmetric slab waveguide with the following parameters:

$$d = 3\,\mu m,\ n_1 = 1.5,\ and\ n_2 = 1.45.$$

Solution

1. For $m = 0$, the cutoff wavelength $\lambda_c = \infty$, meaning there is no cutoff wavelength for this mode.
2. For $m = 1$, the cutoff wavelength $\lambda_c = 2.3\,\mu m$.
3. For $m = 2$, the cutoff wavelength $\lambda_c = 1.15\,\mu m$.

7.4.6 Number of Guided Modes

A slab waveguide supports a specific number of guided modes, determined by factors such as the waveguide's size, wavelength, and refractive index difference between core and cladding. The number of guided modes increases with a higher V-number. In terms of ray optics, the highest-order mode that can be supported corresponds to a wave vector $\bar{k}$ with an incident angle at the critical angle, while the lowest-order mode has a wave vector nearly aligned with the waveguide axis. This relationship can be expressed in terms of the propagation constant β:

$$\beta_{lowest\text{-}order} \approx k_o n_1 \tag{7.81}$$

$$\beta_{highest\text{-}order} \approx k_o n_1 cos\theta_{critical} \approx k_o n_3. \tag{7.82}$$

where k_o is the free-space wavenumber, n_1 is the refractive index of the core, and n_2 is the refractive index of the cladding.

All other modes have to satisfy the the previously derived transcendental equation (Eq. (7.62)) and can be visualized in Fig. 7.11. The approximate number of modes M in terms of the V-number is given by:

$$M \approx \frac{V}{\pi} + \frac{atna(\sqrt{a})}{\pi} \tag{7.83}$$

7.4.7 Orthogonality and Degeneracy of Guided Modes

In an optical waveguide, guided modes are the distinct solutions of the wave equation that propagate along the waveguide with unique transverse field distribution and a propagation constant. These modes are represented by their electric field E_n and magnetic field H_n distributions for the nth mode.

Two modes m and n are said to be orthogonal if the integral of the product of their field components over the waveguide cross section is zero when $m \neq n$. Mathematically, this can be expressed as:

$$\int\int_S \overline{E}_m(x,y) \cdot \overline{E}_n^*(x,y) dx dy = 0 \quad for\ m \neq n$$

where S is the cross-sectional area of the waveguide. For a lossless waveguide, the orthogonality condition for two guided modes can be derived from Maxwell's equations. The orthogonality of the modes ensures that the power carried by different modes does not interfere with each other.

Consider two guided modes with electric fields $\overline{E}_m$ and $\overline{E}_n$ and corresponding propagation constants β_m and β_n. The orthogonality condition can be expressed as:

$$\int\int_S \overline{E}_m(x,y) \times \overline{H}_n^*(x,y) \cdot \overline{a}_z dx dy = \delta_{mn}$$

where $\overline{H}_n$ is the magnetic field of the nth mode, and δ_{mn} is the Kronecker delta function.

Orthogonal modes do not exchange power. This means that the power carried by one mode remains isolated from the power carried by another mode, preventing cross talk between modes. Multiple modes can propagate simultaneously within the same waveguide without interfering with each other, allowing for higher data capacity in communication systems. The orthogonality property simplifies the analysis and design of waveguide structures since each mode can be treated independently when computing the total field distribution. Physically, orthogonality implies that each mode represents a unique and independent solution to the wave equation in the waveguide. This is analogous to the orthogonality of eigenfunctions in quantum mechanics.

In the context of optical waveguides, guided mode degeneracy refers to the phenomenon where two or more modes have the same propagation constant (or eigenvalue) but different field distributions. This degeneracy can occur in various types of waveguides, such as fibers and planar waveguides, and has important implications for the behavior and analysis of the waveguide.

Modes are considered degenerate if they have the same propagation constant β but different transverse field distributions. This means that while they travel with the same phase velocity along the waveguide, their spatial field patterns in the cross-sectional plane are distinct. In waveguides with symmetric structures, modes with orthogonal polarization states (e.g., TE and TM modes) can have the same propagation constant. For example, in a symmetric planar waveguide, the TE_0 and TM_0 modes can be degenerate. A good example of such a degeneracy is the fundamental mode in a circular dielectric fiber. The mode can have two different

7.5 Coupled Mode Theory

electric field polarizations, E_x and E_y, respectively, each of which has the same spatial energy distribution.

7.5 Coupled Mode Theory

In the previous section, we have found that the solutions of the wave equation in slab waveguides result in guided mode solutions. Each mode is a solution to an eigenvalue problem. These guided modes propagate inside the core of the waveguide independently. There is no transfer of energy from one mode to the other.

Coupled mode theory deals with the coupling of spatial distribution modes of different spatial distribution or polarization. Let us represent fundamental eigenmodes as follows:

$$E_a(r) = E_a(x, y)e^{-j\beta_a} \tag{7.84}$$

and

$$H_a(r) = H_a(x, y)e^{-j\beta_a} \tag{7.85}$$

where subscript a is the mode index. When two waveguides are in close proximity, they become coupled. The pair forms supermodes. When two waveguides are in close proximity, they become coupled and exchange power as a function of z. Very often; this leads to a periodic exchange of power between the waveguides.

To simplify the analysis, it will be assumed that the supermodes can be represented as a weighted sum of the individual guided modes. This implies that the modes do not change at all with the introduction of the second guide. In reality, the modes are deformed slightly, but are still coupled.

Let the electric and magnetic fields of guided modes in the first waveguide to be:

$$\overline{E}_1 = \overline{E}_{o,1}(x, y)e^{-j\beta_1 z} \tag{7.86}$$

$$\overline{H}_1 = \overline{H}_{o,1}(x, y)e^{-j\beta_1 z} \tag{7.87}$$

Let the electric and magnetic second of guided modes in the first waveguide be:

$$\overline{E}_2 = \overline{E}_{o,2}(x, y)e^{-j\beta_2 z} \tag{7.88}$$

$$\overline{H}_2 = \overline{H}_{o,2}(x, y)e^{-j\beta_2 z} \tag{7.89}$$

To simplify the analysis, it will be assumed that the supermodes can be represented as a weighted sum of the individual guided modes. This implies that the modes do not change at all with the introduction of the second guide. In reality, the modes are deformed slightly, but are still coupled:

$$\overline{E} = A(z)\overline{E}_1 + B(z)\overline{E}_2, \tag{7.90}$$

$$\overline{H} = A(z)\overline{H}_1 + B(z)\overline{H}_2, \tag{7.91}$$

where A(z) is the amplitude of the first mode, and B(z) is the amplitude of the second mode. When two waveguides are in close proximity, they become coupled and exchange power as a function of z. Assume that Eqs. (7.90) and (7.91) are solutions of Maxwell wave equation. Maxwell curl equations are given by

$$\nabla \times \overline{E} = -j\omega\mu\overline{H}, \tag{7.92}$$

and

$$\nabla \times \overline{H} = j\omega\epsilon\overline{H}. \tag{7.93}$$

To derive the coupled mode equations, we substitute the field expressions into Maxwell's equations and apply the slowly varying envelope approximation (SVEA).

Slowly Varying Envelope Approximation The Slowly Varying Envelope Approximation (SVEA) is an important concept in the analysis of wave propagation, especially in the context of optical waveguides. The approximation simplifies the mathematical treatment of waves whose amplitude varies slowly compared to the wavelength. This allows us to focus on the changes in the amplitude and phase of the wave over longer distances without dealing with the rapid oscillations of the carrier wave. We have applied this technique in Chap. 3 in the derivation of the Paraxial Wave Equation.

The envelope $A(z, t)$ and $B(z, t)$ change slowly with respect to the spatial coordinate z. This implies:

$$\frac{\partial A}{\partial z} \ll k|A| \tag{7.94}$$

and therefore,

$$\frac{\partial^2 A}{\partial z^2} \ll k^2|A|. \tag{7.95}$$

This leads to neglecting second-order derivatives of the amplitudes $A(z)$ and $B(z)$ with respect to z. In the context of coupled mode theory for optical waveguides, the SVEA is used to derive the coupled mode equations.

7.5 Coupled Mode Theory

Let the electric fields in two parallel waveguides be expressed as:

$$E_1(x, y, z, t) = A(z)\psi_1(x, y)e^{-i(\beta_1 z - \omega t)} \tag{7.96}$$

$$E_2(x, y, z, t) = B(z)\psi_2(x, y)e^{-i(\beta_1 z - \omega t)} \tag{7.97}$$

where $A(z)$ and $B(z)$ are the slowly varying amplitudes of the fields in waveguide 1 and waveguide 2, respectively, and $\psi_1(x, y)$ and $\psi_2(x, y)$ are the mode profiles.

To derive the coupled mode equations, we substitute the field expressions into Maxwell's equations and apply the slowly varying envelope approximation. This involves neglecting second-order derivatives of the amplitudes $A(z)$ and $B(z)$ with respect to z z.

The wave equations for the electric fields in the presence of coupling are:

$$\frac{d^2 E_1}{dz^2} + k_0^2 n_1^2 E_1 = -\frac{d^2 E_2}{dz^2} + k_0^2 n_{12}^2 E_2 \tag{7.98}$$

$$\frac{d^2 E_2}{dz^2} + k_0^2 n_2^2 E_1 = -\frac{d^2 E_1}{dz^2} + k_0^2 n_1^2 E_1 \tag{7.99}$$

where and n_1 and n_2 are the refractive indices of the waveguides.

To isolate the coupling terms, we neglect the second-order derivative of A and B and multiply by the conjugate mode profiles and integrate over the cross-sectional area of the waveguides. This yields the generalized coupled mode equations:

$$\frac{dA}{dz} = j\beta_1 A - j\kappa_{12} B \tag{7.100}$$

$$\frac{dB}{dz} = j\beta_1 B - \kappa_{12} A \tag{7.101}$$

where κ_{12} and κ_{21} are the coupling coefficients between the waveguides. These coefficients are generally complex and describe the strength and phase of the coupling.

$$\kappa_{12} = \frac{\omega \epsilon}{2} \int_S \psi_1^* \psi_2 \, dx dy \tag{7.102}$$

$$\kappa_{21} = \frac{\omega \epsilon}{2} \int_S \psi_2^* \psi_1 \, dx dy \tag{7.103}$$

These integrals account for the overlap of the mode fields and the dielectric perturbation that causes the coupling. The generalized coupled mode equations describe how the amplitudes $A(z)$ and $B(z)$ of the modes in the two waveguides evolve along the propagation direction z z due to coupling.

7.6 Optical Codirectional Couplers

Optical codirectional couplers are essential components in integrated photonic circuits and fiber optic communication systems. They are designed to facilitate the transfer of optical power between two or more waveguides that are placed close to each other, allowing for efficient signal processing and routing.

An optical codirectional coupler typically consists of two parallel waveguides positioned close enough so that the evanescent fields of the modes in one waveguide overlap with those in the adjacent waveguide. This proximity allows the optical power to transfer between the waveguides through evanescent coupling. The key principle underlying the operation of these couplers is coupled mode theory, developed in the previous section, which describes how the modes in the coupled waveguides interact.

Consider a two-mode coupling due to a uniform perturbation as in a two-channel directional coupler, which consists of two parallel single-mode waveguides, as shown schematically in Fig. 7.12. This is the case of multiple structure coupling. If the two waveguides are not identical, the directional coupler is not symmetric. Then, in general $\kappa_{21} \neq \kappa_{12}$. In general. If the two waveguides are identical, the directional coupler is symmetric. Then, $\kappa_{21} = \kappa_{12} = \kappa$ and $\beta_1 = \beta_2$. The simplified coupled mode equations are:

$$\frac{dA}{dz} = -j\beta A + j\kappa B \tag{7.104}$$

$$\frac{dB}{dz} = -j\beta B + j\kappa A \tag{7.105}$$

To solve these equations, we eliminate the common phase factor e^{-ijz} by defining new variables:

$$A(z) = \tilde{A}(z)e^{-i\beta z} \tag{7.106}$$

$$B(z) = \tilde{B}(z)e^{-i\beta z} \tag{7.107}$$

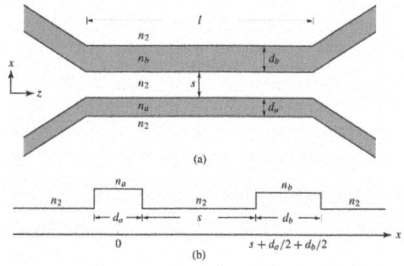

Fig. 7.12 Schematic diagram of (**a**) a two-channel directional coupler of a length l consisting of two parallel waveguides and (**b**) its index profile assuming two step-index waveguides on the same substrate. The coupler is symmetric if $n_a = n_b = n_1$ and $d_a. = d = d$

7.6 Optical Codirectional Couplers

Substituting these into the coupled mode equations, we get the modified coupled mode equation:

$$\frac{d\tilde{A}}{dz} = j\kappa \tilde{B} \tag{7.108}$$

$$\frac{d\tilde{A}}{dz} = j\kappa \tilde{B} \tag{7.109}$$

To solve these second-order equations, we take the derivative of the first equation with respect to z:

$$\frac{d^2\tilde{A}}{dz^2} = j\kappa \frac{d\tilde{B}}{dz} = -\kappa^2 \frac{d\tilde{A}}{dz} \tag{7.110}$$

This results in a second-order differential equation for $\tilde{A}$. The solution to this equation is:

$$\tilde{A}(z) = C_1 cos(\kappa z) + C_2 sin(\kappa z) \tag{7.111}$$

Similarly, we get a solution for $\tilde{B}$ as:

$$\tilde{B}(z) = D_1 cos(\kappa z) + D_2 sin(\kappa z). \tag{7.112}$$

Assume an initial condition such that at $z = 0$, we have $\tilde{A}(0) = A_0$, and $\tilde{B}(0) = B_0$. To find C_2 and D_2, use the first derivative at $z = 0$:

$$\frac{d\tilde{A}}{dz}|_{z=0} = j\kappa \left(D_1 cos(0) + D_2 sin(0) \right) = j\kappa B_0 \tag{7.113}$$

$$\kappa C_2 = j\kappa B_0, \text{ i.e. } C_2 = jB_0 \tag{7.114}$$

Similarly, taking the first derivative of $\tilde{B}$ results in the following:

$$D_2 = jA_0. \tag{7.115}$$

Substituting Eqs. (7.114) and (7.115) into Eqs. (7.111) and (7.112), we get the final solution as:

$$\tilde{A}(z) = A_0 cos(\kappa z) + jB_0 sin(\kappa z) \tag{7.116}$$

$$\tilde{B}(z) = B_0 cos(\kappa z) + jA_0 sin(\kappa z) \tag{7.117}$$

By including the $e^{-j\beta z}$, we get the waveguide mode amplitude as:

$$A(z) = (A_0 cos(\kappa z) + j B_0 sin(\kappa z)) e^{-j\beta z} \qquad (7.118)$$

and

$$B(z) = (B_0 cos(\kappa z) + j A_0 sin(\kappa z)) e^{-j\beta z} \qquad (7.119)$$

These expressions provide the final solution to the coupled mode equations for a codirectional coupler, showing that power oscillates between the two waveguides periodically. The **coupling length** l_c for a symmetric coupler, the distance over which complete power transfer occurs is given by:

$$l_c = \frac{\pi}{2\kappa}. \qquad (7.120)$$

The amplitude of the modes and the power of the guided models in a codirectional coupler are illustrated in Fig. 7.13 using the following parameters: $s = 0.5$ mm, $\kappa = 10^4$, and $\lambda = 1550$ nm.

The coupling length is expected to be:

$$l_c = \frac{\pi}{2\kappa} = \frac{\pi}{2 \times 10^4} = 1.57 \times 10^{-4} \text{ m}.$$

Examining the curves in Fig. 7.13 confirms that this calculated coupling length aligns with the observed values.

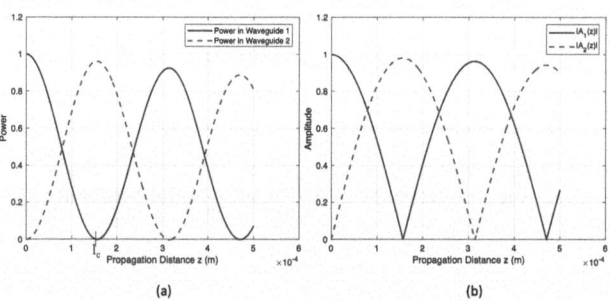

Fig. 7.13 Periodic amplitude and power of the guided modes exchange between two codirectionally coupled modes for (**a**) power exchange a function of z, and (**b**) mode amplitude exchange a function of z. The blue solid line is for waveguid 1, and dotted red line is for waveguide 2

7.7 Summary

This chapter covers the foundational concepts, structures, and applications of optical waveguides, fundamental components for guiding light in photonic systems. The chapter begins by introducing waveguides as structures that confine light within a core surrounded by a cladding, relying on total internal reflection for efficient light transport. Various waveguide structures, including slab, ridge, and rib waveguides, are examined, each with unique geometries tailored for specific applications.

The slab waveguide is presented as a fundamental structure, providing a simplified model to explore the principles of light guidance and mode propagation. Key aspects such as mode confinement, effective refractive index, and dispersion are discussed, with emphasis on the importance of the critical angle in determining guided modes. The wave equation approach further details how electromagnetic waves propagate within these structures, enabling calculation of waveguide parameters like propagation constants and mode profiles.

The chapter also delves into the behavior of guided modes, specifically how light can be confined to discrete modes within the waveguide core while maintaining a specific spatial field distribution. Orthogonality and degeneracy of these modes are explained, emphasizing how they allow for the independent propagation of multiple modes within the same waveguide.

An introduction to coupled mode theory demonstrates how energy can transfer between waveguides in close proximity, forming the basis for optical couplers and other photonic devices. Codirectional couplers are then discussed as an example of practical applications of coupled mode theory, where power exchanges periodically between coupled waveguides through evanescent coupling.

Throughout, the chapter emphasizes practical considerations for designing waveguides, including the role of normalized parameters like the V-number, which determine the number of modes a waveguide can support. This chapter provides with both theoretical and practical tools essential for understanding and designing waveguides.

7.8 Problems

1. A slab waveguide has core and cladding refractive indices $n_1 = 1.6$ and $n_2 = 1.4$, respectively. Find the range of possible propagation constants β for guided modes.
2. Compare the mode confinement between symmetric and asymmetric slab waveguides with core refractive index $n_1 = 1.5$ and cladding indices $n_2 = 1.45$ (symmetric) and $n_3 = 1.4$ (asymmetric).
3. Explain the impact of modal dispersion in a multimode waveguide and calculate the relative group velocities for the fundamental and first higher-order modes given: $\beta_0 = 1.3 \times 10^7 \, \text{m}^{-1}$ and $\beta_1 = 1.29 \times 10^7 \, \text{m}^{-1}$.

4. For a symmetric slab waveguide with core thickness $d = 4\,\mu m$, core refractive index $n_0 = 1.48$, and cladding refractive index $n_2 = 1.45$, calculate the cutoff wavelength λ_c for the first two guided modes.
5. A waveguide has a core refractive index $n_1 = 1.52$, cladding index $n_2 = 1.46$, and thickness $d = 3\,\mu m$ at a wavelength of $\lambda = 1.3\,\mu m$. Calculate the V-number and determine the maximum number of TE modes the waveguide can support.
6. Given two identical waveguides in a codirectional coupler with a coupling coefficient $\kappa = 8 \times 10^3\,m^{-1}$ and initial field amplitudes $A(0) = 1$ and $B(0) = 0$, find the amplitude $A(z)$ and $B(z)$ at a distance $z = 0.1\,mm$.
7. For a symmetric codirectional coupler with a coupling coefficient $\kappa = 1.5 \times 10^4\,m^{-1}$, calculate the coupling length l_c required for complete power transfer between the waveguides.
8. For a slab waveguide with core refractive index $n_1 = 1.5$ and cladding index $n_2 = 1.4$, calculate the effective refractive index n_{eff} for a mode with propagation constant $\beta = 1.47 \times 10^7\,m^{-1}$.
9. Derive the coupled mode equations for two waveguides in close proximity where the field overlap integrals result in coupling coefficients $\kappa_{12} = 5 \times 10^3\,m^{-1}$ and $\kappa 21 = 4.5 \times 10^3\,m^{-1}$.
10. For a codirectional coupler with initial conditions $A(0) = 1$ and $B(0) = 0$, and coupling coefficient $\kappa = 10^4\,m^{-1}$, derive the expressions for $A(z)$ and $B(z)$ over one complete coupling length, and interpret the power distribution between waveguides.

Bibliography

1. Pollock, C. R., & Lipson, M. (2003). Integrated Photonics. Springer Science+Business Media.
2. Liu, J. M. (2016). Principles of Photonics. Cambridge University Press.
3. Snyder, A. W., & Love, J. D. (1983). Optical Waveguide Theory. Springer.
4. Kawano, K., & Kitoh, T. (2001). Introduction to Optical Waveguide Analysis: Solving Maxwell's Equations and the Schrödinger Equation. Wiley.
5. Hunsperger, R. G. (2009). Integrated Optics: Theory and Technology (6th ed.). Springer.
6. Okamoto, K. (2010). Fundamentals of Optical Waveguides (2nd ed.). Academic Press.
7. Marcatili, E. A. J. (1969). "Dielectric Rectangular Waveguide and Directional Coupler for Integrated Optics." Bell System Technical Journal, 48(7), 2071–2102.

Optical Fibers

8.1 Introduction

An optical fiber is a flexible glass or plastic fiber that can transfer light from one end to the other. It is a cylindrical waveguide with a circular cross section, consisting of a core surrounded by a cladding with a lower index of refraction than the core. This structure guides light through total internal reflection (TIR), similar to the optical waveguides discussed in Chap. 7. Optical fiber has revolutionized telecommunications due to its low attenuation, extremely high bandwidth, and low cost.

In this chapter, we will introduce a brief history of optical fiber development. A major challenge in the use of optical fibers for telecommunication has been the production of fibers with low attenuation, which is a primary source of signal degradation.

The revolution in fiber optic communication has been made possible by technological advancements that have resulted in the availability of low-loss silica fibers. The attenuation in a single-mode fiber can be as low as 0.25 dB/km, allowing optical signals to propagate over long distances without the use of repeaters or amplifiers.

We discuss the different types of fibers commonly used in fiber optic communication, as well as the parameters that affect light coupling into the fiber. Additionally, we examine the optical fiber components essential for building a network such as: couplers, splicers, connectors, and cables.

Attenuation and dispersion are the main forms of linear degradation. Signal attenuation is due mainly to absorption and scattering, whereas dispersion can result from the waveguide properties of the fiber or the wavelength dependence of the index of refraction.

In this chapter, we will also examine a critical class of optical devices, namely optical fiber amplifiers. In these devices, the mechanism for attenuation is reversed, and optical gain is realized through an additional pump signal. The most important class of these devices is the erbium-doped fiber amplifier (EDFA) and Raman amplifiers.

8.2 A Brief History of Optical Fiber Development

Optical fibers in one form or another have been known since the discovery of the phenomenon of total internal reflection. Daniel Colladon and Jacques Babinet first demonstrated the guiding of light by refraction, the principle that makes fiber optics possible, in Paris in the early 1840s. John Tyndall included a demonstration of it in his public lectures in London, 12 years later.

Early uses of optical fibers were in industrial and medical applications. Fibers are used to illuminate or image hidden parts of machines or the human body. In the late nineteenth century, a team of Viennese doctors guided light through bent glass rods to illuminate body cavities. Practical applications such as close internal illumination during dentistry followed early in the twentieth century.

In 1953, Dutch scientist Bram van Heel first demonstrated image transmission through bundles of optical fibers with a transparent cladding. That same year, Harold Hopkins and Narinder Singh Kapany at Imperial College in London succeeded in making image-transmitting bundles with over 10,000 fibers and subsequently achieved image transmission through a 75 cm long bundle that combined several thousand fibers. Today, optical fiber bundles are widely used in medical applications such as diagnostics and surgery.

After the invention of the laser in 1960, it became clear that laser beams can be used for communication. The challenge was then, how to bend light around corners. In the 1960, glass-clad fibers had attenuation of about 1 dB/m, which is fine for medical imaging, but much too high for communications.

By 1964, Dr. Charles K. Kao identified a critical and theoretical specification for long-range communication devices, establishing a standard of 10 or 20 dB of light loss per kilometer. He also highlighted the need for a purer form of glass to reduce light loss, for which he earned the Nobel Prize in Physics in 2009.

In the summer of 1970, researchers began experimenting with fused silica, a material capable of extreme purity with a high melting point and a low refractive index. Corning Glass researchers Robert Maurer, Donald Keck, and Peter Schultz invented fiber-optic wire or "optical waveguide fibers," which are capable of carrying 65,000 times more information than copper wire does, through which information carried by a pattern of light waves can be decoded at a destination even a thousand miles away. The team had solved the decibel-loss problem presented by Dr. Kao. The team had developed a single-mode fiber (SMF) with a loss of 17 dB/km at 633 nm by doping titanium into the fiber core. By June of 1972, they invented multimode germanium-doped fibers with a loss of 4 dB/km and much greater strength than titanium-doped fibers. By 1973, John MacChesney developed a modified chemical vapor deposition process for fiber manufacture at Bell Labs, which spearheaded the commercial manufacture of fiber-optic cable.

In April 1977, General Telephone and Electronics tested and deployed the world's first live telephone traffic through a fiber-optic system running at 6 Mbps, in Long Beach, California. They were soon followed by Bell in May 1977, with an optical telephone communication system installed in the downtown Chicago, covering a distance of 1.5 miles.

Further progress resulted in a loss of 0.2 dB/km near the 1.55 μm spectral region by 1979. The availability of low-loss fibers and semiconductor lasers and photodetectors led to a revolution in the field of lightwave technology and started the era of fiber-optic communications, carrying terabits of data per single fiber.

8.3 Types of Optical Fibers

An optical fiber is a cylindrical waveguide consisting of a core surrounded by a cladding layer with a lower refractive index as shown in Fig. 8.1. To protect the fiber from environmental factors, it is coated with a protective jacket.

Optical fibers can be categorized on the basis of their material and/or their structure. Optical fibers can be divided into glass fibers and plastic optical fibers (POFs). Optical attenuation in plastics is much greater than that in glass. For example, materials such as polymethacrylates (PMMA) or polystyrene, commonly commonly used in plastic fibers, exhibit losses on the order of 100–1000 dB/km. However, plastic fibers are mechanically sturdy, easy to manufacture, and inexpensive. As a result, plastic fibers can provide an economic solution for very short-distance applications. Compared with silica-fibers, plastic fibers usually have a much larger dimensions. Glass is an obvious material because of its transparency makes it to be an excellent choice for optical fibers. Silica optical fibers exhibit an attenuation of 0.25 dB/km at a 1.55 μm wavelength, making them the ultimate choice for long-haul communication systems.

Glass optical fibers are made of silica (SiO_2). A significant advantage of silica is that its refractive index can be slightly increased by using dopants such as germanium dioxide (GeO_2) or reduced by boron tetroxide (B_2O_3). This allows the core and cladding refractive indices to be tailored to specific applications.

8.3.1 Step-Index Optical Fiber

The optical fiber structure is essentially made of a core and a cladding with refractive index lower than that of the core. This is the common structure for all types of fibers, as illustrated in Fig. 8.2. The step index refers to the abrupt change in the refractive

Fig. 8.1 Typical construction of an optical fiber

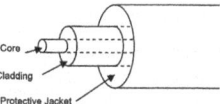

Fig. 8.2 Types of optical fibers and their refractive index profiles

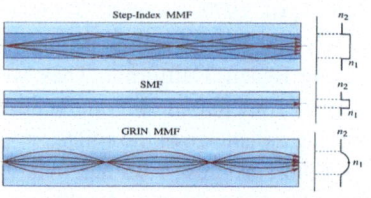

Fig. 8.3 Light confinement through total internal reflection in step-index fibers. Rays for which $\phi < \phi_c$ are refracted out of the core

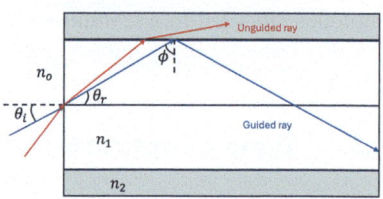

index at the core–cladding interface for both multimode fibers (MMFs) and single-mode fibers (SMFs).

The longitudinal cross section of a step-index fiber is considered to examine the ray path for a light ray impinging of a surface of a fiber (Fig. 8.3). The input ray makes an angle θ_i with the normal to the air–fiber interface. The ray emerges inside the core of the fiber, making an angle θ_r, based on Snell's law:

$$n_o sin\theta_i = n_1 sin\theta_r. \tag{8.1}$$

where n_1 and n_o are the refractive indices of the fiber core and air, respectively. The refracted ray hits the core–cladding interface and refracts again. This ray falls on the core–cladding interface making an angle $\phi = \pi/2 - \theta_r$. However, for all values of the angle ϕ smaller than the critical angle (ϕ_c) some of the light is reflected, and the other part is radiated and refracted into the cladding. For angles of incidence on the core cladding interface where $\phi \geq \phi_c$ the light will experience a total internal reflection (TIR) and no radiated light outside the core. The critical angle is defined with respect to the refractive indices of the core and cladding as follows:

$$sin\phi_c = \frac{n_2}{n_1}, \tag{8.2}$$

where n_2 is the cladding refractive index. Since such reflections occur throughout the fiber length, all rays with $\phi > \phi_c$ remain confined to the fiber core. This is the basic mechanism behind light confinement in optical fibers. Therefore, the incidence angle to the fiber θ_i has a maximum value (θ_{max}) within which the light inside the core satisfies the TIR condition. Thus,

$$n_o sin\theta_{max} = n_1 sin\left(\frac{\pi}{2} - \phi_c\right) = n_1 cos\phi_c = \sqrt{n_1^2 - n_2^2}. \tag{8.3}$$

8.3 Types of Optical Fibers

Fig. 8.4 Ray trajectories in a graded-index fiber

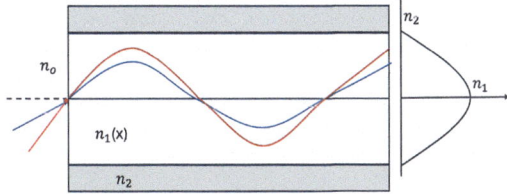

The quantity $n_o \sin\theta_{max}$ is known as the numerical aperture NA of the fiber, represents the light-gathering ability:

$$NA = \sqrt{n_1^2 - n_2^2}. \tag{8.4}$$

The numerical aperture are a very useful measure of the light-collecting ability of a fiber. In silica optical fibers, the refractive indices of the core and the cladding are approximately have the same value, i.e. $n_1 \simeq n_2$. Therefore, if we define Δ as the fractional index change at the core–cladding interface,

$$\Delta = \frac{n_1 - n_2}{n_1}, \tag{8.5}$$

the numerical aperture can be expressed in terms of Δ as::

$$NA = n_1\sqrt{2\Delta}. \tag{8.6}$$

Clearly, the the fractional index change (Δ) should be made as large as possible in order to couple maximum light into the fiber. However, such fibers are not useful for the purpose of optical communications because of the modal dispersion introduced in in Chap. 7 for slab waveguides, which will be introduced later in this chapter in the context of optical fibers. Because of the circular symmetry of the fiber cross section, the maximum angle of incidence (θ_{max}) becomes the subtended angle of a cone within which light rays result in transmitted light along the fiber, as shown in Fig. 8.4. This light acceptance cone becomes very important in the coupling of light into the fiber.

Example A silica optical fiber with a core refractive index of 1.50 and a cladding refractive index of 1.47. Determine: (a) the critical angle at the core–cladding interface, (b) the NA for the fiber, and (c) the acceptance angle in air for the fiber.

(continued)

Solution

(a) The critical angle ϕ_c at the core–cladding interface is given by Eq. (8.2)

$$\phi_c = \frac{n_2}{n_1} = \frac{1.47}{1.5} = 78.5°$$

(b) The numerical aperture is

$$NA = \sqrt{n_1^2 - n_2^2} = \sqrt{1.5^2 - 1.47^2} = 0.3$$

(c) The acceptance angle is

$$\theta_{max} = sin^{-1}\left(\frac{NA}{n_o}\right) = sin^{-1}\left(\frac{0.3}{1}\right) = 17.4°$$

Example A typical fractional index change for an optical fiber designed for long-haul transmission is 1%. Determine:

(a) The numerical aperture, (b) the solid acceptance angle in air if the core index is 1.46, and (c) calculate the critical angle at the core–cladding interface.

Solution

(a) The numerical aperture is

$$NA = n_1\sqrt{2\Delta} = 1.47\sqrt{2 \times 0.01} = 0.21.$$

(b) For small acceptance angles, the solid acceptance angle is

$$\xi = \pi sin^2\theta_{max} = \pi \times NA^2 = 0.13 \text{ rad}.$$

(c) The fractional index change Δ is

$$\Delta = \frac{n_1 - n_2}{n_1} = 1 - \frac{n_2}{n_1} = 1 - \frac{n_2}{1.47} = 0.01$$

(continued)

8.3 Types of Optical Fibers

Therefore,

$$n_2 = 0.99 \times 1.47 = 1.455$$

and the critical angle is

$$\phi_c = sin^{-1}\left(\frac{n_2}{n_1}\right) = sin^{-1}(0.99) = 82°.$$

8.3.2 Graded-Index Fibers

Graded-index fibers (GRINF) are multimode fiber like step-index fibers except that the refractive index of the core is graded such that it peaks in the center reach a minimum at the core–cladding interface, as shown in Fig. 8.4. This approach t is similar to the graded-index lenses we discussed in Chap. 2. Typically, graded-index fibers are designed to have a nearly quadratic decrease using α-$profile$, given by:

$$n(r) = \begin{cases} n_1\left[(1-\Delta)\left(\frac{r}{a}\right)^\alpha\right] & for\ r < a \\ n_1(1-\Delta) = n_2 & for\ r \geq a \end{cases} \quad (8.7)$$

where a is the core radius. A parabolic-index fiber corresponds to $\alpha = 2$.

In step-index fibers, rays traveling closer to the fiber axis travel the shortest distance, whereas rays with greater oblique angle closer to the critical angle travel the longest distance. Consequently, different modes propagate at different velocities, leading to what is known as modal dispersion, a key contributor to pulse broadening as will be discussed in subsequent sections. Graded-index fibers are designed to eliminate or significantly reduce modal dispersion. As illustrated in Fig. 8.4, rays traveling along the axis of the fiber have the slowest velocity because the refractive index is highest at the center. Rays propagating at oblique angles experience a decrease in the refractive index as they move away from the axis, resulting in increased velocity. By carefully designing the refractive index profile, it is possible to ensure that all rays arrive simultaneously at the fiber's output.

8.3.3 Specialty Fibers

Several special optical fibers are designed to mitigate the dispersion encountered in single-mode fibers used for log-haul transmission. Further, we briefly introduce a few of these fibers, which will be discussed in more detail later in this chapter.

Polarization Maintaining Fiber The polarization maintaining fibers (PMFs) are optical fibers engineered to preserve the polarization state of light as it propagates through the fiber. In contrast to standard single-mode fibers, where the polarization state of light can change due to environmental factors such as stress, temperature variations, and fiber bends, PMF is designed to maintain consistent polarization. This is typically achieved by incorporating an elliptical core or impeding stress rods within the cladding creating birefringence. The resulting birefringence splits light into two orthogonal polarization modes that travel at different speeds, thus maintaining the polarization state. PMF is commonly used in applications where maintaining the polarization state is essential, such as in interferometry, fiber optic sensors, quantum key distribution, and in the transmission of polarized light in communication systems.

Dispersion Compensation Fiber Dispersion compensation fibers (DCFs) are designed to mitigate the dispersion effects in optical communication systems, particularly in long-haul fiber-optic links. Dispersion, the phenomenon where different wavelengths of light travel at different speeds through the fiber, leads to pulse broadening, signal degradation, and potential pulse overlap, which can reduce the overall data transmission quality and capacity. The DCF is specifically engineered to compensate for both chromatic and polarization-mode dispersions. The required length of the DCF is considerably shorter than the single-mode fiber (SMF) it compensates for. However, DCF generally introduces higher insertion loss, i.e., it attenuates the signal more than standard fiber.

Dispersion-Shifted Fiber Dispersion-shifted fibers (DSFs) are designed to shift the wavelength at which zero chromatic dispersion occurs, optimizing fiber performance for long-distance, high-capacity telecommunications. By shifting the zero-dispersion wavelength to the 1550 nm region, DSF minimizes chromatic dispersion at the operating wavelength, thereby reducing pulse broadening and signal degradation.

Dispersion Flattened Fiber Dispersion Flattened Fiber (DFF) is a specialized optical fiber engineered to minimize chromatic dispersion over a wide wavelength range. This is achieved by carefully engineering the refractive index profile of the fiber. DFF has a complex refractive index profile, often involving multiple concentric layers with varying refractive indices. The result is a flattened dispersion curve, ensuring low dispersion across a wide range of wavelengths.

8.4 Meridional and Skew Rays

An optical fiber is a waveguide that confines optical propagation within its core. When viewed in a cross section along its axis as shown in Fig. 8.5, it closely resembles a symmetrical slab waveguide as analyzed in Chap. 7. Therefore, the waveguiding condition in Eq. (7.5) is applicable to step-index fibers. The propa-

Fig. 8.5 (a) A meridional ray zig-zags down the fiber, passing through the origin. There is no angular rotation of the ray path as it propagates. (b) A skew ray travels in a spiral path down the fiber. The ray does not go through the origin

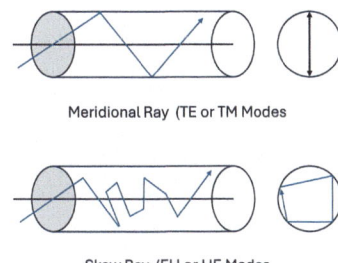

Meridional Ray (TE or TM Modes)

Skew Ray (EH or HE Modes)

Fig. 8.6 Cylindrical coordinate system where any point in space can be given by $P(r, \theta, z)$ coordinates

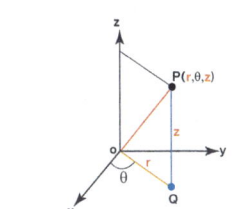

gation modes in a step-index fiber also are similar to those of the symmetric slab waveguide as depicted in Fig. 8.6. Due to the fiber's circular symmetry, these modes are consistent for any longitudinal cross section. For example, if two cross sections perpendicular to each other are taken, the mode profiles will remain identical. Thus, within the fiber every mode has an orthogonal counterpart with similar profile. The light rays considered in our geometrical analysis, which remain in the same plane as they travel down the fiber, are known as **meridional rays**. The modes generated by these rays are the TE (transverse electric) and TM (transverse magnetic) modes.

There are other types of rays that can travel down the fiber but do not stay confined to a plane defined by a longitudinal cross section. These rays are known as **skew rays.** They travel down the fiber in a helical pattern, glancing off the interface as they spiral along the axis, as illustrated in Fig. 8.5. The azimuthal structure is evident from the cyclical path of these rays. The modes generated by skew rays are known as **hybrid modes**, specifically the EH and HE modes.

8.5 Wave Propagation in Step-Index Fibers

In this section, we examine the propagation of light in step-index fibers using Maxwell's equations for electromagnetic waves. These equations were introduced in Chap. 3 and solved in Chap. 7 to derive the optical modes guided in a slab waveguide. Here, we focus on how a step-index fiber supports guided modes, then derive the conditions for single-mode operations, and discuss the properties of single-mode fibers. We begin with the Helmholtz wave equation in cylindrical coordinate system (r, θ, z), as shown in Fig. 8.6.

The wave equation expressed in terms of the field U, which can represent any of the six components of the electric (E_r, E_θ, E_z) or magnetic (H_r, H_θ, H_z) fields, is given by:

$$\nabla^2 U + n^2 k_o^2 U = 0, \tag{8.8}$$

where $n(\omega)$ is the refractive index, and k_o is the wavenumber in free space. The wave equation in cylindrical coordinate system is:

$$\frac{\partial^2 U}{\partial r^2} + \frac{1}{r}\frac{\partial U}{\partial r} + \frac{1}{r^2}\frac{\partial^2 U}{\partial \theta^2} + \frac{\partial^2 U}{\partial z^2} + n^2 k_o^2 U = 0. \tag{8.9}$$

For a step-index fiber of core radius a, the refractive index n is given by:

$$n = \begin{cases} n_1 & for\ r \leq a \\ n_2 & for\ r > a \end{cases}. \tag{8.10}$$

The wave equation, Eq. (8.9), can be solved by the method of separation of variables. Assuming that the field $U(r, \theta, z)$ is of the form:

$$U(r, \theta, z) = R(r)\Theta(\theta)Z(z). \tag{8.11}$$

Substituting Eq. (8.11) into Eq. (8.9) yields:

$$\Theta(\theta)Z(z)\frac{\partial^2 R(r)}{\partial r^2} + \Theta(\theta)Z(z)\frac{1}{r}\frac{\partial R(r)}{\partial r} + R(r)Z(z)\frac{1}{r^2}\frac{\partial^2 \Theta(\theta)}{\partial \theta^2}$$
$$+ R(r)\Theta(\theta)\frac{\partial^2 Z(z)}{\partial z^2} + n^2 k_o^2 R(r)\Theta(\theta)Z(z) = 0. \tag{8.12}$$

Dividing both sides of Eq. (8.12) by $R(r)\Theta(\theta)Z(z)$ gives

$$\frac{1}{R}\frac{\partial^2 R}{\partial r^2} + \frac{1}{R}\frac{1}{r}\frac{\partial R}{\partial r} + \frac{1}{\Theta}\frac{1}{r^2}\frac{\partial^2 \Theta}{\partial \theta^2} + \frac{1}{Z}\frac{\partial^2 Z}{\partial z^2} + n^2 k_o^2 = 0. \tag{8.13}$$

To proceed, we need to rearrange the wave equation into three separate differential equations, each in one of the variables r, θ and z. We know that the field propagates in the z-direction as a a planewave with propagation constant β and that differential equations with circular symmetry have solutions in terms of Bessel functions. The general form differential equation with a Bessel function solution is:

$$\frac{d^2 y}{dx^2} + \frac{1}{x}\frac{dy}{dx} + \left(1 - \frac{m^2}{x^2}\right)y = 0, \tag{8.14}$$

where m is the order of the Bessel function.

8.5 Wave Propagation in Step-Index Fibers

On the basis of these assumptions, Eq. (8.13) can be expressed as the following three differential equations:

$$\frac{d^2 Z}{dz^2} + \beta^2 Z = 0, \tag{8.15}$$

$$\frac{d^2 R}{dr^2} + \frac{1}{r}\frac{dR}{dr} + \left(n^2 k_o^2 - \beta^2 - \frac{m^2}{r^2}\right) = 0, \tag{8.16}$$

$$\frac{d^2 \Theta}{d\theta^2} + m^2 \Theta = 0. \tag{8.17}$$

Equation (8.15) is a simple second-order differential equation with a solution given by

$$Z(z) = e^{-j\beta z} \tag{8.18}$$

which describes a planewave propagating along the fiber axis. The solution of Eq. (8.16) is expressed in Bessel functions. These functions are commonly used to describe the behavior of guided modes, which are oscillatory in the core and decay in the cladding. Therefore, the solution to this differential equation can be expressed as:

$$R(r) = \begin{cases} AJ_m(\kappa r) + BY_m(\kappa r) & \text{for } r \leq a \\ CI_m(qr) + DK_m(qr) & \text{for } r > a \end{cases}, \tag{8.19}$$

where $J_m(x)$ and $Y_m(x)$ are the ordinary Bessel functions of the 1st and 2nd kinds, respectively, each of order m. And $I_m(x)$ and $K_m(x)$ are the modified Bessel functions of 1st and 2nd kind, respectively. The coefficients A, B, C and D are constants to be determined using the boundary conditions at the core–cladding interface. The parameters κ and q are defined as:

$$\kappa^2 = n_1^2 k_o^2 - \beta^2 \tag{8.20}$$

and

$$q^2 = \beta^2 - n_2^2 k_o^2 \tag{8.21}$$

From the analysis of the slab waveguide, we know that a guided wave must have a phase constant β that lies in between those of the core region (β_1) and the cladding (β_2), such that

$$n_1 k_o = \beta_1 > \beta > \beta_2 = n_2 k_o. \tag{8.22}$$

The condition $\beta_2 < \beta$ ensures a decay in the cladding, whereas $\beta < \beta_1$ prevents this decay in the core. Since the refractive index depends on r, a common β for the core and cladding is possible only if Eq. (8.20) holds for the core and Eq. (8.21) holds for the cladding.

> **Examples** Plot the curves for all Bessel functions in Eq. (8.19) as a function $x = 10$ for the values of $m = 0, 1, 2$
>
> **Solution** The following are the plots for Bessel 1st and 2nd kind and Modified Bessel 1st and 2nd kind:

To determine the optical field for a guided mode, we apply the boundary conditions such that the field must be finite at $r = 0$ and decay to zero at $r = \infty$. Since $Y_m(r))$ has a singularity at $r = 0$, as shown in Fig. 8.7, $R(0)$ remains finite only if $B = 0$. Similarly, $R(0)$ vanishes at infinity ($r = \infty$), only if $C = 0$. Therefore, the general solution of Eq. (8.19) for the electric field is reduced to:

$$E_z(r, \theta, z) = \begin{cases} AJ_m(\kappa r)e^{-jm\theta}e^{-j\beta z} & for\ r \leq a \\ DK_m(qr)e^{-jm\theta}e^{-j\beta z} & for\ r > a \end{cases}. \tag{8.23}$$

The same method can be applied to obtain the magnetic field H_z which also satisfies Eq. (8.19). Therefore, the solution for H_z is the same form but with different

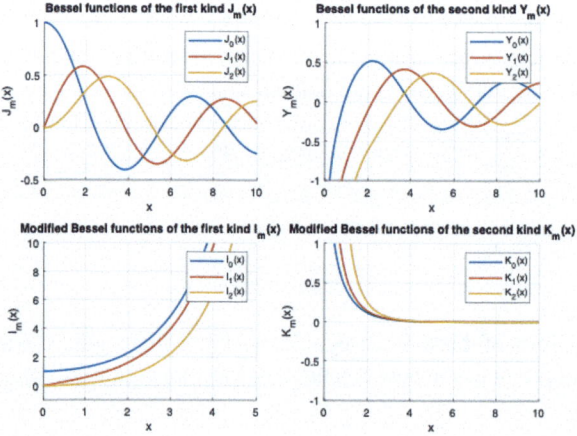

Fig. 8.7 Plots for Bessel functions 1st and 2nd kind and modified Bessel 1st and 2nd kin

8.5 Wave Propagation in Step-Index Fibers

constants B and D, that is,

$$H_z(r, z) = \begin{cases} B J_m(\kappa r) e^{-jm\theta} e^{-j\beta z} & for \ r \leq a \\ C K_m(qr) e^{-jm\theta} e^{-j\beta z} & for \ r > a \end{cases}. \tag{8.24}$$

The remaining four components of the electric and magnetic fields can be derived using Maxwell's equations in terms of E_z and H_z. The boundary conditions require that the tangential field components, E_z, E_θ, H_z, and H_θ, be continuous at the boundary, $r = a$, at the core–cladding interface. These conditions result in the requirement that the θ dependence of E_z be 90° out-of-phase with respect to that of H_z.

Applying the boundary condition that the tangential component of the electric field is continuous across the core–cladding interface, we obtain:

$$A J_m(\kappa a) = D K_m(qa),$$

which leads to:

$$D = A \frac{J_m(\kappa a)}{K_m(qa)}. \tag{8.25}$$

Substituting this result into Eq. (8.23), we obtain the following expression for the z-component of the electric field:

$$E_z = \begin{cases} A J_m(\kappa r) e^{-jm\theta} e^{-j\beta z} & for \ r \leq a \\ A \frac{J_m(\kappa a)}{K_m(qa)} K_m(qr) e^{-jm\theta} e^{-j\beta z} & for \ r > a \end{cases}. \tag{8.26}$$

Example Given the following parameters for an optical fiber:
$a = 2.5\,\mu m$, $\lambda = 1550\,nm$, $k_o = 2\pi/\lambda$, $n_1 = 1.5$, $n_2 = 1.47$. Assume that $\beta = \frac{k_o(n_1 - n_2)}{2}$.
Plot the electric field $E_z(r)$ as a function of $-5\,\mu m \leq r \leq 5\,\mu m$

$$E_z = \begin{cases} J_m(\kappa r) & for \ r \leq a \\ \frac{J_m(\kappa a)}{K_m(qa)} K_m(qr) & for \ r > a \end{cases}$$

for m=0 and m=1.

Solution To plot the electric field $E_z(r)$ for $m = 0$ and $m = 1$, we first calculate the parameters κ and q. Next, we substitute these parameters into the expression for $E_z(r)$ to generate the plot over the range $-5\,\mu m \leq r \leq 5\,\mu m$.

The following is a plot of the electric field (Fig. 8.8):

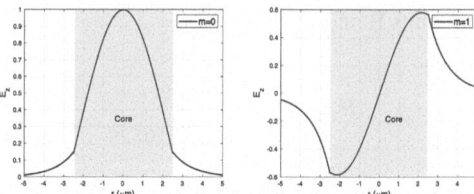

Fig. 8.8 Electric field as a function of the radial distance r for the first two modes ($m=0$ and $m=1$) for the parameters given in the example

In the example, we have assumed a value of the propagation constant β for demonstrative purposes. However, in practice, β should be determined from the solution of the wave equation, as it takes on discrete values rather than defining the guiding modes.

8.6 Guided Modes of the Optical Fiber

In a planar waveguide, as discussed in Chap. 7, the optical field is confined in one dimension, and it supports two kinds of modes: the TE and TM modes. In contrast, the circular cross section of an optical fiber confines the optical field in two dimensions. As a result, the optical fiber supports four kinds of modes. In addition to the TE and TM modes, there are hybrid modes: HE modes, where E_z dominates H_z, and EH modes, where H_z dominates E_z.

In slab waveguides, a single index m is used to identify the mode, such as TE_m or TM_m. However, because a cylindrical waveguide is bounded in two dimensions, two integers, m and n, are required to to specify the modes. Consequently, in a cylindrical waveguide, we therefore refer to modes as TE_{mn} and TM_{mn}. These modes correspond to meridional rays traveling within the fiber. However, hybrid modes where E_z and H_z are nonzero, also occur in a cylindrical waveguide. These modes, resulting from skew ray propagation within the fiber, are designated as HE_{mn} and EH_{mn} depending on whether the components of H_z or E_z contribute more to the transverse field. Therefore, a precise description of the modal fields in a step-index fiber is somewhat complex.

Let us reconsider the field equation given by Eqs. (8.23) and (8.24) and rewrite the fields in terms of the variables r and θ, which yields the following expressions:

$$E_z(r,\theta) = \begin{cases} A_m J_m(\kappa r)\cos(m\theta) & for\ r \leq a \\ D_m K_m(qr)\cos(m\theta) & for\ r > a \end{cases}, \quad (8.27)$$

$$H_z(r,\theta) = \begin{cases} B_m J_m(\kappa r)\sin(m\theta) & for\ r \leq a \\ C_m K_m(qr)\sin(m\theta) & for\ r > a \end{cases}. \quad (8.28)$$

Since E_z is out of phase with H_z enforces the choice of the sine and cosine terms in the field expressions. The constants A_m, B_m, C_m, and D_m are specific to each fiber mode and must be determined accordingly.

8.6 Guided Modes of the Optical Fiber

Alternatively, we could have chosen the θ term to be as follows:

$$E_z(r,\theta) = \begin{cases} A'_m J_m(\kappa r) sin(m\theta) & for\ r \leq a \\ D'_m K_m(qr) sin(m\theta) & for\ r > a \end{cases}, \quad (8.29)$$

$$H_z(r,\theta) = \begin{cases} B'_m J_m(\kappa r) cos(m\theta) & for\ r \leq a \\ C'_m K_m(qr) cos(m\theta) & for\ r > a \end{cases}. \quad (8.30)$$

where the constants A'_m, B'_m, C'_m, and D'_m are also to be determined for a particular fiber mode. Applying the boundary conditions leads to the following eigenvalue function for the allowed values of h and q for the guided modes, which in turn determine the propagation constant β :[1]

$$\left[\frac{J'_m(\kappa a)}{\kappa a J_m(ha)} + \frac{K'_m(qa)}{qa JK(ha)}\right]\left[\frac{n_1^2 J'_m(\kappa a)}{ha J_m(\kappa a)} + \frac{n_2^2 K'_m(qa)}{qa JK(ha)}\right]$$
$$= m^2 \frac{\beta^2}{k_o^2}\left(\frac{1}{\kappa^2 a^2} + \frac{1}{q^2 a^2}\right)^2, \quad (8.31)$$

where J'_m and K'_m are the derivatives of the Bessel functions. Each mode in a circular fiber is characterized by two mode indices m and n. The first index m refers to the angular dependence $cos m\theta$ or $sin m\theta$. The second index n refers to the order of the allowed solutions for eigenvalues h or, equivalently, q. Therefore, m is called the **azimuthal mode index**, or the **angular mode index**, whereas n is called the **radial mode index**. In general, Eq. (8.31) has no closed-form solution so it must be solved numerically.

For the case when m = 0, the first set of solutions for the longitudinal components of the mode fields given in Eqs. (8.27) and (8.28) results in $H_z = 0$, corresponding to TM modes. The second set of solutions given in Eqs. (8.29) and (8.30) results in $E_z = 0$, corresponding to TE modes. Using the following identities of the Bessel functions:

$$J'_m(x) = -J_{m+1}(x) - \frac{m}{x} J_m(x), \quad (8.32)$$

and

$$K'_m(x) = -K_{m+1}(x) - \frac{m}{x} K_m(x). \quad (8.33)$$

---

[1] The derivation of this equation is beyond the scope of this book but can be found in many of the references listed in the end of this chapter.

Equation (8.31) for $m = 0$, either of the two terms on the left side must be equal to zero. Therefore, Eq. (8.31) becomes two separate eigenvalue equations:

$$\frac{J_1(\kappa a)}{\kappa a J_0(\kappa a)} + \frac{K_1(qa)}{qa K_0(qa)} = 0 \quad for\ TE\ modes, \tag{8.34}$$

and

$$\frac{n_1^2 J_1(\kappa a)}{\kappa a J_0(ha)} + \frac{n_2^2 K_1(qa)}{qa K_0(qa)} = 0 \quad for\ TM\ modes. \tag{8.35}$$

Let us rearrange Eq. (8.34) as

$$\frac{J_1(\kappa a)}{\kappa J_0(\kappa a)} = -\frac{K_1(qa)}{qa K(qa)}, \tag{8.36}$$

which is a **transcendental equation** that must be solved graphically to determine the values of the propagation constant β for the allowed guided modes.

Example Determine the values of β for the following parameters:
$a = 5\,\mu\text{m}, \lambda = 1550\,\text{nm}, k_o = 2\pi/\lambda, n_1 = 1.5, n_2 = 1.47$
$k_o^2(n_1^2 - n_2^2) - \kappa^2$
Plot
$\frac{J_1(\kappa a)}{\kappa J_0(\kappa a)} = -\frac{K_1(qa)}{qa K(qa)}$ as a function of κ

Solution Using MATLAB, we obtain the following graph (Fig. 8.9).
The roots the numerical calculation are at $\kappa = 6765.9211\ and\ 12229.4385$ which correspond to the following values of β
$$\beta = \sqrt{n_1^2 k_o^2 - \kappa^2} = 6.0427 \times 10^4\ and\ 5.9562 \times 10^4\,\text{m}^{-1}$$

For $m \geq 1$, the guided modes in a circular fiber are hybrid modes. In these modes both E_z and H_z exist. The hybrid modes HE are the solutions of the eigenvalue equations when E_z dominates H_z, whereas EH modes are the solutions when H_z dominates E_z. For each given m ≥ 1, the eigenvalue equation in Eq. (8.31) yields two sets of solutions, one for HE modes and another for EH modes. All field components can be expressed in terms of E_z and H_z in terms of $J_m(\kappa r)$.

Mode Cutoff In optical fiber waveguides, the propagation of modes is critically dependent on the propagation constant β. When β approaches $n_2 k_o$, the phase velocity of the mode aligns with the speed of light in the cladding. At this juncture,

8.6 Guided Modes of the Optical Fiber

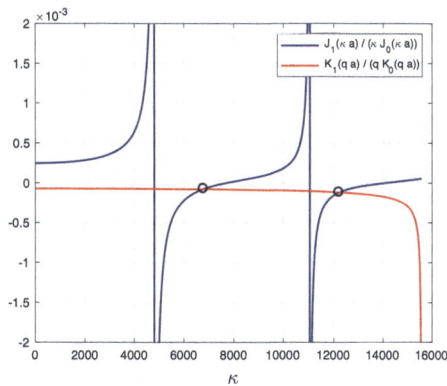

Fig. 8.9 Plotting the transcendental equation Eq. (8.36) as a function of κ. There are two roots of the equation, which are marked with circles. Ignore the vertical lines, since they are artifacts of the calculations

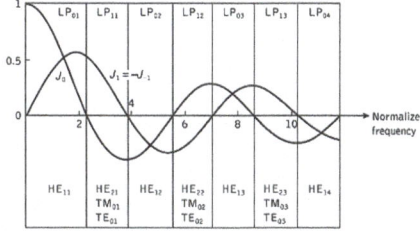

Fig. 8.10 The allowed regions for the LP modes of order $m = 0, 1$ against normalized frequency (V) for a circular optical waveguide with a constant refractive index core (step index fiber). [D. Gloge. Appl. Opt., 10, p. 2552, 1971]

the mode is no longer effectively guided within the core, leading to what is termed as "mode cutoff." In this scenario, the eigenvalue $qa = 0$.

Modes with frequencies below the cutoff, where $\beta < k_o n_2$ are referred to as unguided or radiation modes In these cases, qa becomes imaginary, and solutions to the wave equation describe "leaky modes." These modes often exhibit characteristics of highly lossy guided modes rather than pure radiation modes.

Conversely, as β is increased beyond $k_o n_2$, the power propagated in the cladding diminishes when $\beta = k_o n_1$ and all the power is confined exclusively to the fiber core. This specific range of β values delineates the guided modes within the fiber.

Figure 8.10 illustrates the lower-order modes present in an optical fiber with homogeneous core. Both the linearly polarized (LP) notation and the corresponding traditional (HE, EH, TE and TM) modes notations are depicted. In addition, the Bessel functions J_0 and J_1 are plotted against the normalized frequency (V). The points where these functions cross zero indicate the cutoff points for various modes.

For example, the first zero crossing of J_1 occurs at a normalized frequency $V = 0$, corresponding to the cutoff for the LP_{01} mode. In contrast, the first zero crossing for J_0 is at a normalized frequency $V = 2.405$, yielding a cutoff value $V_c = 2.405$ for the LP_{11} mode. Similarly, the the second zero of J_1 at $V_c = 3.83$ marks the cutoff for the LP_{02} mode.

These observations highlight that fibers can be engineered with specific normalized frequency values to permit only certain modes to propagate. This concept is

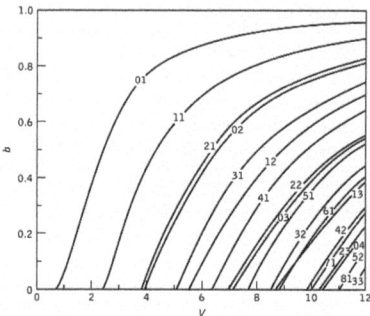

Fig. 8.11 The normalized propagation constant b as a function of normalized frequency V for a number of LP modes. [D. Gloge, Appl. Opt., 10, p. 2552, 1971]

further clarified in Fig. 8.11, which shows the normalized propagation constant b for various LP modes as a function of V. The cutoff normalized frequency V_c, occurring when $\beta = n_2 k_o$, corresponds to $b = 0$.

The cutoff of the normalized frequency is pivotal in determining the single-mode operation of the fiber. Specifically, for single-mode propagation, the normalized frequency V must be below the cutoff value of the second mode. Typically, maintaining $V < 2.405$ ensures single-mode behavior.

The normalized frequency cutoff determines the single-mode operation of the fiber. In the context of multimode optical fibers, the number of supported modes M is approximately given by:

$$M \approx \frac{V^2}{2}. \tag{8.37}$$

This relation underscores the dependence of mode quantity on the square of the normalized frequency, emphasizing the importance of precise control over fiber parameters to achieve desired mode propagation characteristics.

Example A multimode step index fiber with a core diameter of 50 μm and a relative index difference $\Delta = 0.015$ is operating at a wavelength of 0.85 μm. If the core refractive index is 1.48, determine: (a) the normalized frequency for the fiber; and (b) the number of guided modes.

Solution

(a) The normalized frequency is

$$V = \frac{2\pi a n_1}{\lambda}\sqrt{2\Delta} = \frac{2\pi \times 1.48 \times 25 \times 10^{-6}}{0.85 \times 10^{-6}}\sqrt{2 \times 0.015} \approx 23.7$$

(continued)

(b) the number of modes is

$$M \approx \frac{V^2}{2} = 280.$$

Example Estimate the maximum core radius for an optical fiber with the same relative refractive index difference of 1.5 % and core refractive index (1.48) for the fiber to be single mode operating at $\lambda = 0.85\,\mu m$, $1.3\,\mu m$, and $1.55\,\mu m$. Repeat the same as before if $\Delta = 0.001$.

Solution In order for the fiber to be single-mode, its normalized frequency $v \leq 2.405$. Therefore, the maximum core radius is

$$a_{max} = \frac{V\lambda}{2\pi n_1 \sqrt{2\Delta}}$$

The maximum core radii for all the cases aforementioned are summarized in the following table:

Wavelength (λ)	Refractive index difference (Δ)	Maximum core radius (a_{max})
$0.85\,\mu m$	0.015	$1.27\,\mu m$
$0.85\,\mu m$	0.001	$4.9\,\mu m$
$1.3\,\mu m$	0.015	$1.94\,\mu m$
$1.3\,\mu m$	0.001	$7.5\,\mu m$
$1.55\,\mu m$	0.015	$2.3\,\mu m$
$1.55\,\mu m$	0.001	$9\,\mu m$

It is evident from both examples that to achieve a single-mode operation with a maximum V number of 2.405, the single-mode fiber must have a significantly smaller core diameter than a multimode step index fiber. However, it is possible to achieve single-mode operation with a slightly larger core diameter by reducing the relative refractive index difference (Δ) of the fiber. Both of

(continued)

these factors introduce challenges in the design and use of single-mode fibers. The small core diameters pose problems with launching light into the fiber and with field jointing, whereas, the reduced relative refractive index difference presents challenges in the fiber fabrication process. Single-mode fibers used in long-haul telecommunications typically have a core radius of approximately 4.5 μm,, which requires $\Delta \approx 0.004$ for operation at $\lambda = 1.55$ μm.

Example Determine the cutoff wavelength (λ_c) for a step index fiber to exhibit single-mode operation when the core refractive index is 1.46, and the core radius is 4.5 μm with a relative index difference being 0.25%.

Solution The cutoff wavelength is given by

$$\lambda_c = \frac{2\pi a n_1 \sqrt{2\Delta}}{2.405} = \frac{2\pi \times 4.5 \times 10^{-6} \times 1.46 \times \sqrt{2 \times 0.0025}}{2.405} = 1.214\,\mu m$$

Hence, the longest wavelength for this fiber operate as a single-mode is 1214 nm.

8.7 Signal Degradation in Optical Fibers

As optical signals propagate through the fiber, they experience several forms of degradation that ultimately limit the range of a particular fiber link. Signals cannot be transmitted beyond this limit, unless they undergo some form of active reconditioning or optical amplification.

Signal degradation can be divided into the two general types of linear and nonlinear. Linear degradations are those that do not depend on the optical power and affect weak and strong signals equally. In contrast, with nonlinear degradation, on the other hand, the effects become significant at high-power levels. Thus, nonlinear effects limit the maximum power that can be launched into the fiber.

Linear Degradations
- Attenuation: Loss of signal strength due to absorption and scattering.
- Dispersion: Spreading of the signal in time affects bandwidth and transmission quality.

Nonlinear Effects
- Self and Cross-Phase Modulation.

- Four-Wave Mixing.
- Stimulated Raman and Brillouin scattering.

Overall, minimizing signal degradation and improving fiber properties are key to extending the reach and efficiency of optical communication systems. In the following sections, we discuss these transmission characteristic factors and different means of their mitigation.

8.7.1 Attention in Silica Optical Fibers

The most common form of signal degradation in optical fibers is attenuation. The development of low-loss silica fibers was a pivotal breakthrough that enabled the advancement of fiber optic communication. When an optical wave propagates through a lossy medium, its intensity decays exponentially with distance. The power of an optical wave in a fiber is simply the integration of its intensity over the cross section of the fiber. Consequently, the optical power in an optical fiber decreases exponentially with distance. Hence, the power at the end of a fiber of length L is given by:

$$P_{out} = P_{in} e^{-\alpha_f L}, \tag{8.38}$$

where P_{in} is the power of optical signal launched into the fiber, P_{out} is the power at the output of the fiber of length L, and the attenuation α_f is the fiber attenuation factor. P_{in} and P_{out} are measured in watts or milliwatts or microwatts and α_f is per meter or per kilometer or even per centimeter.

In practical engineering applications, it is convenient to express relative changes in quantities using decibels (dB). The attenuation coefficient α is then measured in decibels per meter. For low-loss fibers, the propagation length in a fiber is usually measured in kilometers, and α is conventionally given in decibels per kilometer: Alternatively, Eq. (8.38) can be written in terms of logarithm as follows:

$$10 \log P_{out} = 10 \log P_{in} - \alpha_f L [10 \log(e)] = 10 \log P_{in} - 4.343 \alpha_f L. \tag{8.39}$$

Simplifying, we obtain:

$$10 \log P_{out} = 10 \log P_{in} - \alpha L, \tag{8.40}$$

where α is the fiber attenuation in dB/km, L is the fiber length in km, and the power is in Watts. The power can be measured in decibel (dB), hence Eq. (8.40) becomes:

$$P_{dB}(L) = P_{dB}(0) - \alpha L. \tag{8.41}$$

In optical fiber communication systems, the power emitted by lasers is typically measured in milliwatts (mW). Therefore, calculations are often performed using decibel-milliwatts (dBm), which is defined as follows:

$$P(\text{dBm}) = 10log\,[P(\text{mW})]. \tag{8.42}$$

When power is given in decibel-watts (dB) or decibel-milliwatts (dBm) and the attenuation coefficient α is in decibels per kilometer, Eq. (8.41) accurately describes the relationship between the input and output powers over a given fiber length is given by:

$$P_{out}(\text{dBm}) = P_{in}(\text{dBm}) - \alpha(\text{dB/km})L(\text{km}). \tag{8.43}$$

Example A 50 km long fiber has an attenuation coefficient of (a) 0.5 dB/km at $\lambda = 1.3\,\mu\text{m}$ and (b) 0.3 dB/km at $\lambda = 1.55\,\mu\text{m}$. If 1 mW of optical power at each wavelength is launched into the fiber, what is the output power at each wavelength?

Solution

(a) $P_{out} = 10log(1\text{ mW}) - 50 \times 0.5 = 10\text{ dBm} - 25\text{ dB} = -15\text{ dBm} = 31.6\,\mu\text{W}$

(b) $P_{out} = 10log(1\text{ mW}) - 50 \times 0.3 = 10\text{ dBm} - 15\text{ dB} = -5\text{ dBm} = 0.316\text{ mW}$

The attenuation of optical power propagating in a fiber is primarily is caused by absorption and scattering. In addition, there are losses due to nonlinear optical effects and external effects such as mechanical losses. These loss mechanisms vary, in their effects, but together, they can contribute to the total fiber loss. Since the majority of optical fibers are silica fibers, we discuss the loss mechanisms and their effects on silica fibers. Some of these losses have fundamental limits on the losses fiber experience, whereas others can be reduced and mitigated.

8.7.1.1 Absorption

As the light propagates in the fiber, photon absorption can occur, leading to electron transitions. The bandgap of silica is approximately 8.9 eV, corresponding to a wavelength

$$\lambda = \frac{1.24}{8.9} = 0.139\,\mu\text{m}.$$

Silica becomes opaque at these ultraviolet (UV) wavelengths. The silica bandgap has bandtails that extend to the visible and infrared regions, leading to an absorption tails in these regions.

Another absorption mechanism occurs in the infrared (IR) region and is caused by transitions between different vibrational modes of silica molecules. The fundamental vibrational transition of fused silica results in a strong absorption peak at about 9 µm wavelength. For silica ($S\dot{I}O2$) molecules, electronic resonances occur in the ultraviolet region ($\lambda < 0.4$ µm), whereas vibrational resonances occur in the infrared region ($\lambda > 7$ µm). Due to the amorphous nature of fused silica, these resonances form of absorption bands whose tails extend into the visible region. That intrinsic material absorption for silica in the wavelength range 0.8–1.6 µm is below 0.1 dB/km. In fact, it is less than 0.03 dB/km in the wavelength window 1.3–1.6 µm commonly used fiber communication systems.

Another significant attenuation mechanism is caused by impurities in the silica. Impurity absorption is particularly important in the near infrared region because many impurity ions such as OH^-, form absorption bands in this region where both the electronic and molecular absorption losses of silica glass are very low. Near the peaks of the impurity absorption bands, an impurity concentration as low as one part per billion can contribute to an absorption loss as high as 1 dB/km. The presence of the OH^- radicals are often introduced by water in the silica fibers. These can occur during the manufacturing process or the presence of humidity in the environment.

Thanks to advanced manufacturing technology, the levels of impurities in fibers have been reduced to the point where losses associated with their absorption are negligible, except for the OH^- radical. The most significant absorption peaks related to this radical are at wavelengths of $0.95, 1.25, and\ 1.39$ µm.

8.7.1.2 Scattering

As light propagates in glass, it encounters random local variations in the refractive index due to random defects, material composition fluctuations, and various inhomogeneities. These variations in refractive index cause the light to scatter in random directions contributing to the attenuation of the light in silica fibers.

The dominant attenuation at short wavelengths is caused by scattering, which is known as **Rayleigh scattering**. Rayleigh scattering is in the range of $1/\lambda^4$, meaning that it the silica attenuation decreases rapidly at longer wavelengths. In optical fibers, the losses associated with Rayleigh scattering dominate at shorter wavelengths, particularly below 1 µm. The intrinsic loss of silica fibers due to Rayleigh scattering (α_R) is expressed as:

$$\alpha_R = C/\lambda^4, \qquad (8.44)$$

where the constant C ranges from 0.7–0.9(dB/km) $-$ µm^4, depending on the constituents of the fiber core. The values of C correspond to $\alpha_R = 0.12\ to\ 0.16$ dB/km at $\lambda = 1.55$ µm.

In addition to Rayleigh scattering, there are other nonlinear mechanisms that cause scattering loss. The main mechanisms are stimulated Brillouin scattering (SBS) and stimulated Raman scattering (SRS). These types of scattering are inelastic, meaning that they cause a change in the photon energy of the propagating

Fig. 8.12 Spectral dependence of loss mechanisms and total attenuation in a fiber

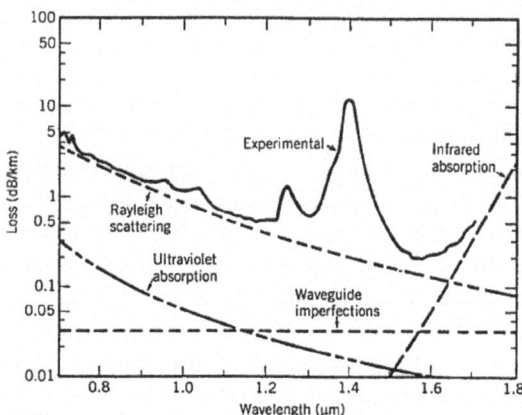

light, unlike Rayleigh scattering, which does not change the frequency of the light. However, both SBS and SRS losses are negligible for low-energy light waves.

A summary of the absorption and scattering losses in silica fibers as a function of wavelength is shown in Fig. 8.12.

8.7.2 Dispersion

Dispersion in optical fibers occurs when different components of a light signal travel at different speeds through the fiber. This effect causes the broadening of the optical pulse as it propagates along the fiber, leading to signal degradation and limiting the bandwidth and transmission distance of the fiber optic communication system.

Figure 8.13 illustrates this dispersion effect. Consider a stream of pulses launched into a fiber as shown in the figure. As these pulses propagate along the fiber, their pulse width increases because of various dispersion mechanisms, whereas their amplitude decreases due to fiber attenuation. Each pulse broadens and begins to overlap with neighboring pulses, eventually becoming indistinguishable at the receiver input. This effect is known as **intersymbol interference** (ISI). As the ISI becomes more pronounced, the error rate on the digital optical channel increases, ultimately limiting the maximum bandwidth that can be achieved. This happens when individual symbols can no longer be clearly distinguished. The error rate is also influenced by signal attenuation along the fiber link and the resulting signal-to-noise ratio (SNR) at the receiver, which will be discussed later in the book.

To avoid overlapping light pulses on an optical fiber link, the digital bit rate B_T must be less than the reciprocal of the broadened (through dispersion) pulse duration (2τ). Thus:

$$B_T \leq \frac{1}{2\tau}, \tag{8.45}$$

where τ is the pulse width.

Fig. 8.13 Digital bit pattern 1011 broadening of light pulses, due to dispersion, as they propagate along an optical fiber: (**a**) pulses at the fiber input, (**b**) pulses at distance L_1 alone the fiber, and (**c**) pulses at a distance $L_2 > L_1$ along the fiber

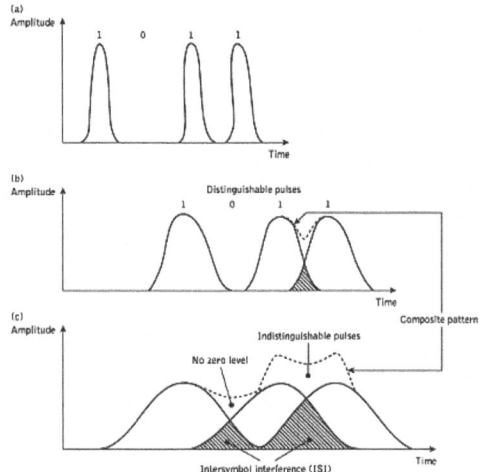

Dispersion in optical fibers arises from several mechanisms, which can be categorized as follows:

1. Modal dispersion occurs in multimode fibers due to differences in group velocity for different modes
2. Chromatic dispersion refers to the dependence of the phase velocity on the frequency of the wave, meaning that the refractive index varies with frequency of the wave. Chromatic dispersion is further divided into:
   (a) **Material Dispersion:** Arising from the wavelength-dependent refractive index of the fiber material, this dispersion results from the fact that different wavelengths travel at different speeds in the fiber material, causing pulse broadening.
   (b) **Waveguide Dispersion**: This is due to the dependence of the effective refractive index on wavelength for the guided modes in a waveguide. It arises because the fraction of light traveling in the fiber core versus the cladding varies with wavelength, impacting the overall speed of the pulse.
3. Polarization mode dispersion.: Occurs because different polarizations travel at different speeds due to small birefringence present in the fiber.

The combined effect of material and waveguide dispersions for a particular mode is known as intramode dispersion. Modal dispersion, also called intermode dispersion, is caused by variations in the propagation constant between different modes and occurs only when more than one mode is excited in a multimode fiber. Importantly, modal dispersion can exist even in the absence of chromatic dispersion. Conversely, if only one mode is excited in a fiber, only intramode chromatic dispersion needs to be considered, even if the fiber is a multimode fiber.

Example A multimode fiber exhibits total pulse broadening of 0.01 μs over a distance of 10 km. Determine: (a) the maximum possible bandwidth on the link assuming no intersymbol interference; (b) the pulse dispersion per unit length; (c) the bandwidth–length product for the fiber.

Solution

(a) The maximum possible optical bandwidth is

$$B_T = \frac{1}{2\tau} = \frac{1}{0.02 \times 10^{-6}} = 50\,\text{MHz}$$

(b) The dispersion per unit length is

$$D = \frac{0.01 \times 10^{-6}}{10} = 1\,\text{ns/km}$$

(c) The bandwidth–length product is

$$B_T L = 50\,\text{MHz} \times 10\,\text{km} = 500\,\text{MHz\,km}$$

Modal Dispersion

Modal dispersion occurs because different modes in a multimode waveguide propagate at different group velocities. In a step-index fiber, the fastest mode (lowest-order mode), can be represented by the axial ray, whereas the slowest mode (highest-order mode) can be represented by the extreme meridional ray. which is incident at the core–cladding interface at the critical angle ϕ_c. The paths taken by these two rays in a step-index fiber are shown in Fig. 8.14. The delay difference between these two rays as they travel through the fiber core provides an estimate of the pulse broadening resulting from intermodal dispersion within the fiber.

Both rays travel at the same velocity within the constant refractive index of the fiber core. The delay difference is directly related to their respective path lengths

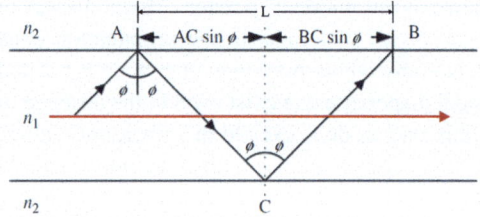

Fig. 8.14 Optical axial and meridian rays paths for the lowest-order and highest-order modes in step-index multimode fiber

8.7 Signal Degradation in Optical Fibers

within the fiber. The time taken for the axial ray to travel between points A and B along a fiber of length L gives the minimum delay time t_{min} given by:

$$t_{min} = \frac{L}{v} = \frac{Ln_1}{c}, \tag{8.46}$$

where n_1 is the refractive index of the core and c is the velocity of light in a vacuum. When the extreme meridional ray travels a distance L along the fiber, the path is longer, following the path $AC+CB$. Hence the maximum delay is given by:

$$t_{max} = \frac{(AC+CB)n_1}{c} = \frac{Ln_1}{c\sin\phi_c}. \tag{8.47}$$

Since the $\sin\phi_c = n_2/n_1$, therefore, the maximum delay is given by:

$$t_{max} = \frac{n_1^2 L}{cn_2}. \tag{8.48}$$

The delay difference δt_{mod} between the highest-order mode and the lowest-order mode is obtained by subtracting Eq. (8.46) from Eq. (8.48). Therefore:

$$\delta\tau_{mod} = t_{max} - t_{min} = \frac{n_1^2 L}{cn_2} - \frac{Ln_1}{c}. \tag{8.49}$$

Simplifying further:

$$\delta\tau_{mod} = \frac{n_1^2 L}{cn_2}\left(\frac{n_1 - n_2}{n_1}\right). \tag{8.50}$$

Alternatively,

$$\delta\tau_{mod} = \frac{n_1^2 L}{cn_2}\Delta, \tag{8.51}$$

where Δ is the refractive index difference. The intermodal pulse broadening of Eq. (8.51) can be expressed in terms of the numerical aperture NA as:

$$\delta\tau_{mod} = \frac{L(NA)^2}{2cn_2} \tag{8.52}$$

This approximate analysis provides an estimate of the maximum pulse broadening due to intermodal dispersion.

Example A 10 km optical link consists of a multimode step-index fiber with a core refractive index of 1.5 and a relative refractive index difference of 1%. Estimate: (a) the pulse broadening due to intermodal dispersion, (b) the maximum bit rate that may be obtained without substantial errors on the link assuming only intermodal dispersion, and (c) the bandwidth–length product.

Solution

(a) The pulse broadening due to intermodal dispersion is

$$\delta \tau_{mod} = \frac{n_1^2 L}{c n_2} \Delta$$

we need to determine the cladding index of refraction n_2

$$\Delta = 0.01 = \frac{n_1 - n_2}{n_1} = \frac{1.5 - n_2}{1.5}$$

$$n_2 = 1.5 - 0.015 = 1.485$$

Therefore,

$$\delta \tau_{mod} = \frac{(1.5)^2 \times 10 \times 10^3}{3 \times 10^8 \times 1.485} = 505 \, \text{ns}$$

(b) The maximum bit rate

$$B_T = \frac{1}{2\delta T} = \frac{1}{2 \times 505 \times 10^{-9}} = 990 \, \text{Mbps}$$

(c) the bandwidth–length product. is

$$B_T \times L = 0.99 \, \text{GHz} \times 10 \, \text{km} = 9.9 \, \text{GHz km}.$$

8.7.2.1 Material Dispersion

Material dispersion is a key factor in the study of optical fibers and other waveguides. Different wavelengths of light travel at different speeds through a material, resulting in a broadening of optical pulses as they propagate over distance. This effect arises from the variations in the material's refractive index with wavelength.

The refractive index of pure silica within the wavelength range between 200 nm and 4μm is described by the empirically fitted Sellmeier equation:

8.7 Signal Degradation in Optical Fibers

$$n^2 = 1 + \frac{0.6961663\lambda^2}{\lambda^2 - (0.0684043)^2} + \frac{0.4079426\lambda^2}{\lambda^2 - (0.1162414)^2} + \frac{0.8974794\lambda^2}{\lambda^2 - (9.896161)^2} \quad (8.53)$$

where λ is in micrometers. The refractive index can be modified by adding dopants to silica, allowing control over the fiber's index profile. The amount of index change depends on the type and concentration of the dopant(s) used. For example, doping with germanium dioxide increases the refractive index. Consequently, the coefficients in the Sellmeier equation actually vary with the composition of the glass.

A pulse of light that contains several light wavelengths experiences different refractive indices and group indices n_g, as each wavelength travels at a phase velocity

$$v_p = \frac{\omega}{k}, \quad (8.54)$$

while the group of wavelengths in the optical pulse travels with a group velocity:

$$v_g = \frac{d\omega}{dk}. \quad (8.55)$$

Since $v = c/n$, then

$$\omega = \frac{c}{n}k. \quad (8.56)$$

To determine the group velocity v_g, differentiate ω with respect to k to get:

$$v_g = \frac{d}{dk}\left(\frac{c}{n}k\right) = \frac{c}{n} - k\left(\frac{c}{n^2}\frac{dn}{dk}\right). \quad (8.57)$$

This can be expressed as:

$$v_g = \frac{c}{n}\left(1 - \frac{k}{n}\frac{dn}{dk}\right) = \frac{c}{n}\left(1 - \frac{k}{n}(-\frac{\lambda}{k}\frac{dn}{d\lambda})\right). \quad (8.58)$$

Simplifying, we get:

$$v_g = \frac{c}{n}\left(1 + \frac{\lambda}{n}\frac{dn}{d\lambda}\right). \quad (8.59)$$

The group velocity as described in Eq. (8.59), is nearly equal to the phase velocity, but is slightly modified by a term proportional to the change in the refractive index with respect to the wavelength. This variation in the refractive index is known as **dispersion**. The phase and group velocities are illustrated in Fig. 8.15.

Fig. 8.15 The refractive index n and group index n_g for pure silica as a function of wavelength

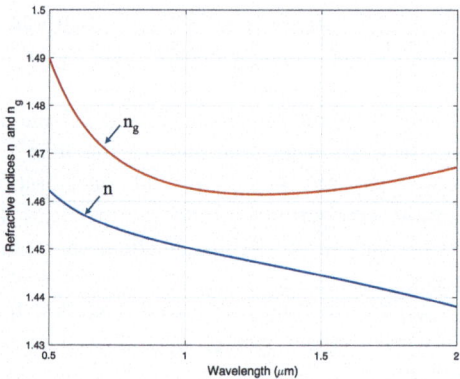

The group index of refraction is given by:

$$n_g = \frac{c}{v_g} = \frac{c}{\frac{c}{n}\left(1 - \frac{k}{n}\frac{dn}{dk}\right)} = n\left(1 - \frac{k}{n}\frac{dn}{dk}\right)^{-1}. \tag{8.60}$$

Since $k = 2\pi/\lambda$, therefore,

$$\frac{dk}{d\lambda} = -\frac{2\pi}{\lambda^2},$$

and

$$\frac{dn}{dk} = \frac{dn}{d\lambda}\frac{d\lambda}{dk} = -\frac{\lambda^2}{2\pi}\frac{dn}{d\lambda} = -\frac{\lambda}{k}\frac{dn}{d\lambda}. \tag{8.61}$$

Substituting Eq. (8.61) into (8.60), we get:

$$n_g = n\left(1 + \frac{\lambda}{n}\frac{dn}{d\lambda}\right)^{-1}. \tag{8.62}$$

Using the binomial expansion, we obtain:

$$n_g \approx n\left(1 - \frac{\lambda}{n}\frac{dn}{d\lambda}\right). \tag{8.63}$$

Considering the nature of the dependency of the refractive index on the wavelength, the group index is indeed larger than the phase index. Therefore, the correct expression for the group index is

$$n_g = n - \lambda\frac{dn}{d\lambda}. \tag{8.64}$$

8.7 Signal Degradation in Optical Fibers

If an optical pulse contains multiple wavelengths, the individual components of this pulse will travel at different group velocities. As a result, these components will reach the end of the fiber at different times, effectively stretching out the pulse duration. This phenomenon is known as **group velocity dispersion** (GVD). Consider an optical pulse with a finite spectral bandwidth, $\delta\lambda$, traveling through an optical fiber. The time required for each component to travel a distance L along the fiber is called the **latency**, which is the product of the group delay, τ_g, and the propagation distance, L is given by:

$$\tau_g = \frac{L}{c} n_g(\lambda).$$

The spectral width of the pulse spans from λ_1 to λ_2 (i.e., $\delta\lambda = |\lambda_2 - \lambda_1|$). Each wavelength component propagates at a slightly different speed. In the time domain, the pulse spread due to material dispersion is given by:

$$\delta\tau_{mat} = \frac{L}{c}\left[n_g(\lambda_2) - n_g(\lambda_1)\right],$$

which simplifies to:

$$= \frac{L}{c}\Delta n_g,$$

and can be further expressed as:

$$\delta\tau_{mat} = \frac{L}{c}\frac{dn_g}{d\lambda}\delta\lambda. \tag{8.65}$$

Taking the first derivative of the group index n_g, Eq. (8.64) with respect to λ and substituting it in Eq. (8.65) we obtain the following for the pulse spread:

$$\delta\tau_{mat} = -\frac{L}{c}\lambda\frac{d^2n}{d\lambda^2}\delta\lambda. \tag{8.66}$$

The term

$$D_m = -\frac{\lambda}{c}\frac{d^2n}{d\lambda^2}, \tag{8.67}$$

where D_m is called the **material dispersion**. Material dispersion is typically measured in units given in $ps/(nm\ km)$, which represents the number of picoseconds by which the pulse spreads as it travels 1 km per nanometer of spectral bandwidth of the optical pulse.

Therefore, the pulse spread due to material dispersion is expressed as:

$$\delta\tau_{mat} = D_m L \delta\lambda. \tag{8.68}$$

If an optical pulse with a spectral width $\delta\lambda$ travels a distance L along an optical fiber, the pulse spread is given by

$$\delta\tau_{mat} = LD_m\delta\lambda. \tag{8.69}$$

8.7.2.2 Waveguide Dispersion

Waveguide dispersion occurs because, in a single-mode fiber, a portion of the light energy propagates in the cladding. Since the index of refraction in the cladding is slightly lower than that in the core, the part of the signal traveling in the cladding moves faster, causing the energy within a pulse to spread. Although waveguide dispersion effects are usually minor, they can become significant in certain cases. For example, in dispersion-compensating fibers, waveguide dispersion is deliberately utilized and maximized to achieve an overall dispersion coefficient that opposes the dispersion in standard fibers.

Figure 8.16 illustrates both material dispersion and waveguide dispersion. The material dispersion is negative for wavelengths $\lambda \lesssim 1.29\,\mu m$ and is positive for wavelengths $\lambda \gtrsim 1.29\,\mu m$. Importantly, the material dispersion for silica fibers is zero at $\lambda = 1.29\,\mu m$. The waveguide dispersion is negative for this wavelength range. The combined effect of waveguide dispersion shifts the material dispersion curve, resulting in chromatic dispersion with zero dispersion at approximately $\lambda \approx 1.45\,\mu m$. These values are provided for demonstration purposes.

The waveguide dispersion is given by:

$$D_{wg} = \left(\frac{n_1 - n_2}{\lambda c}\right) \frac{V d^2(Vb)}{dV^2},$$

where V and b are the normalized parameters of the optical fiber given by

$$V = \frac{2\pi}{\lambda} d \sqrt{n_1^2 - n_2^2},$$

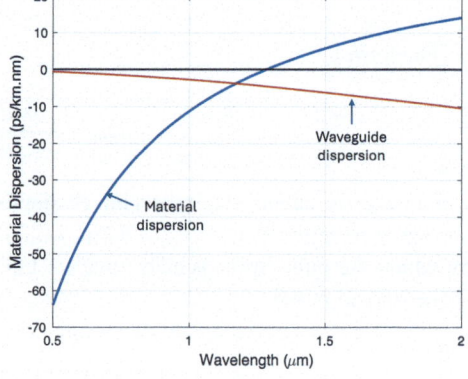

Fig. 8.16 Schematic of the material and waveguide dispersions for a fused silica fiber as a function of the wavelength. The chromatic dispersion is the sum of both dispersions (plot is not to scale)

8.7 Signal Degradation in Optical Fibers

and

$$b = \frac{\beta^2 - k_1^2}{k_1^2 - k_2^2}.$$

8.7.2.3 Total Dispersion

In summary, for multimode fibers, modal dispersion is dominant because chromatic dispersion is relatively small in comparison. In single-mode fibers, where modal dispersion does not occur, chromatic dispersion becomes the limiting factor for the fiber's bandwidth. The total pulse broadening in fibers is generally given by:

$$\delta\tau_{total} = \sqrt{(\delta\tau_{mat} + \delta\tau_{wg})^2 + \delta\tau_{mod}}. \tag{8.70}$$

Example An optical fiber link with the following parameters is operated at $\lambda = 1.55 \, \mu m : n_1 = 1.49, n_2 = 1.47$ and the chromatic dispersion at the operating wavelength is -20 ps/km.nm and the pulse spectral width is $\delta\lambda = 0.5$ nm.

Determine the pulse broadening and bandwidth for the following cases:

(a) Multimode fiber with $L = 1$ km
(b) Single-mode fiber with $L = 1$ km.
(c)) Single-mode fiber with $L = 10$ km.

Solution

(a) the modal dispersion is given by Eq. (8.51) is

$$\delta T = \frac{n_1^2 L}{cn_2}\Delta$$

Finding Δ

$$\Delta = \frac{n_1 - n_2}{n_1} = \frac{1.49 - 1.47}{1.49} = 0.0134$$

Therefore, the pulse broadening

$$\delta\tau_{mod} = \frac{1.49^2 \times 10^3 \times 0.0134}{3 \times 10^8 \times 1.47} = 67.5 \, \text{ns}$$

(continued)

The chromatic dispersion is

$$\delta\tau_{ch} = L|D_{ch}|\delta\lambda = 1 \times 20 \times 0.5 = 2\,\text{ps}$$

The total dispersion

$$\delta\tau_{total} = \sqrt{(\delta\tau_{mat} + \delta\tau_{wg})^2 + \delta\tau_{mod}}$$
$$= \sqrt{(2 \times 10^{-12})^2 + (67.5 \times 10^{-9})^2} \approx 67.5\,\text{ns}$$

and the bandwidth

$$B_T = \frac{1}{2\delta\tau_{total}} = \frac{1}{67.5 \times 10^{-9}} = 14.8\,\text{MHz}.$$

(b) In single-mode fibers, the only dispersion is the chromatic dispersion so the pulse broadening for $L = 1\,\text{km}$ is

$$\delta\tau_{ch} = L|D_{ch}|\delta\lambda = 2\,\text{ps}$$

and the bandwidth

$$B_T = \frac{1}{2\delta\tau_{ch}} = \frac{1}{2 \times 10^{-12}} = 500\,\text{GHz}.$$

(c) In single-mode fiber, the only dispersion is the chromatic dispersion so the pulse broadening for $L = 10\,\text{km}$ is

$$\delta\tau_{ch} = L|D_{ch}|\delta\lambda = 20\,\text{ps}$$

and the bandwidth

$$B_T = \frac{1}{2\delta\tau_{ch}} = \frac{1}{20 \times 10^{-12}} = 50\,\text{GHz}.$$

8.7.2.4 Dispersion Management

As discussed earlier, dispersion is a critical factor that limits the bandwidth and data rate of an optical fiber link. Therefore, reducing or eliminating dispersion is essential for achieving high data rates in long-haul fiber communications, particularly in single-mode fiber links.

Chromatic dispersion can be zero at specific wavelengths, depending on the interplay between material dispersion and waveguide dispersion. Material dispersion is inherent to the fiber material, whereas waveguide dispersion depends on the fiber's structural design. To manage dispersion effectively, specially designed fibers, known as dispersion-shifted fibers (DSF), have been developed to shift the zero-dispersion wavelength. These specialty fibers, briefly described in Sect. 8.3.3, adjust the waveguide properties to achieve minimal dispersion at desired wavelengths.

8.8 Summary

This chapter summarizes the key aspects of optical fiber technology and its impact on telecommunications. The chapter covers the structure and function of optical fibers, emphasizing how the core–cladding configuration enables light propagation through total internal reflection, making fiber optics indispensable in long-distance communication due to low attenuation and high bandwidth. A historical overview outlines major developments, from early discoveries to breakthroughs in low-loss silica fibers, which paved the way for modern fiber-optic systems.

The chapter categorizes optical fibers by material and structure, explaining types such as single-mode and multimode fibers, step-index, graded-index, and specialty fibers like polarization-maintaining and dispersion-compensation fibers. The propagation of light in fibers is analyzed in terms of guided modes, with detailed discussions on the mathematical framework, including numerical aperture, critical angles, and Bessel functions, to describe light confinement and wave propagation.

Signal degradation is explored extensively, identifying both linear effects (like attenuation due to scattering and absorption) and nonlinear effects (such as self-phase modulation and four-wave mixing) that impact signal integrity. Different forms of dispersion—modal, chromatic, and polarization mode—are analyzed, showing their role in pulse broadening and data transmission limits. Techniques to manage these effects, including the use of dispersion-shifted and dispersion-flattened fibers, are introduced.

In summary, this chapter provides foundational knowledge of optical fiber properties, types, and applications, with insights into how engineering advances continue to mitigate signal degradation, allowing for faster and more reliable data transmission over vast distances.

8.9 Problems

1. A silica optical fiber with a core diameter large enough to be considered by ray theory analysis has a core refractive index of 1.50 and a cladding refractive index of 1.47. Determine: (a) the critical angle at the core–cladding interface; (b) the NA for the fiber; (c) the acceptance angle in air for the fiber.
2. For an optical fiber with a core refractive index of 1.495 and cladding index of 1.46, find (a) the numerical aperture, and (b) the maximum acceptance angle in air.

3. The relative refractive index difference for an optical fiber designed for long-haul transmission is 1%. (a) Determine the NA and the solid acceptance angle in air for the fiber when the core index is 1.46. (b) Calculate the critical angle at the core–cladding interface within the fiber.
4. A 5 km multimode step-index fiber has a core refractive index of 1.48 and a relative refractive index difference of 0.01. Estimate the pulse broadening due to modal dispersion.
5. Given the material dispersion coefficient $D_m = 20$ ps/nm-km at 1.55 μm wavelength, calculate the chromatic dispersion-induced pulse spread over a 10 km link for a light source with a spectral width of 2 nm.
6. For an optical fiber with core refractive index 1.45 and cladding index 1.44, find the core radius required for single-mode operation at a wavelengths of 1.3 μm, and 1.55 μm.
7. A multimode step index fiber with a core diameter of 80 μm and a relative index difference of 0.015 is operating at a wavelength of 0.85 μm. If the core refractive index is 1.48. Determine: (a) the normalized frequency for the fiber; (b) the number of guided modes.
8. A standard single-mode fiber has chromatic dispersion of $D = 17$ ps/nm-km at 1550 nm. Determine the required length of a dispersion-compensating fiber with $D = -100$ ps/nm-km to counteract dispersion over a 100 km link.
9. Determine the maximum core diameter for an optical fiber with a relative refractive index difference (0.015) and core refractive index (1.48) to operate as single-mode. It may be assumed that the fiber is operating at the same wavelength (0.85 μm).
10. Determine the cutoff wavelength for a step index fiber to exhibit single-mode operation when the core refractive index and radius are 1.46 and 4.5 μm, respectively, with the relative index difference being 0.0025.

Bibliography

1. G. Agrawal (2009), Optical Fiber Communication Systems, 4th ed., Wiley.
2. D. Marcuse (1991), Theory of Dielectric Optical Waveguides, 2nd ed., Academic Press, San Diego, CA.
3. G. Cancellieri (1991), Single-Mode Optical Fibers, Pergamon Press, Elmsford, NY.
4. J. A. Buck (2004), Fundamentals of Optical Fibers, 2nd ed., Wiley, Hoboken, NJ.
5. J. M. Senior (2009), Optical Fiber Communications: Principles.
6. D. Gloge (1971), "Weakly guiding fibers," Applied Optics, vol. 10, pp. 2252–2258.
7. G. Cancellieri (1991), Single-Mode Optical Fibers, Pergamon Press, Elmsford, NY.
8. J. A. Buck, (2004) Fundamentals of Optical Fibers, 2nd ed., Wiley, Hoboken, NJ.
9. J. Hecht (2005), Understanding Fiber Optics, 5th ed., Pearson.
10. R. G. Hunsperger (2009), Integrated Optics: Theory and Technology, 6th ed., Springer.

Introduction to Quantum Mechanics 9

9.1 Introduction

In the first part of this book, we employed classical physics to explain various aspects of light. Classical physics relies on two fundamental concepts: the notion of a particle, a discrete entity with a well-defined position and momentum governed by Newton's laws of motion, and the concept of an electromagnetic wave, an extended physical entity with a presence at every point in space, characterized by electric and magnetic fields evolving according to Maxwell's laws of electromagnetism. This classical framework offers an orderly description: particle motion laws account for the material world, whereas electromagnetic field laws explain the behavior of light waves that illuminate our surroundings.

However, this elegant theory began to crumble in the late nineteenth century when classical physics failed to explain experimental results related to blackbody radiation and the subsequent photoelectric effect. These shortcomings paved the way for a new theory capable of explaining these phenomena, ushering in a paradigm shift away from the deterministic world of classical physics.

In this chapter, we provide a concise introduction to quantum mechanics, a prerequisite for comprehending the theory required to explore semiconductor optoelectronic devices in subsequent chapters. While we would ideally delve directly into the discussion of these devices, a fundamental understanding of electron behavior in semiconductors under various potential functions is essential to grasp current-voltage characteristics.

Throughout this chapter, we delve into the foundational principles of quantum mechanics, covering critical concepts such as wave-particle duality, superposition, and entanglement. Moreover, we acquaint readers with the mathematical tools underpinning these concepts.

In the following sections, we revisit the experimental results pertaining to blackbody radiation toward the end of the nineteenth century, highlighting the inadequacy of the classical wave theory of light in explaining these observations.

This scientific conundrum spurred the development of a new theory rooted in the quantization of energy, initially proposed by Max Planck and later expanded upon by Albert Einstein. This pivotal development gave rise to the wave-particle duality of light, as previously introduced in Chap. 1.

We subsequently delve into the wave-particle duality of matter, a concept introduced by Louis de Broglie. This duality played a crucial role in formulating the Schrödinger wave equation, a foundational tool for describing the behavior of microscopic particles, particularly electrons. In the following sections, we explore solutions to the Schrödinger wave equation, addressing specific scenarios such as the behavior of single particles within potential wells and atoms.

9.2 Blackbody Radiation

A blackbody is an idealized physical object that absorbs all incident electromagnetic radiation, regardless of its frequency or angle of incidence. The emission coefficient of a blackbody is the highest possible for any given frequency, and it emits thermal radiation with a specific spectrum that depends solely on its temperature; this phenomenon is known as blackbody radiation. While there are no perfect blackbodies in reality, there are many objects that closely approximate one, such as charcoal, stars, and various everyday objects. For example, when charcoal is heated, it emits radiation that initially appears red and gradually shifts toward a yellowish color as the temperature increases. Whereas a blackbody is heated, it emits electromagnetic radiation across a broad range of frequencies.

The exploration of blackbody radiation has played a pivotal role in the evolution of modern physics, marking the inception of quantum mechanics. Classical physics methods fall short in providing an accurate account of the distribution of radiation emitted by a blackbody across various frequencies. This shortfall in classical theory led to a notable theoretical anomaly termed the "ultraviolet catastrophe," wherein the predicted total energy emitted by the blackbody tends to infinity as the frequency of light approaches the ultraviolet spectrum. This divergence between theory and experimental reality was a glaring inconsistency that demanded a new theory.

9.2.1 The Rayleigh-Jeans Radiation Law

Many attempts has been made to predict the radiation of blackbody on the basis of electrodynamic laws known at that time. Let us start by describing the radiation emerging from a small hole in a highly polished cavity as shown in Fig. 9.1.

Let us consider a cavity with dimensions a, b, and d, as illustrated in Fig. 9.1 with a radiating electromagnetic energy through a tiny hole on its surface.. This cavity can be considered as a three-dimensional resonator, as previously discussed in Chap. 6. The energy emitted from this cavity can be assessed through electromagnetic radiation. The cavity can store significant energy at the resonant frequencies

9.2 Blackbody Radiation

Fig. 9.1 A three-dimensional cavity emitting electromagnetic radiation from a tiny hole

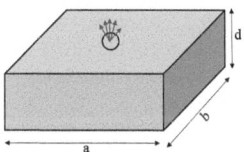

characterized by the following wavenumbers:

$$k^2 = \left(\frac{\omega}{c}\right)^2 = k_x^2 + k_y^2 + k_z^2 = \left(\frac{m\pi}{a}\right)^2 + \left(\frac{p\pi}{b}\right)^2 + \left(\frac{q\pi}{d}\right)^2, \quad (9.1)$$

where m, p and q are positive integers. Therefore, the resonant frequencies, or wavelengths, are discrete and depend on the size of the cavity. Let us assume that the cavity is cubical such that $b = d = a$. Therefore, Eq. (9.1) becomes:

$$k^2 k = \left(\frac{2\pi \nu}{c}\right)^2 = \left(\frac{\pi}{a}\right)^2 \left(m^2 + p^2 + q^2\right), \quad (9.2)$$

hence, the resonant frequencies ν are given by:

$$\nu = \frac{c}{2a}\left[m^2 + p^{2+}q^2\right]^{\frac{1}{2}}, \quad (9.3)$$

and

$$\frac{4a^2\nu^2}{c^2} = m^2 + p^2 + q^2. \quad (9.4)$$

This equation represents a sphere with a radius of $2a\nu/c$. Each resonant frequency corresponds to a point within this sphere. However, due to the positive values of the wavenumbers k_x, k_y, and k_z, the resonant frequencies are confined to one-eighth of the sphere, as shown in Fig. 9.2. It's worth noting that a single set of m, p, and q values represents one mode of vibration. The electromagnetic field has an infinite number of modes (standing waves) in the cavity. The blackbody radiation field is a superposition of planewaves of different frequencies.

The total number of resonances between $\nu = 0$ *and* ν in one-eighth of the sphere with radius $2a\nu/c$ are given by:

$$N(\nu) = \frac{1}{8} \times \frac{4\pi}{3}\left(\frac{2a\nu}{c}\right)^3 = \frac{4\pi \nu^3}{3c^3}a^3, \quad (9.5)$$

Fig. 9.2 Resonant frequencies of the cavity that exist between frequencies v and $v + dv$

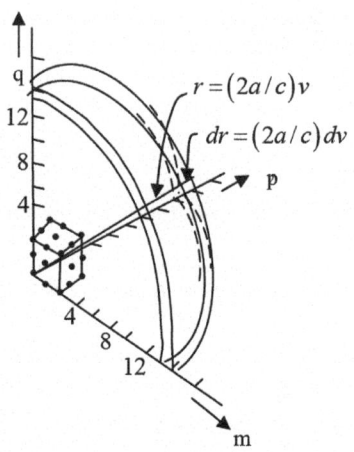

The volume of each mode is $V = a^3$, and the derivative of the number of modes with respect to frequency is given by:

$$\frac{dN}{dv} = \frac{4\pi V v^2}{c^3}. \tag{9.6}$$

Now, if we express these equations in terms of the wavelength λ, we obtain:

$$N(\lambda) = \frac{4\pi V}{3\lambda^3}, \tag{9.7}$$

and

$$\frac{dN}{d\lambda} = \frac{4\pi V}{\lambda^4}. \tag{9.8}$$

According to the kinetic theory of gases, each standing wave, which is equivalent to a harmonic oscillator, contributes energy in the form of kT at equilibrium absolute temperature T. Consequently, the energy per unit volume within the frequency interval v and $v+dv$ in the blackbody spectrum of the cavity at absolute temperature T can be computed as the product of the average energy per standing wave and the total number of standing waves present.

The mode density, which denotes the number of modes per unit volume within the dv frequency interval, can be expressed as follows:

$$g(v)dv = \frac{1}{V}\frac{dN}{dv}dv, \tag{9.9}$$

$$g(v)dv = \frac{8\pi v^2}{c^3}dv. \tag{9.10}$$

9.3 Planck's Radiation Law

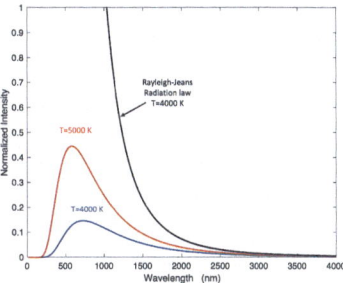

Fig. 9.3 Blackbody radiation as a function of wavelengths for observed radiation (red and blue lines) and the Rayleigh-Jeans radiation law (black line). The plot for the classical method given by the Rayleigh-Jeans law prediction fails at short wavelengths to match the experimental results

According to classical physics, each mode possesses an energy of $k_B T$, where k_B is the Boltzmann constant. The energy density per $d\nu$ is given by:

$$\mathcal{E}(\nu, T) = \left(\frac{8\pi \nu^2}{c^3}\right) k_B T. \tag{9.11}$$

Similarly,

$$\mathcal{E}(\nu, T) = \left(\frac{8\pi}{\lambda^4}\right) k_B T. \tag{9.12}$$

This formula in Eq. (9.12) is known as the **Rayleigh-Jeans** radiation law. We show that it fits the experimental results at low frequencies (long-wavelengths), but it fails to accurately fit the experimental data for high-frequencies (short-wavelengths) emission shown in Fig. 9.3. The Rayleigh-Jeans law, which is based on classical physics, erroneously predicts that a heated body would emit infinite amounts of ultraviolet radiation, a conclusion known as the ultraviolet catastrophe.

9.3 Planck's Radiation Law

The discussion of the photoelectric effect in Chap. 1 offers a crucial insight into the limitations of classical light theory and the emergence of quantum mechanics. This experiment demonstrated that light shining on a metal surface can eject electrons, referred to as photoelectrons. Contrary to the classical physics prediction, which postulates that light of any color or frequency could release an electron from a material if it were sufficiently bright, the experiment reveals a different phenomenon.

It was observed that the maximum speed of the ejected photoelectrons is determined by the light's color, or frequency, rather than its brightness, or intensity.

Fig. 9.4 Kinetic energy of a photoelectron extracted from a material surface illuminated with a light beam of frequency v. The minimum frequency of the photon needed to extract a photoelectron is v_o, which is a function of the work function of the material

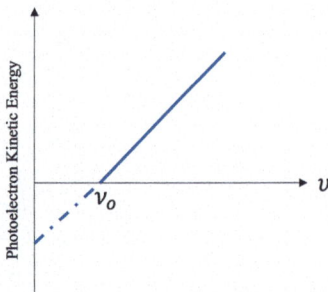

Crucially, a minimum frequency of light is necessary to dislodge electrons, as depicted in Fig. 9.4. Altering the light's brightness affects the quantity of ejected electrons but does not influence their maximum speed.

This disparity with classical theory was instrumental in the development of quantum mechanics. Quantum mechanics accounts for the photoelectric effect by introducing the concept of photons, which are particles of light possessing discrete energy levels.

In 1900, Max Planck addressed the discrepancy between observed blackbody radiation and the classical electromagnetic prediction, known as the Rayleigh-Jeans law. Planck's groundbreaking approach involves proposing that the energy emitted by heated surfaces is not continuous, as classical physics has suggested, but rather is quantized. This means it is released in fixed, discrete amounts, which he termed "quanta." The energy ($\mathcal{E}$) of each quanta correlates with the light frequency (v) and is calculated using Planck's constant (h):

$$\mathcal{E} = hv, \tag{9.13}$$

where

$$h = 6.626 \times 10^{-34} \, Js. \tag{9.14}$$

Planck's hypothesis marked a profound shift from classical theories, where energy was considered a continuous variable with no restrictions on its values. His introduction of quantization, the idea that energy is emitted or absorbed in distinct, fixed quantities, was pivotal in addressing the "ultraviolet catastrophe"—a considerable problem in classical physics.

By proposing that energy is quantized, Planck not only resolved the blackbody radiation problem but also laid the cornerstone for quantum mechanics. This new field revolutionized our understanding of energy and matter, particularly at the microscopic level.

In 1905, building on Planck's theory, Albert Einstein extended the concept of quantization to light energy, suggesting that it is composed of discrete packets or photons. According to Einstein, the photon energy ($\mathcal{E}$) is directly proportional to its

9.3 Planck's Radiation Law

frequency (ν), following the equation $\mathcal{E} = h\nu$. In the context of the photoelectric effect, Einstein determined that a photon with sufficient energy can eject an electron from a material's surface.

The minimum energy necessary to remove an electron is termed the material's work function. Any additional energy from the photon then translates into the kinetic energy ($\mathcal{K}$) of the ejected photoelectron. This theoretical framework has been substantiated through empirical evidence, as depicted in Fig. 9.4.

Therefore, the photoelectric effect not only reveals the quantized nature of photon energy but also highlights the notion that light possess particle-like attributes. This understanding forms a fundamental aspect of the quantum theory of light.

In his analysis of blackbody radiation, Max Planck focused on the energy per mode of the electromagnetic spectrum. He introduced the initial quantum hypothesis, which posited that electromagnetic (EM) energy at a specific frequency (ν) can exist only in discrete amounts, specifically as multiples of a quantum of energy, expressed as $h\nu$. According to this hypothesis, energy values between $h\nu$ and $2h\nu$ are not possible—energy increments occur in these fixed quantum steps.

Planck then revisited the principles of Boltzmann statistics to explain the distribution of these energy quanta among the different modes. In this context, the relative probability for a particular energy state $\mathcal{E}_j$ is given by the Boltzmann factor: $e^{-\mathcal{E}_j/k_B T}$, where k_B is Boltzmann's constant and T is the absolute temperature.

The Boltzmann constant is given by:

$$k_B = 1.380649 \times 10^{-23} \text{ J/K}.$$

The Boltzmann factor indicates that the likelihood of a system occupying a state with energy $\mathcal{E}_j$ decreases exponentially with increasing energy and inversely depends on the temperature.

Planck's integration of quantum hypotheses with classical statistical mechanics was crucial in deriving his law for blackbody radiation, which accurately described the observed radiation spectrum and resolved the inconsistencies of classical theories. This groundbreaking work not only explained the behavior of blackbody radiation but also paved the way for the development of quantum mechanics.

To derive Planck's radiation law, we start by determining the average energy of all the modes emitted by the blackbody. The energy emitted by each mode is $\mathcal{E}_m = m h \nu$, where $m = 0, 1, 2, 3, \ldots$ is the mode number. Therefore,

$$< \mathcal{E} > = \frac{e^{-0/k_B T}(0) + e^{-h\nu/k_B T}(h\nu) + e^{-2h\nu/k_B T}(2h\nu) + \ldots}{e^{-0/k_B T} + e^{-h\nu/k_B T} + e^{-2h\nu/k_B T} + \ldots}, \tag{9.15}$$

which can be represented by the infinite sums as:

$$< \mathcal{E} > = \frac{\sum_{m=0}^{\infty} m h\nu e^{-m h\nu/k_B T}}{\sum_{m=0}^{\infty} e^{-m h\nu/k_B T}}. \tag{9.16}$$

where $e^{-mh\nu/k_B T}$ is the relative probability of energy $\mathcal{E}_m$ and the sum $\sum_{m=0}^{\infty} e^{-mh\nu/k_B T}$ is the sum of all relative probabilities, which is the sum of infinite geometric series. Let $\xi = h\nu/k_B T$, so the sum of all relative probabilities is given by:

$$\sum_{0}^{\infty} e^{-mh\nu/k_B T} = \sum_{0}^{\infty} e^{-m\xi} = \frac{1}{1 - e^{-\xi}}. \tag{9.17}$$

Hence, the average energy of all modes is given by:

$$<\mathcal{E}> = \frac{h\nu}{e^{h\nu/k_B T} - 1}. \tag{9.18}$$

Consequently, by modifying Eq. (9.11), the energy density of the blackbody radiation can be expressed as:

$$g(\nu) = \frac{8\pi \nu^2}{c^3} \frac{h\nu}{e^{h\nu/k_B T} - 1} \text{ J/m}^3, \tag{9.19}$$

This is known as the **Planck's radiation law** which gives the energy density of the blackbody radiation.

We can express blackbody radiation in terms of wavelength by initially recognizing that radiation is equivalent in terms of wavelength or frequency, as follows:

$$g(\nu)d\nu = g(\lambda)d\lambda. \tag{9.20}$$

We then derive the expression for $g(\lambda)$ as:

$$g(\lambda) = g(\nu) \frac{d\nu}{d\lambda}. \tag{9.21}$$

Considering that $\nu = c/\lambda$, we find that:

$$\frac{d\nu}{d\lambda} = -\frac{c}{\lambda^2}, \tag{9.22}$$

where the negative sign indicates the inverse proportionality between the wavelength and frequency. Consequently, we can omit the negative sign in our derivation of blackbody radiation in terms of wavelength. Thus, we have:

$$g(\lambda) = \frac{8\pi \nu^2}{c^3} \frac{h\nu}{e^{h\nu/k_B T} - 1} \times \frac{c}{\lambda^2}.$$

Therefore, **Planck's radiation law** terms of the wavelength can be expressed as:

9.3 Planck's Radiation Law

Fig. 9.5 Blackbody radiation energy density emitted as a function of wavelength and temperature

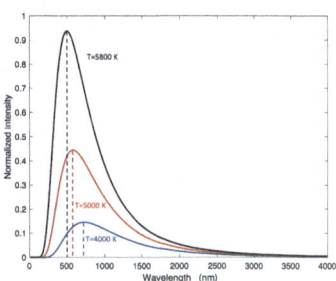

$$g(\lambda) = \frac{8\pi hc}{\lambda^5} \frac{1}{e^{hc/\lambda k_B T} - 1}. \tag{9.23}$$

The energy density for a blackbody radiation is plotted using Eq. (9.23) in Fig. 9.5. The plots match the obsessed the radiation of a typical blackbody. The plots indicate that the peak wavelength for the blackbody radiation shifts to shorter wavelengths as the temperature increases. The peak wavelength as a function of temperature is given by **Wien's displacement law** as follows:

$$\lambda_p = \frac{2.898 \times 10^{-3} \text{ m.K}^\circ}{T}. \tag{9.24}$$

Example Calculate the peak wavelength for the human body ($T = 310$ K), molten iron ($T = 1800$ K), and the sun ($T = 5800$ K).

Solution The peak wavelength for the human body is:

$$\lambda_p = \frac{2.898 \times 10^{-3}}{310} = 9.35 \,\mu\text{m},$$

for the molten iron is:

$$\lambda_p = \frac{2.898 \times 10^{-3}}{1800} = 1.62 \,\mu\text{m},$$

and for the sun is:

$$\lambda_p = \frac{2.898 \times 10^{-3}}{5800} = 500 \,\text{nm}.$$

Therefore, the only peak wavelength in the visible region of the spectrum is that of the sun, while the other two are in the infrared region.

Example Calculate the emission intensity of the sun in a 20 nm region centered at 500 nm.

Solution The energy density over $\Delta\lambda = 20$ nm of the radiation is given by:

$$g(\lambda)\Delta\lambda = \frac{8\pi hc}{\lambda^5} \frac{\Delta\lambda}{e^{hc/\lambda k_B T} - 1} = 8.7 \times 10^{-3} \text{ J/m}^3.$$

Therefore the intensity emitted at the surface is:

$$I(\lambda) = g(\lambda)\Delta\lambda c = 337 \text{ W/cm}^2.$$

9.4 Quantum Mechanics and the Schrödinger Equation

Planck's success in explaining blackbody radiation through the quantization of light was a pivotal moment in physics, leading to the concept of wave-particle duality for photons. This discovery raised profound questions about the nature of matter at the microscopic scale, especially during the early twentieth century. Scientists begun to delve deeper into the structure and behavior of atoms.

Niels Bohr, a key figure in this exploration, proposed a model of the atom that significantly differed from earlier models. According to Bohr, an atom consists of a heavy, positively charged nucleus surrounded by negatively charged electrons. These electrons orbit the nucleus, much like planets orbiting the sun, but their behavior is governed by a set of quantum rules that Bohr postulated.

Additionally, Louis de Broglie extended the concept of wave-particle duality beyond photons. He proposed that electrons also exhibit this duality, behaving both as particles and as waves. This revolutionary idea represents a significant step forward in quantum mechanics, providing a new way to understand the behavior of particles at the atomic scale. De Broglie's hypothesis laid the groundwork for the development of wave mechanics, an essential component of quantum theory.

9.4.1 De Broglie Wave-Particle Duality

Louis de Broglie proposed a groundbreaking concept in physics: electrons possess a dual nature, exhibiting both particle and wave-like behaviors. This duality, however, requires empirical validation, akin to the photoelectric and Compton effects which demonstrate the dual nature of photons. The notion of particles and waves being distinct yet complementary is central to understanding this duality. This implies that an object cannot exhibit its wave and particle characteristics simultaneously.

9.4 Quantum Mechanics and the Schrödinger Equation

Instead, the nature revealed—whether wave-like or particle-like—depends on the specific experimental setup.

For example, the double-slit experiment vividly demonstrates the wave aspect of electrons. In this experiment, electrons passing through two adjacent slits create an interference pattern characteristic of waves. On the other hand, the photoelectric effect showcases the particle nature of the electrons. Here, electrons are emitted from a material when it absorbs photons, behaving distinctly as particles. This duality forms a cornerstone of quantum mechanics, illustrating the complex and often counterintuitive nature of the microscopic world.

Louis de Broglie, pioneering quantum theory, posited that particles such as electrons possess wave-like characteristics. This idea was revolutionary, suggesting that particles could exhibit behavior akin to electromagnetic waves. A fundamental aspect of this theory is the concept of a particle's wavelength.

To understand this, consider the momentum p of a particle, expressed as:

$$p = mv, \tag{9.25}$$

where m is the particle mass and v is its velocity. This formula is rooted in classical mechanics, providing a bridge to quantum mechanics.

Einstein's renowned equation:

$$E = mc^2. \tag{9.26}$$

This equation links a particle's energy E with its mass m and the speed of light c. This equation was a cornerstone in the development of modern physics, revealing the profound relationship between mass and energy.

For photons, which are particles of light, the momentum is given by:

$$p = \frac{E}{c} = \frac{h\nu}{c} = \frac{h}{\lambda}, \tag{9.27}$$

where h is Planck's constant, ν is the frequency of the photon, and λ is its wavelength. This relationship underscores the wave-particle duality inherent in photons.

De Broglie extended this concept to all particles, including electrons, proposing that the wavelength λ of a particle could be determined using the formula:

$$\lambda = \frac{h}{p}, \tag{9.28}$$

where λ is known as **de Broglie wavelength,** a critical concept in quantum mechanics that reveals the wave-like properties of particles. This wavelength is inversely proportional to the particle's momentum, illustrating how particles exhibit both particle and wave-like behaviors, a fundamental principle of quantum mechanics.

Example What is the wavelength of an electron traveling at 50,000 m/s and the mass of the electron $m_e = 9.1 \times 10^{-31}$ kg?

Solution The electron wavelength is:

$$\lambda = \frac{h}{p} = \frac{h}{m_e v} = \frac{6.626 \times 10^{-34}}{(9.1 \times 10^{-31} \text{kg})(5 \times 10^4 \text{m/s})}$$

$$\lambda = 14.6 \text{ nm}.$$

How about the wavelength of horse ($m = 600$ kg) runs at a speed of $v = 80$ km/h?

$$\lambda = \frac{6.626 \times 10^{-34}}{(600 \text{kg})(80 \times 1000/3600) \text{ m/s}} = 5 \times 10^{-38} \text{ m}.$$

9.4.2 Schrödinger Equation Origins

Erwin Schrödinger's seminar at the University of Vienna, which focused on Louis de Broglie's work and the concept of wave-particle duality, highlighted a crucial question in early quantum theory: if electrons exhibit wave-like properties, what is the wave equation that governs their behavior? This question marked a pivotal moment in the development of quantum mechanics.

Schrödinger's pursuit of a wave equation for electrons necessitated bridging the gap between classical mechanics, which describes particles through energy and momentum, and traditional wave theory, which is characterized by frequency and wavelength. The challenge was to create a framework that unified these seemingly disparate aspects of physical reality.

For an electron, or any quantum particle, the duality is evident in its possession of both a wavelength, denoted as λ, and a frequency, ν. These are given by the de Broglie relations:

$$\lambda = \frac{p}{h} = \frac{mv}{h},$$

and

$$\nu = \frac{E}{h}.$$

9.4 Quantum Mechanics and the Schrödinger Equation

The energy of the particle E is given by the sum of its kinetic and potential energies, i.e.,

$$E = KE + PE. \tag{9.29}$$

The particle kinetic energy is given by:

$$KE = \frac{1}{2}mv^2 = \frac{p^2}{2m}, \tag{9.30}$$

and if the potential energy is given by $V(x, t)$, then the total energy of the particle

$$E = \frac{p^2}{2m} + V. \tag{9.31}$$

The force F exerted on a particle by the potential energy gradient is given by:

$$F = -\frac{\partial V}{\partial x}. \tag{9.32}$$

This relationship is fundamental in classical mechanics, linking the force acting on a particle to the spatial variation in its potential energy.

The principle of linearity, which is central to classical wave theory, states that the sum of individual solutions to a wave equation also forms a valid solution. This concept is exemplified in the Helmholtz wave equation used in electromagnetism, where the superposition principle allows for the addition of wave solutions. Schrödinger aimed to incorporate this principle into the quantum domain.

To achieve this, Schrödinger introduced the wave function, $\Psi(x, t)$, as a fundamental descriptor of a quantum particle's state. The wave function, a complex-valued function of position and time, encapsulates the probabilistic nature of quantum particles.

Let the complex wave function be represented by $\Psi(x, t)$. If $\Psi_1(x, t)$ and $\Psi_2(x, t)$ are different solutions of the wave equation, then

$$\Psi(x, t) = a\Psi_1(x, t) + b\Psi_2(x, t) \tag{9.33}$$

is also a solution. Schrödinger's postulate that a constant potential leads to a sinusoidal solution with constant frequency and wavelength in the derived wave equation is a key aspect of quantum mechanics. In this scenario, where the potential energy $V(x)$ is constant, the force F experienced by a particle in this potential field can be examined.

When the potential $V(x)$ is constant, it implies that it does not vary with position x or time t. Therefore, the force that is related to the spatial derivative of the potential is given by:

$$F = -\frac{\partial V}{\partial x} = 0$$

This indicates a constant potential, i.e., the particle experiences no force, and thus its momentum remains constant. Additionally, since the force is zero, the time derivative of the potential is also zero, as the potential does not change over time.

9.5 Derivation of the Schrödinger Wave Equation

The derivation of the Schrödinger wave equation is a pivotal moment in quantum mechanics, integrating classical mechanics concepts with wave theory. Let us break down this derivation step by step.

The energy E of a particle can be expressed in terms of its momentum p and potential energy V, the energy equation becomes:

$$E = \frac{p^2}{2m} + V = \frac{(h/\lambda)^2}{2m} + V.$$

Therefore,

$$E = \frac{h^2}{2m\lambda^2} + V. \tag{9.34}$$

By introducing the reduced Planck constant $\hbar = \frac{h}{2\pi}$, the angular frequency $\omega = 2\pi \nu$, and recalling that the wavenumber $k = 2\pi \lambda$ (from Chap. 3), the energy equation can be rewritten as:

$$E = \frac{\hbar^2 k^2}{2m} + V(x,t) = \hbar \omega. \tag{9.35}$$

This equation is known as the **Einstein-De Broglie** equation.

The classical wave equations in one spatial dimension is generally of the form:

$$\frac{\partial^2 U(x,t)}{\partial x^2} + \frac{1}{v^2}\frac{\partial^2 U(x,t)}{\partial t^2} = 0. \tag{9.36}$$

This represents the wave equation for a wave $U(x,t)$ traveling with speed v. As developed in Chap. 3, the Helmholtz wave equation solution for a monochromatic wave is of the form:

$$U(x,t) = A e^{j(\omega t - kx)}, \tag{9.37}$$

the second derivative with respect to x and t result in factors of k^2 and ω^2, respectively. Here, the factor j is the imaginary unit, and $j^2 = -1$. Given the energy equation, Eq. (9.35), has the terms k^2 and ω, Schrödinger inferred that the quantum wave equation should have a second derivative with respect to x but only a first derivative with respect to t. Thus, he proposed a wave equation for the wave function $\Psi(x, t)$ in the form:

$$\alpha \frac{\partial^2 \Psi(x,t)}{\partial x^2} + V\Psi(x,t) = \beta \frac{\partial \Psi(x,t)}{\partial t}, \qquad (9.38)$$

where α and β are constants to be determined.

To determine the constants α and β in the proposed wave equation for a quantum particle, we can substitute a trial solution and analyze the resulting equation. Let us consider the trial solution $\Psi(x, t) = cos(kx - \omega t)$ and substitute it into the wave equation (9.38); we obtain:

$$\alpha[-k^2 \cos(kx - \omega t)] + V \cos(kx - \omega t) = \beta \omega \sin(kx - \omega t),$$

simplifying this equation it becomes:

$$(\alpha k^2 - V) \cos(kx - \omega t) + \beta \omega \sin(kx - \omega t) = 0. \qquad (9.39)$$

For this equation to hold true, the coefficients of both $cos(kx - \omega t)$ and $sin(kx - \omega t)$ must be zero. Initially, setting $\beta = 0$ and $\alpha = k^2/V$ may seem appropriate, but this does not align with the quantum mechanical relation in Eq. (9.35).

Therefore, we reconsider β and set it as $\beta = j\hbar$, and $\alpha = -\frac{\hbar^2}{2m}$. This choice leads to the following wave equation:

$$-\frac{\hbar^2}{2m} \frac{\partial^2 \Psi}{\partial x^2} + V\Psi = j\hbar \frac{\partial \Psi}{\partial x}. \qquad (9.40)$$

This equation is a form of the **time-dependent Schrödinger equation**, which is fundamental to quantum mechanics. The equation governs the evolution of the wave function $\Psi(x, t)$ over time, encapsulating the particle's energy and its wave-like properties.

9.6 Solving Schrödinger Equation

The general form of the three-dimensional time-dependent Schrödinger wave equation is given by:

$$j\hbar \frac{\partial \Psi(r,t)}{\partial t} = -\frac{\hbar^2}{2m} \left(\nabla^2 + V(r,t) \right) \Psi(x,t), \qquad (9.41)$$

where

$$\nabla^2 = \frac{\partial^2}{\partial x^2} + \frac{\partial^2}{\partial y^2} + \frac{\partial^2}{\partial z^2}.$$

The wave function $\Psi(x, t)$ is a complex scalar function. Max Born proposed that the square of the absolute value of the wave function, $|\Psi(x, t)|^2$, represents the probability density of finding the particle at a position x at time t. This interpretation is a cornerstone of quantum mechanics, introducing probabilistic elements to the description of particle behavior.

Solving the Schrödinger equation (SE) is fundamental in understanding quantum mechanics. Let us look at the steps involved in solving the time-dependent Schrödinger equation in one dimension and interpret the physical significance of the solution.

The time-dependent Schrödinger equation in one dimension is expressed as:

$$j\hbar \frac{\partial \Psi(x, t)}{\partial t} = -\frac{\hbar^2}{2m} \frac{\partial^2 \Psi(x, t)}{\partial x^2} + V(x, t)\Psi(x, t). \tag{9.42}$$

If the potential energy V is independent of time and space, then a harmonic solution of SE resembles a solution for the Helmholtz equation as shown in Chap. 3. Let the complex wave function be represented by:

$$\Psi(x, t) = A e^{j(\omega t - kx)} \tag{9.43}$$

where A is the amplitude, ω is the angular frequency, and k is the wavenumber. By substituting this solution into the Schrödinger equation and simplifying, we obtain:

$$j\hbar \frac{\partial}{\partial t} A e^{j(\omega t - kx)} = \frac{-\hbar^2}{2m} \frac{\partial^2}{\partial x^2}(A e^{j(\omega t - kz)}) + V A e^{j(\omega t - kz)},$$

simplifying,

$$-\omega \hbar \Psi = \frac{-\hbar^2}{2m}(-jk)^2 \Psi + V\Psi. \tag{9.44}$$

By divide both sides of the equation by Ψ, we obtain:

$$\hbar \omega = E = \frac{\hbar^2 k^2}{2m} + V. \tag{9.45}$$

This equation is the Einstein-De Broglie equation as shown previously, Eq. (9.35), which relates the total energy to the kinetic energy and potential.

In mathematics and quantum mechanics, complex relations are often simplified using operator notation. For instance, the differential operation can be represented by an operator $\mathcal{L}$, such that:

$$f(x) = \frac{d}{dx}\{g(x)\}.$$

This can be rewritten such that we replace the derivative by an operator $\mathcal{L}$, so that:

$$f(x) = \mathcal{L}\{g(x).$$

The Hamiltonian operator $\hat{H}$ is a fundamental concept in quantum mechanics, encapsulating the total energy (kinetic plus potential) of the system. It is defined as:

$$\hat{H} = -\frac{\hbar^2}{2m}\frac{\partial^2}{\partial x^2} + V(x) \tag{9.46}$$

Thus, the Schrödinger equation, Eq. (9.42), can be elegantly rewritten using the Hamiltonian operator:

$$j\hbar\frac{\partial}{\partial t}\Psi = \hat{H}\Psi. \tag{9.47}$$

This form of the equation highlights the role of the Hamiltonian as the generator of time evolution in quantum systems. The use of the Hamiltonian operator in the Schrödinger equation not only simplifies the mathematical representation but also provides deep insights into the dynamics of quantum systems. It allows for the analysis of complex quantum systems by examining the properties of the Hamiltonian, facilitating a broader understanding of quantum phenomena.

9.7 Time-Dependent Schrödinger Wave Equation

In analyzing quantum mechanical systems, we are often interested in solving the wave equation at a particular instant in time. To do this, we will need to modify the wave equation to be independent of the time variable. We start by assuming that the space and time behaviors of the quantum articles are independent of each other. Thus, the wave function can be represented by the product of two functions such that:

$$\Psi(x,t) = \psi(x)\phi(t). \tag{9.48}$$

We can use the separation of variables technique to separate the differential equation into two independent equations. Substituting the function given in Eq. (9.48) into the Schrödinger wave equation leads to:

$$j\hbar \frac{\partial}{\partial t} \psi(x)\phi(t) = -\frac{\hbar^2}{2m}\frac{\partial^2}{\partial x^2}\psi(x)\phi(t) + V(x)\psi(x)\phi(t). \tag{9.49}$$

Dividing both sides by $\psi(x)\phi(t)$ results in the partial differential equation:

$$j\hbar \frac{1}{\phi(t)}\frac{\partial}{\partial t}\phi(t) = -\frac{\hbar^2}{2m}\frac{1}{\psi(x)}\frac{\partial^2}{\partial x^2}\psi(x) + V. \tag{9.50}$$

Rearranging the equation, we obtain:

$$j\hbar \frac{1}{\phi(t)}\frac{\partial \phi(t)}{\partial t} = \frac{1}{\psi(x)}\left[-\frac{\hbar^2}{2m}\frac{\partial^2 \psi(x)}{\partial x^2} + V\psi(x)\right]. \tag{9.51}$$

Since the right-hand side of the equation does not depend on t and the left-hand side of the equation does not depend on x, then the common value of both sides of the equation cannot depend on either x nor t. Therefore, the common value must be a constant. Let this constant be K. Therefore, we can write two equations as follows:

$$j\hbar \frac{1}{\phi(t)}\frac{\partial \phi(t)}{\partial t} = K, \tag{9.52}$$

a time-dependent equation, Eq. (9.52), and

$$\frac{1}{\psi(x)}\left[-\frac{\hbar^2}{2m}\frac{\partial^2 \psi(x)}{\partial x^2} + V\psi(x)\right] = K \tag{9.53}$$

a space-dependent equation, Eq. (9.53). The constant K is called the separation constant. The separation of variables technique converts a partial differential equation into two ordinary differential equations.

Let us first consider the time-dependent equation and rearrange so that:

$$\frac{d\phi(t)}{dt} = -j\frac{K}{\hbar}\phi(t), \tag{9.54}$$

rearranging it further we obtain:

$$d\phi(t) = -j\frac{K}{\hbar}\phi(t)dt. \tag{9.55}$$

This is a first-order linear differential equation. Integrating both sides gives:

$$\int d\phi(t) = -j\frac{K}{\hbar}\int \phi(t)dt.$$

9.7 Time-Dependent Schrödinger Wave Equation

Therefore, this integral leads to the following solution:

$$\phi(t) = e^{\frac{-jKt}{\hbar}}. \tag{9.56}$$

We can rewrite the solution using Euler's identity to obtain:

$$\phi(t) = \cos\left(\frac{Kt}{\hbar}\right) - j\sin\left(\frac{Kt}{\hbar}\right),$$

Upon substituting $\hbar = \frac{h}{2\pi}$, the solution becomes:

$$\phi(t) = \cos\left(\frac{2\pi Kt}{h}\right) - j\sin\left(\frac{2\pi Kt}{h}\right). \tag{9.57}$$

The function $\phi(t)$ is an oscillatory function in time with angular frequency ω given by:

$$\omega = \frac{2\pi K}{h} = 2\pi \nu,$$

hence,

$$K = h\nu = E. \tag{9.58}$$

Therefore the separation constant K is equal to the energy E. Therefore the function $\phi(t)$ can be expressed in the final form:

$$\phi(t) = e^{-j\frac{E}{\hbar}t}. \tag{9.59}$$

This represents the time evolution part of the wave function in quantum mechanics and reflects the fact that the separation constant K in the wave equation corresponds to the energy E of the quantum system.

This solution is significant because it illustrates how the energy of a quantum system directly influences the time evolution of the system's wave function. The oscillatory nature of $\phi(t)$ reflects the fundamental principle of wave-particle duality in quantum mechanics.

Let us now return to the space-dependent equation by subsisting for K by E in Eq. (9.53) to obtain:

$$\frac{1}{\psi(x)}\left[-\frac{\hbar^2}{2m}\frac{\partial^2 \psi(x)}{\partial x^2} + V\psi(x)\right] = E. \tag{9.60}$$

Multiplying both sides by $\psi(x)$ leads to:

$$\left[-\frac{\hbar^2}{2m} \frac{\partial^2}{\partial x^2} + V \right] \psi(x) = E\psi(x). \tag{9.61}$$

The square bracket is the Hamiltonian operator as we defined it previously. This equation can be expressed using the Hamiltonian operator $\hat{H}$ as:

$$\hat{H}\psi(x) = E\psi(x). \tag{9.62}$$

Here, E represents the eigenvalues of ψ when $\hat{H}$ operates on the eigenfunctions ψ. The Hamiltonian operator $\hat{H}$ encapsulates the total energy (kinetic and potential) of the quantum system. The equation $\hat{H}\psi(x) = E\psi(x)$ is an eigenvalue problem, where $\psi(x)$ represents the eigenfunctions and E are the corresponding eigenvalues. This formulation is the core of the time-independent Schrödinger equation.

Equation (9.62) is known as the time-independent Schrödinger equation. The solution of this equation determines the spatial dependence of the wave function $\psi(x)$, which is known as an eigenfunction. After we determine the eigenfunctions, we can rewrite the complex wave function in the following form:

$$\Psi(x, t) = \psi(x) e^{-j \frac{E}{\hbar} t}. \tag{9.63}$$

The probability density function of the quantum particle is given by:

$$\Psi \Psi^* = |\Psi|^2,$$

$$= \left(\psi e^{-j \frac{E}{\hbar} t} \right) \left(\psi^* e^{+j \frac{E}{\hbar} t} \right) = \psi \psi^* = |\psi|^2 \tag{9.64}$$

The absolute value square of the eigenfunction, $|\psi|^2$, determines the probability of finding a quantum particle at a location x in the space domain. Therefore, the probability of finding a quantum particle in all space, i.e., in $-\infty \leq x \leq \infty$, should equal to unity, i.e.,

$$\int_{-\infty}^{\infty} |\psi|^2 dx = 1. \tag{9.65}$$

This means that the quantum particle exists in space, but it is diffused and/or it is non-localized.

One major difference between classical and quantum mechanics is that in classical mechanics, the position of a particle or body can be determined precisely, whereas in quantum mechanics, the position of a particle is found in terms of a probability. This is the implication of another quantum mechanical principle known as the **Heisenberg uncertainty principle**, which states that it is impossible to simultaneously describe with absolute accuracy the position and momentum of a quantum particle. If the uncertainty in the momentum is Δp and the uncertainty in the position is Δx, then the uncertainty principle is stated as:

$$\Delta x \Delta p \geq \hbar. \tag{9.66}$$

One consequence of the uncertainty principle is that we cannot, for example, determine the exact position of an electron. Instead we determine the probability of finding an electron at a particular position, which is given by $\psi \Delta x$.

9.7.1 Dirac Bra-Ket Notation

Dirac's bra-ket notation is a powerful mathematical notation introduced by the physicist Paul Dirac for quantum mechanics. It is a convenient and concise way of representing quantum states and operations on them. This notation allows for a very elegant and compact representation of quantum states and operations, such as inner- and outer-products in linear algebra. A breakdown of its key components:

1. **Ket**: Represented as $|\psi\rangle$, it is a column vector in a complex vector space. It represents a quantum state.
2. **Bra**: Represented as $\langle\phi|$, it is a row vector and the conjugate transpose of a ket. If $|\phi\rangle$ is a ket, then $\langle\phi|$ is its corresponding bra. The conjugate transpose involves taking the transpose of the vector and then taking the complex conjugate of each element.
3. **Bra-Ket or InnerProduct**: The notation $\langle\phi|\psi\rangle$ represents the inner product between two states. It is a complex number that provides important physical information, such as the probability amplitude for transitions between states in a quantum system. The inner product of linearly independent functions ψ_1 and ψ_2 is given by:

$$<\psi_1|\psi_2> = \int_{-\infty}^{+\infty} \psi_1^* \psi_2 dx. \tag{9.67}$$

4. **Outer Product**: Written as $|\psi\rangle\langle\phi|$, this represents a matrix formed from a ket and a bra. In quantum mechanics, these outer products are often used to represent operators.

We use the bra-ket notation to describe very important operations in quantum mechanics.

Let us suppose that a quantum system is described by a set of wave functions $(\psi_1, \psi_2, \psi_3 \ldots)$ each representing a quantum state of the system in a Hilbert space. In a Hilbert space, the wave functions are typically orthogonal to each other. This orthogonality means that the inner product between any two different wave function is zero. If the wave functions is expressed as:

$$\psi = a\psi_1 + b\psi_2 + c\psi_3 + \ldots \tag{9.68}$$

Mathematically, for two wave functions ψ_i and ψ_j, their orthogonality is expressed as:

$$\langle \psi_i | \psi_j \rangle = 0 \ for \ i \neq j. \tag{9.69}$$

We can write the inner product of these wave functions as:

$$<\psi_n | \psi_m> = \delta_{n,m}, \tag{9.70}$$

where $\delta_{n,m}$, is the Kronecker delta defined as:

$$\delta_{n,m} = \begin{cases} 0 & if\ n \neq m \\ 1 & if\ n = m \end{cases}. \tag{9.71}$$

Therefore, the bra-ket of the wave function that is not equal to zero is hen $n = m$, i.e.,

$$<\psi_n | \psi_n> = 1.$$

Therefore, $|\psi>$ is a vector in Hilbert space, whereas $\psi(x)$ the projection of that vector on to the familiar space. The probability density function can be represented as:

$$p(x) = \psi^* \psi = |\psi(x)|^2, \tag{9.72}$$

where $p(x)$ at some instant (t) is the probability density of finding a particle between x and $(x + dx)$. The probability, P, of finding a particle between x_1 and x_2 is given by:

$$P_{x_1 \leq x \leq x_2} = \int_{x_1}^{x_2} |\psi(x)|^2 dx,$$

as shown in Fig. 9.6.

Therefore, if a solution to the Schrödinger equation is given by ϕ_1, then:

$$<\phi_1 | \phi_1> = \int_{-\infty}^{\infty} |\phi_1|^2 dx = q. \tag{9.73}$$

and we can normalize the solution by setting $q = 1$.

9.7 Time-Dependent Schrödinger Wave Equation

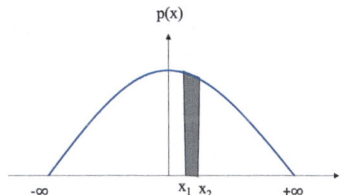

Fig. 9.6 The probability of finding a quantum particle between x_1 and x_2

9.7.2 The Superposition Principle

The superposition principle is a fundamental concept in quantum mechanics, reflecting one of the most striking differences between classical and quantum physics. It states that if a quantum system can be in multiple states, then it can also be in a combination (or superposition) of those states. This principle is critical in understanding the behavior of quantum systems. Quantum systems can be described by the superposition of a set of wave functions. The superposition principle asserts that if a quantum system has multiple possible states, represented by wave functions, the actual state of the system can be any linear combination (or superposition) of these states. Mathematically, if ψ_1 and ψ_2 are two possible states of a system, then a valid state of the system can also be $a\psi_1 + b\psi_2$, where a and b are complex coefficients.

The double-slit experiment for waves was discussed in Chap. 3, resulting in an interference pattern that is a property of waves. Using the same setup, but now using a stream of electrons as the source, followed by two slits labeled S_L and S_R results in a similar pattern. The wave function of the electron at the observation screen is given by:

$$\psi = a_L \psi_L + a_R \psi_R, \tag{9.74}$$

where ψ_L and ψ_R are the wave functions of the electron at the left or right slits, respectively, and a_L and a_R are complex coefficients. The electron before reaching the observation screen cab pass either through slit S_L, or slit S_R, or through both of them at the same time. The double slit experiment of the electron resulted in a similar interference pattern as that for the interference of light waves. Therefore, a quantum particle can be in several places at the same time (Fig. 9.7).

9.7.3 Expectation Values

We have previously defined the Hamiltonian operator as:

$$\hat{H} = -\frac{\hbar^2}{2m}\frac{\partial^2}{\partial x^2} + V).$$

Several operators are used in the mathematical analysis of quantum systems. We introduce here a number of those that are used in this book.

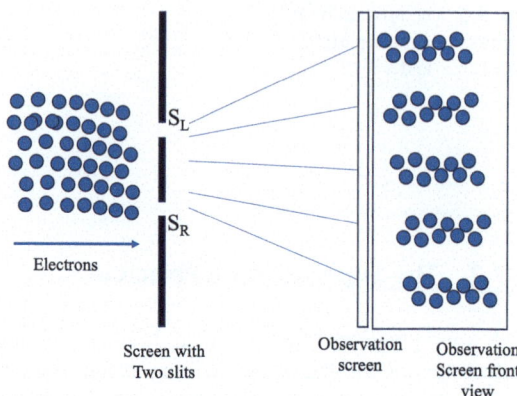

Fig. 9.7 Electrons double slit experiment showing that electrons interfere like wave

1. The position operator $\hat{x}$ and is given by the relation:

$$\hat{x}\psi = \hbar\psi. \tag{9.75}$$

2. The momentum operator $\hat{p}$ which is given by the following relation:

$$\hat{p}\psi = -j\frac{\partial}{\partial x}\psi(x). \tag{9.76}$$

3. The kinetic energy operator $\hat{T}$ which is given by the following relation:

$$\hat{T}\psi = -\frac{\hbar^2}{2m}\frac{\partial^2}{\partial x^2}\psi. \tag{9.77}$$

We can rewrite the momentum operator by using a solution of the Schrödinger wave equation to be:

$$\hat{p}_x\psi(x) = -j\frac{\partial}{\partial x}\left(e^{j(kx-\omega t)}\right) = -\hbar k\psi. \tag{9.78}$$

Therefore, $\hbar k$ are the eigenvalues of the momentum operator, similar to E, which are the eigenvalues of the Hamiltonian operator. We can express the Hamiltonian operator as follows:

$$\hat{H} = \hat{T} + \hat{V}, \tag{9.79}$$

where $\hat{V}$ is the potential operator.

The probability of finding a particle somewhere in space is 100%. However, where is it most likely to find the particle? Its position using a position operator

$\hat{x}$ is given by:

$$\int \psi^*(x)\hat{x}\psi(x)dx = x_0$$

x_0 is an expected value that is the most likely location of the particle. We define the expectation value using the bra-ket notation as:

$$<\hat{x}> = <\psi|\hat{x}|\psi> = \int_{-\infty}^{\infty} \psi^*\hat{x}\psi(x)dx. \qquad (9.80)$$

Similarly, the expectation value of the momentum is given by:

$$<\hat{p}> = <\psi|\hat{p}|\psi> = \int_{-\infty}^{\infty} \psi^*\hat{p}\psi(x)dx,$$

or

$$<\hat{p}> = \int_{-\infty}^{\infty} \psi^*(-j\frac{\hbar\partial}{\partial x})\psi(x)dx,$$

simplifying we obtain:

$$<\hat{p}> = -j\int_{-\infty}^{\infty} \psi^*\frac{\partial \psi}{\partial x}dx. \qquad (9.81)$$

The expectation value of the kinetic energy of a quantum particle is given by:

$$\begin{aligned}<\psi|\hat{T}|\psi> &= \int_{-\infty}^{\infty} \psi^*(-\frac{\hbar^2}{2m})\frac{\partial^2}{\partial x^2}\psi dx \\ &= -\frac{\hbar^2}{2m}\int_{-\infty}^{\infty} \psi^*\frac{\partial^2 \psi}{\partial x^2}dx\end{aligned} \qquad (9.82)$$

9.8 Application of Schrödinger Wave Equation

Applying Schrödinger's wave equation across different examples with various potential functions is important for understanding the techniques involved in solving this fundamental differential equation in quantum mechanics. As we delve into these examples, we will gain insights into electron behavior under various potentials. This understanding lays the groundwork for later chapters where we explore semiconductor properties. The behavior of electrons, dictated by the wave function solutions of Schrödinger's equation, directly influences the electronic and optical properties of semiconductors, which are pivotal in modern electronics. Thus, these examples represent a step toward connecting abstract quantum mechanics with practical applications in materials science and engineering.

9.8.1 Quantum Particle in Free Space

As a first example, we consider an electron in free space where no force is applied to and examine its behavior as a solution to the Schrödinger equation. Let us rewrite the time-independent Schrödinger equation given by Eq. (9.61) as:

$$\left[-\frac{\hbar^2}{2m}\frac{\partial^2}{\partial x^2} + V\right]\psi(x) = E\psi(x),$$

rearranging terms we obtain:

$$\frac{\partial^2 \psi}{\partial x^2} + \frac{2m}{\hbar^2}[E - V(x)]\psi = 0. \qquad (9.83)$$

For the electron to move freely in space, then we can take $V(x) = 0$. Therefore, the wave equation for a free electron is expressed as:

$$\frac{d^2\psi}{dx^2} + \frac{2mE}{\hbar^2}\psi = 0. \qquad (9.84)$$

This a typical ordinary second order differential equation similar to the Helmholtz electromagnetic wave equation. We use similar approach to solving this equation. We assume the following solution:

$$\psi(x) = Ae^{j\left(\frac{\sqrt{2mE}}{\hbar}x\right)} + Be^{-j\left(\frac{\sqrt{2mE}}{\hbar}x\right)}, \qquad (9.85)$$

or

$$\psi(x) = Ae^{jkx} + Be^{-jkx}, \qquad (9.86)$$

where

$$k = \sqrt{\frac{2mE}{\hbar^2}}, \qquad (9.87)$$

which is called the wavenumber.

To obtain the solution for the time-dependent Schrödinger, wave equation we must include the time-dependent portion of the solution given by Eq. (9.59) to obtain:

$$\phi(t) = e^{-j\left(\frac{E}{\hbar}\right)t} = e^{-j\omega t}.$$

9.8 Application of Schrödinger Wave Equation

Hence the complete solution for the wave function is given by:

$$\Psi(x,t) = Ae^{j(kx-\omega t)} + Be^{-j(kx+\omega t)}. \tag{9.88}$$

Hence, the wave function solution for an electron in free-space is a traveling wave. The first term of Eq. (9.88) represents a planewave traveling in the $+x$ direction, whereas the second term represents a planewave traveling in the $-x$ direction. The coefficients A and B are determined on the basis of the boundary conditions of the problem under consideration.

Assume that the coefficient $B = 0$.

The wave function for a particle traveling in the $+x$ direction is given by:

$$\Psi(x,t) = Ae^{j(kx-\omega t)}. \tag{9.89}$$

The energy-momentum relationship is given by:

$$E = \frac{1}{2}mv^2 = \frac{p^2}{2m}, \tag{9.90}$$

therefore,

$$k = \sqrt{\frac{2mE}{\hbar^2}} = \sqrt{\frac{p^2}{\hbar^2}} = \frac{p}{\hbar}, \tag{9.91}$$

or

$$p = \hbar k. \tag{9.92}$$

Additionally, since the de Broglie wavelength is given by:

$$\lambda = \frac{h}{p} = \frac{2\pi\hbar}{p} = \frac{2\pi}{k},$$

or

$$k = \frac{2\pi}{\lambda}. \tag{9.93}$$

Therefore, a particle that has a well-defined energy, has a well-defined momentum, and a well-defined wavelength. This relationship exemplifies the particle-wave duality of the electron.

The probability density function is given by:

$$\Psi(x,t)\Psi^*(x,t) = AA^* = |A|^2, \tag{9.94}$$

which is a constant independent of position. A free particle with a well-defined momentum can be found anywhere with equal probability. This result is in agreement with the Heisenberg uncertainty principle in which a precise momentum implies an undefined position.

9.8.2 Particle in an Infinite Potential Well

In this section we present the classical problem of a particle in an infinite quantum. This case will define a specific boundary condition that is used in determining the coefficients A and B of the wave function solution.

The wave equation is given by:

$$\frac{\partial^2 \psi}{\partial x^2} + \frac{2m}{\hbar^2}[E - V(x)]\psi = 0.$$

The potential $V(x)$ is defined as in Fig. 9.8. The boundary conditions in the three regions is as follows:

1. Region I: $V(x) = \infty$, $\psi(x) = 0$ and $p(x) = 0$
2. Region II: $V(x) = 0$, $\psi(x) \neq 0$ and $p(x) \neq 0$
3. Region III: $V(x) = \infty$, $\psi(x) = 0$ and $p(x) = 0$

Hence, in regions I and III, the wave function and the probability density function are equal to zero. The solution for the Schrödinger wave equation in region II can be written as:

$$\psi(x) = A\sin(kx) + B\cos(kx). \tag{9.95}$$

Applying the boundary condition at $x = 0$ where $\psi(0) = 0$, leads to the following wave function:

$$\psi(0) = A\sin(0) + B\cos(0),$$

Therefore,

$$B = 0. \tag{9.96}$$

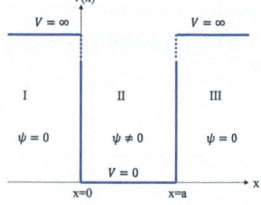

Fig. 9.8 The boundary conditions for an infinite potential well

9.8 Application of Schrödinger Wave Equation

Thus, in region II the wave function is given by:

$$\psi(x) = A sin(kx). \tag{9.97}$$

To determine the coefficient A, we will apply the boundary condition at $x = a$ which leads to:

$$\psi(a) = A\sin(ka) = 0.$$

Therefore,

$$ka = n\pi,$$

where m is a positive integer, i.e., $n = 1, 2, 3, \ldots$ and

$$k = \frac{n\pi}{a}. \tag{9.98}$$

Therefore, the wave function for a particle in an infinite potential well is given by:

$$\psi(x) = A\sin\left(\frac{n\pi x}{a}\right). \tag{9.99}$$

The coefficient A can be found from the normalization boundary condition:

$$<\psi|\psi> = \int_0^a \psi^*(x)\psi(x)dx = 1, \tag{9.100}$$

$$<\psi|\psi> = \int_0^a A^2 \sin^2(\frac{n\pi x}{a})dx = A^2 \int_0^a \frac{1-\cos(2n\pi x/a)}{2}dx$$

$$= \frac{A^2}{2}[\int_0^a dx - \int_0^a \cos^2(\frac{2n\pi x}{a})dx]$$

Therefore,

$$<\psi|\psi> = \frac{aA^2}{2} - A^2\left[\frac{\sin(2ka)}{4k} - \frac{\sin(0)}{4k}\right] = 1.$$

However, since $ka = m\pi$, then $sin(2ka) = 0$.
Therefore

$$\frac{aA^2}{2} = 1,$$

or

$$A = \sqrt{\frac{2}{a}}. \tag{9.101}$$

Finally, the time-independent wave solution is given by:

$$\psi(x) = \sqrt{\frac{2}{a}} \sin(kx). \tag{9.102}$$

However, since $ka = n\pi$, then ψ is dependent on the integer n, therefore, there are n independent wave functions:

$$\psi_n(x) = \sqrt{\frac{2}{a}} \sin(k_n x), \tag{9.103}$$

so these are the eigenfunctions that satisfy Schrödinger equation and the boundary condition of the infinite potential well:

$$\psi_n(x) = \sqrt{\frac{2}{a}} \sin\left(\frac{n\pi x}{a}\right). \tag{9.104}$$

This solution represents the electron in the infinite potential well and is a standing wave solution. While we have shown in the previous section that the free electron was represented by a traveling wave, now the bound particle is represented by a standing wave.

The wavenumber k was defined by two equations, Eqs. (9.91) and (9.98), so by equating both of these equations, we obtain:

$$k = \sqrt{\frac{2mE}{\hbar^2}} = \frac{n\pi}{a},$$

so that the total energy E must have discrete values that depend on the value of the integer n. Therefore, the total energy of the electron bound in an infinite potential well is given by:

$$E_n = \frac{n^2 \pi^2 \hbar^2}{2ma^2}. \tag{9.105}$$

These energy levels are the eigenvalues of the wave function and exist as discrete values. Consequently, the electron energy levels are not continuous but discrete, scaling with n^2. For instance, if the first energy level an electron can occupy is E_1, the subsequent three energy levels are:

$$E_2 = 4E_1, \ E_3 = 9E_1, \ and \ E_4 = 16E_1. \tag{9.106}$$

9.8 Application of Schrödinger Wave Equation

This result signifies that the particle's energy is quantized, meaning that it can only take on specific discrete values. This quantization is in stark contrast to classical physics, which predicts continuous energy values for particles. The discrete nature of these energies leads to quantum states, a topic that will be explored in more depth in this and subsequent chapters. The quantization of energy for a bound particle is a pivotal concept, especially in understanding the properties of semiconductor devices, as will be demonstrated in the following chapters

Example Consider an electron in an infinite potential well with a width of 5 Å. Determine: the first four energy levels of an electron with a mass, $m = 9.1 \times 1$ kg and electron charge, $q = 1.6 \times 10^{-19}$ C.

Solution The first energy level is:

$$E_1 = \frac{\pi^2 \hbar^2}{2ma^2} = \frac{(1.054 \times 10^{-34})^2 (3.1416)^2}{2 \times 9.1 \times 10^{31} \times (5 \times 10^{-10})^2} = 2.4 \times 10^{-19} \text{ J},$$

$$E_1 = \frac{2.4 \times 10^{-19}}{1.6 \times 10^{-19}} = 1.5 \text{ eV}.$$

Therefore,

$E_2 = 4 \times 1.5 = 6 \text{ eV}, \ E_3 = 9 \times 1.5 = 13.5 \text{ eV}, \ and \ E_4 = 16 \times 1.5 = 24 \text{ eV}.$

Figure 9.9a shows the first four allowed energies (E_n) for the electron in the infinite potential well, and Fig. 9.9b,c shows the corresponding wave functions ψ_n and probability density functions $|\psi_n|^2$.

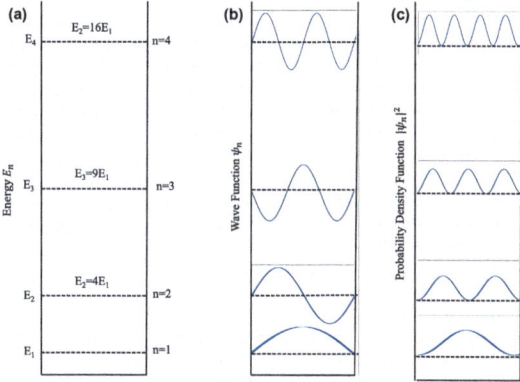

Fig. 9.9 Electron in an infinite potential well: (**a**) four lowest discrete energy levels, (**b**) corresponding wave functions, and (**c**) corresponding probability density functions

The momentum of the particle in an infinite potential well is given by Eq. (9.81)

$$<\psi|\hat{p}|\psi> = \int_{-\infty}^{\infty} \psi^* \hat{p}\psi(x)dx.$$

or

$$<\psi|\hat{p}|\psi> = -j\hbar \int_0^a \psi^* \frac{\partial \psi}{\partial x}dx.$$

Substituting for ψ from Eq. (9.104), we obtain:

$$<\psi|\hat{p}|\psi> = -j\hbar \int_0^a \sqrt{\frac{2}{a}} \sin\left(\frac{n\pi x}{a}\right) \frac{d}{dx}\left(\sqrt{\frac{2}{a}} \sin\left(\frac{n\pi x}{a}\right)\right) dx, \quad (9.107)$$

$$<\psi|\hat{p}|\psi> = -j\frac{2n\pi\hbar}{a^2} \int_0^a \sin\left(\frac{n\pi x}{a}\right) \cos\left(\frac{n\pi x}{a}\right) dx,$$

$$= -j\frac{2n\pi\hbar}{a^2}\left[\sin\left(\frac{n\pi a}{a}\right)\cos\left(\frac{n\pi a}{a}\right) - \sin(0)\cos(0)\right] = 0.$$

Therefore, the momentum of the particle is equal to zero.

Next, let us find the expectation value of the particle. Using Eq. (9.82), we obtain:

$$<\psi_1|\hat{T}|\psi_1> = \int_0^a \sqrt{\frac{2}{a}} \sin(k,x)\left(-\frac{\hbar^2}{2m}\frac{\partial^2}{\partial x^2}\right)\sqrt{\frac{2}{a}} \sin(k,x)dx$$

$$= \frac{(\hbar k_1)^2}{am} \int_0^a \sin(k_1 x)\sin(k_1 x)dx \quad (9.108)$$

$$= \frac{(\hbar k_1)^2}{am}\left[\frac{a}{2} - \frac{\sin(2k_1 a)}{4k_1}\right]$$

$$<\psi|\hat{T}|\psi> = \frac{(\hbar n\pi/a)^2}{am}(a/2)$$

thus,

$$<\psi|\hat{T}|\psi> = \frac{(\hbar n\pi)^2}{2ma^2} \quad (9.109)$$

Therefore the particle has no momentum but has a kinetic energy.

Reexamining the wave function solution, let us express $sin(k_n x)$ into an exponential form:

$$\psi_n(x) = \sqrt{\frac{2}{a}} \sin(k_n x) = \sqrt{\frac{2}{a}}\left(\frac{e^{jkx} - e^{-jk_n x}}{2j}\right).$$

Next, we introduce the time-dependent portion to construct the complex wave function:

$$\Psi_n(x, t) = \sqrt{\frac{2}{a}} \frac{1}{2j} \left(e^{j(\omega t - k_n x)} - e^{j(\omega t + k_n x)} \right)' \tag{9.110}$$

Therefore, the wave function $\Psi_n(x, t)$ is a superposition of two waves: one propagating in the $+x$ direction and the other in the $-x$ direction, resulting in a standing wave. This implies that the particle is simultaneously travels in both directions. As a result, the particle exhibits no net movement, which is indicative of zero momentum. This standing wave pattern is a clear manifestation of the particle's confined state in quantum mechanics.

9.9 Hydrogen Atom

In the previous section, we have described one-dimensional potential energy functions and solved Schrödinger's time-independent wave equation to find the probability density function for the position of a particle. Now, turning to the hydrogen atom, a one-electron system, offers an intriguing shift to a more complex scenario. The classical view, as per Bohr's theory, describes the electron as orbiting a heavy, positively charged proton, but quantum mechanics provides a more subtle understanding.

The potential function in this case arises from the Coulomb attraction between the proton and the electron. It is expressed as:

$$V = \frac{q_1 q_2}{4\pi \varepsilon_o r} = -\frac{e^2}{4\pi \varepsilon_o r}, \tag{9.111}$$

where r is Bohr's radians, e is the electron charge, and ε_o is the permittivity of free space. This potential is spherically symmetric and introduces a three-dimensional problem when approached in quantum mechanics, necessitating the use of spherical coordinates.

Now we attempt to solve the Schrödinger equation for a Hydrogen atom. The electron occupies a spherical orbital around the atom. We start with the time-independent Schrödinger equation given by:

$$\hat{H}\psi = E\psi.$$

The three-dimensional Hamiltonian operator given in Eq. (9.41) takes the following form:

$$\hat{H} = -\frac{\hbar^2}{2m} \nabla^2 + V. \tag{9.112}$$

Since V is a function of r, then we use the spherical coordinate, (r, θ, φ) system instead of the Cartesian coordinate system, (x, y, z), where:

$$r \geq 0, 0 \leq \theta \leq \pi, \text{ and } 0 \leq \varphi \leq 2\pi \tag{9.113}$$

and the Laplacian (∇^2) operator in spherical coordinates is given by:

$$\nabla^2 = \frac{1}{r^2}\frac{\partial}{\partial r}(r^2 \frac{\partial}{\partial r}) + \frac{1}{r^2 \sin\theta}\frac{\partial}{\partial \theta}(\sin\theta \frac{\partial}{\partial \theta}) + \frac{1}{r^2 \sin\theta}\frac{\partial^2}{\partial \varphi} \tag{9.114}$$

The wave function in the spherical coordinate system is $\psi(r, \theta, \varphi)$, and the Schrödinger equation is expressed in spherical coordinate system as:

$$\frac{1}{r^2}\frac{\partial}{\partial r}(r^2 \frac{\partial \psi}{\partial r}) + \frac{1}{r^2 \sin\theta}\frac{\partial}{\partial \theta}(\sin\theta \frac{\partial \psi}{\partial \theta}) + \frac{1}{r^2 \sin^2\theta}\frac{\partial^2 \psi}{\partial \varphi^2}$$
$$-\frac{e^2 \psi}{4\pi \varepsilon_o r} + \frac{2m_o}{\hbar^2}(E - V)\psi(r, \theta, \varphi) = 0. \tag{9.115}$$

The solution to Eq. (9.115) can be determined by the separation-of-variables technique. We will assume that the solution to the time-independent wave equation can be written in the form:

$$\psi(r, \theta, \varphi) = R(r)\Theta(\theta)\Phi(\varphi). \tag{9.116}$$

Substituting for $\psi(r, \theta, \varphi)$ in Eq. (9.115) and multiplying both sides by $\frac{\sin^2\theta}{R\Theta\Phi}$, we obtain the following:

$$\frac{\sin^2\theta}{R}\frac{\partial}{\partial r}(r^2 \frac{\partial R}{\partial r}) + \frac{\sin\theta}{\Theta}\frac{\partial}{\partial \theta}(\sin\theta \frac{\partial \Theta}{\partial \theta}) + \frac{1}{\Phi}\frac{\partial^2 \Phi}{\partial \varphi^2} + \frac{2m_o r^2 \sin^2\theta}{\hbar^2}(E - - \frac{e^2}{4\pi \varepsilon_o r}) = 0. \tag{9.117}$$

The third term in this equation is a function of Φ only, whereas the other terms are either a function of r or θ. Hence, we can write:

$$\frac{1}{\Phi}\frac{\partial^2 \Phi}{\partial \varphi^2} = -m^2,$$

or

$$\frac{d^2 \Phi}{d\varphi^2} + m^2 \Phi = 0, \tag{9.118}$$

where m is a separation variable. The solution of the second-order ordinary differential equation is given by:

$$\Phi(\varphi) = Ae^{jm\varphi}. \tag{9.119}$$

9.9 Hydrogen Atom

Since φ and $\varphi + 2\pi$ represent a single point in space, we must have:

$$Ae^{jm\varphi} = Ae^{jm(\varphi+2\pi)},$$

this happens only for $m = 0, \pm 1, \pm 2, \pm 3, \ldots$

By incorporating the separation-of-variables constant, we can further separate the variables and generate two additional separation-of-variables constants l and n, which are both integers. The integer n is called the principal quantum number, m is called the magnetic quantum number, and l is the orbital quantum number. These quantum numbers follow the following selection rules:

$$n \geq 0, 0 \leq l \leq n \text{ and } -l \leq m \leq l. \tag{9.120}$$

The energy eigenvalues for a hydrogen atom are given by:

$$E_n = -\frac{m_o e^4}{32\pi^2 \varepsilon_o^2 \hbar^2} \left(\frac{1}{n^2}\right), \tag{9.121}$$

or

$$E_n = -\frac{13.6}{n^2} eV. \tag{9.122}$$

The observation in Fig. 9.10 about the quantized energy levels in a hydrogen atom is an important aspect of quantum mechanics. Like the case of a particle in an infinite potential well, the energy levels in a hydrogen atom are quantized, but they follow a different pattern due to the nature of the potential and the three-dimensional aspects of the problem.

In the hydrogen atom, the energy levels are determined by the principal quantum number n. However, unlike the infinite potential well where the energy levels are proportional to n^2, in the hydrogen atom, the energy levels are inversely proportional to n^2. The negative sign in Eq. (9.122) indicates that these are bound states with energies less than zero, representing the energy needed to remove the electron from the atom (ionization). As n increases, the energy levels become closer together, which is evident in the graphical representation you mentioned. This pattern is a direct consequence of the $1/n^2$ dependence, which is a characteristic feature of the hydrogen-like atoms.

Fig. 9.10 The quantized energy levels for a hydrogen atom

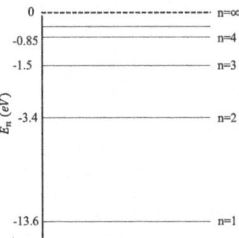

Example Calculate the first four quantized enemy levels for a hydrogen atom.

Solution The quantized energy as a s function of the principle quantum number is:

$$E_n = -\frac{13.6}{n^2}\,\text{eV}.$$

For $n = 1, 2, 3$ and 4, the quantized energy levels are:

$$E_1 - 13.6\,\text{eV},\ E_2 = -\frac{13.6}{4} = -3.4\,\text{eV};\ E_3 = 1.5\,\text{eV}\ and\ E_4 = 0.85\,\text{eV}.$$

The energy at the surface of the material is:

$$E_\infty = 0\,\text{eV}.$$

9.10 Particle in a Finite Potential Well

In this section we present the classical problem of a particle in an infinite quantum. This case will define a set of boundary conditions that are used in the wave equation. Consider the potential well shown in Fig. 9.11, where:

$$-L/2 \leq x \leq L/2. \tag{9.123}$$

The boundary conditions for a particle in a finite potential well, as shown in Fig. 9.11 are:

1. $V(x) = 0$ for $-L/2 \leq x \leq L/2$ in region II.
2. $V(x) = V_o$ for $x < L/2$ in region I.

Fig. 9.11 The boundary conditions for a particle in a finite potential well

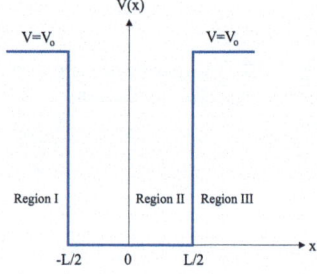

9.10 Particle in a Finite Potential Well

3. $V(x) = V_o$ for $x > L/2$ in region III.
4. The wave function $\psi(x)$ and its first derivative with respect to x are continuous at $x = \pm L/2$. This requirement ensures that the wave function is physically realistic, as abrupt changes or discontinuities in $\psi(x)$ or its derivative would imply infinite momentum or energy, which is not possible.

The solution for the wave function is expected to be symmetric and $\psi(x)$ has no discontinuities. The Schrödinger time-independent wave equation is expressed as:

$$\frac{\partial^2 \psi(x)}{\partial x^2} = \frac{2m(V(x) - E)}{\hbar^2} \psi(x).$$

Let us consider the solution of the Schrödinger equation ((SE) in the different regions of the potential well.

Region I This region is defined by $x \leq 0$, and the potential is $V(x) = V_o$. The SE can be rewritten as:

$$\frac{\partial^2 \psi(x)}{\partial x^2} = \frac{2m(V_o - E)}{\hbar^2} \psi(x). \quad (9.124)$$

Let

$$\alpha^2 = \frac{2m(V_o - E)}{\hbar^2}. \quad (9.125)$$

Substituting Eq. (9.125) into Eq. (9.124), the wave equation becomes:

$$\frac{\partial^2 \psi(x)}{\partial x^2} = \alpha^2 \psi(x).$$

The solution of this second-order differential equation is given by:

$$\psi(x) = De^{-\alpha x} + Be^{\alpha x}, \quad (9.126)$$

where α is a real quantity since $V(x) > E$. The wave function approaches zero as x approaches $-\infty$, therefore, the coefficient $D = 0$.

Therefore the wave equation solution in Region I becomes:

$$\psi(x) = Be^{\alpha x}. \quad (9.127)$$

Region III In this region $V = V_o$, which is similar to region I, hence, the solution for the wave equation is:

$$\psi(x) = Ae^{-\alpha x} + Ce^{\alpha x},$$

the wave function approaches zero as x approaches ∞. Therefore the coefficient $C = 0$.

Therefore, the wave equation solution in region III reduces to:

$$\psi(x) = Ae^{-\alpha x}. \tag{9.128}$$

The wave function takes the form of a decaying exponential in terms of x.

Region II In this region $V = 0$, the wave equation becomes:

$$\frac{\partial^2 \psi}{\partial x^2} = -k^2 \psi, \tag{9.129}$$

where

$$k = \sqrt{\frac{2mE}{\hbar^2}}.$$

The solution for Eq. (9.129) is:

$$\psi(x) = Ccos(kx) + Dsin(kx). \tag{9.130}$$

Coefficients A, B, C and D are detriment by the boundary conditions listed above. The solution assumes both **even-parity**, i.e., even function solutions are given by the *cosine* term, and anti-symmetric **odd-parity solutions are** given by the *sine* term.

9.10.1 Even-Parity Solutions

The solutions of the Schrödinger equation for a particle in a finite potential well are separated into even and odd-parity states, and we need only consider positive values of x (which could be inferred from the potential). With the application of the boundary conditions, the even-parity solutions in region II are given by:

$$\psi_e(x) = Ccos(kx) \tag{9.131}$$

The first boundary condition to apply is that the wave function is continuous across the boundary. Hence at $x = L/2$,

$$Ae^{-\alpha L/2} = Ccos(kL/2). \tag{9.132}$$

The continuity of the wave function derivative results in the following:

$$\alpha Ae^{-\alpha L/2} = -kCsin(kl/2). \tag{9.133}$$

9.10 Particle in a Finite Potential Well

Dividing Eq. (9.133) by Eq. (9.132) results in the following:

$$\frac{\alpha}{k} = tan\left(\frac{kL}{2}\right). \tag{9.134}$$

Introduce the variable

$$\xi = \frac{kL}{2}, \tag{9.135}$$

and

$$\chi = \frac{\alpha L}{2}. \tag{9.136}$$

Substituting Eqs. (9.135) and (9.136) into Eq. (9.134) leads to:

$$\chi = \xi tan(\xi). \tag{9.137}$$

This is a transcendental equation that can be solved either graphically or numerically. Figure 9.12 depicts the graphical solution of the transcendental equation. The points of the intersection between χ and $\xi tan(\xi)$, labeled by circles, are the eigenvalues of the solution of the Schrödinger wave equation. In the context of the specified potential V_o, there are four even-parity quantized bound energy states. The number of these bound states depends on the depth of the potential well, with more states being available in deeper wells, for larger values V_o.

9.10.2 Odd-Parity Solutions

The second set of solutions are the odd-parity solutions where the wave function is given by:

$$\psi_o(x) = Dsin(kx). \tag{9.138}$$

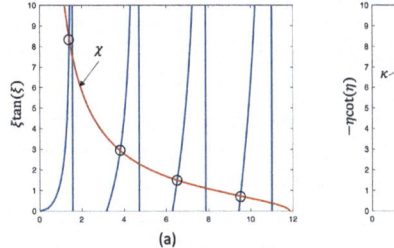
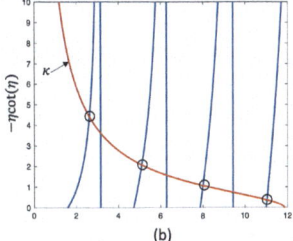

Fig. 9.12 Graphical solution for the transcendental equation (**a**) for even-parity quantized bound states, and (**b**) for odd quantized bound states

The first boundary condition to apply is that the wave function is continuous across the boundary. Hence at $x = L/2$,

$$Ae^{-\alpha L/2} = D\sin(kx)). \tag{9.139}$$

The continuity of the wave function derivative results in the following:

$$-\alpha Ae^{-\alpha L/2} = kD\cos(kl/2). \tag{9.140}$$

Dividing Eq. (9.140) by Eq. (9.139) results in the following

$$-\frac{\alpha}{k} = \cot\left(\frac{kL}{2}\right). \tag{9.141}$$

Introduce the variable

$$\eta = \frac{kL}{2}, \tag{9.142}$$

and

$$\kappa = \frac{\alpha L}{2}. \tag{9.143}$$

Substituting Eqs. (9.142) and (9.143) into Eq. (9.141) leads to:

$$\kappa = -\eta \cot(\eta). \tag{9.144}$$

This is a transcendental equation that can be solved either graphically or numerically. Figure 9.13 depicts the graphical solution of the transcendental equation. The points of the intersection between κ and $\eta \cot(\eta)$, labeled by circles, are the eigenvalues of the solution of the Schrödinger wave equation. In the context of the specified potential V_o, there are four odd-parity quantized bound energy states. The number of these bound states depends on the depth of the potential well, with more states being available in deeper wells, for larger values V_o.

9.10.3 The Wave Function for Even- and Odd-Parity Solutions

The wave functions for even- and odd-parity solutions in a finite potential well are detailed in Eqs. (9.132) and (9.139). Inside the well, these solutions are periodic, reflecting the quantized nature of the particle's states. Outside the well, the wave functions exhibit exponential decay, indicating a decrease in the probability of finding the particle; however, unlike in an infinite potential well, there remains a nonzero probability of the particle's presence outside the well. This characteristic is a distinctive feature of finite potential wells.

Fig. 9.13 The wave functions for even- and odd-parity solutions in a finite potential well

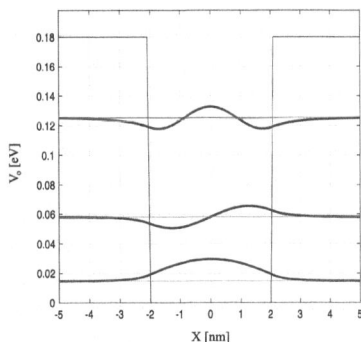

Figure 9.13 visually represents the wave functions for two even-parity states and one odd-parity state, for specified values of the potential V_o and the well width L. These illustrations demonstrate the contrasting behaviors of the wave functions inside and outside the well.

A key difference between finite and infinite potential wells is highlighted here: In a finite potential well, the particle has a finite probability of existing outside the well's boundaries, whereas in an infinite potential well, the particle is strictly confined within the walls of the potential well.

The quantization of the wavenumber and energy in a finite potential well is given by:

$$k_n = \frac{n\pi}{L}, \tag{9.145}$$

and the quantized energy is given by:

$$E_n = \frac{n^2 \pi^2 \hbar^2}{2mL^2}. \tag{9.146}$$

These equations underscore the quantized nature of energy and momentum in quantum systems, a fundamental departure from classical physics. The ability of the particle to exist outside the well, albeit with diminishing probability, is a manifestation of quantum tunneling, a phenomenon with no counterpart in classical mechanics.

9.11 Summary

In this chapter, we transitioned from classical to quantum physics, exploring the revolutionary principles that underpin modern physics. We began by highlighting the limitations of classical theories in explaining phenomena like blackbody radiation and the photoelectric effect, which led to the development of quantum mechanics. Key concepts such as wave-particle duality, quantization of energy, and the Schrödinger wave equation were introduced, forming the basis of our understanding of microscopic particle behavior.

We delved into Planck's and Einstein's contributions, showing how quantization resolved the "ultraviolet catastrophe" and explained the photoelectric effect. The Schrödinger equation, a cornerstone of quantum mechanics, was derived and applied to scenarios like free particles, infinite potential wells, and the hydrogen atom, illustrating the principles of energy quantization and probabilistic behavior of particles.

The chapter culminated with practical applications of quantum mechanics, setting the stage for discussions on semiconductor behavior and optoelectronic devices. These insights form the foundation for bridging theoretical quantum mechanics with real-world technological advancements in later chapters.

9.12 Problems

1. Derive the Rayleigh-Jeans law for blackbody radiation and explain its failure at high frequencies (the "ultraviolet catastrophe"). How does Planck's quantization hypothesis resolve this issue?
2. Calculate the peak wavelength for blackbody radiation at temperatures of 300 K, 2000 K, and 6000 K using Wien's displacement law. Compare your results to the visible spectrum and discuss the implications.
3. A metal has a work function of 2.5 eV. Determine the threshold frequency of light required to eject an electron. If the incident light has a frequency of 1.5×10^{15} Hz, calculate the maximum kinetic energy of the emitted electrons.
4. Calculate the de Broglie wavelength of:
   (a) An electron moving at 10^5 m/s.
   (b) A baseball ($mass = 0.145$ kg) moving at 40 m/s. Compare the results and discuss their significance.
5. A particle in a double-slit experiment produces an interference pattern. Derive the expression for the probability density if the wave function is a superposition of two waves passing through each slit.
6. Derive the time-dependent wave function for a particle in an infinite potential well. Show how the time-dependent term affects the probability density of the particle. Does the probability density depend on time? Explain your findings in the context of quantum mechanics.
7. For a particle confined in a potential well of width 1 nm, calculate the first three energy levels. Discuss how the energy levels change with increasing the width of the well.
8. For a particle in an infinite potential well, calculate the probability of finding the particle between $x = 0$ and $x = L/2$ in the $n = 1$ and $n = 2$ states. Interpret your results.
9. Describe the qualitative differences between the energy eigenvalues and wave functions for a finite potential well compared to an infinite potential well. Solve

for the ground state energy of a finite potential well with $V_0 = 5$ eV, $L = 1$ nm, and $m = 9.1 \times 10^{-31}$ kg.
10. A particle is confined in a one-dimensional box of width 2 nm. Estimate the uncertainty in the particle's momentum. Use the uncertainty principle to estimate the minimum kinetic energy of the particle.

Bibliography

1. Griffiths, D. J., & Schroeter, D. F. (2018). Introduction to Quantum Mechanics (3rd ed.). Cambridge University Press.
2. Shankar, R. (2014). Principles of Quantum Mechanics (2nd ed.). Springer.
3. Sakurai, J. J., & Napolitano, J. (2020). Modern Quantum Mechanics (3rd ed.). Cambridge University Press.
4. Feynman, R. P., Leighton, R. B., & Sands, M. (2011). The Feynman Lectures on Physics, Vol. 3: Quantum Mechanics. Basic Books.
5. Cohen-Tannoudji, C., Diu, B., & Laloë, F. (1977). Quantum Mechanics (Vol. 1). Wiley-Interscience.
6. Schiff, L. I. (1968). Quantum Mechanics (3rd ed.). McGraw-Hill.
7. Planck, M. (1901). "On the Law of Distribution of Energy in the Normal Spectrum." Annalen der Physik, 4(3), 553–563.
8. Bohr, N. (1913). "On the Constitution of Atoms and Molecules." Philosophical Magazine Series 6, 26(153), 1–25.
9. Dirac, P. A. M. (1928). "The Quantum Theory of the Electron." Proceedings of the Royal Society of London. Series A, 117(778), 610–624.
10. Landau, L. D., & Lifshitz, E. M. (1981). Quantum Mechanics: Non-Relativistic Theory (Course of Theoretical Physics, Vol. 3). Butterworth-Heinemann.

Semiconductor Energy Bands 10

10.1 Introduction

Semiconductors form the cornerstone of modern electronic and photonic devices. Their unique ability to absorb and emit light is fundamental in the design of various devices such as light sources, photodetectors, and optical modulators. These applications and their underlying principles will be introduced in subsequent chapters.

To fully grasp these phenomena, a fundamental understanding of the structure and behavior of semiconductor materials is essential. At the heart of these properties is the occupation and transition of electrons between energy levels.

In the previous chapter, we solved the Schrödinger wave equation for isolated particles in potential wells and single-electron atoms. We found that electrons can take only discrete values of energy; that is, the energies are quantized. Building on this foundation, we now turn our attention to the behavior of electrons in solids, which comprise a large number of atoms. In such environments, the discrete quantum states of electrons in isolated atoms give way to energy bands. These bands are a result of the overlap of atomic orbitals in solids and play a pivotal role in determining the electronic and optical properties of materials.

Understanding the formation and characteristics of these energy bands is crucial. This phenomena explains not only the fundamental electrical properties of semiconductors but also their interactions with light. This chapter aims to bridge the gap between the quantum mechanical concepts addressed previously and their application in the macroscopic world of semiconductor physics. It provides the necessary theoretical framework to understand the operation and design of a myriad of semiconductor-based devices.

10.2 Conductors, Semiconductors, and Insulators

Materials can be broadly classified into three distinct groups on the basis of their electrical conductivity:

1. **Conductors**: These materials exhibit very high electrical conductivity and correspondingly low resistivity. Common conductors include metals such as copper, gold, and aluminum. The high conductivity of these materials is due to the abundance of free electrons that can move easily through the material.
2. **Semiconductors**: Semiconductors are characterized by moderate intrinsic electrical conductivity, which is significantly lower than that of conductors. However, their conductivity greatly enhanced by doping them with specific impurities. Silicon and germanium are prime examples of semiconductor materials. The unique ability of semiconductors to alter their conductivity through doping is the cornerstone of modern electronic devices.
3. **Insulators**: Insulators possess very high electrical resistivity, making them poor conductors of electricity. This property is intrinsic to the material and cannot be significantly altered. Examples of insulating materials include wood, diamond, and various plastics. The high resistivity of these materials is due to the lack of free charge carriers, making it difficult for an electric current to flow through them.

The fundamental differences between these categories of materials arise from their energy band structures, a topic we explore into in detail. At the core of electrical conductivity is the concept of band theory, which explains the availability and mobility of charge carriers, primarily electrons, within these materials. Conductors have partially filled or overlapping energy bands, allowing electrons to move freely and conduct electricity. Semiconductors have a small bandgap, which can be bridged by thermal energy or doping, enabling controlled conductivity. Insulators, on the other hand, have a large bandgap, making it difficult for electrons to gain enough energy to contribute to electrical conduction.

In the remaining sections of this chapter, we address the band structure of semiconductors and develop a theory to determine the carrier density in these energy bands.

10.3 Splitting of Energy Levels

When single-electron atoms, such as the hydrogen atom, are examined, we observe that electrons are restricted to discrete energy levels. This observation is crucial for understanding the electronic structure of atoms. A key principle governing this behavior is the **Pauli exclusion principle**, which asserts that no two electrons can occupy the same quantum state simultaneously.

Consider, for example, two one-electron atoms that are initially far apart. As illustrated in Fig. 10.1, each atom has its lowest energy level, denoted as $n = 1$,

10.3 Splitting of Energy Levels

Fig. 10.1 Energy level splitting: (**a**) two atoms far from each other have $n = 1$ energy levels, (**b**) the atoms brought closer so that their $n = 1$ energy level splits into two levels, and (**c**) three atoms are brought closer so their $n = 1$ energy level splits into three levels

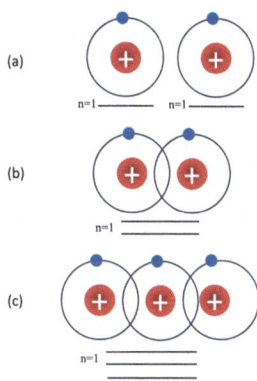

occupied by an electron. When we bring these two atoms closer together, the electric fields of their nuclei begin to interact with the electrons of the neighboring atom. Due to the Pauli exclusion principle, these electrons cannot occupy the same energy level. Consequently, the original $n = 1$ energy level splits into two distinct levels, a phenomenon clearly depicted in the figure.

By extending our examination to a crystal composed of N atoms, each with a single electron, we observe an interesting process: the initial $n = 1$ energy level splits into N closely spaced energy levels. This process ultimately leads to the formation of what we term as "energy bands." To grasp the scale of the spacing between these energy states within a band, consider a crystal with $N = 10^{20}$ single-electron atoms. If the total width of the allowed energy band at the equilibrium interatomic distance is 1 eV, then the spacing between individual energy levels within this band is approximately 10^{-20} eV. This minuscule energy difference means that, for all practical purposes, we can consider the energy distribution within the allowed band as quasi-continuous.

Energy level splitting in a crystalline structure continues until the atoms reach their characteristic interatomic spacing. At this juncture, the splitting stabilizes due to the physical limitations imposed by repulsive forces between atoms; they cannot become any closer. This limitation in atomic proximity sets a definitive boundary for the extent of energy level splitting. As a result, the energy levels form distinct bands, each comprising a multitude of closely spaced energy states. The formation of these energy bands, a direct consequence of the restricted movement of atoms within the crystal lattice, is pivotal for understanding the electronic properties of the materials.

Figure 10.2 provides a schematic representation of how three energy states split into allowed bands of energies in a crystalline structure. The regions between these energy bands represent forbidden states, which electrons cannot occupy.

The concept of energy bands is central to understanding the electrical properties of materials, particularly semiconductors. In a large crystal, the energy levels become so closely packed that they essentially form a continuous band. The behavior of electrons within these bands, as well as the gaps between them, that

Fig. 10.2 Schematic of energy level splitting for three energy levels $n = 1, 2$, and 3.

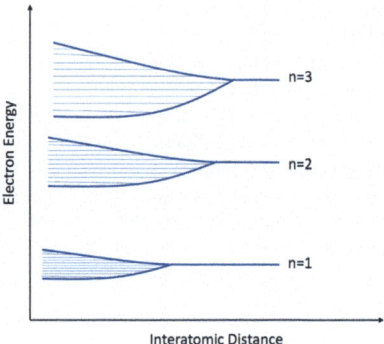

fundamentally determines whether a material acts as a conductor, a semiconductor, or an insulator.

10.4 Periodic Potential Wells in Crystals

In revisiting the concept of a one-electron atom, let us consider its potential and energy characteristics:

1. The potential (V) is inversely proportional to the distance from the positively charged nucleus.
2. Given that the electron is negatively charged, its energy E is negative, indicating an attraction to the nucleus.
3. At an infinite distance from the nucleus, both V and E approach zero, resembling the state of a free electron.

These relationships for an isolated one-electron atom are depicted in Fig. 10.3.

Most semiconductors such as silicon exhibit a crystalline structure. Consider a one-dimensional periodic lattice in such a crystal. The overlapping potentials of the atoms in this lattice form a periodic potential $V(x)$, as illustrated in Fig. 10.4.

The task of solving the Schrödinger wave equation for a complex potential profile, such as the one shown in Fig. 10.4, is quite challenging due to its intricacy. To simplify this problem, we adopt the **Kronig-Penney model,** a more manageable representation of the periodic potential function in a one-dimensional single-crystal lattice, as depicted in Fig. 10.5. This model requires solving the Schrödinger wave equation in each region of the potential.

More intriguing solutions emerge when the electron energy E is less than the potential barrier V_o, corresponding to an electron being bound within the crystal. In this scenario, electrons are confined within the potential wells, yet there is a possibility for tunneling between adjacent wells. This phenomenon of tunneling is a quintessential quantum mechanical effect and is particularly pertinent in the context of periodic potentials.

10.4 Periodic Potential Wells in Crystals

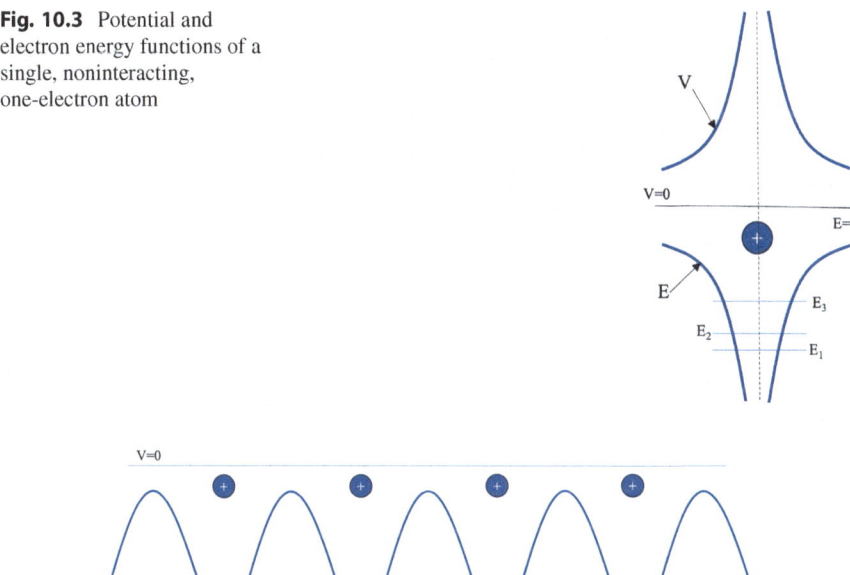

Fig. 10.3 Potential and electron energy functions of a single, noninteracting, one-electron atom

Fig. 10.4 Periodic potential for a one-dimensional single crystal

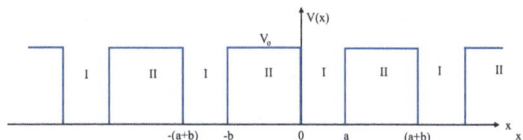

Fig. 10.5 The one-dimensional periodic potential function of the Kronig-Penney model approximating the potential well-shown in Fig. 10.4

The Kronig-Penney model, despite being an idealized representation of a one-dimensional crystal lattice, effectively demonstrates many key aspects of the quantum behavior of electrons in a periodic lattice. It sheds light on how electrons are distributed in energy bands and how they can move across these bands, a concept that is central to the understanding of the electrical properties of semiconductors. This model serves as a crucial stepping stone in understanding the complex quantum mechanics behind the behavior of electrons in the crystalline structures of semiconductor materials.

To solve the Schrödinger wave equation within the framework of the Kronig-Penney potential model, we turn to **Bloch theorem**. This theorem provides a crucial insight: the eigenstates ψ of the Hamiltonian ($\hat{H}$) in a periodic potential can be expressed as the product of a planewave and a function that shares the periodicity of the crystal lattice. Accordingly, the eigenfunction takes the form:

$$\psi(x) = u_k(x)e^{jkx}, \tag{10.1}$$

where $u_k(x)$ is the Bloch function, characterized by the periodicity of the lattice period $(a + b)$, and k is a constant of motion, whose value is determined.

The complete solution of the time-dependent Schrödinger wave equation, referring back to Chap. 9, Eq. (9.62), is represented as:

$$\Psi(x, t) = \psi(x)\phi(t) = \psi(x) = u_k(x)e^{jkx}e^{-jEt/\hbar}. \tag{10.2}$$

This formulation results in a traveling-wave solution, indicative of the motion of an electron in a single-crystal material. The amplitude of this traveling wave, described by the periodic Bloch function $u_k(x)$, modulates the planewave, whereas the parameter k, known as the wavenumber, is integral to defining the wave's spatial characteristics.

It is important to note that the inclusion of the Bloch function in this solution reflects the fundamental nature of electrons in crystalline materials: their behavior is governed by the periodic potential of the lattice.

The time-independent portion of Schrödinger's wave equation can now be rewritten as:

$$\frac{\partial^2 \psi(x)}{\partial x^2} + \frac{2m}{\hbar^2}[E - V(x)]\psi(x) = 0. \tag{10.3}$$

If we consider region I in Fig. 10.5 where $V(x) = 0$, for $0 < x < a$ and substitute from Eq. (10.1) in Eq. (10.3), we obtain the following:

$$\frac{d^2 u_{k1}(x)}{dx^2} + j2k\frac{du_{k1}(x)}{dx} - \left(k^2 - \frac{2mE}{\hbar^2}\right)u_{k1}(x) = 0, \tag{10.4}$$

where $u_{k1}(x)$ is the amplitude of the wave function in region I.

Let

$$\alpha^2 = \frac{2mE}{\hbar^2}, \tag{10.5}$$

then time-dependent Schrödinger wave equation of Eq. (10.4) becomes:

$$\frac{d^2 u_{k1}(x)}{dx^2} + j2k\frac{du_{k1}(x)}{dx} - \left(k^2 - \alpha^2\right)u_{k1}(x) = 0. \tag{10.6}$$

This wave equation has a solution in the form of two planewaves traveling in opposite directions:

$$u_{k1}(x) = Ae^{-j(k-\alpha)} + Be^{j(k-\alpha)}. \tag{10.7}$$

10.4 Periodic Potential Wells in Crystals

In region II ($-b < X < 0$), in which $V(x) = V_o$, the time-dependent Schrödinger wave equation of Eq. (10.4) becomes:

$$\frac{d^2 u_{k2}(x)}{dx^2} + j2k\frac{du_{k2}(x)}{dx} - \left(k^2 - \alpha^2 + \frac{2mV_o}{\hbar^2}\right)u_{k2}(x) = 0, \quad (10.8)$$

where $u_{k2}(x)$ is the amplitude of the wave function in region II.

Let

$$\beta^2 = \alpha^2 - \frac{2mV_o}{\hbar^2}, \quad (10.9)$$

substituting Eq. (10.9) in the Schrödinger wave equation in region II Eq. (10.8) results in the following wave equation:

$$\frac{d^2 u_{k2}(x)}{dx^2} + j2k\frac{du_{k2}(x)}{dx} - \left(k^2 - \beta^2\right)u_{k2}(x) = 0. \quad (10.10)$$

This wave equation has a solution in the form of two planewaves traveling in opposite directions

$$u_{k2}(x) = Ae^{-j(k-\beta)} + Be^{j(k-\beta)}. \quad (10.11)$$

These solutions need to be considered within the boundary conditions:
The wave functions and their first derivatives are continuous, i.e.,

$$u_{k1}(0) = u_{k2}(0),$$

$$\frac{du_{k1}}{dx}\bigg|_{x=0} = \frac{du_{k2}}{dx}\bigg|_{x=0},$$

$$u_{k1}(a) = u_{k2}(-b),$$

and

$$\frac{du_{k1}}{dx}\bigg|_{x=a} = \frac{du_{k2}}{dx}\bigg|_{x=b}.$$

Applying these four boundary conditions to Eq. (10.7) and Eq. (10.11) and working through the algebra to obtain an equation that is more susceptible to a graphical solution and thus will illustrate the nature of the results, let the potential barrier width $b \to 0$ and the barrier height such that the product bV_o remains finite. We obtain:

$$p\left(\frac{sin(\alpha a)}{\alpha a}\right) + cos(\alpha a) = cos(ka), \quad (10.12)$$

where

$$p = \frac{mV_oba}{\hbar^2}. \tag{10.13}$$

Since α as given by Eq. (10.5) is a function of electron energy E, Eq. (10.13) provides a relation between E and K. For the case when $V_o = 0$, i.e., the free-electron case, we find from Eq. (10.12) that $k = \alpha$. Since the potential is equal to zero the electron has kinetic energy only, therefore:

$$\alpha = \sqrt{\frac{2mE}{\hbar^2}} = \sqrt{\frac{2m(\frac{1}{2}mv^2)}{\hbar^2}} = \frac{p}{\hbar} = k, \tag{10.14}$$

where p is the particle momentum. Let the let-hand-side of Eq. (10.12) to be $f(\alpha a)$; then,

$$f(\alpha a) = cos(ka). \tag{10.15}$$

The function $f(\alpha a)$ values can be larger than $+1$ and less than -1 as a function of αa but the right-hand-side of Eq. (10.12) varies between ± 1. Therefore, for Eq. (10.12) to be valid, the function $f(\alpha a)$ must be bounded between -1 and $+1$. The right-hand side of Eq. (10.12) must lie between -1 and $+1$ and, therefore, so must the left-hand side. The plot shows that Eq. (10.5) is only satisfied for certain ranges of values of αa and hence only certain ranges of values of the energy. The regions of allowed energy, called bands, are separated by regions where there are no allowed energies (forbidden regions) called bandgaps. Figure 10.6 shows the allowed values of $f(\alpha a)$ and the allowed values of αa in the shaded areas. Also shown in the figure are the values of $ka = n\pi/a$, where n is an integer that makes the RHS of Eq. (10.15) equal ± 1, which correspond to the allowed values of $f(\alpha a)$.

The parameter α is related to the total energy E of the particle through Eq. (10.5). A plot of the energy E of the particle as a function of the wavenumber k can be generated from Fig. 10.6. Figure 10.7 shows this plot and thus shows the concept of allowed energy bands for the particle propagation in the crystal lattice.

Fig. 10.6 Plot of the function $f(\alpha a)$. The shaded areas show the allowed values of (αa) corresponding to real values of ka

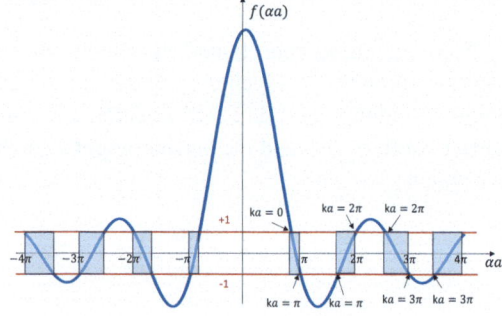

10.4 Periodic Potential Wells in Crystals

Fig. 10.7 The $E - k$ diagram generated from Fig. 10.6. The allowed energy bands and forbidden energy bandgaps are indicated

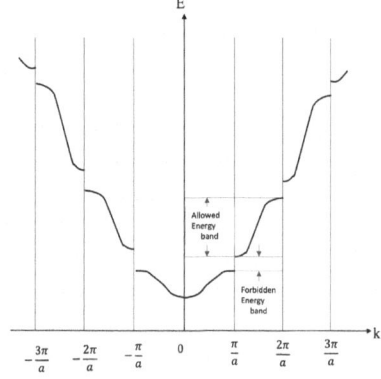

Since the energy E has discontinuities, we also have the concept of forbidden energies for the particles in the crystal.

Example Determine the width of a forbidden energy band (in eV) that exists at $ka = \pi$ (see Fig. 10.8). Assume that the coefficient $p = 10$ and that the potential width is $a = 5 \text{Å}$.

Solution Combining Eq. (10.12), we have:

$$p\left(\frac{sin(\alpha a)}{\alpha a}\right) + cos(\alpha a) = cos(ka),$$

at $ka = \pi$ and $p = 10$, we obtain:

$$10\left(\frac{sin(\alpha a)}{\alpha a}\right) + cos(\alpha a) = -1,$$

Energy gap for the values of αa when $ka = \pi$. The figure shows that $\alpha_1 a = \pi$, and the second value is where $\pi < \alpha_2 a < 2\pi$.

$$\alpha_1 = \frac{\pi}{a} = \sqrt{\frac{2mE_1}{\hbar^2}}.$$

Therefore,

$$E_1 = \frac{\pi^2 \hbar^2}{2ma^2} = \frac{\pi^2 \times (1.054 \times 10^{-34})^2}{2 \times (9.11 \times 10^{-31}) \times (5 \times 10^{-10})^2} = 2.41 \times 10^{-19} \text{ J}.$$

(continued)

To accurately determine the value for α_2, we utilized the MATLAB code employed in plotting the diagram shown in Fig. 10.8. Our focus was on identifying the intersection point where $f(\alpha a) = -1$ and $ka = \pi$. From this analysis, we find that at this specific intersection, the value of $\alpha_2 a$ is equal to 5.32. Therefore,

$$\alpha_2 = \frac{5.32}{a} = \sqrt{\frac{2mE_1}{\hbar^2}},$$

$$E_2 = \frac{5.32^2 \hbar^2}{2ma^2} = \frac{5.32^2 \times (1.054 \times 10^{-34})^2}{2 \times (9.11 \times 10^{-31}) \times (5 \times 10^{-10})^2} = 6.6 \times 10^{-19}$$

Thus, the bandgap energy is given by:

$$E_g = E_2 - E_1 = (6.6 - 2.41) \times 10^{-19} = 4.19 \times 10^{19} \text{ J}.$$

or

$$E_g = \frac{4.13 \times 10^{-19} \text{ J}}{1.6 \times 10^{-19} \text{ C}} = 2.69 \text{ eV}.$$

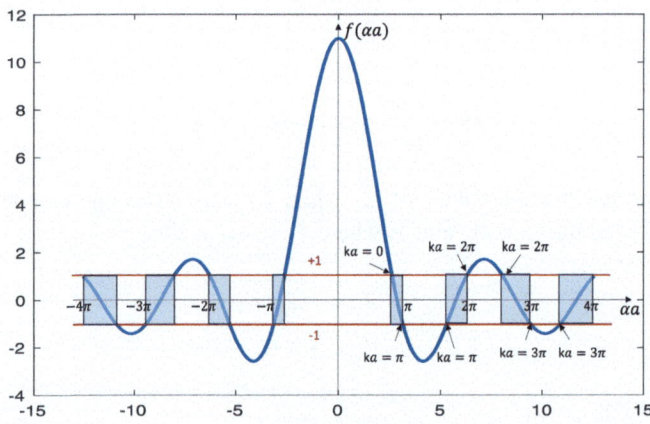

Fig. 10.8 The plot for $f(\alpha a)$ to determine the energy bandgap when $ka = \pi$ and the parameter $p = 10$ used in the example

10.5 The $E - K$ Diagram and the Energy Bands

In this section, we delve into the intricate world of the $E - K$ diagram and energy bands, which are pivotal tools in solid-state physics and semiconductor theory. The $E - K$ diagram, also known as the energy-momentum diagram, is fundamental for visualizing and understanding the behavior of electrons within a crystal lattice. It graphically represents the relationship between the energy (E) of electrons and their wave vector (k), revealing the quantum nature of these particles in solid materials.

The energy bands of semiconductors are categorized into two main types: the valence band, (VB), which is predominantly occupied by electrons at lower energy levels, and the conduction band (CB), where higher energy levels reside.

The $E - k$ diagram can be divided into three distinct zones:

1. Periodic Zone: This zone reflects the periodic nature of the crystal lattice and its influence on electron behavior.
2. Reduced Zone: The reduced zone simplifies the representation of electron states by folding the diagram into the first Brillouin zone, the shaded region in Fig. 10.9.
3. Extended Zone: This zone extends the representation to higher energy levels, showcasing the relationship between energy and momentum over multiple Brillouin zones.

In the remainder of this chapter, our focus will be on the reduced zone of the $E - k$ diagram, which offers a simplified yet comprehensive view of the electron behavior in crystals.

For each value of the wave vector k, there exists a corresponding solution to the Schrödinger equation. As a result, each k value is linked to a unique allowed energy state. Due to the periodic boundary conditions inherent in a crystal, k is quantized, leading to discrete energy levels E. The wave vector k serves as the propagation vector, playing a crucial role in determining the behavior of electrons within the crystal lattice.

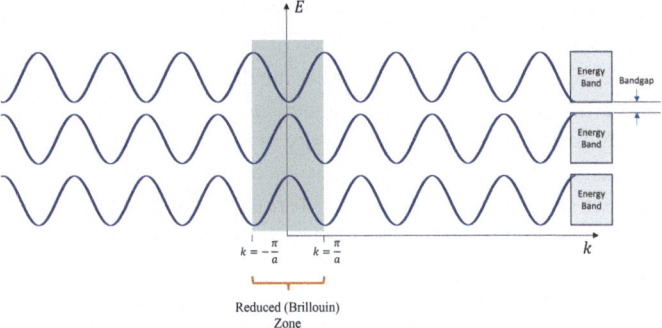

Fig. 10.9 The E-k diagram for the extended region including the energy bands and bandgaps. The reduced (Brillouin) region is shaded

Fig. 10.10 (a) Conduction and valence bands, (b) the $E - k$ diagram of the conduction and valence bands of a semiconductor at $T = 0\,\text{K}°$, and (c) at $T > 0\,\text{K}°$

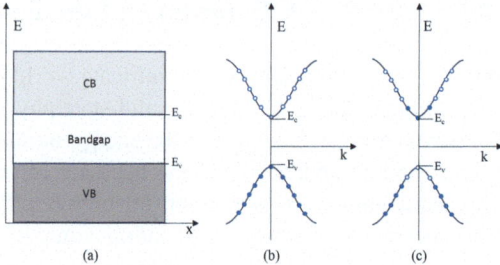

The magnitude of this wave vector $|\bar{k}|$ is related to the de Broglie wavelength Λ by the equation:

$$|\bar{k}| = \frac{2\pi}{\Lambda}. \tag{10.16}$$

The momentum p of an electron in the crystal is expressed as $p = \hbar k$. This relationship defines what is termed the crystalline momentum, blending the quantum mechanical nature of electrons with the periodic structure of the crystal lattice.

Energy bands in semiconductors, as depicted in Fig. 10.10a, are divided into a valence band (VB) and a conduction band (CB), separated by a forbidden region known as the bandgap. At zero degree Kelvin ($T = 0\,\text{K}°$), the VB is completely filled with electrons, meaning all its energy levels are occupied. Conversely, the CB is empty, meaning that all its energy levels are available. This state is illustrated in Fig. 10.10b, where the $E - k$ diagram is used to demonstrate such a condition. At room temperature ($T > 0\,\text{K}°$), electrons in the VB gain sufficient energy to transition to the lower energy states in the CB, leaving behind empty states (holes) in the VB, as shown in Fig. 10.10c.

10.6 Density of the Quantum States

To understand the flow of carriers between energy states in semiconductor devices, it is essential to first determine the number of electrons and holes available for conduction. The current in semiconductors is attributed to the flow of these charge carriers, making the quantification of available carriers a critical step. This number is fundamentally linked to the available energy or quantum states in the material.

The Pauli exclusion principle states that each quantum state can be occupied by only one electron. When discussing the formation of energy bands, which are comprised of allowed and forbidden energy levels, we noted that these bands consist of discrete energy levels. To accurately calculate the concentrations of electrons and holes, we must determine the number of these allowed energy states per unit volume, a concept known as the "density of states." This density is described by a function indicating the number of electronic states per unit energy at each energy level and varies based on the energy and material properties.

10.6 Density of the Quantum States

In quantum mechanics, the quantum states are the solutions to the Schrödinger equation that satisfy boundary conditions in k-$space$, where k is the wave vector.

In the realm of quantum mechanics, the quantum states are the permissible solutions of Schrödinger wave equation that satisfy the boundary conditions in k-$space$, where k is the wave vector.

Let $\rho(k)$ be the density of states in k-space and $\rho(E)$ be the density of states in the energy-space. Thus, $\rho(k)dk$ is the number of states between k and $k + dk$ per unit volume of the material. Similarly, $\rho(E)dE$ is the number of states between E and $E + dE$ per unit volume of the material.

The total number of states in both k-$space$ and energy-space must be equal, i.e.,

$$\rho(k)dk = \rho(E)dE. \tag{10.17}$$

The energy distribution of electrons in these bands can be approximated parabolically. In the CB, the energy E is given by:

$$E = E_c + \frac{\hbar^2 k^2}{2m_c}, \tag{10.18}$$

where E_c is the energy at the bottom of the CB and m_c is the effective mass of an electron in the CB.

Similarly, in the VB, the energy E is described by:

$$E = E_v - \frac{\hbar^2 k^2}{2m_v}, \tag{10.19}$$

where E_v is the energy at the top of the VB and m_v is the effective mass of an electron in the VB.

The kinetic energy of an electron is expressed as:

$$\frac{1}{2}mv^2 = \frac{p^2}{2m}. \tag{10.20}$$

The effective mass of the electron in both the CB and VB is given by:

$$m_{c,v} = \frac{\hbar^2}{(\frac{\partial^2 E}{\partial k^2})}. \tag{10.21}$$

This formulation of effective mass is crucial, as it encapsulates the modified electron behavior due to the periodic potential of the crystal lattice (Table 10.1).

Table 10.1 The electron effective mass and electron and hole mobilities of some semiconductor materials

Material	m_o	m_c	m_v	μ_e	μ_h
Silicon	9.1×10^{-31} kg	0.98	0.49	1450	505

10.6.1 Quantum States in Block Semiconductor Material

The time-independent Schrödinger equation for a particle in three-dimensional space is given by:

$$\nabla^2 \psi(x, y, z) + k^2 \psi(x, y, z) = 0, \tag{10.22}$$

where $\psi(x, y, z)$ is the wave function. This wave function can be expressed as a product of three separate functions, one for each dimension:

$$\psi(x, y, z) = \psi_x(x)\psi_y(y)\psi_z(z). \tag{10.23}$$

Upon substituting this wave function into the SE and dividing both sides by $\psi_x(x)\psi_y(y)\psi_z(z)$, and considering that $k^2 = k_x^2 + k_y^2 + k_z^2$, we arrive at the following equation:

$$\frac{\partial^2 \psi_x}{\partial x^2} + \frac{\partial^2 \psi_y}{\partial y^2} + \frac{\partial^2 \psi_z}{\partial z^2} + k_x^2 \psi_x(x) + k_y^2 \psi_y(y) + k_z^2 \psi_z(z) = 0, \tag{10.24}$$

This leads to the derivation of three separate differential equations, each representing a one-dimensional problem:

1.
$$\frac{\partial^2 \psi_x}{\partial x^2} + k_x^2 \psi_x(x) = 0, \tag{10.25}$$

2.
$$\frac{\partial^2 \psi_y}{\partial y^2} + k_y^2 \psi_y(y) = 0, \text{ and} \tag{10.26}$$

3.
$$\frac{\partial^2 \psi_z}{\partial z^2} + k_z^2 \psi_z(z) = 0. \tag{10.27}$$

10.6 Density of the Quantum States

Each of these equations can be solved in a manner similar to the one-dimensional cases we have previously discussed. The solutions to these equations provide the quantum states for each dimension.

In a block of material, the total number of quantum states available to an electron is equal to the number of allowed k values. These states form the foundation for understanding the quantum mechanical behavior of particles in three-dimensional systems.

According to Bloch's theorem, the wave function for an electron in a crystal lattice is represented as:

$$\psi(r) = u_k e^{j\underline{k}\cdot\underline{r}}, \quad (10.28)$$

where u_k is a periodic function, known as the Bloch function, corresponding to each cell within the lattice. The probability of finding an electron is given by $|\psi|^2$.

The wave vector is composed of its three components along the Cartesian coordinates:

$$\underline{k} = k_x \underline{a_x} + k_y \underline{a_y} + k_z \underline{a_z}. \quad (10.29)$$

Given any point in space represented by the vector $\underline{r} = x\underline{a_x} + y\underline{a_y} + z\underline{a_z}$, we have:

$$\underline{k}\cdot\underline{r} = k_x x + k_y y + k_z z. \quad (10.30)$$

Hence the wave function can be expressed as:

$$\psi(x, y, z) = u_k e^{j(k_x x + k_y y + k_z z)}. \quad (10.31)$$

If we consider an electron confined within a box of semiconductor bulk material with dimensions L_x, L_y and L_z, the probability of finding the electron outside this box is zero. Hence, the wave function ψ will be periodic within the box and forms a standing wave, similar to the solutions for an infinite quantum well. For the lowest energy state E_1, the length of the potential well L corresponds to half the wavelength of the particle in this state, implying that $\Lambda = 2L$. The roundtrip phase of the wave function must be a multiple of 2π, i.e., m

$$2\underline{k}\cdot\underline{r} = N2\pi, \quad (10.32)$$

where N is a positive integer ($N = 1, 2, 3, \ldots$).

Consequently, the roundtrip phase shift between each two-parallel sides of the box to form standing waves is given by::

$$2k_x L_x = m2\pi, \quad m = 1, 2, 3, \ldots, \quad (10.33)$$

$$2k_y L_y = p2\pi, \quad p = 1, 2, 3, \ldots, \quad (10.34)$$

and

$$2k_z L_z = q2\pi, \quad q = 1, 2, 3, \ldots \tag{10.35}$$

Hence, the three wavenumbers are expressed as:

$$k_x = (\frac{\pi}{L_x})m, \; k_y = (\frac{\pi}{L_y})p, \; and \; k_z = (\frac{\pi}{L_z})q, . \tag{10.36}$$

The amplitude of the wave vector is as follows:

$$|\underline{k}| = (k_x^2 + k_y^2 + k_z^2)^{1/}, \tag{10.37}$$

and

$$k^2 = \frac{2mE}{\hbar^2}. \tag{10.38}$$

According to Eq. (10.37) the values of k are quantized and discerned such that the corresponding energy values E.

The $\underline{k}$ vector can assume any value in a sphere, and $\rho(k)$ is the number of states between k and $k + dk$.

Example Consider one-dimensional case where $L_x = 1$ mm, and the interatomic distance for silicon is 5A°.

Determine the number of possible values that k can take.

Solution The smallest value of the wavenumber k_x is:

$$k_{x-min} = \frac{\pi}{1 \text{ mm}} = 10^3 \pi \text{ m}^{-1}.$$

The largest value of k_x for a crystal with an interatomic distance a is given by:

$$k_{x-max} = \frac{\pi}{a}.$$

Therefore, the maximum value of k_x is:

$$k_{x-max} = \frac{\pi}{a} = \frac{\pi}{5 \times 10^{-10}} = 2\pi \times 10^{+9} \text{ m}^{-1}$$

(continued)

10.6 Density of the Quantum States

There are approximately k_{x-max}/k_x discrete values of k_x exist, i.e.,

$$\frac{k_{x-max}}{k_{x-min}} = \frac{2\pi \times 10^{-9}}{10^3 \pi} = 2 \times 10^6.$$

Therefore, this is indicating that k can take millions of values.

10.6.2 Number of k Values Between k and $k + dk$

To determine the number of k values, we analyze the volume occupied by one state in the k-$space$. Consider a two-dimensional space, and visualize the allowed quantum states in k-space as a function of k_x and k_y, where each point represents an allowed state corresponding to various integral values of m and p. Both positive and negative values of k_x, k_y or k_z have the same energy and represent the same energy state. The density of quantum states is determined by considering only the positive one-eighth of the spherical k-$space$ (Fig. 10.11).

The distance between two quantum states in the k_x direction is given by:

$$k_{x+1} - k_x = (m+1)\frac{\pi}{L_x} - m\frac{\pi}{L_x} = \frac{\pi}{L_x}. \tag{10.39}$$

Therefore, the volume of each allowed quantum state in the three-dimensional k-$space$ is:

$$V = (\frac{\pi}{L_x})(\frac{\pi}{L_y})(\frac{\pi}{L_z}). \tag{10.40}$$

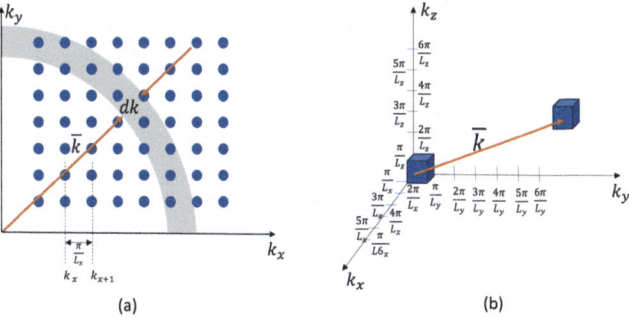

Fig. 10.11 (**a**) A two-dimensional array of allowed quantum states in k space (indicated by blue circles). (**b**) Three-dimensional k-$space$ showing the volume of each quantum state by a blue box

Considering that each point in the sphere is shared by four cubes and that each point on the surface of the cube is shared by eight cubes, the volume of one-eighth of a sphere in k-$space$ is $4\pi k^2/8$. Thus, the total number of quantum states in one-eighth of the sphere in k-$space$ is given by:

$$\frac{\pi k^2 dk/2}{(\frac{\pi}{L_x})(\frac{\pi}{L_y})(\frac{\pi}{L_z})} \times 2 \tag{10.41}$$

where the factor of 2 accounts for two possible electron spins per state.

Therefore, the number of quantum states per unit volume of the material between k and $k+dk$ is:

$$\rho(k)dk = \frac{\pi k^2 dk}{\pi^3} \frac{(L_x L_y L_z)}{L_x L_y L_z} = \frac{k^2}{\pi^2} dk, \tag{10.42}$$

simplifying, we obtain:

$$\rho(k) = \frac{k^2}{\pi^2}. \tag{10.43}$$

This is the density of states in the momentum, $k-$space.

The density of states in the energy space (E-$space$), between energies E an $dE+dE$, is related to the density of states in k-$space$ as follows:

$$\rho(k)dk = \rho(E)dE, \tag{10.44}$$

hence

$$\rho(E) = \rho(k)\frac{dk}{dE}. \tag{10.45}$$

To determine the derivative of E with respect to k, we start by the total energy relation in the conduction band given by:

$$E = E_c + \frac{\hbar^2 k^2}{2m_c}, \tag{10.46}$$

which leads to:

$$\frac{dE}{dk} = \frac{\hbar^2 k}{m_c}. \tag{10.47}$$

Rewriting this energy equation in terms of the momentum k, we obtain:

$$k = (E - E_c)^{1/2} (\frac{2m_c}{\hbar^2})^{1/2}. \tag{10.48}$$

10.6 Density of the Quantum States

Therefore, the density of states in the energy space is:

$$\rho(E) = \rho(k)/\frac{dE}{dk} = \frac{k^2}{\pi^2}/(\frac{\hbar^2 k}{m_c}) = \frac{k m_c}{\pi^2 \hbar^2}, \qquad (10.49)$$

Consequently, the density of states in the energy space in the conduction band becomes:

$$\rho_c(E) = \frac{m_c}{(\pi \hbar)^2}(E - E_c)^{1/2}(\frac{2m_c}{\hbar^2})^{1/2}, \qquad (10.50)$$

or more explicitly:

$$\rho_c(E) = \frac{1}{2\pi^2}(\frac{2m_c}{\hbar^2})^{3/2}(E - E_c)^{1/2}. \qquad (10.51)$$

where the density of states starts is zero at the edge of the conduction band and increases with the energy difference $E - E_c$.

Similarly, the density of states in the valence band (VB), $\rho_v = (E)$, is given by:

$$\rho_v(E) = \frac{1}{2\pi^2}(\frac{2m_v}{\hbar^2})^{3/2}(E_v - E)^{1/2}, \qquad (10.52)$$

where the density of states starts at zero at the edge of the valence band and increases with the energy difference $E_v - E$. The energy density of states for both the conduction and valence bands is shown in Fig. 10.12 as a function of energy. It is important to note that allowed quantum states are absent within the forbidden energy band, hence the density of states, $\rho(E) = 0$ for energies between the valence band maximum E_v and the conduction band minimum E_c, i.e., $\rho(E) = 0$ for $E_v < E < E_c$. Figure 10.12 also illustrates the density of quantum states as a function of energy.

Fig. 10.12 (a) The density of states in the conduction and in the valence bands as a function of energy and (b) the $E - k$ diagram for a bulk semiconductor material

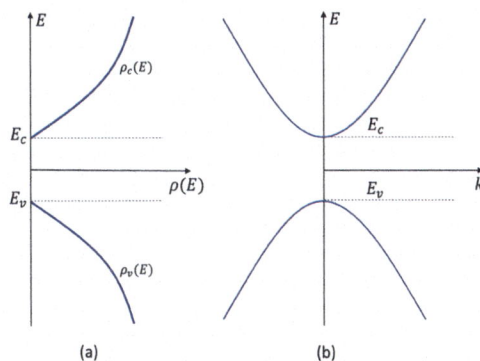

In scenarios where the effective masses of electrons and holes are equal, the density of states functions for the conduction band, $\rho_c(E)$, and the valence band, $\rho_v(E)$, exhibit symmetry about the midpoint of the energy gap, which lies midway between E_v and E_c. This symmetry is a direct consequence of the equal effective masses, leading to a mirrored behavior of the density of states in the conduction and valence bands around this median energy level.

10.6.3 Quantum Wells

In bulk semiconductor materials, as previously shown, the number of allowed quantum states within the energy bands can reach millions. This count varies based on the material's dimensions, specifically $L_x, L_y,$ and L_z. Reducing one of these dimensions, e.g., L_z, increases the spacing between the k_z values, i.e., increasing the spaces between allowed energy states in the *k-space*.

Quantum wells, which are thin layers of semiconductor materials, constrain the number of allowed quantum states to a few discrete levels. This is achieved by reducing one dimension of the semiconductor to a size comparable to the interatomic spacing, thus confining particles, typically electrons, to move in only two dimensions. This confinement leads to the emergence of new energy states.

For example, let $L_z = 20\,\text{Å}$ (angstroms). Then, the k_z values for different integers q are given by:

$$k_{zq} = \frac{q\pi}{L_z}, \tag{10.53}$$

yields specific values:

$$k_{z1} = \frac{\pi}{20 \times 10^{-10}} \cong 1.5 \times 10^9 \text{ m}^{-1}, \tag{10.54}$$

$$k_{z3} = \frac{3\pi}{20 \times 10^{-10}} = 4.5 \times 10^9, \tag{10.55}$$

$$k_{z4} = \frac{4\pi}{20 \times 10^{-10}} = \frac{\pi}{5 \times 10^{-10}} = \frac{\pi}{a}, \tag{10.56}$$

where a represents the interatomic spacing for silicon, which is five Angstrom. Therefore, for the given $L_z = 20\,\text{A}°$, thus, k_z has only four allowed quantum states.

In bulk semiconductors, the tip of the $\bar{k}$ can occupy any position within the k-space sphere. However, in quantum wells, with k_z assuming only a few specific values—four in the example above—the tip of the k-vector is restricted to disks corresponding to each k_z value, as illustrated in Fig. 10.13.

10.6 Density of the Quantum States

Fig. 10.13 The allowed quantum states (denoted by blue circles) for a quantum well exist in a disk at $k_z = q\pi/L_z$

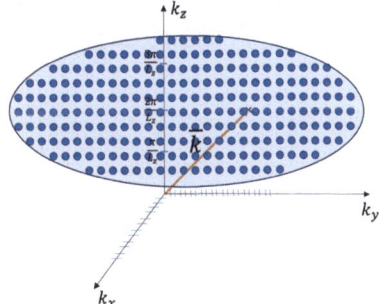

Thus, in quantum wells, the tip of the k-vector for $q = 2m$ must lie within the disk where:

$$k_z = \frac{2\pi}{L_z}.$$

To derive a formula for the density of states in quantum wells, we must determine the number of allowed states between k and $k+dk$. The amplitude of the wave vector is given by

$$k_T^2 = k_x^2 + k_y^2, \quad (10.57)$$

which is the transverse component of the wavenumbers that are in the disc determined by k_z as shown in Fig. 10.13.

Thus, the energy in terms of the three components of the wavenumber can be expressed as:

$$E = E_c + \frac{\hbar^2 k_x^2}{2m_c} + \frac{\hbar^2 k_y^2}{2m_c} + \frac{\hbar^2 k_z^2}{2m_c}, \quad (10.58)$$

and hence, the energy in the conduction band corresponding to the disc where $k_z = q\pi/L_z$ is given by:

$$E = E_c + E_q + \frac{\hbar^2 k_T^2}{2m_c}, \quad (10.59)$$

where

$$E_q = \frac{q^2 \hbar^2 k_z^2}{2m_c}, \quad (10.60)$$

and q is a positive integer, as was shown before.

Therefore the first allowed energy state in the conduction band are determined by the value k_z when $q = 1$. Hence,

$$E_{c1} = E_c + E_1 + \frac{\hbar^2 k_T^2}{2m_c}. \quad (10.61)$$

Similarly,

$$E_{c2} = E_c + E_2 + \frac{\hbar^2 k_T^2}{2m_c}, \quad (10.62)$$

and so on for the other allowed energy states. Consequently, the allowed energy states in the valence band are shifted downward by the values E_q. Therefore, the first energy level in the valence band is given by:

$$E_{v1} = E_v - E_{1'} - \frac{\hbar^2 k_T^2}{2m_v}, \quad (10.63)$$

where

$$E_{1'} = \frac{\hbar^2 k_z^2}{2m_v}. \quad (10.64)$$

The $E - k$ diagram and density of states for quantum well devices is shown in Fig. 10.14. It is clear from the figure that the energy gap for the quantum well is given by:

$$E_{g-qw} = E_{c1} - E_{v1} = (E_c - E_v) + (E_1 + E_{1'}), \quad (10.65)$$

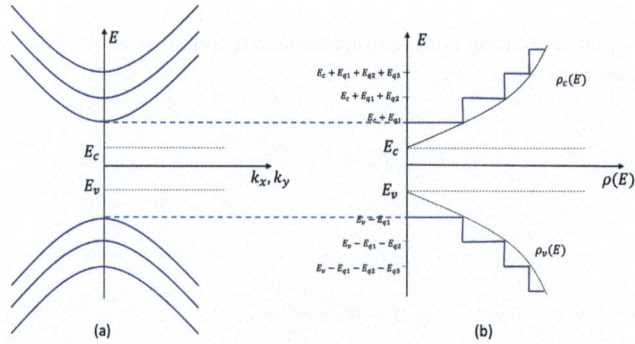

Fig. 10.14 The $E - k$ diagram and density of states for a quantum well structure

10.6 Density of the Quantum States

hence,

$$E_{g-qw} == E_g + \frac{\hbar^2 k_z^2}{2}\left(\frac{1}{m_c} + \frac{1}{m_v}\right), \tag{10.66}$$

where E_g is the energy gap for a bulk semiconductor.

Next let us find the number of states between k and $k + dk$, i.e., find

$$\rho(k)dk = \rho(k_T)dk_T. \tag{10.67}$$

These are the number of states in annular in the disc at a specific value for k_z as shown in Fig. 10.15. Each point, which represents a quantum state, is surrounded by four squares.

The number of quantum states between k and $k_T + dk_T$ is:

$$\frac{2 \times 1/4 \times 2\pi k_T dk_T}{(\pi/L_x)(\pi/L_y)}.$$

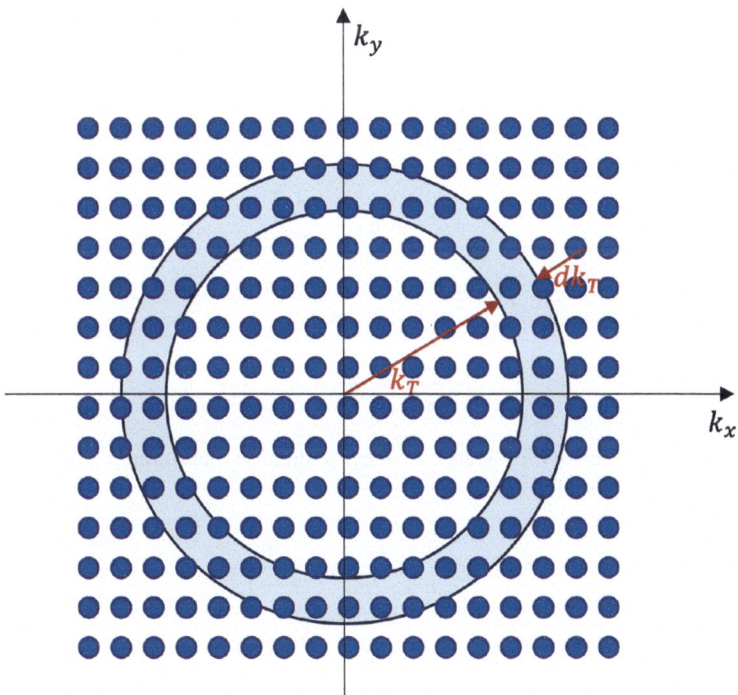

Fig. 10.15 A two-dimensional array of allowed quantum states in the k_T-*space* showing the states in the annular representing the quantum states between k_T and $k_T + dk_T$

Additionally, we count the number of quantum states in a quarter of the annular area and account for both spins of the electron.

The density of states is equal to the number of allowed states per unit volume, which is given by:

$$\rho(k)dk = [\frac{2 \times 1/4 \times 2\pi k_T dk_T}{(\pi/L_x)(\pi/L_y)}]/L_x L_y L_z. \tag{10.68}$$

Therefore, the density of states for a quantum well is:

$$\rho(k)dk = \frac{k_T}{\pi L_z}dk_T. \tag{10.69}$$

However, since the density of states is the same in the *k-space* and in the *E-space*, then

$$\rho(E)dE = \rho(k)dk = \rho(k_T)dk_T, \tag{10.70}$$

simplifying we obtain:

$$\rho(E) = \frac{\rho(k)}{(dE/dk)}. \tag{10.71}$$

The relationship between *E* and k_T for $q = 1$ is given by:

$$E = E_c + E_1 + \frac{\hbar^2 k_T^2}{2m_c}. \tag{10.72}$$

The derivative of *E* with respect to k_T is:

$$\frac{dE}{dk_T} = \frac{\hbar^2 k_T}{m_c}. \tag{10.73}$$

Therefore, the density of states in the conduction band for a quantum well semiconductor is given by:

$$\rho_c(E) = \frac{k_T/\pi L_z}{\hbar^2 k_T/m_c} = \frac{m_c}{\hbar^2 \pi L_z}. \tag{10.74}$$

Similarly, the density of states in the valence band for a quantum well semiconductor is given by:

$$\rho_v(E) = \frac{k_T/\pi L_z}{\hbar^2 k_T/m_v} = \frac{m_v}{\hbar^2 \pi L_z}. \tag{10.75}$$

The density of states for both the conduction and valence bands are shown in Fig. 10.14.

Hence, in a quantum well the density of states is independent of energy, whereas, for bulk semiconductors, it depends on energy as we have shown before where:

$$\rho_c(E) = \frac{1}{2\pi^2}\left(\frac{2m_c}{\hbar^2}\right)^{3/2}(E - E_c)^{1/2}. \quad (10.76)$$

10.7 Semiconductor Carrier Density

The carrier concentration depends on the number of available states that the electrons can occupy and the likelihood of electrons occupying these available states. To quantify this, consider N_c as the total number of available states within the conduction band, spanning energy levels from E_c to E_2, as shown in Fig. 10.16. This can be mathematically represented as:

$$N_c = \int_{E_c}^{E_2} \rho_c(E)dE, \quad (10.77)$$

where $\rho_c(E)$ denotes the density of states in the conduction band. This density of states, for a bulk semiconductor, can be expressed as:

$$N_c = \int_{E_c}^{E_2} \rho_c(E)dE = \frac{1}{2\pi^2}\left(\frac{2m_c}{\hbar^2}\right)^{3/2} \int_{E_c}^{E_2} (E - E_c)^{1/2}dE. \quad (10.78)$$

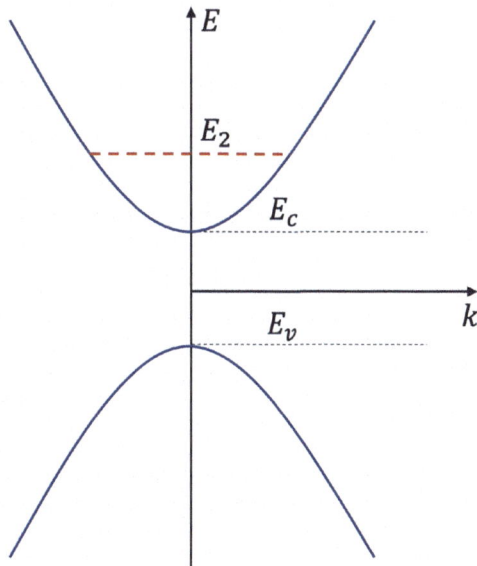

Fig. 10.16 $E - K$ diagram used to calculate N_c between two energy levels E_c and E_2

Integrating Eq. (10.78) over the energy range from E_c to E_2 gives the total number of available state N_c as:

$$N_c = \frac{1}{3\pi^2}\left(\frac{2m_c}{\hbar^2}\right)^{3/2}(E - E_c)^{3/2}. \tag{10.79}$$

This equation highlights the dependence of the carrier concentration on the effective mass of the charge carriers (m_c) and the energy difference between the conduction band minimum and the specific energy level ($E_2 - E_c$). Importantly, note that this relationship provides insight into how the physical properties of a semiconductor material, such as carrier effective mass, influence its electrical properties, including the carrier concentration.

> **Example** Calculate the number of carriers (N_c) for gallium arsenide (GaAs) given that $E_2 = E_c + 0.1\,\text{eV}$, with the effective mass of the charge carriers (m_c) being 0.067 times the rest mass of an electron (m_o), where $m_o = 9.1 \times 10^{-31}$ kg.
>
> **Solution**
>
> $$N_c = \frac{1}{3\pi^2}(\frac{2m_c}{\hbar^2})^{3/2}(0.1\text{eV})^{3/2}$$
>
> $$N_c = \frac{1}{3\pi^2}\left(\frac{2 \times 0.067 \times 9 \times 10^{-31}}{(1.05 \times 10^{-34})^2}\right)^{3/2}(0.1 \times 1.6 \times 10^{-19})^{3/2},$$
>
> Therefore,
>
> $$N_c \simeq 10^{18}\,\text{cm}^{-3}.$$

10.7.1 Fermi-Dirac Distribution

The Fermi-Dirac distribution is a cornerstone concept in solid-state physics, providing a statistical framework for the behavior of electrons in materials. It hinges on two key elements: the density of states, $\rho(E)$, and the occupancy probability, $n(E)$. The density of states represents the number of quantum states available per unit volume per unit energy at a specific energy level E, which is applicable to both the conduction and valence bands. Conversely, $n(E)$ quantifies the density of charge carriers per unit volume per unit energy that are likely to occupy these available states. However, the inherent statistical nature of quantum mechanics

10.7 Semiconductor Carrier Density

introduces a probability that a given state may remain unoccupied. This likelihood is characterized by the Fermi-Dirac distribution function, $f(E)$, which predicts the occupancy probability of a quantum state at a certain energy level by an electron or a hole at a specific temperature.

Mathematically, the Fermi-Dirac distribution is given by:

$$f(E) = \frac{1}{e^{(E-E_f)/k_B T} + 1}, \tag{10.80}$$

where E is the energy level of the state, E_f is the Fermi energy, k_B is the Boltzmann constant, and T is the absolute temperature. This function delineates how the occupancy of energy states by electrons evolves with temperature, electron energy, and the Fermi energy.

At absolute zero temperature ($T = 0°K$), the Fermi-Dirac distribution function simplifies to a step function, where all states with energy less than the Fermi energy (E_f) are fully occupied ($probability = 1$) and all states with energy greater than E_f are empty ($probability = 0$). As the temperature increases, the distribution at energies near E_f becomes "smeared out" due to the thermal excitation of electrons, allowing some electrons to occupy states above E_f. Figure 10.17 depicts the Fermi function for $0°K$ and $300°K$ for bulk intrinsic semiconductor material.

10.7.2 Fermi Energy

Fermi energy is a fundamental in the behavior of charge carriers in semiconductors. This refers to the energy level at which the probability of finding an electron is 50% at absolute zero temperature ($0°K$).

In solids, electrons fill the available energy states in accordance with the Pauli exclusion principle, which dictates that no two electrons can simultaneously occupy the same quantum state simultaneously. At absolute zero temperature, all the

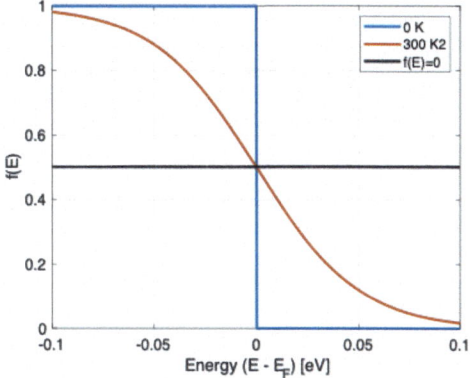

Fig. 10.17 The Fermi-Dirac distribution function at $0\,K$ and $300\,K$

electrons in the solid sequentially fill the lowest available energy states, up to a maximum energy level known as the **Fermi energy,** denoted by E_f.

The Fermi energy is a characteristic property of a material and is determined by the electron concentration of the material and the nature of its energy bands. For metals, the Fermi energy lies within the conduction band, meaning that there are electrons at the Fermi level that can be excited to higher energy states by even small thermal or electric fields, contributing to the electrical and thermal conductivity of the material. For semiconductors and insulators, the Fermi level typically lies within the bandgap, the energy range in which no electron states exist, which significantly affects their electrical properties.

10.7.3 Carrier Density

In the study of semiconductor physics, understanding carrier density is vital for analyzing the behavior of charge carriers within a material. The carrier concentration at an energy level E, denoted as $n(E)$, is determined by multiplying the density of available states, $\rho(E)$, by the probability that an electron or hole occupies these states, represented by $f(E)$. This is represented by the formula:

$$n(E) = \rho(E)f(E) \text{ and } p(E) = \rho(E)[1 - f(E)]. \tag{10.81}$$

Figure 10.18 shows the carrier concentration in intrinsic, n-doped, and p-doped semiconductors. For intrinsic semiconductors, the Fermi level is located midway between the valence and conduction bands. In n-doped semiconductor, the Fermi level shifts closer to the conduction band, whereas in p-doped semiconductors, the Fermi level moves closer to the valence band.

The total carrier concentration across a spectrum of electron energies can be expressed as an integral over energy:

$$n = \int n(E)dE = \int \rho(E)f(E)dE. \tag{10.82}$$

At absolute zero (0°K), the valence band is full while the conduction band is empty. At temperature when $T > 0°K$, some of the electrons transit to the conduction band and their concentration will be given by:

$$n = \int_{E_c}^{\infty} \rho_c(E)f(E)dE. \tag{10.83}$$

Substituting Eq. (10.76) into Eq. (10.83) results in the total number of carriers in the conduction band to be:

$$n = \frac{1}{2\pi^2}\left(\frac{2m_c}{\hbar^2}\right)^{\frac{3}{2}} \int_{E_c}^{\infty} (E - E_c)^{\frac{1}{2}} \frac{1}{e^{(E-E_f)/k_B T} + 1} dE. \tag{10.84}$$

10.7 Semiconductor Carrier Density

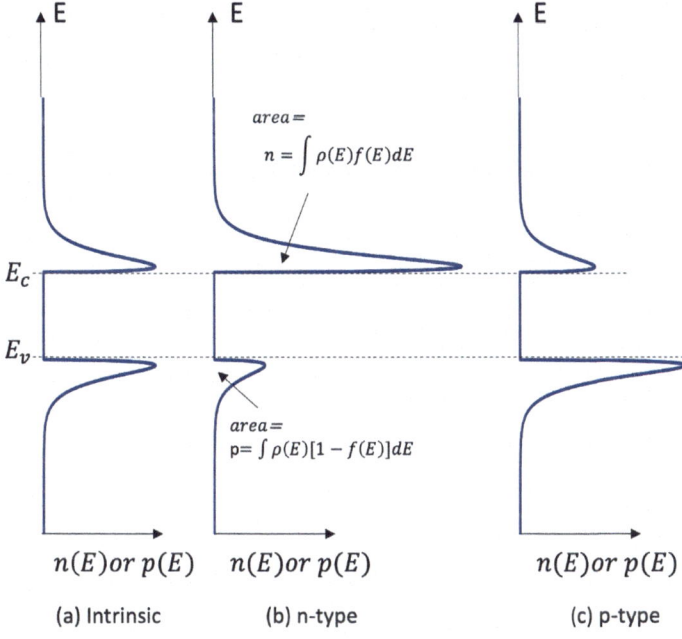

Fig. 10.18 Carrier density as a function of energy for (**a**) intrinsic semiconductors, (**b**) n-type doped semiconductors, and (**c**) p-type doped semiconductors

To determine this integral, we use a change of variable such that:

$$\frac{E - E_c}{k_B T} = \eta, \tag{10.85}$$

so that

$$\frac{dE}{d\eta} = k_B T, \tag{10.86}$$

and let

$$\frac{(E - E_f)}{kT} = \eta_F, \tag{10.87}$$

Making these changes of variables reduces the integral in Eq. (10.84) to the following:

$$n = \frac{1}{2\pi^2} \left(\frac{2m_c}{\hbar^2}\right)^{3/2} \int_0^\infty \frac{(k_B T)^{\frac{3}{2}} \eta^{\frac{1}{2}}}{e^{(\eta - \eta_f)} + 1} d\eta. \tag{10.88}$$

Therefore, the total number of carriers in the conduction band is:

$$n = \frac{1}{2\pi^2}\left(\frac{2m_c}{\hbar^2}\right)^{3/2}(k_BT)^{\frac{3}{2}}\int_0^\infty \frac{\eta^{\frac{1}{2}}d\eta}{e^{(\eta-\eta_f)}+1}. \tag{10.89}$$

We can use the well-known integral **Fermi-Half integral** given by:

$$F_{1/2}(\eta_f) = \int_0^\infty \frac{\eta^{\frac{1}{2}}d\eta}{\exp(\eta-\eta_f)+1}. \tag{10.90}$$

Determining the previous integral for n (Eq. 10.89) using the Fermi-Half integral, we obtain the following expression for the carrier concentration:

$$n = \frac{1}{2\pi^2}\left(\frac{2m_c}{\hbar^2}\right)^{3/2}(k_BT)^{\frac{3}{2}}F_{1/2}(\eta_f). \tag{10.91}$$

The Fermi-Half integral has no analytical solution, so we use its numerical values.

10.7.4 Boltzmann Approximation

The Boltzmann approximation is a significant concept in statistical mechanics and is particularly useful in the study of semiconductors. It simplifies the description of the distribution of particles over energy states in a system at a given temperature, making it easier to understand and predict the behavior of systems with many particles. Simplifying the Fermi-Dirac integral, Eq. (10.84) is very useful in determining the carrier concentration.

There are approximate solutions to Fermi-Dirac integral. For doped semiconductors where E_f approaches E_c. Let us consider the case where

$$E - E_f = 0.1 \text{ eV},$$

and since at 300 K° $k_BT = 0.026$ eV, then in this case

$$e^{(E-E_f)/k_bT} \simeq e^4 \simeq 55,$$

which is much larger than 1. Consequently, $E - E_f \gg k_BT$, the Fermi-Dirac integral, for doped semiconductor, can be approximated as:

$$f(E) = \frac{1}{e^{(E-E_f)/k_BT}+1} \approx e^{-(E-E_f)/k_BT}. \tag{10.92}$$

This is the Boltzmann approximation for the Fermi-Dirac integral.

10.7 Semiconductor Carrier Density

Using this Boltzmann approximation results in the charge carrier concentration being given by:

$$n = \frac{1}{2\pi^2}\left(\frac{2m_c}{\hbar^2}\right)^{\frac{3}{2}} \int_{E_c}^{\infty} (E - E_c)^{\frac{1}{2}} e^{-\left(\frac{E-E_f}{k_B T}\right)} dE \tag{10.93}$$

Use the change of variables such that:

$$\frac{E - E_c}{k_B T} = \eta, \tag{10.94}$$

therefore,

$$\frac{dE}{d\eta} = k_B T \text{ and } dE = k_B T d\eta, \tag{10.95}$$

and

$$\frac{E - E_F}{k_B T} = \frac{(E - E_c) + (E_c - E_F)}{k_B T} = \eta + \eta_f, \tag{10.96}$$

where

$$\eta_f = \frac{E_c - E_f}{k_B T}. \tag{10.97}$$

Substituting the variables η and η_F in Eq. (10.93), the carrier concentration can be represented by:

$$n = \frac{1}{2\pi^2}\left(\frac{2m_c}{\hbar^2}\right)^{\frac{3}{2}} (k_B T)^{\frac{3}{2}} \int_0^{\infty} \eta^{\frac{1}{2}} e^{-(\eta+\eta_f)} d\eta, \tag{10.98}$$

simplifying we obtain:

$$n = \frac{1}{2\pi^2}\left(\frac{2m_c k_B T}{\hbar^2}\right)^{\frac{3}{2}} e^{-\eta_f} \int_0^{\infty} \eta^{\frac{1}{2}} e^{-\eta} d\eta. \tag{10.99}$$

The integral $\int_0^{\infty} \eta^{\frac{1}{2}} e^{-\eta} d\eta$ is very similar to the integral Gamma function ($\Gamma(z)$) where $z = 3/2$, i.e.,

$$\Gamma\left(\frac{3}{2}\right) = \frac{1}{2}\Gamma\left(\frac{1}{2}\right) = \frac{\sqrt{\pi}}{2}. \tag{10.100}$$

Therefore, the electron carrier density is given by:

$$n = 2\left(\frac{mk_BT}{2\pi\hbar^2}\right)^{\frac{3}{2}} e^{\left(\frac{E_f-E_c}{k_BT}\right)}. \tag{10.101}$$

Let

$$N_c = 2\left(\frac{m_c k_B T}{2\pi\hbar^2}\right)^{3/2}. \tag{10.102}$$

Therefore, the electron carrier densities is given by:

$$n = N_c e^{(E_f-E_c)/k_B T} \tag{10.103}$$

Let

$$N_v = 2\left(\frac{m_v k_B T}{2\pi\hbar^2}\right)^{3/2}. \tag{10.104}$$

the hole carrier density is:

$$p = 2\left(\frac{m_v kT}{2\pi\hbar^2}\right)^{\frac{3}{2}} e^{\left(\frac{E_v-E_f}{k_BT}\right)}, \tag{10.105}$$

Therefore, the product of the carrier densities is given by:{

$$np = N_c N_v e^{-E_g/k_B T}. \tag{10.106}$$

For intrinsic semiconductors the density of holes is equal to that of electrons, i.e.,

$$n = p = n_i, \tag{10.107}$$

and

$$n_i^2 = np. \tag{10.108}$$

Therefore, the intrinsic concentration is:

$$n_i = \sqrt{N_c N_v} e^{-E_g/2k_B T}. \tag{10.109}$$

Intrinsic carrier concentrations are given in Table 10.2 for various semiconductors along with other relevant parameters.

10.7 Semiconductor Carrier Density

Table 10.2 Several semiconductor materials and their relevant parameters

Parameter	E_g	m_c	m_v	N_c	N_v	n_i
Unit	(eV)	(m_o)	(m_o)	$(10^{18}\,\text{cm}^{-3})$	$(10^{18}\,\text{cm}^{-3})$	cm^{-3}
Si	1.12	1.18	0.55	32.2	10.2	7×10^9
Ge	0.66	0.22	0.34	2.6	5.0	1×10^{13}
GaAs	1.42	0.063	0.52	0.40	9.41	2×10^6
InP	1.34	0.079	0.60	0.56	11.6	1×10^7

Example Calculate the values of N_c and N_v for silicon.

Solution

$$N_c = 2\left(\frac{m_c k_B T}{2\pi \hbar^2}\right)^{3/2}$$

$$= 2\left(\frac{1.18 \times 9.11 \times 10^{-31} \times 300 \times 1.38 \times 10^{-23}}{(6.626 \times 10^{-34})^2}\right)^{3/2}$$

$$N_c = 32.1 \times 10^{18}\,\text{cm}^{-3}$$

and

$$N_v = 2\left(\frac{m_v k_B T}{2\pi \hbar^2}\right)^{3/2}$$

$$= 2\left(\frac{0.55 \times 9.11 \times 10^{-31} \times 300 \times 1.38 \times 10^{-23}}{(6.626 \times 10^{-34})^2}\right)^{3/2}$$

$$10.2 \times 10^{18}\,\text{cm}^{-3}$$

In the case of intrinsic semiconductors when $n = p$ that leads to:

$$N_c e^{(E_f - E_c)/k_B T} = N_v e^{(E_v - E_f)/k_B T}, \tag{10.110}$$

substituting for N_c and N_v from Eq. (10.79) that results in:

$$(m_c)^{\frac{3}{2}} e^{(E_f - E_c)/k_B T} = (m_v)^{\frac{3}{2}} e^{(E_v - E_f)/k_B T}, \tag{10.111}$$

$$e^{(E_f-E_c)/k_BT} = \left(\frac{m_v}{m_c}\right)^{3/2} e^{(E_v-E_f)/k_BT}. \tag{10.112}$$

Taking the natural log of both sides, we obtain:

$$2\frac{E_f}{k_BT} = \frac{3}{2}ln\left(\frac{m_v}{m_c}\right) + \frac{E_c + E_v}{k_BT}. \tag{10.113}$$

Let us label the Fermi energy for the intrinsic semiconductor by E_{f_i} Eq. (10.113) becomes:

$$E_{f_i} = \frac{E_c + E_v}{2} + \frac{3}{4}ln\left(\frac{m_v}{m_c}\right). \tag{10.114}$$

When $m_v = m_c$, the Fermi energy is midway between E_c and E_v so that:

$$E_{f_i} = \frac{E_c + E_v}{2}. \tag{10.115}$$

In the case of highly doped materials, the Boltzmann approximation becomes invalid, and the Fermi energy levels for such materials can be approximated by the Joyce-Dixon approximation given by:

$$E_{fc} = E_c + k_BT\left[ln\left(\frac{n}{N_c}\right) + \frac{1}{\sqrt{8}}\frac{n}{N_c} - \left(\frac{3}{16} - \frac{\sqrt{3}}{9}\right)\left(\frac{n}{N_c}\right)^2 + \cdots\right], \tag{10.116}$$

for electrons and by

$$E_{fv} = E_v - k_BT\left[ln\left(\frac{p}{N_v}\right) + \frac{1}{\sqrt{8}}\frac{p}{N_v} - \left(\frac{3}{16} - \frac{\sqrt{3}}{9}\right)\left(\frac{p}{N_v}\right)^2 + \cdots\right], \tag{10.117}$$

for holes.

The Joyce-Dixon approximation is typically applied for semiconductors with high doping levels where the Fermi level moves into the energy bands.

10.8 Quasi-Fermi Levels

For an intrinsic non-doped semiconductor at equilibrium there is a single Fermi level. However, under non-equilibrium conditions, such as those induced by an external electric bias or light illumination, the distribution of electrons and holes can deviate significantly from the equilibrium state. This results in two Fermi levels

10.8 Quasi-Fermi Levels

on for the electrons, E_{fc}, and one for the holes, E_{fv}. These are known as the quasi-Fermi levels.

In n-doped and p-doped semiconductors can be deduced by taking the natural log of Eqs. (10.105) and (10.106), which results in the Fermi energies for the n-doped (E_{f_n}) and the Fermi energies for the p-doped (E_{f_p}) being given by:

$$E_{fc} = E_c + k_B T \ln\left(\frac{n}{N_c}\right), \qquad (10.118)$$

and

$$E_{fv} = E_v - k_B T \ln\left(\frac{p}{N_v}\right). \qquad (10.119)$$

The quasi Fermi level for electrons indicates the energy level where the probability, the Fermi-Dirac function $f(E_{fc})$, of finding an electron is 0.5. Similarly, the quasi Fermi level for holes indicates the energy level where the probability, the Fermi-Dirac function $f(E_{fv})$, of finding a hole is 0.5. These are demonstrated in Fig. 10.19.

This separation of quasi-Fermi levels is a key factor in the operation of semiconductor devices, influencing their electrical properties and behavior under non-equilibrium conditions.

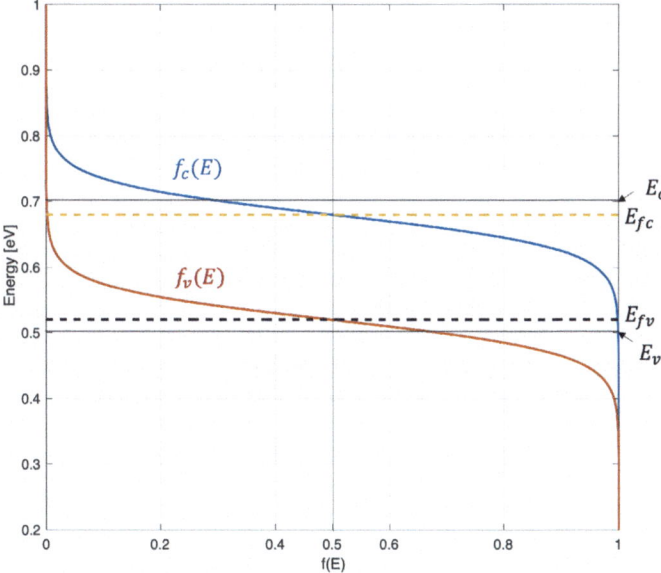

Fig. 10.19 Fermi-Dirac function $f(E)$ for quasi-Fermi levels. In this plot, the following values were used: $E_c = 0.7$, $E_{fn} 0.68$, $E_{fv} 0.52$, and $E_v = 0.5$ eVs

At thermal-equilibrium, the transition of electrons and holes happens due to the thermalization process. At the steady state, the same number of electrons and holes make transitions between CB and VB. These transitions can be intraband, i.e., within the CB or the VB. These transitions are very fast with lifetimes in the range of $\tau \approx 10^{-12} - 10^{-13} s$. On the other hand, the interband transitions between the CB and VB, which can result in the emission of a photon, have lifetimes in the range $10^{-8} - 10^{-9} s$.

Two Fermi functions are used in a quasi-Fermi levels that are expressed as:

$$f_c(E) = \frac{1}{e^{\frac{E-E_{fc}}{kT}}}, \tag{10.120}$$

and

$$f_v(E) = \frac{1}{e^{\frac{E-E_{fv}}{k_B T}}}. \tag{10.121}$$

Under this approximation the electron density is:

$$n = N_c e^{\frac{(E-E_c)}{kT}}. \tag{10.122}$$

At quasi-equilibrium, the carrier densities for electrons and holes are given by:

$$n = N_c e^{\frac{E_{fc}-E_c}{k_B T}}, \tag{10.123}$$

and

$$p = N_v e^{\frac{E_v-E_{fv}}{k_B T}}. \tag{10.124}$$

Therefore,

$$np = N_c N_v e^{\frac{-E_g}{k_B T}} e^{\frac{E_{fc}-E_{fv}}{k_B T}}. \tag{10.125}$$

For intrinsic semiconductors given by Eq. (10.109), we obtain:

$$N_c N_v e^{\frac{-E_g}{k_B T}} = n_i^2. \tag{10.126}$$

Therefore,

$$np = n_i^2 e^{\frac{E_{fc}-E_{fv}}{k_B T}}. \tag{10.127}$$

This gives rise to several conditions for pn-junction devices:

10.8 Quasi-Fermi Levels

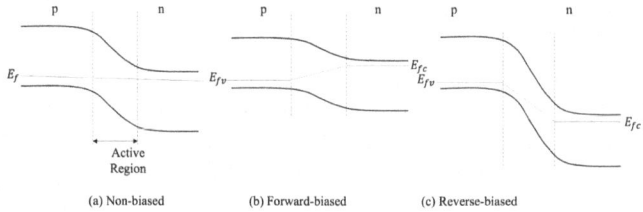

Fig. 10.20 pn-junctions under several biasing conditions

1. Unbiased pn-junction $np = n_i^2$.
2. Forward biased pn-junction $E_{F_c} - E_{F_v} > 0 \rightarrow np > n_i^2$.
3. Reverse biased pn-junction $(E_{F_c} - E_{F_v}) < 0 \rightarrow np < n_i^2$.

These conditions are shown in Fig. 10.20 for pn-junctions.

Example An acceptor-doped silicon, p-type, at $N_A = 1.0 \times 10^{17}\,\text{cm}^{-3}$. Determine the quasi-Fermi levels when excess carriers are injected at $n' = p' = 8 \times 10^{14}\,\text{cm}^{-3}$. Use required parameters from Table 10.2 are used.

Solution

(a) No injection current:

$$p_o = 1.0 \times 10^{17} = N_v e^{-\frac{E_f - E_v}{k_B T}}.$$

take the natural log of both sides to get:

$$ln\left(\frac{N_v}{p_o}\right) k_B T = E_f - E_v,$$

Therefore,

$$E_f - E_v = 0.026 \times ln\left(\frac{1.84 \times 10^{19}}{1.0 \times 10^{17}}\right) 0.136\,\text{eV},$$

and

$$n_o = \frac{n_i^2}{p_o} = \frac{(7 \times 10^9)^2}{1.0 \times 10^{17}} = 490\,\text{cm}^{-3}.$$

(continued)

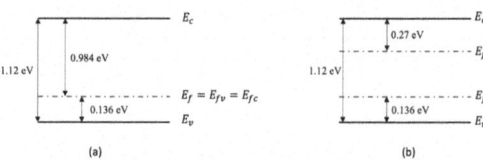

Fig. 10.21 Fermi energy levels for (**a**) no current injection and (**b**) current injection

(b) In the presence of injection current:

$$p = p_o + p' = 1.0 \times 10^{17} + 8 \times 10^{14} \cong 1 \times 10^{17}$$

So $E_f - E_v$ does not change for the holes:

$$n = n_o + n' = 490 + 8 \times 10^{14} \simeq 8 \times 10^{14}.$$

$$E_c - E_{fc} = k_B T ln\left(\frac{N_c}{n}\right) = 0.026 \times ln\left(\frac{2.8 \times 10^{19}}{8 \times 10^{14}}\right) = 0.27\,\text{eV}.$$

Therefore, the Fermi energy changes only for the minority carriers, the electrons. These results are illustrated in Fig. 10.21.

10.9 Summary

This chapter delves into the foundational principles of semiconductor physics, focusing on the structure and behavior of energy bands. It provides a comprehensive framework for understanding the electronic and optical properties of semiconductors, which are essential for modern electronic and photonic devices. Semiconductors are materials with electrical conductivity that lies between conductors and insulators, with properties that can be significantly modified through doping. The quantum mechanical principles underlying the formation of energy bands are introduced, stemming from the quantization of energy levels in isolated atoms.

As atoms form a crystalline lattice, their discrete energy levels split and merge into bands. These bands, separated by energy gaps known as bandgaps, define whether a material behaves as a conductor, semiconductor, or insulator. This transition from discrete energy levels in atoms to continuous energy bands in crystals is central to understanding the electronic properties of materials. The periodic potential in crystalline structures further influences these energy bands. The Kronig-Penney model, a simplified approach, is used to explain the formation of these bands and the behavior of electrons in a periodic lattice.

The relationship between electron energy and wave vector is visualized using the energy-momentum ($E - k$) diagram. This diagram illustrates the conduction and valence bands, separated by the forbidden bandgap. At absolute zero, the valence band is fully occupied, and the conduction band is empty. With increasing temperature, electrons in the valence band can gain sufficient energy to transition into the conduction band, leaving behind holes. The concept of the density of states quantifies the number of available quantum states at each energy level. The density of states in the conduction and valence bands plays a critical role in determining carrier concentrations and current flow in semiconductors.

Carrier concentrations are described using Fermi-Dirac statistics, which predict the probability of occupancy of energy states by electrons at a given temperature. At thermal equilibrium, this distribution helps calculate the electron and hole densities. For doped semiconductors, approximations such as the Boltzmann distribution simplify these calculations when the Fermi energy lies far from the band edges. In reduced-dimensional structures like quantum wells, the density of states becomes altered, confining electrons to discrete energy levels due to spatial restrictions.

In non-equilibrium conditions, such as those created by external biases or illumination, quasi-Fermi levels for electrons and holes arise. These levels play a critical role in determining the behavior of semiconductor devices, such as p n pn-junctions under different biasing conditions. This chapter concludes by bridging quantum mechanical concepts with their practical applications, forming the theoretical foundation for designing and understanding semiconductor devices, including lasers, light-emitting diodes (LEDs), and photodetectors.

10.10 Problems

1. Consider a crystal of 10^{20} atoms with an energy band width of 1 eV. Calculate the spacing between individual energy levels within this band. Explain why the energy levels are considered quasi-continuous.
2. Sketch an $E - k$ diagram for a semiconductor and label the conduction band minimum, valence band maximum, and the bandgap. Explain how the bandgap influences the material's optical and electrical properties.
3. Derive the expression for the density of states in the conduction band, $\rho_c(E)$, and show how it depends on the effective mass of the electron and the energy above the conduction band minimum.
4. For intrinsic silicon at 300 K, calculate the intrinsic carrier concentration (n_i) given the density of states ($Nc = 3.22 \times 10^{19}$ cm^{-3}), bandgap ($E_g = 1.12$ eV, and effective masses. Assume the Boltzmann approximation holds.
5. Plot the Fermi-Dirac distribution function at $T = 0$ K and $T = 300$ K. Discuss the significance of the Fermi energy and how the function changes with temperature.

6. Using the Boltzmann approximation, derive an expression for the electron concentration in a doped semiconductor where E_f lies close to E_c. Verify the approximation's validity for $E_f - E_c = 0.05$ eV at 300 K.
7. Consider a quantum well structure with a well width of 15 nm. Calculate the first three allowed energy states in the conduction band, assuming the effective mass of an electron is $0.067\, m_0$, where m_0 is the free electron mass.
8. Using the Kronig-Penney model, calculate the bandgap for a semiconductor with a potential barrier height $V_0 = 10$ eV, barrier width $b = 5$ A°, and well width $a = 5$ A°.
9. Show how the intrinsic carrier concentration (n_i) of a semiconductor changes with temperature. Derive the exponential dependence of n_i on temperature, and explain its implications for semiconductor performance at high temperatures.
10. Determine the values of n_i, N_c, and N_v for the semiconductors listed in Table 10.2.

Bibliography

1. R. F. Pierret (1996), Semiconductor Device Fundamentals, 2nd Edition.
2. S. M. Sze and K. K. Ng (2006), Physics of Semiconductor Devices, 3rd Edition.
3. C. Kittel (2004), Introduction to Solid State Physics, 8th Edition.
4. J. Singh (2001), Semiconductor Devices: Basic Principles, 1st Edition.
5. B. G. Streetman and S. K. Banerjee (2004), Solid State Electronic Devices, 7th Edition.
6. P. Y. Yu and M. Cardona (2010), Fundamentals of Semiconductors: Physics and Materials Properties, 4th Edition.
7. Kasap, S. O. (2012). Optoelectronics & Photonics: Principles & Practices (2nd ed.). Pearson.

Optoelectronic Semiconductors 11

Optoelectronic semiconductors play a pivotal role in modern technology, enabling the conversion of electronic signals into optical signals and vice versa. These materials are at the heart of numerous devices that have become integral to our daily lives, ranging from laser diodes and light-emitting diodes (LEDs) to photodetectors and solar cells and imaging displays. Optoelectronic semiconductors utilize the unique properties of semiconducting materials to manipulate and control the flow of photons and electrons. One of the defining characteristics of optoelectronic semiconductors is their ability to absorb, emit, or modulate light. To achieve this goal, semiconductor materials are carefully engineered and often combined in heterostructures to tailor their optical and electronic properties to specific applications. The choice of semiconductor materials and their combinations is critical in optimizing the performance and efficiency of optoelectronic devices. Numerous material combinations are commonly used in the design of optoelectronic devices, each presenting distinct advantages. These materials can be classified into two main categories: elemental semiconductors from Groups IV and VI in the periodic table, including silicon (Si) and germanium (Ge) and compound semiconductors such as gallium-arsenide ($GaAs$), indium-phosphide (InP), and gallium-nitride (GaN). The following represent some of the most widely used semiconductors in optoelectronic devices:

1. Silicon (Si): Silicon is widely used in the electronics industry, but its indirect bandgap has traditionally limited its use in optoelectronics. However, advanced Si-$based$ materials and structures have been developed for applications such as photovoltaics, optical modulators, and silicon photonics.
2. Gallium Arsenide($GaAs$): GaAs is a versatile semiconductor material known for its high electron mobility and direct bandgap, which allows for efficient emission and detection of light in the near-infrared range. It is extensively used in applications such as laser diodes and photodetectors for optical communication.

3. Indium Phosphide (InP): InP is another semiconductor with a direct bandgap that falls within the optical communication spectrum. Owing to its optical properties, it is commonly employed in high-speed optoelectronic devices, including photodetectors and lasers, due to its excellent optical properties.
4. Gallium Nitride (GaN): GaN is known for its wide bandgap, making it suitable for blue and ultraviolet LEDs and high-power electronic devices. It has revolutionized lighting technology and is essential for high-efficiency LEDs and laser diodes.

11.1 How to Choose the Proper Material?

Selecting the appropriate material hinges on the specific application at hand. For light sources, materials with a direct bandgap are favored because of their high efficiency in converting electrons into photons. In contrast, for detectors, indirect bandgap materials may be suitable in certain scenarios. Another pivotal factor in material selection is the wavelength of the emitted light, which is inherently linked to the bandgap of the material.

The following formula establishes a connection between the bandgap size (Eg) and the wavelength (λ) of the emitted light:

$$h\nu = E_g, \tag{11.1}$$

which can be further expressed as:

$$h\frac{c}{\lambda} = E_g.$$

Upon substituting the values of the constants c and h, we obtain:

$$6.2626 \times 10^{-34} \frac{3 \times 10^8}{\lambda} = E_g.$$

In this equation, the bandgap energy (Eg) is expressed in Joules, and the wavelength (λ) is in meters.

To facilitate practical calculations, we can convert these units (eV = Joules/(1.60219×10^{-19})), leading to the following formula:

$$\lambda = \frac{1.24}{E_g}, \tag{11.2}$$

where the wavelength (λ) is measured in micrometers (μm) and the bandgap (Eg) measured in electron-volts (eV).

Table 11.1 lists examples of several semiconductors along with their bandgap energies, and the corresponding wavelengths at which these materials can emit photons.

11.2 Direct and Indirect Bandgap Semiconductors

Table 11.1 Bandgap energy and emission wavelengths for typical semiconductors

Semiconductor	Bandgap (E_g)	Emission wavelength (λ)
Silicon (Si)	1.12 eV	1.11 µm (infrared)
Gallium arsenide (GaAs)	1.43 eV	0.87 µm (near infrared)
Indium phosphide (InP)	1.35 eV	0.92 µm (near infrared)
Gallium nitride (GaN)	3.4 eV	0.36 µm (ultraviolet/blue)
Zinc sulfide (ZnS)	3.6 eV	0.34 µm (ultraviolet)

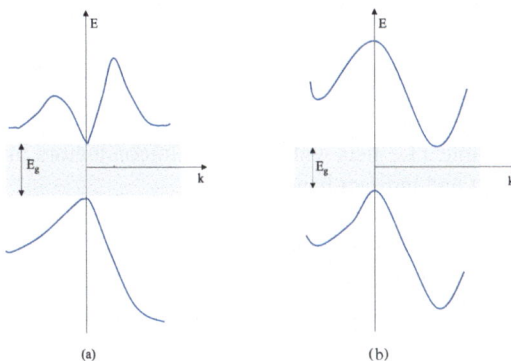

Fig. 11.1 Direct and indirect bandgap semiconductors: (**a**) GaAs direct bandgap ($E_g = 1.4$ eV) and (**b**) Si indirect bandgap ($E_g = 1.1$ eV)

To design devices that emit light at different wavelengths, one can explore the use of ternary (three-material alloys) or quaternary (four-material alloys) semiconductors, as discussed in detail later in this chapter.

11.2 Direct and Indirect Bandgap Semiconductors

In the previous chapter, we discussed the band structures of semiconductor materials. The energy bands are determined by the density of states and demonstrated through the $E - k$ diagram. The energy gap is determined by the energy difference between the minimum of the conduction band and maximum of the valence band. In direct bandgap semiconductors, the maximum of the valence band and the minimum of the conduction band occur at the same momentum in the Brillouin zone. This alignment allows electrons to transit from the conduction band to the valence band with a change in momentum. This results in an efficient emission of photons. On the other hand, indirect bandgap semiconductors have their maximum valence band and minimum conduction band at different momentum in the Brillouin zone. Because of this mismatch, direct transitions between these bands are not allowed without a change in momentum. In these materials, for an electron to transit from the conduction band to the valence band, a change in energy and momentum is needed. Hence, the emission of photons becomes very inefficient. Figure 11.1 shows the direct and indirect bandgaps for GaAs and Si, respectively.

Table 11.2 List of direct and indirect bandgap semiconductors and their relevant parameters

Semiconductor	Bandgap (eV)	Bandgap type	Refractive	Lattice constant (A)	Wavelength (nm)
InAs	0.38	D	3.51	6.06	3263
Ge	0.66	I	4	5.65	1879
Si	1.12	I	3.5	5.43	1107
InP	1.35	D	3.1	5.87	919
GaAs	1.42	D	3.3	5.65	873
AlAs	2.16	I	3.2	5.66	574
GaP	2.26	I	3.3	5.45	549
AlP	2.45	I	3	5.46	506

Table 11.2 lists widely used semiconductors in optoelectronic devices that have direct and indirect bandgaps.

11.3 Bandgap Modification

The semiconductor materials shown in the previous table can be used to make devices that operate at the specified wavelengths. Many practical applications require wavelengths that are very different from those provided by devices made from $IV-VI$ group materials. The first generation, in the 1980s, of optical fiber communication systems used GaAs lasers that operate at around 0.8 μm. However, at that time it was known that optical fibers have much lower attenuations at 1.3 and 1.55 μm. To design devices that operate at these wavelengths, there was a need to modify the bandgaps of these materials. This led to the creation of ternary and quaternary materials.

An example of a ternary-based semiconductor material is the $Al_xGa_{1-x}As$. The fraction x defines the ratio of Al in this alloy. If $x = 0$, then we obtain a binary alloy $GaAs$. In this alloy $GaAs$ is the substrate. Similarly, if the fraction $x = 1$, then we obtain a binary alloy $AlAs$. Thus, by changing the ratio x, the bandgap will change. leading to a different wavelength of emission on the basis of the formula given above, $\lambda = 1.24/E_g$.

An example of a quaternary semiconductor alloy id $In_{1-x}Ga_xAs_yP_{1-y}$, where two elements are from group IV and two element are from group VI. In this case if $x = 0$ and $y = 0$, we obtain an InP binary alloy, which can be the substrate for this alloy. Additionally, we can obtain $In_xGa_{1-x}As_{1-y}P_y$. Now, if $x = 1$ and $y = 0$, we obtain $GaAs$ is the substrate.

To understand the selection of the composition factor x on the formation of the ternary semiconductor $Al_xGa_{1-x}As$, consider when $x = 0$, the substrate is $GaAs$, and when $x = 1$, we get $AlAs$. Both of these alloys have about similar lattice constants as shown in Table 11.2. $GaAs$ and $AlAs$ exhibit both direct and indirect bandgaps, as illustrated in Fig. 11.2. At $x = 0$, the $GaAs$ has a direct bandgap,

11.3 Bandgap Modification

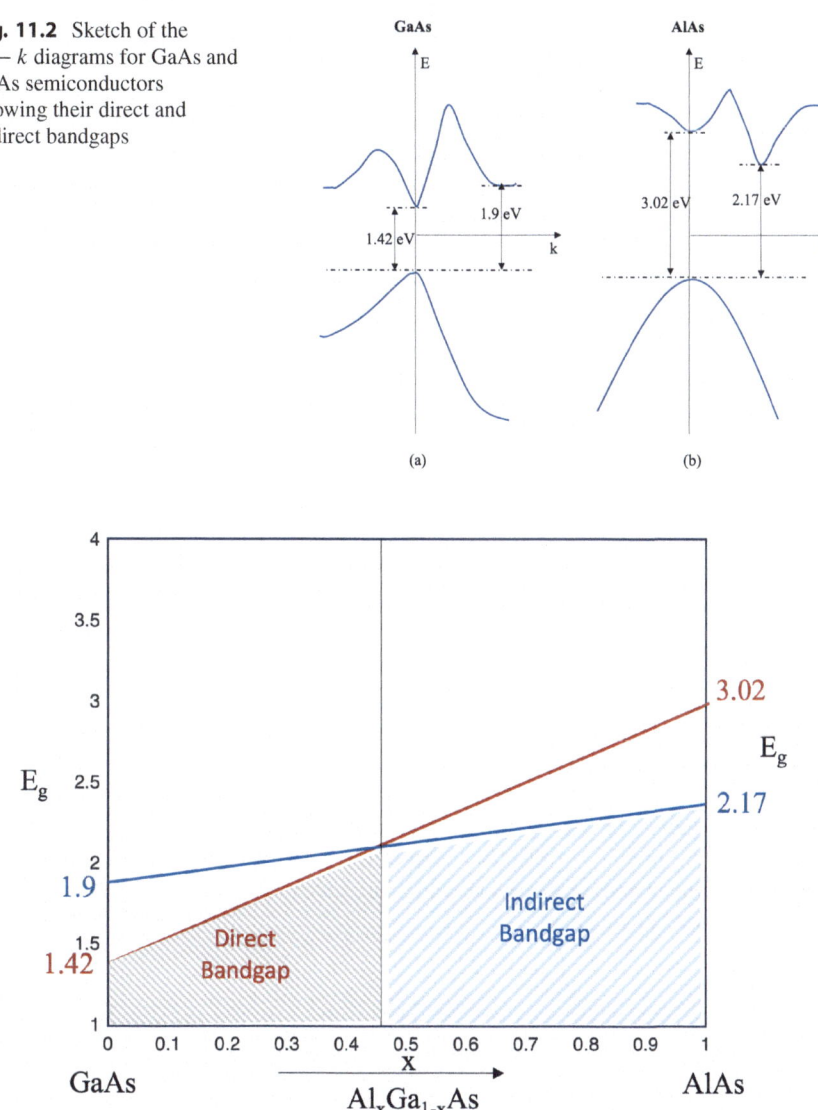

Fig. 11.2 Sketch of the $E - k$ diagrams for GaAs and AlAs semiconductors showing their direct and indirect bandgaps

Fig. 11.3 Dependence of the formation of the ternary semiconductor $AlGaAs$ on the composition factor x

and at $x = 1$, $AlAs$ has an indirect bandgap. Consequently, with increasing x, the alloy $AlGaAs$ transitions from a direct to an indirect bandgap semiconductor beyond a critical value of x. This transition is depicted in Fig. 11.3, where the red line connects the direct bandgap of $GaAs$ at 1.42 eV to that of $AlAs$ at 3.02 eV. Simultaneously, the blue line represents the transition from the indirect bandgap of $GaAs$ at 1.9 eV to that of the $AlAs$ at 2.17 eV. The intersection of both lines determines the critical composition value of x at which the semiconductor transitions from having a direct bandgap to an indirect bandgap. In the example shown in Fig. 11.3, the value of $x \simeq 0.45$. Therefore, at this point, the ternary alloy is denoted by $Al_{0.45}Ga_{0.55}As$, showing that the selection of x is crucial for achieving the desired wavelength for specific applications.

Ternary and quaternary semiconductors are extensively used in the design of optoelectronic devices. To determine the carrier transition involved in the generation and absorption of photons, we develop the theoretical representation of the optical density of states, in the following section.

11.4 Optical Joint Density of States

In the previous chapter, we derived a formula for the density of states (number of allowed states per unit volume per energy interval) in the conduction band (ρ_c) and in the valence band (ρ_v). The case of electron transitions between the conduction and valence bands that can result in the absorption or emission of photons is described by the **optical joint density of states** ($OJDS$). It describe the density of optical states available to a photon within a material or structure as a function of photon energy or frequency. It gives the number of states (one in the CB and one in the VB) per unit volume per unit photon energy interval where the energy difference equals $h\nu$. For a given energy ($h\nu$), there are several pairs of ($E_2 - E_1$). Therefore, there are a number of pairs of states available for a photon of energy ($h\nu$) to be absorbed or edited. These are given by the joint density of states. The OJDS provides information about the number of photon states that can interact with a specific frequency or energy of light within a given material.

Consider the number of electrons and holes per unit volume in the conduction band and valence band, respectively, which are given by:

$$n = \int n(E) \, dE = \int \rho_c(E) f(E) \, dE, \qquad (11.3)$$

and

$$p = \int \rho_v(E) [1 - f(E)] \, dE. \qquad (11.4)$$

These states are available in both bands. Consider, as shown in Fig. 11.4, the absorption of a photon by an electron at energy level E_1 to transition to energy level E_2 in the conduction band, leaving behind a hole. The energy of the photon must satisfy the following:

$$h\nu = E_2 - E_1. \qquad (11.5)$$

11.4 Optical Joint Density of States

Fig. 11.4 Interband transition resulting in the emission or absorption of a photon

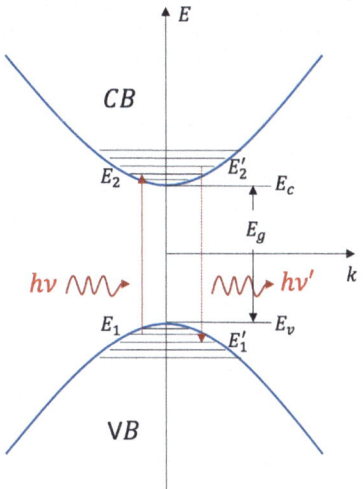

Similarly, an electron in energy level E_2' can make a transition to energy level E_1' results in an emission of a photon with energy:

$$h\nu' = E_2' - E_1'. \tag{11.6}$$

In both cases $h\nu > E_g$ and $h\nu' > E_g$. The number of these transitions does not depend on either E_1 or E_2, but rather on the difference $E_2 - E_1$. Therefore, the optical joint density of states ($OJDS$) is independent of the individual energy states E_1 and E_2 but rather depends on their energy difference, $E_2 - E_1$.

For each of these energy states (E_1 and E_2), several values exist depending on the direction of the wavevector k. The energy state E_2 in the conduction band is given by:

$$E_2 = E_c + \frac{\hbar^2 k^2}{2m_c}, \tag{11.7}$$

whereas the energy state E_1 in the valence band is given by:

$$E_1 = E_v - \frac{\hbar^2 k^2}{2m_v}. \tag{11.8}$$

Therefore, the energy difference, $h\nu = E_2 - E_1$, can be expressed as:

$$h\nu = E_2 - E_1 = (E_c - E_v) + \frac{\hbar^2 k^2}{2}\left(\frac{1}{m_c} + \frac{1}{m_v}\right). \tag{11.9}$$

Here, $E_c - E_v$ represents the energy bandgap (Eg), and let

$$\frac{1}{m_c} + \frac{1}{m_v} = \frac{1}{m_r}, \tag{11.10}$$

where m_r is the reduced mass. Therefore, the energy of the photon can be written as:

$$h\nu = E_g + \frac{\hbar^2 k^2}{2m_r}. \tag{11.11}$$

This equation can be further expressed in terms of momentum as:

$$k^2 = \frac{2m_r}{\hbar^2} \left(h\nu - E_g \right). \tag{11.12}$$

The energy state E_2 in the conduction band can be expressed as:

$$E_2 = E_c + \frac{\hbar^2}{2m_c} \left[\frac{2m_r}{\hbar^2} \left(h\nu - E_g \right) \right], \tag{11.13}$$

which simplifies to:

$$E_2 = E_c + \frac{m_r}{m_c} \left(h\nu - E_g \right), \tag{11.14}$$

and

$$E_2 - E_c = \frac{m_r}{m_c} \left(h\nu - E_g \right). \tag{11.15}$$

The only variables in this equation are E_2 and ν. Therefore, for each energy state in the conduction band, E_2, there is one value for the photon frequency ν, with a one-to-one correspondence.

We have shown in the previous chapter that $\rho_c(E_2)dE_2$ to be the number of states per unit volume available for electron occupation. Therefore, we can relate this to the number of photons with energies between $h\nu$ and $h(\nu + d\nu)$ that can be emitted or absorbed. Therefore

$$\rho_{opt}(\nu)d\nu = \rho_c(E_2 dE_2, \tag{11.16}$$

where the function $\rho_{opt}(\nu)$ is known as the optical joint density of states, which can be expressed as:

$$\rho_{opt}(\nu) = \rho_c(E_2) \frac{dE_2}{d\nu}. \tag{11.17}$$

11.4 Optical Joint Density of States

Differentiating Eq. (11.15) results in the following:

$$\rho_{opt}(\nu) = \rho_c(E_2)\frac{hm_r}{m_c}. \qquad (11.18)$$

The density of states in the conduction band derived in the last chapter is given by:

$$\rho_c(E_2) = \frac{1}{2\pi^2}(\frac{2m_c}{\hbar^2})^{3/2}(E_2 - E_c)^{1/2}. \qquad (11.19)$$

Substituting Eq. (11.19) into Eq. (11.18) for the density of states, we obtain:

$$\rho_{opt}(\nu) = \frac{1}{2\pi^2}(\frac{2m_c}{\hbar^2})^{3/2}(E_2 - E_c)^{1/2}\frac{hm_r}{m_c}. \qquad (11.20)$$

Using Eq. (11.15) for $E_2 - E_c$, we obtain:

$$\rho_{opt}(\nu) = \frac{1}{2\pi^2}(\frac{2m_c}{\hbar^2})^{3/2}\left[\frac{m_r}{m_c}(h\nu - E_g)\right]^{1/2}\frac{hm_r}{m_c}, \qquad (11.21)$$

Therefore, the optical joint density of states is given by:

$$\rho_{opt}(\nu) = \frac{1}{\pi\hbar^2}(2m_r)^{3/2}(h\nu - E_g)^{1/2}. \qquad (11.22)$$

Figure 11.5 illustrates the $OJDS$ as a function of energy $(h\nu)$, and it has values only if $h\nu > E_g$. One of the advantages of the optical joint density of states that it provides information about the number of possible transition paths between the conduction band (CB) and valence band (VB) without depending on the specific values of E_1 or E_2.

Fig. 11.5 The optical joint density of states $(\rho_{opt}(\nu))$ as a function of photon energy

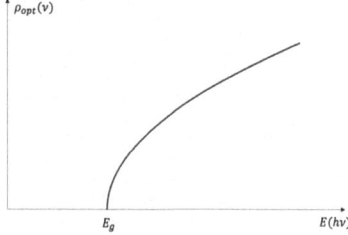

11.5 Probabilities of Emission and Absorption

Emission requires an electron at E_2 and a hole in E_1 for the electron to transit from the CB to valence band losing an energy $h\nu = E_2 - E_1$ in the form of a photon. Therefore, the probability of emission (f_e) is given by:

$$f_e = f(E_2)[1 - f(E_1)]. \tag{11.23}$$

Similarly, the probability of absorption (f_a) is given by

$$f_a = f(E_1)[1 - f(E_2)] \tag{11.24}$$

Hence, a net emission requires that f_e must be greater than f_a, i.e., $f_e > f_a$.

Case 1: Thermal Equilibrium In this case, there is no external source of energy to move the Fermi level closer to the conduction band, which results in the condition that absorption exceeds emission, i.e.,

$$f_a >> f_e. \tag{11.25}$$

Case 2: Quasi Equilibrium This is the case of a doped semiconductor, which leads to the existence of two Fermi energies E_{fc} and E_{fv}. Substituting the Fermi-Dirac function, the probability of emission in this case of quasi-equilibrium being given by:

$$f_e = \left[\frac{1}{e^{(E_2-E_{fc})/k_BT}+1}\right]\left[1 - \frac{1}{e^{(E_1-E_{fv})/k_BT}+1}\right], \tag{11.26}$$

$$f_e = \left[\frac{1}{e^{(E_2-E_{fc})/k_BT}+1}\right]\left[\frac{e^{(E_1-E_{fv})/k_BT}}{e^{(E_1-E_{fv})/k_BT}+1}\right]. \tag{11.27}$$

Similarly, the probability of absorption in this case of quasi-equilibrium is given by:

$$f_a = \left[\frac{1}{e^{(E_1-E_{fv})/k_BT}+1}\right]\left[\frac{e^{(E_2-E_{fc})/k_BT}}{e^{(E_2-E_{fc})/k_BT}+1}\right]. \tag{11.28}$$

For probability of emission exceeding that of absorption, the following condition must be satisfied when Eqs. (11.27) and (11.28) are used:

$$e^{(E_1-E_{fv})/k_BT} > e^{(E_2-E_{fc})/k_BT}, \tag{11.29}$$

i.e.,
$$(E_1 - E_{fv}) > (E_2 - E_{fc}). \tag{11.30}$$

Rearranging the terms of the inequality yields:
$$E_{fc} - E_{fv} > E_2 - E_1 = h\nu. \tag{11.31}$$

Therefore, the condition of emission is:
$$E_{fc} - E_{fv} > E_g. \tag{11.32}$$

This condition requires that the Fermi levels are located within the conduction and valence bands for doped semiconductor to have net emission. This is the condition is sufficient for the operation of semiconductor light sources.

11.6 Rates of Emission and Absorption

The rates of emission and absorption refer to the processes by which electrons transition between different energy states in the conduction and valence bands of semiconductors. These processes are fundamental for understanding the behavior of electrons and holes (electron vacancies), as they transition between energy states within semiconductor materials. Emission and absorption processes play a crucial roles in the operation of various optoelectronic devices, including lasers, light-emitting diodes (LEDs), and photodetectors.

Emission in semiconductors occurs when an electron transitions from a higher energy state to a lower energy state, typically resulting in the release of energy in the form of a photon. The emission can take two forms:

1. Spontaneous Emission: In spontaneous emission, an electron in an excited state (conduction band) spontaneously drops to a lower energy state (valence band) and emits a photon. The rate of spontaneous emission, r_{sp}, is governed by the radiative recombination rate, which depends on the material's properties and the transition probability.
2. Stimulated Emission: In stimulated emission, an incoming photon with energy matching the energy difference between the excited and lower energy states can trigger the emission of additional photons with the same energy, direction, and phase.

Absorption involves the process where an electron in the valence band absorbs energy (typically in the form of a photon) and transitions to a higher energy state in the conduction band. This process is responsible for the semiconductor's ability to absorb light and convert it into an electrical current, which is the basis for optical photodetectors.

In certain semiconductors, the emission and absorption processes encompass direct transitions between the valence and conduction bands. However, in semiconductors with indirect bandgaps, these processes involve indirect transitions, necessitating both energy transition and momentum conservation.

The rate of emission in semiconductors is a consequence of the recombination of an electron and a hole. Recombination processes in semiconductors describe how charge carriers, specifically electrons and holes, combine and return to lower energy states. These processes typically occur through two primary mechanisms: radiative recombination and nonradiative recombination.

Radiative recombination is a process in which an electron in the conduction band recombines with a hole in the valence band, and as a result, a photon is emitted. This photon carries away the energy difference between the electron and hole energy states. Nonradiative recombination is a process in which electrons and holes recombine without emitting photons. Instead of emitting light, the excess energy is dissipated as heat or transferred to lattice vibrations (phonons).

The rate of emission is the number of emissions per unit volume per unit time. The rate of spontaneous emission r_{sp} is proportional to the product of the optical joint density of state and the probability of emission, P_e, i.e.,

$$r_{sp} = A\rho_{opt}(v)f_e. \tag{11.33}$$

Similarly, the rate of stimulated emission is:

$$r_{st} = B\rho_{opt}(v)f_e(v)u(v). \tag{11.34}$$

The spontaneous emission coefficient (A) depends on the radiative recombination lifetime, i.e.,

$$A = \frac{1}{\tau_r}, \tag{11.35}$$

where τ_r is the radiative recombinations lifetime. Let T be the total number of recombinations and let τ_{nr} be the nonradiative recombinations lifetime. Therefore,

$$T = \frac{1}{\tau} = \frac{1}{\tau_r} + \frac{1}{\tau_{nr}}, \tag{11.36}$$

where τ is the recombination lifetime.

In direct bandgap semiconductors, $\tau_r \approx \tau_{nr}$, whereas, in the case of indirect bandgaps, $\tau_{nr} \gg \tau_r$.

In the cases of absorption and stimulated emission, the material is illuminated with stream of photons with energy equal to or exceeding the bandgap. The energy density of illumination is given by:

11.6 Rates of Emission and Absorption

$$u(\nu) = \frac{nh\nu}{V},\qquad(11.37)$$

where n is the number of photons and V is the volume of the material illuminated. The rates of emission and absorption can be expressed as:

$$r_{sp} = A\rho_{opt}(\nu)f_e(\nu),\qquad(11.38)$$

$$r_{st} = B\rho_{opt}(\nu)f_e(\nu)u(\nu),\qquad(11.39)$$

and

$$r_{ab} = B\rho_{opt}(\nu)f_e(\nu)u(\nu).\qquad(11.40)$$

Let us now consider each of these processes separately in the following subsections.

11.6.1 Rate of Spontaneous Emission

The spontaneous emission depends on the optical joint density of states and the probability of emission, which can be given by:

$$r_{sp} = \frac{1}{\tau_r}\frac{1}{\pi\hbar^2}(2m_r)^{3/2}(h\nu - E_g)^{1/2}f_e(\nu),\qquad(11.41)$$

where the probability of emission is given by:

$$f_e = f(E_2)[1 - f(E_1)],\qquad(11.42)$$

$$f_e = \left[\frac{1}{e^{(E_2-E_f)/k_BT}+1}\right]\left[\frac{e^{(E_1-E_f)/k_BT}}{e^{(E_1-E_f)/k_BT}+1}\right].\qquad(11.43)$$

At room temperature $k_BT \approx 0.026$, in the case of thermal equilibrium $E_2 - E_f \gg k_BT$, and $E_f - E_1 \gg k_BT$ but it is a negative quantity sinc $E_f > E_1$, therefore,

$$e^{(E_2-E_f)/k_BT} \gg 1,\qquad(11.44)$$

and

$$e^{(E_1-E_f)/k_BT} \ll 1.\qquad(11.45)$$

The probability of emission in thermal equilibrium is reduced to the following:

$$f_e \simeq e^{-(E_2-E_f)/k_BT} \times e^{(E_1-E_f)/k_BT}, \qquad (11.46)$$

therefore,

$$f_e = e^{-(e_2-E_1)/k_BT} = e^{-h\nu/k_BT}, \qquad (11.47)$$

which is very small.

Following the same procedure for the quasi-equilibrium case, the probability of emission is given by:

$$f_e \simeq e^{-(E_2-E_{fc})/k_BT} \times e^{(E_1-E_{fv})/k_BT}. \qquad (11.48)$$

Therefore,

$$f_e \simeq e^{-h\nu/k_BT} e^{(E_{fc}-E_{fv})/k_BT}. \qquad (11.49)$$

The first term of this equation is small but the second term is very large. Therefore, the ratio of the probability of emission in quasi-equilibrium to that at thermal equilibrium is given by:

$$\frac{(f_e)_{quasi-equilibrium}}{(f_e)_{therma-equilibrium}} = \frac{e^{-h\nu/k_BT} e^{(E_{fc}-E_{fv})/k_BT}}{e^{-h\nu/k_BT}} = e^{(E_{fc}-E_{fv})/k_BT}. \qquad (11.50)$$

Therefore, the probability of emission in the quasi-equilibrium case is greater than that at the thermal equilibrium.

The spontaneous emission in thermal equilibrium can be considered by substituting Eq. (11.47) into Eq. (11.41) to obtain the following expression for r_{sp} in thermal equilibrium:

$$r_{sp}(\nu) = \frac{1}{\tau_r} \frac{1}{\pi \hbar^2} (2m_r)^{3/2} (h\nu - E_g)^{1/2} e^{-h\nu/k_BT}, \qquad (11.51)$$

rearranging some of the terms in Eq. (11.51) to obtain:

$$r_{sp}(\nu) = \frac{1}{\tau_r} \frac{1}{\pi \hbar^2} (2m_r)^{3/2} e^{-E_g/k_BT} (h\nu - E_g)^{1/2} e^{-(h\nu-k_g)/k_BT}, \qquad (11.52)$$

then perform a change of variables s such that:

$$D_o = \frac{1}{\tau_r} \frac{1}{\pi \hbar^2} (2m_r)^{3/2} e^{-E_g/k_BT}, \qquad (11.53)$$

$$\kappa = h\nu - E_g, \qquad (11.54)$$

11.6 Rates of Emission and Absorption

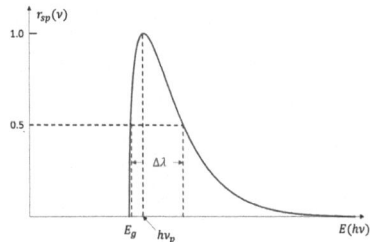

Fig. 11.6 The rate of spontaneous emission is plotted as a function of the photon energy ($h\nu$)

and

$$p = \frac{1}{k_B T}. \tag{11.55}$$

By substituting these changes in Eq. (11.52), the spontaneous emission can be expressed in the following simplified form:

$$r_{sp}(\nu) = D_o \kappa^{1/2} e^{-p\kappa}. \tag{11.56}$$

The rate of spontaneous emission (r_{sp}) given by Eq. (11.56) is shown in Fig. 11.6. The amplitude of the r_{sp} value is zero for all photon energies ($h\nu$) less than the bandgap energy (E_g). Beyond this threshold r_{sp} increases reaching a peak at $h\nu_p$, after which it decays exponentially following $e^{-p\kappa}$.

The maximum photon energy ($h\nu_p$) at of the spontaneous emission curve is determined by taking the first derivative of Eq. (11.56) with respect to $h\nu$ and setting the derivative to zero. This procedure yields the expression for the peak photon energy as follows:

$$h\nu_p = E_g + \frac{1}{2}k_B T. \tag{11.57}$$

Consequently, the peak wavelength of the spontaneous emission rate can be expressed as:

$$\lambda_p = \frac{2hc}{2E_g + k_B T}, \tag{11.58}$$

This formula reveals that the peak wavelength is inversely related to the sum of twice the bandgap energy and the temperature-dependent term $k_B T$.

Therefore, the peak frequency ν_p is a function of the bandgap energy and temperature, rendering it highly sensitive to temperature changes in the material. The rate of spontaneous emission is plotted in Fig. 11.7 as a function of photon energy for various temperatures: $T = 300$, 400 and 500 K. Clearly, the peak frequency (ν_p) increases with the increase of the semiconductor temperature.

Fig. 11.7 Spontaneous emission rate as a function of photon energy at various temperatures

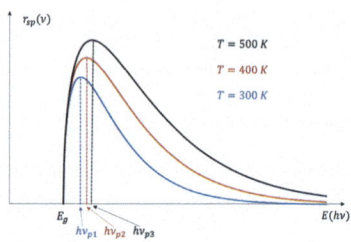

The spontaneous emission rate has a spectral width, full-width at half-magnitude ($FWHM$), $\Delta\lambda$, as shown in Fig. 11.6 and is given by:

$$\Delta\lambda \simeq 1.45 k_B T \lambda_p^2, \tag{11.59}$$

The units of constants in this equation are as follows: λ and $\Delta\lambda$ in μm and $k_B T$ is in eV.

11.6.2 Rats of Absorption and Stimulated Emission

In this section, we explore the processes of absorption and stimulated emission in semiconductor materials, focusing on their interaction with a light beam composed of photons. During absorption, electrons in the valence band absorb photons whose energy exceeds the bandgap energy, causing these electrons to transition to the conduction band. The rate of absorption (r_{ab}) can be described by Eq. (11.40). In the case of stimulated emission, an external photon stimulates an electron in the conduction band to transition back to the valence band, emitting a photon that possesses the same energy, polarization, and direction of propagation as the stimulating photon. The formula for the stimulated emission rate (r_{st}) is provided in Eq. (11.39).

Consider an optical beam with intensity I_{in} impinging on a box of a bulk semiconductor material characterized by a cross-sectional area S and length L as shown in Fig. 11.8. As the photons travel through the semiconductor, they can either be absorbed or initiate stimulated emission. Consequently, an output beam I_{out} emerges from the semiconductor. Depending on the balance between absorption and stimulated emission within the material, this output beam can be either attenuated or amplified.

To analyze this, process considers a very thin slice of the semiconductor with thickness dz and a volume Sdz along the path of the beam inside the box. The intensity of the beam entering this slice is $I_\nu(z)$, and the intensity of the beam exiting from this slice of material is $I_\nu(z + dz)$.

The number of stimulated emitted photons per unit time in the volume element Sdz is:

11.6 Rates of Emission and Absorption

Fig. 11.8 Schematic of a box of semiconductor material showing the input and output beams of light as it goes through absorption and stimulated emissions

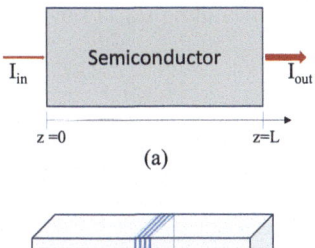
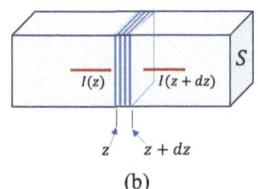

$$N_{st} = r_{St}(v)Sdz, \quad (11.60)$$

and the stimulated energy generated per unit time in the volume element Sdz is:

$$E_{st} = r_{St}(v)Sdzhv. \quad (11.61)$$

Similarly, the energy absorbed per unit time in the element Sdz is

$$E_{ab} = r_{ab}(v)Sdzhv. \quad (11.62)$$

The units of intensity, I_v, are Joules/unit time/unit area. Therefore, the net energy entering/ unit time entering the semiconductor slice, Sdz, is:

$$E(z) = I_v(z)S \quad (11.63)$$

and the optical energy leaving the slice at $z + dz$ is:

$$E(z + dz) = I_v(z + dz)S. \quad (11.64)$$

Therefore, the net energy generated in the volume Sdz is:

$$E_{gen} = [I_v(z+dz) - I_v(z)]S, \quad (11.65)$$

$$E_{gen} = (r_{st} - r_{ab})Sdzhv. \quad (11.66)$$

However, the intensity at the end of slice $I_v(z+dz)$ can be expressed in terms of the rate of change along the z-axis as:

$$I_v(z+dz) = I(z) + \frac{\partial I_v}{\partial z}dz. \quad (11.67)$$

Substituting Eq. (11.67) into Eq. (11.65), we obtain the following:

$$[I_v(z) + \frac{\partial I_v}{\partial z}dz - I_v(z)]S = (r_{st} - r_{ab})Sdzhv. \tag{11.68}$$

Therefore, the rate of change of the light beam intensity along the z-axis, and by using Eqs. (11.39) and (11.40) for the rates of stimulated emission and absorption, is given by:

$$\frac{\partial I_v}{\partial z} = (r_{st} - r_{ab})hv = B\rho_{opt}(v)[f_e(v) - f_a(v)]u_v hv. \tag{11.69}$$

The optical beam energy density $u(v)$ is related to the intensity through the following formula:

$$u(v) = \frac{I_v}{v}, \tag{11.70}$$

where v is the volume and the Einstein $A \& B$ coefficient as developed in Chap. 9 are given by

$$\frac{A}{B} = \frac{8\pi h v^3}{v^3}. \tag{11.71}$$

However, since coefficient A is given by Eq. (11.35), therefore, the coefficient B becomes:

$$B = \frac{Av^3}{8\pi h v^3} = \frac{v^3}{\tau_r 8\pi h v^3}. \tag{11.72}$$

Substituting Eq. (11.72) into Eq. (11.69), we obtain the following for the intensity rate of change along the semiconductor bar:

$$\frac{\partial I_v}{\partial z} = \frac{v^2}{8\pi v^2} \frac{\rho_{opt}(v)}{\tau_r}[f_e(v) - f_a(v)]I_v, \tag{11.73}$$

which can be expressed as:

$$\frac{dI_v}{dz} = \gamma(v)I_v, \tag{11.74}$$

where γ is the gain coefficient and it is given by:

$$\gamma(v) = \frac{v^2}{8\pi v^2} \frac{\rho_{opt}(v)}{\tau_r}[f_e(v) - f_a(v)]. \tag{11.75}$$

11.6 Rates of Emission and Absorption

Therefore, the intensity of the optical beam at any point z along the semiconductor material is the solution of the first-degree order differential equation, Eq. (11.74), given by:

$$I_\nu(z) = I_\nu(0)e^{\gamma(\nu)z}, \tag{11.76}$$

where $I(0)$ is the beam intensity at $z = 0$. Therefore, the optical beam intensity at the output at $z = L$ then becomes:

$$I_\nu(L) = I_\nu(0)e^{\gamma L}. \tag{11.77}$$

The intensity $I_\nu(z)$ at any point at z increases (if $\gamma > 0$) or decreases (if $\gamma < 0$) exponentially as a function of z along the semiconductor as shown in Fig. 11.9.

Substituting for the optical joint density of states Eq. (11.22) in the fain coefficient ($\gamma(\nu)$) given by Eq. (11.74) results in

$$\gamma(\nu) = \frac{(c/n)^2}{8\pi\nu^2}\frac{1}{\tau_r}(2m_r)^{3/2}\frac{1}{\pi\hbar^2}(h\nu - E_g)^{1/2}f_g(\nu), \tag{11.78}$$

where $f_g(\nu)$ is the Fermi inversion factor which is given by:

$$f_g(\nu) = f_e(\nu) - f_a(\nu). \tag{11.79}$$

The gain of a semiconductor element of volume SL is given by:

$$Gam = \frac{I_\nu(L)}{I_\nu(0)} = e^{\gamma L}. \tag{11.80}$$

The gain will be positive if:

$$f_e(\nu) - f_a(\nu) > 0, \tag{11.81}$$

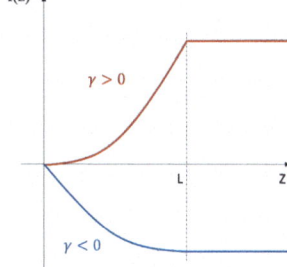

Fig. 11.9 The input-output relationship for the beam intensity in cases of both gain and attenuation

and the gain will be negative, i.e., attenuation, if:

$$f_e(v) < f_a(v). \tag{11.82}$$

The semiconductor material becomes a gain medium (Eq. (11.81)) and requires the quasi-equilibrium condition as given by Eq. (11.31). Therefore, the gain amplification condition for all frequencies is given by:

$$E_g < hv < (E_{fc} - E_{fv}), \tag{11.83}$$

determines the bandwidth of the optical amplifier.

Example Determine the central frequency the bandwidth of a $GaAs$ ($E_g = 1.42$ eV) under a quasi-equilibrium condition such that:

$$E_{fc} - E_c = 0.1 \text{ eV},$$

and

$$E_v - E_{fv} = 0.1 \text{ eV}.$$

Solution The center frequency of the gain spectrum is given by:

$$v_o = \frac{c}{\lambda_o} = \frac{c}{(1.24/E_g)} = \frac{3 \times 10^8 \text{ m/s}}{(1.24/1.42 \, \mu\text{m})} = 3.4 \times 10^{14} \text{ Hz}.$$

The bandwidth of the amplifier v_{bw} is given by:

$$\frac{E_g}{h} < v_{bw} < \frac{E_{fc} - E_{fv}}{h}$$

$$v_1 < v_{bw} < v_2.$$

Therefore, the bandwidth is:

$$v_{bw} = v_2 - v_1,$$

$$\Delta E = hv_b = 0.2 \text{ eV}$$

$$v_{bw} = \frac{0.2 \times 1.6 \times 10^{-19}}{6.6 \times 10^{-34}}$$

$$Bandwidth = 50 \times 10^{12} \text{ Hz}.$$

11.7 Semiconductor Gain Medium

A semiconductor is considered a gain medium if its gain coefficient (γ) is positive. The gain medium is an essential component for optical amplifiers and lasers. It defines the characteristics of these devices. One of the important characteristics of such a medium is its gain profile as a function of frequency (wavelength). The gain coefficient given in Eq. (11.78) depends on the material properties, temperature, photon energy ($h\nu$), and the Fermi inversion factor. The gain coefficient as derived in the previous section, Eq. (11.78), is given by:

$$\gamma(\nu) = \left[\frac{(c/n)^2}{2\tau_r}(2m_r)^{3/2}\right]\frac{(h\nu - E_g)^{1/2}}{(h\nu)^2}f_g(\nu),$$

which can rewritten as:

$$\gamma(\nu) = K\frac{(h\nu - E_g)^{1/2}}{(h\nu)^2}f_g(\nu), \quad (11.84)$$

where the factor K is a material dependent constant given by:

$$K = \frac{(c/n)^2}{2\tau_r}(2m_r)^{3/2}. \quad (11.85)$$

In the case of thermal equilibrium at 0 K (absolute zero), the Fermi function $f(E_2) = 0$ and $f(E_1) = 1$. Consequently, $f_9(\nu) = -1$, leading to the gain coefficient to becoming as follows:

$$\gamma(\nu) = -K\frac{(h\nu - E_g)^{1/2}}{\nu^2}f_0(\nu), \quad (11.86)$$

where $f_0(\nu)$ is $f_g(\nu)$ at 0 K. In this case, since the gain coefficient $\gamma < 0$, the medium becomes an attenuating medium. Therefore, the absorption (attenuation) coefficient is determined by:

$$\alpha(\nu) = K\frac{(h\nu - E_g)^{1/2}}{\nu^2} = \frac{(c/n)^2}{8\pi^2\hbar^2\tau_r}(2m_r)^{3/2}\frac{(h\nu - E_g)^{1/2}}{\nu^2}f_0(\nu). \quad (11.87)$$

This coefficient characterizes the extent to which the medium absorbs the intensity of the incident radiation as a function of the frequency of the incident beam.

The gain and attenuation coefficients are shown in Fig. 11.10 as a function of the photon energy ($h\nu$). The effect of dividing the gain coefficient by $h\nu$ is very limited because of the narrow bandwidth of the gain medium as shown in the previous example. The gain coefficient has a value only for photon energy between

Fig. 11.10 The gain and attenuation coefficients for a semiconductor medium

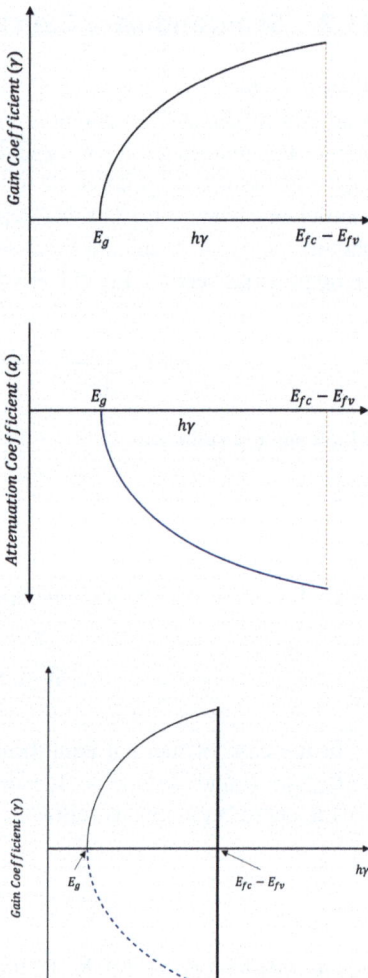

Fig. 11.11 The gain coefficient ($\gamma(\nu)$) for a range of photon energies $h\nu > E_{fc} - E_{fv}$ where the gain medium becomes an absorber following the attenuation coefficient shown as a blue dashed line

E_g and ($E_{fc} - E_{fv}$) beyond which it becomes zero. The semiconductor becomes an absorber for all $h\nu > (E_{fc} - E_{fv})$, as shown in Fig. 11.11.

The effect of temperature on the gain coefficient is shown in Fig. 11.12. At $T = 0\,\text{K}$, the gain exhibits an abrupt change at $h\nu = E_{fc} - E_{fv}$ where the medium abruptly becomes an absorption medium. Whereas, for temperatures $T > 0\,\text{K}$, the gain has a gradual transition, reaching zero at $h\nu = E_{fc} - E_{fv}$ before the medium becomes an attenuation medium. Additionally, the peak gain decreases as the temperature increases.

Fig. 11.12 Effect of temperature on the gain coefficient

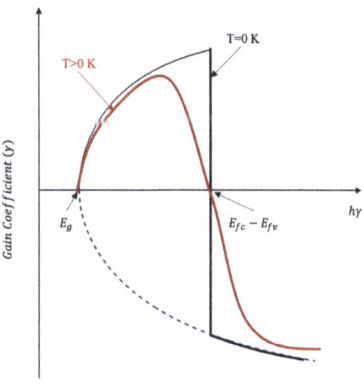

11.8 Summary

This chapter explored the fundamental principles and materials behind optoelectronic semiconductors, which are crucial for modern photonics applications such as LEDs, laser diodes, photodetectors, and solar cells. Key semiconductor materials, including silicon, gallium arsenide, indium phosphide, and gallium nitride, were discussed in the context of their bandgap properties and applications.

The chapter explained the differences between direct and indirect bandgap semiconductors, emphasizing their impact on photon emission and absorption efficiency. Bandgap engineering was introduced as a method to tailor materials for specific wavelengths using ternary and quaternary alloys.

Further, the concepts of Optical Joint Density of States and its role in photon interactions were discussed. Probabilities and rates of emission and absorption processes, including spontaneous and stimulated emission, were analyzed mathematically to show their dependence on factors like temperature, material properties, and Fermi energy levels.

Finally, the chapter discussed semiconductor gain media, highlighting conditions for optical amplification versus attenuation and the effects of temperature on gain coefficients. These principles underscore the role of optoelectronic semiconductors in enabling advanced photonic technologies.

11.9 Problems

1. Calculate the built-in potential of a pn-junction at room temperature (300 K) with doping concentrations $N_A = 10^{18}\,\mathrm{cm}^{-3}$ and $N_D = 10^{16}\,\mathrm{cm}^{-3}$. Assume intrinsic carrier concentration $ni = 1.5 \times 10^{10}\,\mathrm{cm}^{-3}$.
2. Determine the width of the depletion region for a pn-junction with uniform doping ($N_A = 10^{17}\,\mathrm{cm}^{-3}$, $N_D = 10^{15}\,\mathrm{cm}^{-3}$) and a built-in potential of 0.7 V. Use the permittivity of the material as $\varepsilon = 11.7\varepsilon_0$.

3. For a forward-biased pn-junction diode, calculate the reduction in the potential barrier when a voltage of 0.3 V is applied. Explain its effect on carrier injection across the junction.
4. An LED has a radiative recombination lifetime of 100 ns and a non-radiative recombination lifetime of 50 ns. Calculate the internal quantum efficiency (η_i).
5. A $GaAs$ LED with a bandgap energy of 1.42 eV emits photons with an internal quantum efficiency of 0.8. If a current of 50 mA is applied, calculate the optical power output at the junction.
6. Determine the peak wavelength and full-width at half-maximum (FWHM) for a $GaAs$ LED operating at 300 K. Use the bandgap energy of 1.42 eV.
7. For a Fabry-Perot laser diode, calculate the threshold current if the transparency carrier concentration is 1.2×10^{18} cm^{-3}, and the active region dimensions are $200\,\mu m \times 10\,\mu m \times 1\,\mu m$. Assume the carrier lifetime $\tau = 2$ ns.
8. A semiconductor optical amplifier (SOA) has a gain coefficient of 300 cm^{-1}. Calculate the total gain of the device if its length is 500 μm. Express the result in dB.
9. An LED has a refractive index of $n = 3.5$, and the surrounding medium has a refractive index of $n_o = 1.0$. Calculate the fraction of photons that escape the surface, considering total internal reflection.
10. Discuss the mechanism of wavelength tuning in a tunable laser and its importance for dense wavelength-division multiplexing ($DWDM$) systems. Provide an example of a practical scenario where tunable lasers are indispensable.

Bibliography

1. Sze, S. M., & Ng, K. K. (2006). Physics of Semiconductor Devices (3rd ed.). Wiley.
2. Kasap, S. O. (2020). Principles of Electronic Materials and Devices (4th ed.). McGraw-Hill.
3. Coldren, L. A., Corzine, S. W., & Mašanović, M. L. (2012). Diode Lasers and Photonic Integrated Circuits (2nd ed.). Wiley.
4. Wilson, J., & Hawkes, J. F. B. (1998). Optoelectronics: An Introduction (3rd ed.). Prentice Hall.
5. Bhattacharya, P. (1996). Semiconductor Optoelectronic Devices (2nd ed.). Pearson.
6. Schubert, E. F. (2006). Light-Emitting Diodes (2nd ed.). Cambridge University Press.
7. Chuang, S. L. (2009). Physics of Photonic Devices (2nd ed.). Wiley.
8. Kressel, H., & Butler, J. K. (1977). Semiconductor Lasers and Heterojunction LEDs. Academic Press.
9. Agrawal, G. P., & Dutta, N. K. (1993). Semiconductor Lasers. Springer.
10. Saleh, B. E. A., & Teich, M. C. (2019). Fundamentals of Photonics (3rd ed.). Wiley.

Semiconductor Light Sources

12

Light is emitted in semiconductor materials as a result of electron transitions from the conduction band to recombine with a hole in the valence band, thereby releasing energy in the form of a photon. This emission can occur through either spontaneous or stimulated emission, as discussed in the previous chapter. Semiconductor light sources such as light-emitting diodes ($LEDs$) are based on spontaneous emission, whereas laser diodes (LDs) utilize the property of stimulated emission. These light sources are typically based on pn-junction diodes under a forward bias condition. The electric current in the device injects electrons and creates an excess carrier concentration, leading to radiative recombination and the subsequent emission of photons. This chapter provides a comprehensive overview of their operation, characteristics, and emerging trends. We discuss their theory of operation, input-output relationships, output characteristics, and device structures.

12.1 pn-Junction Diodes

Semiconductor light sources are pn-junctions of direct-bandgap materials such as GaAs and InGaAs. In this section, we introduce the basic principles of pn-junction diodes both homogeneous and heterostructure-based devices.

12.1.1 Unbiased pn-Junction

A pn-junction is formed by combining *p-type* and *n-type* semiconductor materials, which have holes and electrons as their majority carriers, respectively. When these materials contact each other to form a pn-junction diode, diffusion of majority carriers across the junction occurs. Near the junction, the p-type material loses holes, leaving behind negatively charged ions, whereas the n-type material loses electrons, also leaving behind positively charged ions. This movement of carriers generates

Fig. 12.1 The unbiased *pn-junction* (**a**) charge carriers, (**b**) energy band structure, (**c**) built-in electric field, and (**d**) built-in potential

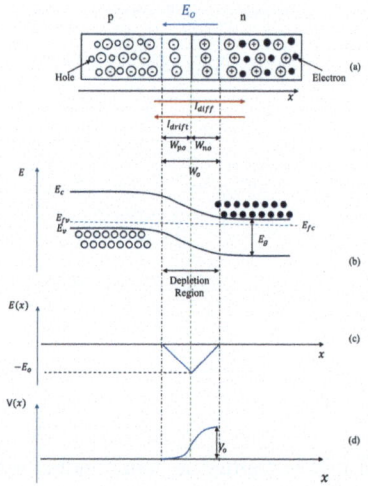

a diffusion current (I_{diff}) flowing from the p-side to the n-side. Simultaneously, the accumulation of charged ions on either side of the junction establishes a built-in electric field, oriented from the n-side to the p-side. This electric field results in a drift current (I_{drift}). The interplay between these currents eventually reaches equilibrium, balancing the flow of carriers, i.e.,

$$I_{diff} = I_{drift}. \tag{12.1}$$

The unbiased *pn-junction* in Fig. 12.1 illustrates the charge carriers, energy band structure, and built-in electric field as well as the drift and diffusion currents at steady state. Note that we must, under equilibrium conditions (e.g., no applied bias or photo-excitation), have the product of the charge densities $pn = n_i^2$ everywhere. The width of the depletion region is W_o.

It is clear that depleted ions create an internal electric field $E(x)$ with a peak value $-E_o$ at the junction interface (Fig. 12.1c), from positive ions to negative ions, that is, in the $-x$ direction, which causes the holes to drift back into the p-region and electrons back into the n-region. This field drives the holes in the opposite direction to their diffusion. The electric field E_o, as shown in Fig. 12.1a, imposes a drift force on holes in the $-x$ direction whereas the hole diffusion flux is in the $+x$ direction. A similar situation also applies for electrons with an electric field, which causes electrons to drift against diffusion from *n-* to the *p-region*. It is apparent as increasing number of holes diffuse toward the right and electrons move toward the left, the internal field around the junction interface increases until an equilibrium is eventually reached when the rate of holes diffusing toward the right is just balanced by the rates of holes drifting back to the left, driven by the field E_o. The electron diffusion and drift fluxes will also be balanced in equilibrium. At this point the current flow reaches a steady state.

12.1 pn-Junction Diodes

The electric field creates an internal built-in potential $V(x)$ related to the electric field through the following relationship:

$$E = -\frac{dV}{dx}. \qquad (12.2)$$

The built-in potential $V(x)$ is an integral of Eq. (12.2). By assuming that $V(x) = 0$ in the *p-region* away from the junction, the potential increases in the depletion region reaching its peak value V_o at the edge of the depletion region in the $n - type$ as shown Fig. 12.2d. The depletion region extends from approximately $x = -W_{po}$ on the *p-side* to $x = W_{no}$ on the *n-side*, and the subscripts indicate the *p-* or *n-side* and open circuit. The total width of the depletion region is $W_o = W_{po} + Wn_o$. For uniformly doped *p-* and *n-regions*, let the acceptor atoms density be given by N_a for the *p-region* and the donor atoms density be given by N_d for the *n-region* of the junction. Therefore, the total charge on the left-hand side must equal to that on the right-hand side of the junction for overall charge neutrality, so that:

$$N_a W_{po} = N_d W_{no}. \qquad (12.3)$$

The built-in electric field can be shown to be expressed as:

$$E_o = -\frac{eN_a W_{po}}{\epsilon} = -\frac{eN_d W_{no}}{\epsilon}, \qquad (12.4)$$

where $\epsilon = \epsilon_o \epsilon_r$ is the permittivity of the semiconductor material. The built-in potential is equal to the total area under the triangle in Fig. 12.1c, so:

$$V_o = -\frac{1}{2} E_o W_o = \frac{eN_a N_d W_o^2}{2\epsilon (N_a + N_d)}. \qquad (12.5)$$

Using the Boltzmann statistics relating the carrier densities at both sides of the junction, we can derive the following expression for the built-in potential:

$$V_o = \frac{k_B T}{e} \ln\left(\frac{N_a N_d}{n_i^2}\right), \qquad (12.6)$$

$$V_o = V_{th} \ln\left(\frac{N_a N_d}{n_i^2}\right), \qquad (12.7)$$

where V_{th} is known as the thermal voltage and it is equal to 26 mV at room temperature (300 K). Therefore the built-in potential is a function of the doping concentration and temperature.

12.1.2 Forward Biased pn-Junction

Consider the case in which a DC voltage source (battery) with a voltage V is connected across a *pn-junction* so that the positive terminal of the battery is attached to the *p-side* and the negative terminal to the *n-side* (forward bias). The applied voltage causes a current I to flow through the diode, which increases the number of majority carriers on both sides of the junction. This in turn reduces the width of the depletion region to W which in turn reduces the built-in electric field to $E_o - E$ as shown in Fig. 12.2a. The energy band diagram demonstrates that the energy difference between the bands on the *n*- and *p-sides* are reduced allowing easier flow of carriers across the junction, as shown in Fig. 12.2b. The flow of carriers across the junction results in electron-hole recombination. This process is fundamental to n light emissions. The negative polarity of the supply reduces the potential barrier V_o by V. The reason is that the bulk regions outside the depletion region have high conductivities, due to the plenty of majority carriers in the bulk, in comparison with the depletion region in which there are mainly immobile bound ions. Thus, the applied voltage drops mostly across the depletion width W. The applied bias V now directly opposes V_o. The potential barrier against diffusion therefore becomes reduced to $(V_o - V)$, as illustrated in Fig. 12.2c.

Consider the process of the flow of majority and minority carriers in the *pn-junction*. In the *p-region*, the injected electrons diffuse toward the positive terminal. As they diffuse, they recombine with some of the many holes in this region. Those holes that are lost by recombination can be readily replenished by the positive terminal of the battery connected to this side. The current due to the diffusion of electrons on the p-side can be maintained by the supply of electrons from the n-side, which itself can be replenished by the negative terminal of the battery. It is clear that an electric current can be maintained through a *pn-junction* under forward bias and that the current flow seems to be due to the diffusion of minority carriers. There is, however, some drift of majority carriers as well. When a hole (minority carrier) is injected into the *n-side*, it diffuses in this region, inasmuch as

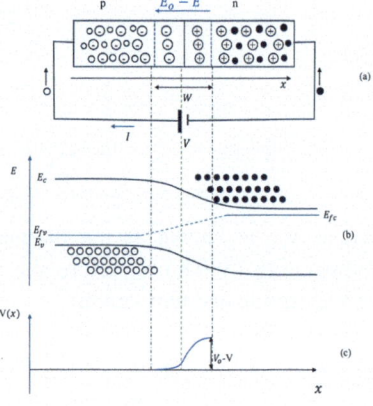

Fig. 12.2 A *pn-junction* under forward-bias condition

there is very little electric field in the neutral n-region resulting in drift. Eventually, the hole will recombine with an electron (majority carrier). The average time it takes for a hole (minority carrier) to recombine with an electron on the n-side is called the minority carrier recombination time or lifetime τ_h. The reciprocal $1/\tau_h$ is the mean probability per unit time that a hole will recombine and disappear. This recombination can be either radiative (editing a photon) or a nonradiative, as will be discussed later in light source operations.

12.1.3 Reverse Biased pn-Junction

Reverse biasing a *pn-junction* increases the width of the depletion region by pulling majority carriers away from the junction. This expansion of the depletion region not only increases the resistance of the device but also enhances its capacitance. The reverse current in this scenario is typically very minimal. In reverse bias, as depicted in Fig. 12.3, the applied voltage $V_o + V_r$ is dropped across the resistive depletion region, intensifying the electric field to $E_o + E$. This increased field causes the depletion zones on both the *p-* and *n-sides* to widen, exposing more ionized dopants, as illustrated in Fig. 12.3a. Consequently, there is a substantial enhancement in the electric field within the depletion region due to the reverse bias.

The movement of electrons in the n-region toward the positive terminal of the battery is halted because of the absence of an electron supply on the n-side. Similarly, the p-side cannot provide electrons to the n-side as it possesses very few electrons. Nonetheless, a slight reverse current arises from two factors: the applied voltage increases the built-in potential barrier, as shown in Fig. 12.3c, and the electric field in the depletion region exceeds the internal built-in field. This condition causes a few holes near the depletion region on the *n-side* to be extracted and swept across to the *p-side* by the field. This minimal current is sustained by

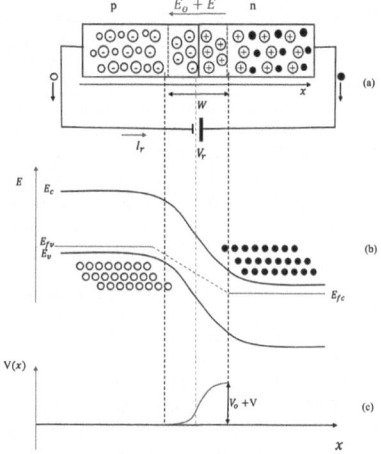

Fig. 12.3 A *pn-junction* under reverse-biased conditions

the diffusion of holes from the bulk of the *n-side* to the boundary of the depletion region.

12.2 Light-Emitting Diodes

Electroluminescence in thermal equilibrium is the phenomenon where a material emits light when it is in a state of thermal equilibrium and an external electric field is applied to it. This phenomenon is a fundamental principle underlying the operation of light-emitting diodes (*LEDs*). At room temperature the thermally excited electron and hole concentrations are so small that the generated photon flux is very small. The photon emission rate can be increased by using external means to increase excess electron-hole pairs such that recombine and result in the emission of photons. As discussed in the previous chapter, light can be due to spontaneous or stimulated emissions. Light-emitting diodes operate on the basis of spontaneous emission that is created by forward biasing a *pn-junction* made of a semiconductor with a direct-bandgap such as *GaAs*. *LEDs* can be based on homojunction structures. A homojunction is a semiconductor interface that occurs between layers of similar semiconductor materials that have the same bandgaps but different doping to create the *p-* and *n-sides* of the junction. *LEDs* can also be based on a heterojunction structure. A heterojunction is the region of the interface between two dissimilar semiconductors with different bandgaps.

12.2.1 Homojunction Light-Emitting Diodes

A homojunction LED is a pn-junction diode whose *n-* and *p-sides* are made of the same semiconductor but are doped differently. When the *LED* is forward biased, the recombination of electron-hole pairs occurs resulting in the emission of a photons with energy $h\nu \approx E_g$. Figure 12.1b illustrates the energy band diagram of an unbiased *pn-junction* device. In this state, the band diagram shows uniform Fermi levels, E_{fc} on the *n-side* and E_{fv} on the *p-side*, a condition necessary for equilibrium when no bias applied.

There exists a potential energy barrier, eV_o, between the edge of the conduction band on the *n-side* (E_c) and the conduction band edge on the *p-side*, represented as $\Delta E_c = eV_o$, where V_o is the built-in voltage. When a forward bias V is applied, this voltage primarily dropped across the depletion region, the most resistive part of the device. Consequently, the built-in potential V_o is reduced to $V_o - V$, facilitating the diffusion or injection of electrons from the *n-side* into the *p-side* as depicted in Fig. 12.4. The recombination of the electron-hole pairs leads to the spontaneous emission of photons, which disperse in various directions through the material, as illustrated in Fig. 12.4. The photon emission predominantly occurs within the depletion region of the device.

12.2 Light-Emitting Diodes

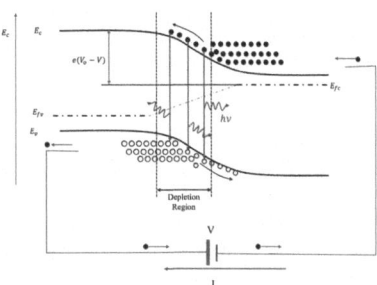

Fig. 12.4 The applied bias potential V reduces V_o and thereby allows electrons to diffuse and be injected, into the $p\text{-}side$. Recombination around the junction of the electrons on the $p\text{-}side$ leads to spontaneous photon emission

In a forward-biased device, the Fermi levels E_{fv} and E_{fc} are separated by the potential energy eV, such that:

$$\Delta E_f = E_{fc} - E_{fn} = eV. \tag{12.8}$$

Under these quasi-Fermi level conditions, the electron and hole concentrations can be expressed as:

$$n = N_c e^{-(E_c - E_{fc})/k_B T}, \tag{12.9}$$

and

$$p = N_v e^{-(E_{fv} - E_v)/k_B T}, \tag{12.10}$$

respectively.

In an LED, photon emission occurs via the transition of an electron from the conduction band to an empty state (hole) in the valence band. The recombination of an electron-hole pair can be either radiative, resulting in photon emission, or nonradiative, involving a phonon. The energy of the emitted photon corresponds to the bandgap energy, $h\nu = E_g$.

However, as photons exits the active (depletion) region of the LED, they may be reabsorbed within the device creating an electron-hole pair. This happens when a photon, possessing sufficient energy, excites an electron in the valence band to transition to the conduction band, leaving behind a hole. As a result, a significant portion of the emitted photons are absorbed within the device, and only a small fraction emerges as useful light.

This inefficiency in light generation in a homojunction $pn\text{-}junction$ diodes motivated the development of single and double heterojunction devices, which enhance light generation and emission efficiency.

12.2.2 Heterojunction Light-Emitting Diodes

The homojunction pn light-emitting diode (LED), while foundational in the development of light-emitting technologies, has several limitations that have led to its classification as an inefficient light source. These drawbacks are primarily due to its structural and material properties:

1. In a homojunction LED, the active region, where the light is generated, coincides with the depletion region, and the close proximity of the *p*- and *n-sides* of the diode increases the probability of nonradiative recombinations occurring.
2. The lack of confinement mechanism in the homojunction structure causes carriers to diffuse outside the light-emitting region before they recombine to emit a photon.
3. A uniform semiconductor material across the device with the same bandgap causes the emitted photon to be absorbed before exiting the device.

These are some of the drawbacks of homojunction LEDs. Heterojunction diodes are engineered to overcome these limitations. In a single heterojunction device, the p- and n-sides are made from materials with different bandgaps. This design ensures that if photons are generated predominantly in a material with a smaller bandgap, they will not be absorbed as they travel through the other side of the device, which has a larger bandgap. A double heterojunction diode consists of three layers, where the middle layer is composed of a material with an active region with a smaller bandgap than the two outer layers. This configuration enhances carrier confinement in the active region, reducing recombination losses and improving the efficiency of photon generation and emission. One advantage of such structures is the use of the optical confinement since materials with high bandgaps have lower indices of refraction.

A typical structure of a double heterojunction LED is demonstrated as shown in Fig. 12.5. A double-heterostructure device based on two heterojunctions between

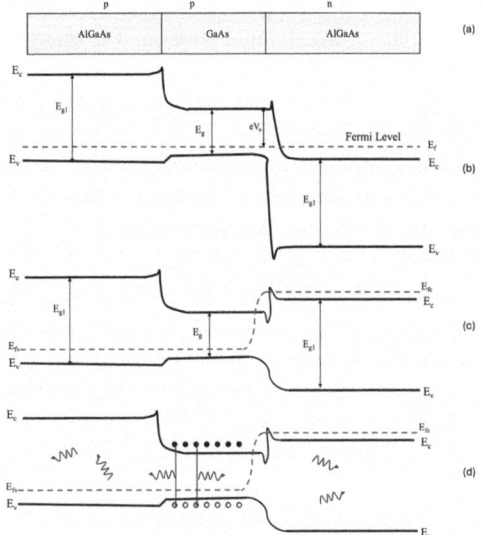

Fig. 12.5 A heterojunction light-emitting diode: (**a**) device structure, (**b**) energy band diagram for an unbiased device, (**c**) energy band diagram for a forward biased device, and (**d**) electron-hole recombination and emission of photons in the narrow bandgap

different semiconductor crystals with different bandgaps. Specifically, the semiconductors involved are $AlGaAs$ ($Al_xGa_{1-x}As$) with $E_g \approx 2\,eV$ and $GaAs$ with $E_g \approx 1.4\,eV$. The p-p-n double heterostructure in Fig. 12.5a has a p-n heterojunction between p-GaAs and n-AlGaAs and another heterojunction between p-AlGaAs and P-GaAs. The p-GaAs region is a thin layer, typically a fraction of a micron, and it is lightly doped. The energy band diagram of an unbiased LED is shown in Fig. 12.5b. In this case, the Fermi level is continuous throughout the device. There is a potential energy barrier eV_o for electrons in the CB of $n-AlGaAs$ against diffusion into $p - GaAs$. There is a bandgap change between p-GaAs and p-GaAs that prevents electrons in the CB of $p - GaAs$ from passing into CB of $p - AlGaAs$.

When a forward bias voltage V is applied to the device, the band structure changes as shown in Fig. 12.5c. The pending of the Fermi level by a potential energy eV reduces the barrier between $n - AlGaAs$ and $p - GaAs$ layers, thus facilitating the diffusion of electrons into the p-GaAs layer. The p-AlGaAs layer, with its wide bandgap acts as a confinement layer that restricts the injected electrons to the p-GaAs layer. The recombination of these injected electrons with the holes present in the p-GaAs layer leads to the spontaneous emission of photons, as shown in Fig. 12.5d. Since the bandgap E_g of $AlGaAs$ is greater than that of $GaAs$, the emitted photons are not reabsorbed as they traverse the active region, enabling them to reach the surface of the device without being absorbed.

12.3 The LED Characteristics

Light-emitting diodes are semiconductor devices that convert electrical energy into light. Understanding their operation and characteristics is crucial for their development and application. The outputs of these devices are characterized by their energy conversion efficiency, wavelength spectrum, temperature, and material dependence.

12.3.1 Current to Optic Power Relation

The light-emitting diode under forward bias of voltage V leads to a steady-state carrier concentration expressed as:

$$\left. \begin{array}{l} n = n_o + \Delta n \\ p = p_o + \Delta p \end{array} \right\}, \quad (12.11)$$

where Δn the carrier density of the excess electrons injected by the voltage source, whereas Δp is the carrier density of the excess holes injected. Since the pumping is performed by forward biasing the $pn\text{-}junction$, therefore,

$$\Delta n = \Delta p. \quad (12.12)$$

The excess electron-hole pairs recombine at a rate given by:

$$\frac{1}{\tau} = \frac{1}{\tau_r} + \frac{1}{\tau_{nr}} \tag{12.13}$$

where τ is the total (radiative and nonradiative) electron-hole recombinations time under steady-state conditions, τ_r is the radiative recombinations lifetime, and τ_{nr} is nonradiative recombinations lifetime. The rate of generation of photons (R) rate must balance the recombination rate, i.e.,

$$R = \Delta n/\tau \text{ pairs/cm}^3., \tag{12.14}$$

Rearranging the previous equation, we obtain:

$$\Delta n = R\tau \text{ pairs/cm}^3. \tag{12.15}$$

Let the recombination coefficient be $r = 1/\tau$. Thus, the pumping rate is given by:

$$R = r\Delta n. \tag{12.16}$$

Only radiative recombinations generate photons: hence, we define the internal quantum efficiency (η_i) as:

$$\eta_i = \frac{r_r}{r} = \frac{\tau}{\tau_r}. \tag{12.17}$$

where r_r is the radiative recombinations coefficient. Therefore, the internal quantum efficiency depends on the radiative and nonradiative recombinations lifetimes. The radiative and nonradiative lifetimes and typical inner quantum efficiencies for various semiconductor materials are given in Table 12.1.

Since the radiation of photons takes place inside the active region, if the volume of the active region is V, therefore, the total injected carrier, RV, results in the following expression for the photon flux (number of photons emitted per second):

$$\phi = \eta_i RV = \eta_i \frac{\Delta n}{\tau} V, \tag{12.18}$$

Table 12.1 Recombinations lifetimes and internal quantum efficiencies

Semiconductor	τ_r	τ_{nr}	τ	η_i
Si	10 ms	100 ns	0.25 ns	2.5×10^{-6}
GaAs	100 ns	100 ns	50 ns	0.5
GaN	10 ns	0.1 ns	0.1 ns	0.01

12.3 The LED Characteristics

simplifying,

$$\phi = \frac{V \Delta n}{\tau_r}. \quad (12.19)$$

The internal photon flux (ϕ) of an LED is proportional to the carrier-pair injection ratio (R). For a forward-biased device with a current i, the photon flux is expressed as:

$$\phi = \eta_i \frac{i}{q}. \quad (12.20)$$

where q is the electron charge and i/q is the number of electrons injected in the active region per second. Consequently, the generated optic power generated within the depletion region is given by:

$$P_{opt} = \eta_i \frac{i}{d} h\nu, \quad (12.21)$$

or equivalently

$$P_{opt} = \eta_i \left(\frac{h\nu}{q}\right) i. \quad (12.22)$$

From this relationship, the electric current (i) and the emitted optic power (P_{opt}) are linearly related. Figure 12.6 illustrates a typical input (i)-output (P_{opt}) relationship for an LED; this relationship remains linear for low injection currents; however, at higher currents, the relationship becomes nonlinear, and the output power saturates.

The diode basically acts as a resistance (R_d) under high current conditions; it generates heat proportional to the electrical power. This heating reduces the efficiency of photon generation, as nonradiative recombination increases and fewer electrons contribute to photon emission, ultimately decreasing the optical power output.

Fig. 12.6 Current-optic power characteristic for a typical light-emitting diode

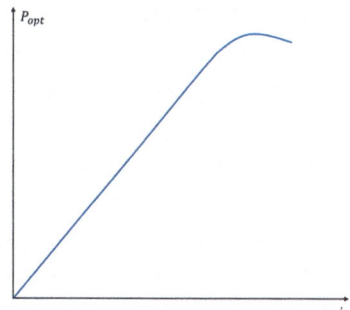

For an active region with dimensions l (length), w (width), and d (thickness), consider a voltage source applied to the surface lw. The current density of the injected current is then given by:

$$J = \frac{i}{lw}. \tag{12.23}$$

Equating Eqs. (12.18) and (12.19), we obtain:

$$\Delta n = \frac{J\tau}{qd}. \tag{12.24}$$

Example For an LED made of GaAs where $\tau_r = 100$ ns and $\tau_{nr} = 100$ ns. A forward bias current $i = 100$ mA is used to pump the diode. The devise dimensions are $2 \times 100 \times 10\,\mu$m. Determine the optic power generated at the depletion region.

Solution The total recombination lifetime is

$$\frac{1}{\tau} = \frac{1}{\tau_{nr}} + \frac{1}{\tau_r} = \frac{1}{100} + \frac{1}{100} = \frac{1}{50},$$

therefore, $\tau = 50$ ns.
The carrier injection rate is:

$$R = \frac{i/e}{V} = \frac{100 \times 10^{-3}/1.626 \times 10^{-19}}{2 \times 10 \times 100 \times 10^{-18}} = 3.12 \times 10^{-32}$$

The access carrier density is given by:

$$\Delta n = R\tau = 50 \times 10^{-9} \times 3.12 \times 10^{32} = 1.56 \times 10^{25}$$

The internal quantum efficiency is:

$$\eta_i = \frac{\tau}{\tau_r} = \frac{50}{100} = 0.5$$

Therefore, the photon flux is:

$$\phi = \eta_i \frac{i}{e} = 3.015 \times 10^{17} \text{ phptons/S}.$$

(continued)

12.3 The LED Characteristics

For GaAs the wavelength of emission is:

$$\lambda = \frac{1.24}{1.42} = 0.8 \, \mu m.$$

Therefore, the output optical power is:

$$P_{opt} = \phi h \nu = 3.015 \times 10^{17} \times 6.626 \times 10^{-34} \times \frac{3 \times 10^8}{0.87 \times 10^{-6}},$$

$$P_{opt} = 70 \, mW.$$

12.3.2 Spectral Response

The energy of an emitted photon from an LED is not strictly equal to the bandgap energy E_g. This deviation raises because electrons in the conduction band and holes in the valence band are distributed across a range of energies, following Fermi-Dirac statistics. As a result, the emitted light spans a spectrum rather than a single wavelength. $LEDs$ emit light through the process of spontaneous emission of photons, which is influenced by these energy distributions.

The rate of spontaneous emission as derived in the previous chapter is given by:

$$r_{sp}(\nu) = \frac{1}{\tau_r} \rho_{opt}(\nu) f_e(\nu), \tag{12.25}$$

where $\rho_{opt}(\nu)$ is the optical joint density of states, which is the number of photons per second per unit volume and is expressed as:

$$\rho_{opt}(\nu) = \frac{(2m_r)^{3/2}}{\pi \hbar^2} \sqrt{h\nu - E_g}, \tag{12.26}$$

where m_r is the reduced mass of the electron, which is given by:

$$m_r = \frac{m_c m_v}{m_c + m_v}. \tag{12.27}$$

Using the Fermi-Dirac function, the probability of emission is expressed as:

$$f_e(\nu) = f_c(E_2)[1 - f_v(E_1)], \tag{12.28}$$

where E_2 is the energy state in the conduction band and E_1 is the energy level in the valence band between which transition of an electron takes place as shown in

Fig 12.6. These energies are given by:

$$E_2 = E_c + \frac{m_r}{m_c}(h\nu - E_g), \tag{12.29}$$

and

$$E_1 = E_2 - h\nu. \tag{12.30}$$

The emitted flux (ϕ) emitted over all frequencies is given by:

$$\phi = V \int_0^\infty r_{sp}(\nu) d\nu, \tag{12.31}$$

where V is the volume of the active region. Integrating this expression using the method of variable substitution and applying the value of the Fermi integral, the total photon flux is derived as:

$$\phi = \frac{V(m_r)^{3/2}}{\sqrt{2}\pi^{3/2}\hbar^3 \tau_r}(k_B T)^{3/2} e^{(E_{fc} - E_{fv} - E_g)/k_B T}. \tag{12.32}$$

Equation (12.18) shows that the photon flux (ϕ) is directly proportional to the pumping rate (R). Hence, increasing the pumping rate R leads to an increase in Δn, which in turn increases E_{fc} and decreases E_{fv}.

Even in the case of low injection, the following conditions still hold:

$$E_{fc} - E_c \gg k_B T, \tag{12.33}$$

and

$$E_v - E_{fv} \gg k_B T. \tag{12.34}$$

The peak value of the frequency of the emitted intensity, as derived in the previous chapter, is:

$$h\nu_p = E_g + \frac{1}{2}k_B T, \tag{12.35}$$

hence,

$$\nu_p = \frac{E_g + \frac{1}{2}k_B T}{h}. \tag{12.36}$$

12.3 The LED Characteristics

The peak wavelength (λ_p) is given by:

$$\lambda_p = \frac{ch}{E_g + \frac{1}{2}k_B I} \tag{12.37}$$

Similarly, the linewidth ($FMHM$) of spectral intensity is given by:

$$\Delta \lambda \simeq 1.8 \lambda_p^2 k_B T / hc, \tag{12.38}$$

such that the units for $k_B T$ is in eV and λ_p is in μm. And since

$$\Delta \nu = \Delta \lambda \frac{\lambda^2}{c}, \tag{12.39}$$

the negative sign in the last expression is omitted because, for the same signal, as frequency increases, wavelength decreases (their product equals the speed of light).

Example Find $\Delta \nu$ and $\Delta \lambda$ at $T = 300\,^\circ$K for $GaAs$.

Solution The peak wavelength is given by:

$$\lambda_p = \frac{ch}{E_g + \frac{1}{2}k_B T}$$

Substituting using the given parameters, we obtain:

$$\lambda_p = \frac{3 \times 10^8 \times 6.626 \times 10^{-34}}{1.42 \times 1.6 \times 10^{-19} + 0.5 \times 300 \times 1.381 \times 10^{-23}}$$

$$= 8.527 \times 10^{-7}\,\text{m} = 0.85\,\mu\text{m}$$

$$\nu_p = \frac{c}{\lambda_p} = \frac{3 \times 10^8}{8.527 \times 10^{-7}} = 3.516 \times 10^{14}\,\text{Hz}$$

$$\Delta \nu = 1.8 \times 300 \times 1.381 \times 10^{-23} / 6.626 \times 10^{-34} = 1.15 \times 10^{13}\,\text{Hz}$$

$$\Delta \lambda = 1.8 \lambda_p^2 k_B T / hc = \frac{1.8 \times (8.527 \times 10^{-7})^2 \times 300 \times 1.381 \times 10^{-23}}{6.626 \times 10^{-34} \times 3 \times 10^8}$$

$$\Delta \lambda = 2.729 \times 10^{-8}\,\text{m} = 0.027\,\mu\text{m}.$$

(continued)

> If we use a different formula for $\Delta\lambda$ which was given before, we obtain:
>
> $$\Delta\lambda \simeq 1.45\lambda_p^2 k_B T = 1.45 \times (0.8527)^2 \times 0.026 = 0.027\,\mu m,$$
>
> which is the same.

The spectral response of an LED is heavily dependent on temperature. Since the rate of spontaneous emission is a function of temperature, changes in temperature result in variations in both peak wavelength λ_p and spectral width $\Delta\lambda$. The rate of spontaneous emission as a function of wavelength is plotted in Fig. 12.7 for temperatures 10, 30, and 60 °C.

The following key observations are made from Fig. 12.7a, where bandgap energy (E_g) is considered to be independent of temperature:

1. The peak of the rate of spontaneous emission increases with temperature due the presence of the exponential term $e^{-E_g/k_B T}$ included in the term D_o in Eq. (12.53) as given by:

$$D_o = \frac{1}{\tau_r}\frac{1}{\pi\hbar^2}(2m_r)^{3/2}e^{-E_g/k_B T}. \tag{12.40}$$

2. The peak wavelength (λ_P) decreases with increasing temperature following Eq. (12.37),
3. The spectral width ($\Delta\lambda$) broadens with temperature as indicated by Eq. (12.38).

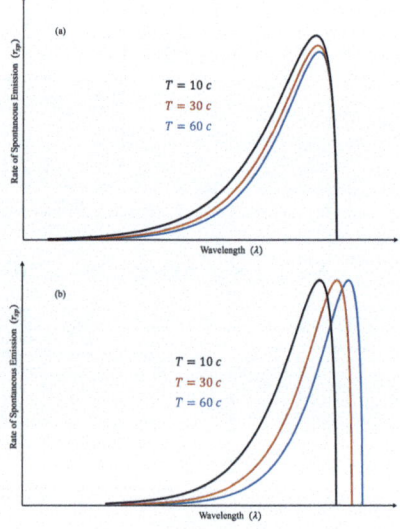

Fig. 12.7 Rate of spontaneous emission as a function of wavelength at various temperatures. Units are arbitrary where (**a**) assuming E_g is assumed to be independent of temperature and (**b**) applied values for (E_g) as a function of temperature, as given by Eq. (12.41). Amplitudes are normalized

12.3 The LED Characteristics

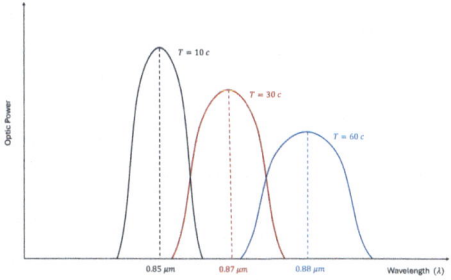

Fig. 12.8 Schematic of the experimental results for the emitted optical power as a function of wavelength ay various operating temperatures, not to scale

In contrast Fig. 12.8 presents experimental results for the optical power emitted by a GaAs LED at the same temperatures as theoretical results in Fig. 12.7. These experimental results reveal a significantly different emission dependence on temperature. As the temperature increases, both λ and $\Delta\lambda$ increase, while the emitted optical power decreases.

This discrepancy between the theoretical and experimental results raises from the assumption in the theoretical model that the bandgap energy (E_g) is independent of temperature. In reality, though, while the temperature dependence of E_g is very small, it significantly impacts the exponential terms in the rate of spontaneous emission, causing noticeable changes in the emission characteristics. The following empirical relation describes the temperature dependence of E_g:

$$E_g = 1.52 - \frac{5.405 \times 10^{-4} \times T^2}{T + 204}, \tag{12.41}$$

where $E_g(T = 0\,\text{K}) = 1.52\,\text{eV}$ and T is in degree Kelvin.

When this temperature dependent on this E_g is incorporated in the computation of the rate of spontaneous emission, the results as shown in Fig. 12.7b align more closely with experimental observations. The peak of the emission spectrum decreases with increasing temperature, and the upper wavelength of the emission shifts to longer wavelengths. Published experimental results in Fig. 12.8 confirm the trends seen in Fig. 12.7a, demonstrating the critical role of accounting for the temperature dependence of E_g.

Example Determine the bandgap energy for GaAs at $T = 10, 27, 30$ and $60\,°\text{C}$, and determine the peak wavelength and spectral width for theses temperature.

Solution The temperatures given in degree Kelvin are 283, 300, 303, and 333 °K.

(continued)

Using Eqs. (12.37), (12.38), and (12.41), we obtain the following results:

Temperature (K)	E_g(eV)	λ_p (μm)	$\Delta\lambda$ (nm)
283	1.4311	0.867	32.7
300	1.4235	0.872	33.1
303	1.4221	0.873	33.15
383	1.4081	0.881	33.8

The question that arises when considering the experimental results is: Does the spectrum's shape align more closely with the theoretical or experimental curves? The theoretical spectrum sharply rises at $h\nu = E_g$ and tapers off as the photon energy $h\nu$ increases beyond the curve's peak. Figure 12.9 illustrates the energy distributions of electrons and holes in the conduction band (CB) and valence band (VB) for a p-type semiconductor.

The electron concentration, $n(E)$, as a function of energy in the CB is given by $\rho_c(E)f(E)$, where $\rho_c(E)$ is the density of states in the CB and $f(E)$ is the Fermi-Dirac function, representing the probability of finding an electron in a state with energy E. Similarly, there is an energy distribution for holes, $p(E)$, in the VB, which is significantly larger than $n(E)$. The $E - k$ diagram for a typical direct bandgap semiconductor (e.g., $GaAs$) is shown in Fig. 12.9. Since the hole concentration is much greater, the rate of recombination depends primarily on the concentration of injected electrons.

The electron concentration in the CB as a function of energy is asymmetrical, peaking above E_c. When an electron at E_c recombines with a hole at E_v, as shown by transition 1 in Fig. 12.9, a photon is emitted with an energy $h\nu_1 = E_c - E_v = E_g$.

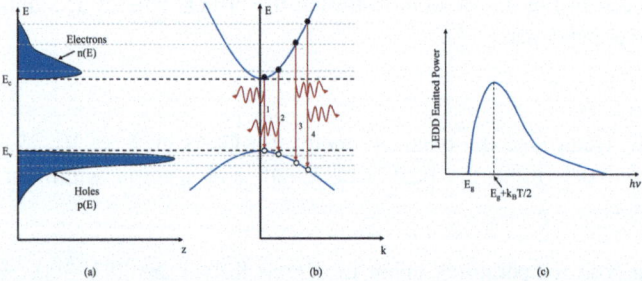

Fig. 12.9 (a) Energy distribution of electrons in the CB and holes in the VB. (b) A simplified energy-momentum ($E - k$) diagram showing direct recombination transitions where momentum (k) is conserved. (c) Relative emitted intensity as a function of photon energy based on transitions depicted in (b).

12.3 The LED Characteristics

However, because there are relatively few electrons and holes at the band edges, such recombination events are infrequent, and the emitted light intensity from type 1 transitions is small.

Transition 2, involving the largest electron concentration (the peak in $n(E)$), emits a photon with $h\nu_2 > h\nu_1$. These transitions are more frequent due to the higher $n(E)$, resulting in higher emitted intensity compared to type 1. Transition 3, with photon energy $h\nu_3 > h\nu_2$, involves electrons higher up in the CB, where $n(E)$ is smaller than for transition 2. Thus, type 3 transitions occur less frequently, leading to reduced emission intensity. Similarly, transition 4 emits photons with energy $h\nu_4 > h\nu_3$, but because $n(E)$ is even smaller, such transitions are rare. As a result, the emission intensity rises to a maximum and then falls as $h\nu$ increases, as depicted in Fig. 12.9.

The experimentally observed LED output spectrum depends not only on the semiconductor material and dopant concentrations but also on the structure of the pn-junction diode. The spectrum in Fig. 12.7 represents an idealized case, excluding the effects of heavy doping on the energy bands. In a heavily doped n-type semiconductor, the donor concentration is so high that the donor wave functions overlap, forming a narrow impurity band centered at E_d, slightly below and overlapping with the conduction band. This overlap effectively lowers E_c, reducing the minimum emitted photon energy to below E_g, depending on the doping level.

Additionally, excitonic transitions, which occur at energies below Eg, further modify the spectrum, causing it to taper off. These transitions explain why the output spectrum for heavily doped semiconductors resembles the experimental spectrum shown in Fig. 12.8. These effects illustrate the influence of doping and recombination dynamics on the spectral response of $LEDs$.

12.3.3 External Quantum Efficiency

As outlined in Sect. 12.3.1, the internal quantum efficiency of a light-emitting diode (LED) is defined as the ratio of radiative recombinations to the total recombinations. The photons generated in the LED's active region are emitted in all directions. For these photons to contribute to external light output, they must traverse from the active region to the device's exterior. The external quantum efficiency is determined by the ratio of photons that successfully escape the device to the total number of photons generated at the junction. This efficiency is impacted by three primary loss mechanisms:

1. Photon reabsorption within the semiconductor: Emitted photons may be reabsorbed, generating an electron-hole pair. This typically occurs when the photon energy matches the semiconductor's bandgap.
2. Reflection at the semiconductor-air interface: A fraction of photons are reflected back into the semiconductor when encountering the interface.
3. Total internal reflection: Photons striking the surface at angles exceeding the critical angle are subject to total internal reflection and fail to exit the device.

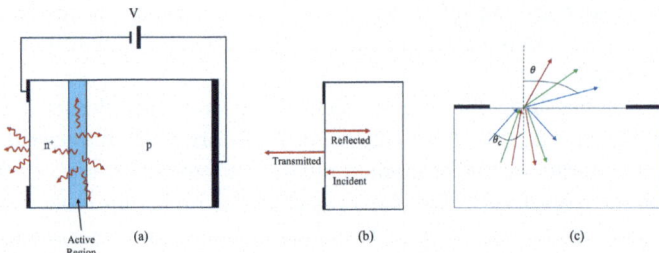

Fig. 12.10 Illustrating the loss mechanisms: (**a**) a schematic of an LED showing emitted photons traversing the device, (**b**) losses due to device surface reflection, and (**c**) losses due to total internal reflection

Minimizing these losses is crucial and requires careful consideration in the design and structural planning of the LED.

Figure 12.10 illustrates loss mechanisms mentioned above. Figure 12.10a demonstrates the losses caused by absorption. The light emitted in the active region traverses along many paths. Along these paths it can be absorbed by the semiconductor generating electron-hole pairs. This can happen in very short distances, with few micrometers. The attenuation experienced by a ray traveling a distance d from the active region to the front surface of the device can be represented by the coefficient:

$$a_1 = e^{-\alpha d}, \tag{12.42}$$

where α is the material absorption coefficient. Figure 12.10b depicts reflection losses at the interface between the semiconductor and the external medium. The refractive indices of the semiconductor and the outside medium are denoted as n_1 and n_2, respectively. For normal incidence, the reflection loss is expressed as:

$$a_2 = \left(\frac{n_2 - n_1}{n_2 + n_1}\right)^2. \tag{12.43}$$

Figure 12.10c illustrates the reflection and transmission of light rays making acute angles with the normal to the interface. These rays emerge at angles larger than the angles of incidence. When the incidence angle is equal to or greater than the critical angle ($\theta_c = \sin^{-1}(n_1/n_2)$), the light undergoes total internal reflection. The emitted light forms a cone shape outside the device, and the area of the spherical cap at the top of this cone is given by:

$$A = \int_0^{\theta_c} 2\pi r \sin(\theta) r d\theta = 2\pi r^2 (1 - \cos\theta_c), \tag{12.44}$$

12.3 The LED Characteristics

where r is the radius of the sphere. The area of the entire sphere is $4\pi r^2$. Thus, the ratio of the spherical cap area to the total sphere area is:

$$a_3 = \frac{A}{4\pi r^2} = \frac{1}{2}(1 - \cos \Theta_c) = \frac{1}{2}\left(1 - \sqrt{1 - n_2^2/n_1^2}\right) \simeq \frac{n_2^2}{4n_1^2}. \tag{12.45}$$

Consequently, the total external quantum efficiency η_e is expressed as:

$$\eta_e = a_a a_2 a_3 = \frac{n_1^2}{4n_2^2}\left(\frac{n_2 - n_1}{n_2 + n_1}\right)^2 e^{-\alpha d}. \tag{12.46}$$

The external photon flux considering the eternal quantum efficiency is:

$$\phi_o = \eta_e \phi = \eta_e \eta_i \frac{i}{q}. \tag{12.47}$$

Let the LED extraction efficiency be defined as:

$$\eta_{ext} = \eta_e \eta_i, \tag{12.48}$$

therefore, the LED output optical power be given by:

$$P_{opt} = \phi_o h\nu = \eta_{ext} h\nu \frac{i}{q}. \tag{12.49}$$

12.3.4 LED Responsivity

Another important characteristic of and LED is its responsivity, (R_e) which is defined as the ratio between the output optic power and the input current, measured in watts per ampers (W/A).Therefore, the responsivity can be determined by:

$$R_e = \frac{P_{opt}}{i} = \frac{h\nu\phi_o}{i} = \eta_{ext}\frac{h\nu}{q}. \tag{12.50}$$

Substituting for $h\nu$ and q in the above equation, the responsivity expression becomes:

$$R_e = \eta_{ext}\frac{1.24}{\lambda_o}. \tag{12.51}$$

This formula provides a direct link between material properties and device efficiency, offering a precise estimation of LED performance. It highlights the influence of quantum efficiency, extraction efficiency, and material characteristics on the $LED's$ optical output.

Example Determine the emitted optical power and responsivity for an LED made of $GaAs$ with the following parameters:

- Wavelength, $\lambda_o = 0.87\,\mu m$,
- Internal quantum efficiency, $\eta_i = 0.9$
- Refractive index of GaAs, $n1 = 3.5$.
- Refractive index of air, $n_2 = 1.0$
- Material absorption coefficient, $\alpha = 100\,cm^{-1}$
- Thickness of the active region, $d = 5\,\mu m$.
- Input current, $i = 20\,mA$.

Solution Absorption loss:

$$a_1 = e^{-\alpha d} = e^{-100 \times 0.0005} = e^{-0.05} \approx 0.9512.$$

Reflection loss:

$$a_2 = \left(\frac{n_2 - n_1}{n_1 + n_2}\right)^2 = \left(\frac{3.5 - 1}{3.5 + 1}\right)^2 \approx 0.3086.$$

Total internal reflection loss (a_3):

$$a_3 = \frac{n_1^2}{4n_2^2} = \frac{1}{4 \times (3.5)^2} \approx 0.082.$$

External quantum efficiency (η_e):

$$\eta_e = a_1 a_2 a_3 = 0.9512 \times 0.3086 \times 0.082 \approx 0.024.$$

Extraction efficiency (η_{ext}):

$$\eta_{ext} = \eta_e \eta_i, = 0.024 \times 0.9 = 0.022$$

Output optical power (P_{opt}):

$$P_{opt} = \phi_o h\nu = \eta_{ext} h\nu \frac{i}{q} = 62.7\,mW.$$

Improving the External Quantum Efficiency of LED Improving the external quantum efficiency (EQE) of light-emitting diodes ($LEDs$) is crucial for enhancing their performance, particularly in terms of the light output per electrical input. The following are some of the methods used including:

12.3 The LED Characteristics

1. **Antireflection coating:** This technique involves the application of a thin multi-layer optical filter on the surface of the LED. The purpose of the antireflection (AR) coating is to minimize the reflection of light at the interface between the semiconductor and external media. By reducing the amount of light is reflected back into the LED, more light can escape from the surface, thus improving the EQE. AR coatings are designed to be effective at the wavelength of the emitted light and are often used in high-performance $LEDs$.
2. **Spherical done to remove the total internal reflection TIR:** Total internal reflection (TIR) occurs when light rays strike a boundary at an angle greater than the critical angle causing all the incident light to be reflected back into the device, rather than transmitted through the surface. Incapsulating the LED in a spherical dome can alter the path of these rays, reducing the occurrence of TIR by changing the geometry of the $LED's$ surface. This dome acts as a lens, redirecting more light rays in a manner that allows them to exit the LED, thereby increasing the light output. Figure 12.11a illustrates this LED design.
3. **Roughening the surface of the LED:** Surface roughening is another effective method to increase the EQE of $LEDs$. By texturing the surface of the LED, the roughened textures create multiple scattering points, which disrupt the path of light within the device. This disruption reduces the TIR by providing more angles for light to escape the surface LED. Surface roughening can be achieved through various techniques such as etching, sandblasting, or applying a rough coating.
4. **Optical confinement:** This technique involves designing the LED structure to confine light more effectively within the active region where light generation occurs. By confining the light, the probability of photon reabsorption is reduced, and more photons are directed toward the output surface of the LED. Techniques for optical confinement include the use of reflective coatings, altering the shape of the emitting layer, or integrating optical waveguides that guide the light toward the surface.

Other techniques such as photon recycling, plasmonic enhancements, and the use of quantum well structures have been proposed. Each of these techniques addresses specific optical challenges in the design of LEDs, and often, a combination of these

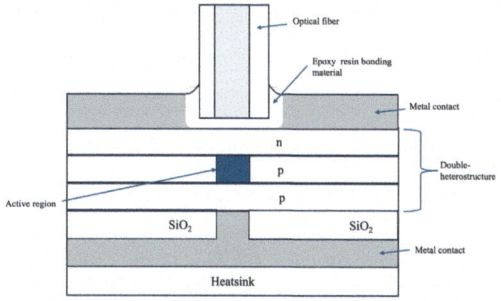

Fig. 12.11 Schematic of a typical structure of a surface-emitting LED. In this case, the output is coupled directly into an optical fiber

methods is used to achieve the best performance in terms of efficiency and light output.

12.3.5 LED Response Time

The response time of a light-emitting diode (LED) is a critical parameter in various applications, particularly in communication systems where it impacts the bandwidth of the system. The response time is typically defined as the time it takes for an LED to transition from on to off (or vice versa), which directly affects how quickly the LED can modulate light in response to electrical signals. The information can be transmitted as an analog or digital signal. The LED is essentially a forward-biased *pn-junction* diode in which minority charges are injected into an active recombination region. To be able to modulate the output of the device, it must be possible to modulate the injected carriers. A key issue in the device speed is the time taken to extract the charge. This time is controlled by the carrier recombination time.

The response time of an LED is determined primarily by two factors:

1. **Charge and Discharge of the Capacitance**: The depletion region of the diode acts as a capacitance. This inherent capacitance that needs to be charged or discharged when switching states. The time constant associated with this process is related to the RC (resistance-capacitance) time constant of the LED and its driving circuit.
2. **Recombination and Generation Rates**: The physical processes of electron-hole recombination and generation in the semiconductor material of an LED also affect the response time. The faster these carriers can recombine or be generated, the quicker the LED can respond.

The rise time of the LED is affected by the capacitance and resistance of junction. The bandwidth of an LED, in the context of data transmission, is influenced by its response time. The bandwidth determines the maximum rate at which data can be transmitted via the LED as a light source in optical communication systems. Mathematically, the bandwidth (B) of an LED can be approximated as:

$$B = \frac{1}{2\pi\tau} \text{ Hz}, \quad (12.52)$$

and

$$\frac{1}{\tau} = \frac{1}{\tau_r} + \frac{1}{\tau_{nr}}. \quad (12.53)$$

Typical rise time for LED is in the range of 1 to 50 ns, depending on the material, doping level and size.

12.3.6 LED Device Structures

Thus far, in our discussions, light-emitting diodes have been described as simple pn-junctions to facilitate an understanding of their fundamental operations and some key characteristics. However, these theoretical structures alone are not practical for manufacturing or suitable for the diverse applications for which LEDs are intended. Modern LEDs are fabricated via advanced semiconductor technologies that enable mass production. These devices are constructed on a stable semiconductor substrate where various specifically doped layers are deposited to form the active junctions and electrical contacts. The architecture of LEDs is meticulously designed to minimize optical and electrical losses, enhancing efficiency and performance. There are two principal device structures commonly used in the industry:

1. Surface-Emitting Light-Emitting Diodes ($SLED$): In this design, the semiconductor layers are stacked vertically. The photons generated in the active region travel perpendicularly to this region and are emitted directly from the surface of the device. This configuration is particularly advantageous for applications that require requiring high-intensity illumination and are readily compatible with fiber optic technology for effective light transmission.
2. Edge-Emitting Light-Emitting Diodes (ELEDs): In these devices, the photons propagate parallel to the plane of the active region before exiting the device at one of its edges. The design of the active region in ELEDs includes features that ensure optical confinement within a single-mode optical waveguide, optimizing the directional output of light which is crucial for applications such as laser diodes and high-speed fiber-optic communications.

12.3.6.1 Surface-Emitting LEDs

The construction of surface light-emitting diode is detailed in Fig. 12.11, which demonstrates the vertical stacking of semiconductor layers—essential for steering photon emissions directly from the surface. This structural configuration optimizes light output and is particularly beneficial in applications requiring precise control of light directionality and minimal spatial intrusion, such as in sensor technologies and certain display lighting systems. The specific device depicted in Fig. 12.11 uses a double heterojunction diode design.

The electrical contacts in SLEDs are designed so that the active region is confined to a specific area beneath the optical fiber contact. This targeted design limits photon emission primarily to this defined area. However, it is important to note that the light emitted can still display isotropic characteristics that photons may be emitted in various directions rather than being tightly confined. Despite this, the photons that do travel vertically are effectively captured and guided into the optical fiber, leveraging the fiber's numerical aperture to enhance efficient light coupling.

This precise control of the emission area combined with strategic fiber alignment enhances the device's efficiency in coupling light into fiber-optic systems, making SLEDs highly suitable for applications that require efficient, directed light transmission that is critical. These characteristics make SLEDs versatile for not only general illumination but also for advanced optical applications such as fiber-optic data transmission and high-resolution displays.

12.3.6.2 Edge-Emitting LEDs (ELEDs)

The typical design of an edge-emitting light-emitting diode (ELED) is depicted in Fig. 12.12. This configuration utilizes a double heterojunction diode, where the active region is restricted to a rectangular cross-sectional area along the device. Electrons are injected via electrical contacts and flow through this confined area. In this heterojunction structure, the active region is composed of a material with a smaller bandgap and a higher index of refraction than the surrounding materials, enhancing the optical confinement.

The dimensions of the active region are carefully selected to ensure that it functions as a single-mode optical waveguide. This design confines the emitted photons to travel longitudinally along the waveguide and emerge from both edges of the device. Such a configuration significantly reduces absorption losses, as photons do not travel through the bulk material but are instead guided within the active layer. Additionally, the edges of the $ELED$ are engineered to minimize reflection losses. The single-mode characteristic of the waveguide facilitates a smaller angle of incidence relative to the normal to the surface of the edges, avoiding total internal reflection (TIR) losses.

ELEDs are predominantly employed in optical communications, where their design allows for direct integration with optical fibers. Figure 12.13 shows an ELED packaged with an optical fiber pigtail, illustrating how these devices can be effectively coupled to fiber-optic systems for efficient light transmission. This capability makes ELEDs essential in high-speed, high-efficiency optical communication applications.

12.3.7 White Light-Emitting Diodes

White light covers a broad spectrum of radiation spanning the visible range from 400 to 700 nm. Light-emitting diodes (WLED) typically emit light within a narrow

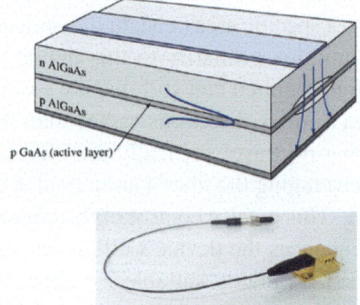

Fig. 12.12 Double heterojunction edge-emitting light-emitting diode (ELED)

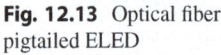

Fig. 12.13 Optical fiber pigtailed ELED

12.3 The LED Characteristics

band of approximately 30–50 nm around a specific wavelength (λ_p), which results in the emission of distinctly colored light such as red, blue, or green. Historically, LEDs were primarily used as indicators or for communication purposes. However, advancements in LED technology have now made them a predominant source of illumination. In display technology, the challenge of rendering light to appear white to the human eye was met by mixing the primary colors red, green, and blue (RGB). This method of color mixing allows for the production of white light, which is essential for a wide range of applications from general illumination to displays. Over the past few decades, while LEDs have continued to be used for indicators and communications, the emergence of white LEDs has shifted their role to become almost the main source of lighting. Here, how white LEDs work:

1. Phosphor-Coated Blue LED: The most common type of white LED uses a blue LED chip coated with a yellow phosphor. When the blue light from the LED chip strikes the yellow phosphor layer, part of the blue light is converted to yellow light. The combination of this yellow light with the remaining blue light results in light that appears white to the human eye. The specific shade of white can be adjusted by changing the composition of the phosphor and the intensity of the blue LED. This method is illustrated in Fig. 12.14a.
2. RGB LEDs: Another method involves the use of red, green, and blue (RGB) LEDs in a single package. By carefully controlling the output of each colored light, white light can be produced through color mixing. This method allows for adjustable color temperature and can produce a wide range of colors, including various shades of white (Table 12.2). This RGB approach is common in applications that require dynamic color control, such as in display technologies and smart lighting systems. This method is illustrated in Fig. 12.14b.

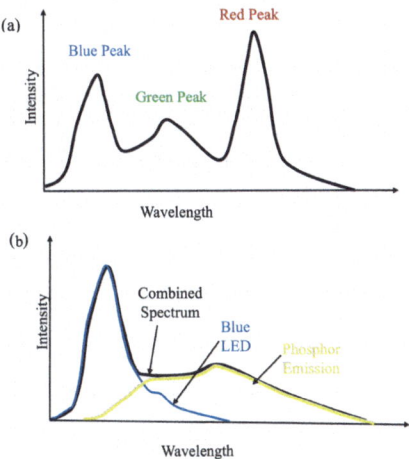

Fig. 12.14 The white light spectrum generated by (**a**) red, green, and blue LEDs and (**b**) blue LED and yellow phosphorus emulsions

Table 12.2 LEDs that emit RGB colors

Semiconductor	Emitted color	Wavelength
AlGaAs	Red	$610 < \lambda < 760$
AlgaN or GaN	Green	$500 < \lambda < 570$
InGaN	Blue	$460 < \lambda < 500$

12.4 Semiconductor Optical Amplifiers (SOAs)

Since the advent of optical fiber communication in the early 1980s, there has been a persistent need for signal regeneration in long-haul systems. As we discussed in Chap. 9, optical signals traveling through fibers suffer attenuation, which limits the effective range of a fiber link to approximately 50 km. Traditionally, to extend this range, the optical signal is converted to an electrical signal at repeater stations and then retransmitted on the next fiber link. This process, particularly in extensive and costly networks such as undersea links that can span up to 10,000 km, added considerable complexity and cost.

To address these challenges, the development of an optical alternative to electrooptic conversion at repeater stations has become a priority. Semiconductor optical amplifiers were developed to fulfill this role by directly amplifying optical signals. An *SOA* operates on the principle of stimulated emission to enhance incoming light. It comprises a semiconductor material that acts as the gain medium, Chap. 11. When direct current is applied, this semiconductor is "pumped" into a higher-energy state. As the incoming light traverses this energized medium, it stimulates the semiconductor atoms, which are already excited, to emit additional photons. These photons are coherent with the incoming light, thereby effectively amplifying the original signal. This advancement significantly streamlines the optical transmission process, reducing the need for costly and complex electrooptic conversions and supporting the expansion of global telecommunications infrastructure.

Figure 12.15 presents a schematic of an optical amplifier characterized by a gain (G), where the input optical power is P_i and the output optical power is P_o. For the output power to exceed the input power ($P_o > P_i$), an external energy source must compensate for this power deficit. In semiconductor optical amplifiers (SOAs), this energy is supplied by the pumping of charge carriers in a forward-biased pn-junction diode.

In Sect. 11.7, we discussed that under specific conditions where the Fermi inversion factor (f_g) equals 1, the semiconductor becomes a gain medium. The gain coefficient under these conditions is described by Eq. (11.78), rewritten here for clarity:

$$\gamma(\nu) = \frac{(c/n)^2}{8\pi \nu^2} \frac{1}{\hbar^2 \tau} (h\nu - E_g)^{1/2} [f_c(E_2) - f_v(E_1)]. \tag{12.54}$$

Figure 12.16 shows the energy-momentum ($E - k$) diagram for stimulated emission under conditions of quasi-Fermi equilibrium. This condition means that

12.4 Semiconductor Optical Amplifiers (SOAs)

Fig. 12.15 A schematic for an optical amplifier

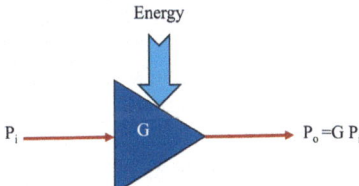

Fig. 12.16 The $E - k$ diagram for a gain medium satisfying the conditions for net stimulated emission

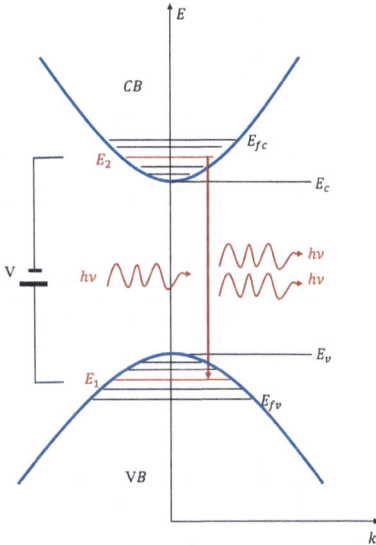

a photon interacting with the semiconductor results in the emission of two photons that share the same energy, direction of propagation, and polarization. The $E - k$ diagram demonstrates the conditions required for net emission in this medium, which are:

1. The energy states in the conduction band (CB) are filled up to the quasi-Fermi level E_{fc}, whereas energy states are available in the valence band (VB) from E_{fv} and E_v for electron transitions from the conduction band.
2. The energy levels E_2 and E_1 must reside within the conduction band and the valence band, respectively, such that $E_c < E_2 < E_{fc}$ and $E_{fv} < E_1 < E_v$. These specific transitions are the only ones that result in gains.
3. The energy a photon emits is $h\nu = E_2 - E_1$.

Therefore, the photon energy must satisfy the following condition:

$$E_g < h\nu < E_{fc} - E_{fv}, \tag{12.55}$$

This equation not only determines the operation but also sets the bandwidth of the amplifier.

In the operation of a semiconductor optical amplifier, an excess carrier concentration, Δn, is injected into the active region to facilitate electron-hole recombination. At low levels of Δn, absorption predominates over emission, resulting in the medium absorbing more light than it emits. However, as Δn increases, the medium transitions through a state of transparency and eventually becomes a gain medium when the carrier concentration surpasses a certain threshold. Thus, the gain coefficient γ is proportional to the carrier injection rate Δn and can be expressed as follows:

$$\gamma \propto \frac{\Delta n}{\Delta n_T} - 1 \tag{12.56}$$

where Δn_T is the transparency carrier concentration, i.e., the carrier concentration that causes the gain coefficient to be zero, so the medium is neither absorbing nor amplifying. The proportionality factor in this relationship depends on the material's inherent losses. The peak value of the gain coefficient is as follows:

$$\gamma_p = \alpha_a \left(\frac{\Delta n}{\Delta n_T} - 1 \right) = \alpha_a \left(\frac{J}{J_T} - 1 \right), \tag{12.57}$$

where α_a is the material absorption coefficient and J_T is the transparency current density. This relationship is depicted in Fig. 12.17a.

The gain coefficient, $\gamma(\nu)$, is directly proportional to the excess carrier concentration (Δn). Additionally, the Fermi levels E_{fc} and E_{fv} are functions of the carrier concentration, as described by the Joyce-Dixon formula in Eqs. (10.16) and (10.17). Consequently, the cutoff photon energy for the gain, represented by $E_{fc} - E_{fv}$, increases as a function of the excess carrier concentration. This relationship is depicted in Fig. 12.17b. As Δn increases, both the gain and the bandwidth of the amplifier also increase. Typically, the carrier concentration for GaAs lies within the range of 10^{18} cm^{-3}.

Fig. 12.17 (a) The optical amplifier peak gain coefficient as a function of excess carrier density and (b) the gain versus photon energy curves for a variety of carrier injections

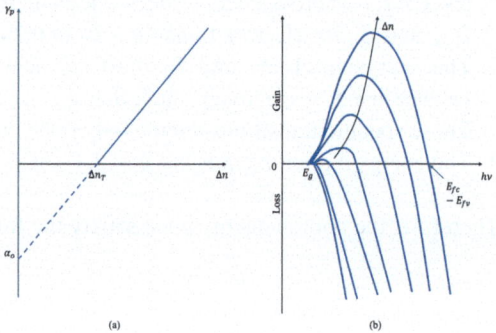

12.4 Semiconductor Optical Amplifiers (SOAs)

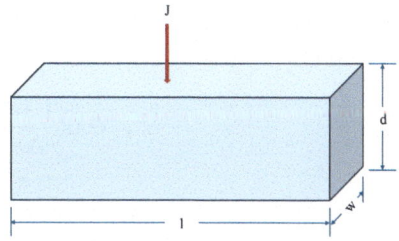

Fig. 12.18 The gain medium volume (active region)

Considering a device with dimensions d, l, and w as shown in Fig. 12.18, the excess carrier injection is calculated by:

$$\Delta n = \frac{\tau}{qdlw}, \qquad (12.58)$$

where τ is the carrier lifetime and q is the charge of an electron. The carrier density related to the injected current density J is given by:

$$\Delta n = \frac{\tau}{qd} J. \qquad (12.59)$$

Consequently,

$$\frac{\Delta n}{\Delta n_T} = \frac{J}{J_T} = \frac{i}{i_T}.$$

Therefore, the expression for the peak gain coefficient then becomes:

$$\gamma_p = \alpha_a \left(\frac{\tau}{qd \Delta n_T} J - 1 \right). \qquad (12.60)$$

This formulation allows for a clear understanding of how changes in current density affect the properties of the semiconductor optical amplifier.

Example In an InGaAsP SOA:

$$\Delta n_T \simeq 1.25 \times 10^{18}\,\text{cm}^{-3},$$

$$\alpha = 600\,\text{cm}^{-1},$$

and let the excess carriers $\Delta n = 1.4 \Delta n_T$ determine the peak gain coefficient and the total gain for an SOA with length $l = 350\,\mu\text{m}$.

(continued)

Solution

$$\gamma_p = 600(\frac{1.4\Delta n}{\Delta n_T} - 1) = 240 \text{ cm}^{-1}$$

The total gain

$$G = e^{\gamma_p l} = e^{240 \times 350 \times 10^{-4}} = 4447,$$

or

$$G = 10 \log 4447 = 36.5 \text{ dB}.$$

Pumping of SOA In a semiconductor optical amplifier, pumping can be achieved by electron-hole injection through a forward-biased pn-junction. Let the volume of the active region be $ldw = lA$, where A is the surface area of the active region and l is its length. Therefore, the pumping rate is given by:

$$R = \frac{i}{qlA} \; S^{-1} \text{ cm}^{-3}, \tag{12.61}$$

similarly,

$$R = \frac{J}{ql} \tag{12.62}$$

where J is the current density in A/cm^2 and excess carrier concentration is:

$$\Delta n = \tau R = \frac{\tau}{qlA}i = \frac{\tau}{ql}J. \tag{12.63}$$

Therefore, the peak gain is:

$$\gamma_p \simeq \alpha(\frac{J}{J_T} - 1), \tag{12.64}$$

where

$$J_T = \frac{ql}{\eta_i \tau_r}\Delta n_T, \tag{12.65}$$

and since the internal quantum efficiency is:

$$\eta_i = \frac{\tau}{\tau_r}. \tag{12.66}$$

12.4 Semiconductor Optical Amplifiers (SOAs)

Therefore, the transparency current density is given by:

$$J_T = \frac{ql}{\tau}\Delta n_T. \tag{12.67}$$

Example In a $GaAsP$ SOA, the following parameters are given:

$$\tau_r = 2.5\,\text{ns},$$

$$\eta_i = 0.5,$$

$$\Delta n_T = 1.25 \times 10^8, \alpha_o = 600\,\text{cm}^{-1},$$

and

$$d = 2\,\mu\text{m}, l = 200\,\mu\text{m}, w = 10\,\mu\text{m}.$$

Determine the peak gain coefficient and the total gain if $i = 700\,\text{mA}$.

Solution

$$J_T = \frac{1.626 \times 10^{-19} \times 2 \times 10^{-4}}{0.5 \times 2.5 \times 10^{-9}} \times 1.25 \times 10^{18} = 3.2 \times 10^4\,\text{A/cm}^2$$

Since $i = 700\,\text{mA}$, thus,

$$J = \frac{700\,\text{mA}}{200 \times 10\,\mu\text{m}^2},$$

therefore,

$$J = 3.5 \times 10^4\,\text{A/cm}^2.$$

Thus, the peak gain coefficient is given by:

$$\gamma_p = 600 \left(\frac{3.5 \times 10^4}{3.2 \times 10^4} - 1 \right) = 56.25\,\text{cm}^{-1},$$

and the peak gain of the SOA is:

$$G = e^{\gamma_p l} = e^{56.25 \times 200 \times 10^{-4}} \simeq 3.$$

Example An optical amplifier made of $In_{0.7}G_{0.3}aAs_{0.6}P_{0.4}$ has the following parameters:

$$\eta_i = \frac{\tau}{\tau_r}, \eta_i = 0.5, \tau_r = 2.0\,\text{ns}, \Delta n_T = 1.2 \times 10^{18}\,\text{cm}^{-3}$$

$$\alpha = 600\,\text{cm}^{-1}, q = 1.6 \times 10^{-19}\,\text{C}, 1 = 300\,\mu\text{m}, \text{ and } w = 200\,\mu$$

1. Determine the transparency current for this laser diode.
2. Determine the transparency current by changing the sizes of the active region such that $d = 0.2\,\mu\text{m}$, and $w = 10\,\mu$

Solution

$$J_T = \frac{qd}{\eta_i \tau_r} \Delta n_T$$

$$J_T = 40\,\text{kA/cm}^2$$

$$i_T = J_T \times \omega \times l = J_T \times 300 \times 200 \times 10^{-8}\,\text{A}$$

$$i_T = 24\,\text{A}$$

This is extremely high. Impractical
(2) The threshold current becomes $i_T = 120\,\text{mA}$.

The previous example demonstrates that for homojunction laser diode in order to reach transparency conditions for the active region, the injected current needs to be 24 A, which is impossible to run the LD on this DC current; otherwise the device will melt because of the heat caused by this large current. This led to the development of the heterojunction laser diodes which require a small transparency current i_T. Double heterostructure laser diodes provide many advantages including (1) carrier confinement, (2) optical confinement, (3) low loss, and (4) bandgap flexibility.

A typical double heterojunction of a semiconductor optical amplifier is illustrated in Fig. 12.19. This device consists of a heterojunction pn-diode comprising a *GaAs* layer sandwiched between AlGaAs layers. The refractive index of the *GaAs* layer is greater than that of the surrounding *AlGaAs* layers, which forms an optical waveguide, which provides optical power confinement, as detailed in Chap. 7.

In the active region of the SOA, the optical field profile is mostly confined to the waveguide, although its tails extend beyond the active region boundaries. As a result, only the portion of the optical field within the active region experiences

12.4 Semiconductor Optical Amplifiers (SOAs)

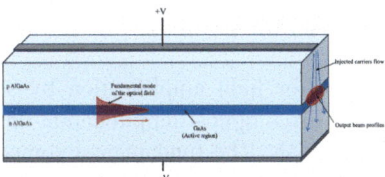

Fig. 12.19 Semiconductor optical amplifier structure for a heterojunction pn-diode illustrating the optical confinement within the fundamental mode of the optical field

Fig. 12.20 Packaged semiconductor optical amplifier illustrating the electric contacts and input and output optical fibers, made by InPhenix Inc.

amplification. The extent of this confinement is quantified by the confinement factor, Γ, which depends on the height (d) of the active region and the refractive indices of the materials involved.

The end facets of the SOA are coated with antireflection coatings to minimize reflections at the input and output ports. In practical applications, these facets are often directly butt-coupled to single-mode optical fibers to facilitate efficient light transfer. Figure 12.20 shows a packaged SOA, demonstrating how these components are integrated into a functional unit ready for deployment in optical systems.

Semiconductor optical amplifiers can be utilized in various applications where amplification of optical signals is a critical system design requirement. In optical communication systems, SOAs can extend the reach of optical links and also function as preamplifiers in optical receiver circuits. However, incorporating SOAs into wavelength division multiplexing (WDM) systems presents several challenges:

1. Nonuniform Gain Profile: The gain of an SOA is not consistent across its operational bandwidth. This nonuniformity can lead to differential amplification for (WDM) channels, resulting in signal imbalance and increased cross talk between channels.
2. Limited Bandwidth: Typically, spanning only approximately 30 nm, the bandwidth of an SOA is insufficient to cover the entire C-band, which ranges from 1530 nm to 1565 nm. This limitation prevents uniform amplification across all channels in the C-band. These issues necessitate careful consideration and adjustment when integrating SOAs into WDM systems.

These issues will be addressed in more detail later in the optical communications system chapter.

12.5 Semiconductor Lasers Diodes

The acronym LASER stands for light amplification by stimulated emission of radiation, a principle that involves optical amplification on the basis of quantum mechanical properties of materials. The fundamental component at the heart of a laser diode (LD) is a semiconductor optical amplifier, which functions as the gain medium.

To generate an optical signal, however, the laser requires additional components that extend beyond the SOA. Analogous to an electronic oscillator, which typically consists of an amplifier combined with a feedback mechanism and a resonant circuit (such as an RC or RLC circuit), a laser diode also comprises three essential elements:

1. **Gain Medium:** This is the active region where light amplification occurs, provided by the semiconductor material in the SOA. When excited by external energy, this medium amplifies light through the stimulated emission of photons.
2. **Pumping Source**: This component supplies energy to the gain medium to excite the electrons within the semiconductor. In semiconductor lasers, pumping is typically electrical, where current is injected into the device to stimulate the gain medium.
3. **Optical Feedback**: Necessary for sustained laser operation, this is achieved through an optical resonator. For a semiconductor laser using an SOA, the resonator is typically formed by cleaving the semiconductor crystal to create two parallel, highly reflective surfaces at either end of the gain medium. These surfaces act as mirrors, forming what is known as a Fabry-Perot resonator, previously detailed in Chap. 3. This resonator structure enables the light to oscillate back and forth through the gain medium, resulting in repeated amplification.

In edge-emitting laser diodes, as will be discussed later in this section, the cleaved edges of the semiconductor chip form these parallel mirrors, effectively creating a cavity that supports standing waves of light. This arrangement allows for efficient light amplification and the creation of a coherent light output, which is essential for applications ranging from fiber-optic communication to barcode scanning.

By integrating these components, semiconductor laser diodes efficiently convert electrical energy into a coherent light output, leveraging the stimulated emission processes within the semiconductor gain medium facilitated by the SOA.

In a semiconductor optical amplifier, photons are multiplied via stimulated emission. In standard SOA configurations, photons are confined in the dimensions transverse to the waveguide but are allowed to escape from the ends of the waveguide. Now, we shift our focus to optical cavities where photons are confined in all three dimensions and retained within the cavity for extended periods. This confinement allows photons to multiply through stimulated emission, resulting in a significant accumulation of photons inside the cavity. Typically, SOAs have an antireflective coating at these facets to minimize reflections. By altering these

12.5 Semiconductor Lasers Diodes

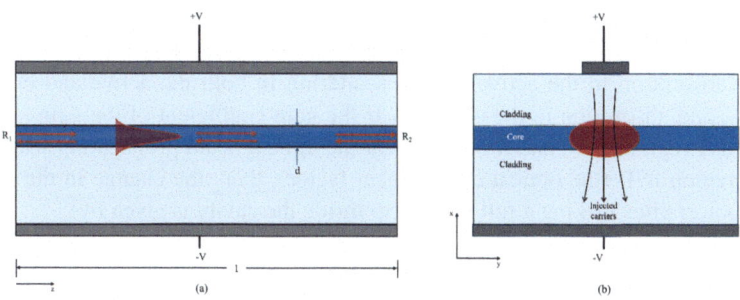

Fig. 12.21 Schematic of a cross section of a double heterostructure semiconductor optical amplifier

coatings to high-reflectivity (HR) optical coating, the resulting optical cavity formed is known as a Fabry-Perot cavity, as detailed in Chap. 5.

12.5.1 Laser Dynamics

Consider a cross section along an SOA with highly reflective facets, as depicted in Fig. 12.21. The reflective properties of the two facets transform the device into an optical cavity. A forward-biased voltage facilitates the injection of electrons and holes into the device, creating a quasi-Fermi equilibrium. Initially, the injected carriers lead to spontaneous emission of photons, but owing to continuous carrier injection, these photons stimulate further emission in the active region. As these photons traverse the active region, their numbers increase because of the gain provided by the active medium.

The active region, composed of $GaAs$, has a higher refractive index than the surrounding $AlGaAs$ layers. This difference in refractive indices creates an optical waveguide, with the $GaAs$ serving as the core and the $AlGaAs$ serving as the cladding. Typically, the thickness of the active region (d) is approximately 0.1 μm, ensuring that the waveguide remains single-mode along the x-$axis$ and thus enhances the optical confinement. The device is designed such that the upper electrode is just a few micrometers wide, confining carrier injection and limiting the extent of the active region along the y-$axis$. As the optical field propagates in the active region, the fundamental mode remains mostly within the active region, but some extends outside. Only the optical field inside the active region will experience significant gain.

The optical cavity introduces losses both in the active region and at the reflective surfaces. These losses (α_r) were derived in Chap. 6 and are given by Eq. (6.81), rewritten here as:

$$\alpha_r = \alpha_s + \frac{1}{2l} \ln\left(\frac{1}{R_1 R_2}\right),$$

where R_1 and R_2 are the reflectances of both facets of the cavity and where α_s is the material absorption coefficient. These losses in semiconductor lasers are free carrier absorption in the active region, scattering in both the active and cladding regions, and absorption in the cladding. If the gain coefficient of the gain medium (the active region) is γ, and the fraction of the optical beam propagating within the active region is Γ (the optical confinement factor); then, the change in the optical beam power after making a full round trip inside the cavity is given by:

$$\Delta P = R_1 R_2 e^{(\Gamma \gamma - \alpha_r)2l}. \tag{12.68}$$

When γ is relatively small, such that:

$$R_1 R_2 e^{(\Gamma \gamma - \alpha_r)2l} \ll 1, \tag{12.69}$$

then, any photons generated by spontaneous emission in the cavity will eventually escape or be absorbed. This occurs through one of the following mechanisms: transmission out of the cavity through one of the facets, absorption within the cavity due to material losses (α_s), or attenuation caused by modal losses (Γ).

However, if the gain γ is increased sufficiently such that $R_1 R_2 e^{(\Gamma \gamma - \alpha)2l}$ approaches unity, a critical condition emerges. At this point, the photon losses due to cavity facet transmission and waveguide absorption are exactly compensated by the gain in photon count from stimulated emission after each roundtrip within the cavity. This condition is expressed as:

$$R_1 R_2 e^{(\Gamma \gamma - \alpha_r)2l} = 1. \tag{12.70}$$

This is the **lasing condition** required for continuous laser operation. When this condition is satisfied, photons originating from spontaneous emission can build up significantly inside the cavity. This buildup marks the transition to laser operation, where light amplification occurs through the process of stimulated emission of radiation.

This mechanism underpins the fundamental operation of lasers and highlights how changes to the structure and materials in semiconductor optical amplifiers (SOAs) can enhance their performance and lead to the development of highly efficient laser systems.

The gain coefficient of the active medium $\gamma(\nu)$, Eq. (11.52), is given by:

$$\gamma(\nu) = D_o(h\nu - Eq)^{1/2} e^{-(h\nu - Eg)/k_B T}. \tag{12.71}$$

If the gain exceeds the losses, gain saturation occurs, which moderates the gain to maintain the balance:

$$\Gamma \gamma = \alpha_r,$$

12.5 Semiconductor Lasers Diodes

or explicitly,

$$\Gamma D_o(h\nu - Eq)^{1/2} e^{-(h\nu-Eg)/k_B T} = \alpha_r = \alpha_s + \frac{1}{2l}\ln\left(\frac{1}{R_1 R_2}\right), \quad (12.72)$$

where Γ is the confinement factor, α_s represents material absorption losses, l l is the cavity length, and R_1 and R_2 are the reflectivities of the cavity facets.

This balance ensures a steady-state operation of the laser diode, where the photon generation through stimulated emission compensates for losses, resulting in continuous laser output.

Example Calculate the total losses for a $GaAs$ cavity given the following parameters:
$\alpha_s \sim 20\,\text{cm}^{-1}, l = 300\,\mu\text{m}\; n = 3.6$.

Solution Assuming both facets were cleaved to provide a reflective surface then

$$R_1 = R_2 = \left(\frac{n-1}{n+1}\right)^2 = \left(\frac{3.6-1}{3.6+1}\right)^2 \cong 0.32,$$

therefore,

$$\alpha_n = \frac{1}{2l}\ln\left(\frac{1}{R_1 R_2}\right) = \frac{1}{2\times 300}\ln\left(\frac{1}{032\times 032}\right) \simeq 38\,\text{cm}^{-1}$$

$$\alpha_r \cong 58\,\text{cm}^{-1}.$$

12.5.2 Gain Threshold

The steady-state operation of the laser diode requires the gain to be equal to the total loss. This gain is the threshold gain γ_{th}. When the reflections from both sides of the cavity is the same, i.e., $R_1 = R_2 = R$, the threshold gain condition can be expressed as:

$$\Gamma\gamma_{th}(\nu) = \alpha_s - \frac{ln(R)}{l}. \quad (12.73)$$

This is different from the transparency condition when the cavity suffers no gain nor los, i.e.,

$$\Gamma\gamma_T = 0. \quad (12.74)$$

Fig. 12.22 The cavity gain is plotted as a function of frequency below and above the threshold

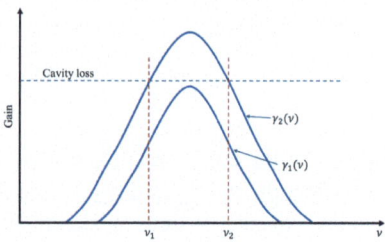

In Fig. 12.22 the gain is plotted as a function of frequency (v). We consider two cases: the gain $\gamma_1(v)$ is just below the cavity loss so the lasing condition is not satisfied and the photons emitted are lost in the cavity. When the gain is increased ($\gamma_2(v)$), the gain exceeds the loss for all frequencies between v_a and v_2. In this range of frequencies the lasing condition is satisfied, and the laser emits a steady optical beam outside the cavity.

12.5.3 Laser Diode Longitudinal Modes

The reflective end facets of the laser diode's active region act as an optical resonator. The resonant frequencies of this cavity are determined by ensuring that the phase shift for a wave completing a full round trip equals multiples of 2π. Mathematically, this is expressed as:

$$k_o n \times 2l = q2\pi, \tag{12.75}$$

where k_o is the wavenumber in free space, n is the refractive index of the active region, and l is the cavity length.

Consequently, the resonant frequencies are given by:

$$v_q = q \frac{c}{2nl}. \tag{12.76}$$

The free spectral range (FSR), which is the frequency difference between two adjacent longitudinal modes, is:

$$v_F = \frac{c}{2nl}. \tag{12.77}$$

These resonant frequencies define the longitudinal modes of the laser diode. These modes are separated by the free spectral range and are the only frequencies that can be sustained within the active region. The laser diode achieves lasing only in the range where the gain exceeds the losses, specifically between v_1 and v_2, as illustrated in Fig. 12.22. The laser emits light at the resonant frequencies within this range and with amplitudes corresponds to the gain at each mode.

12.5 Semiconductor Lasers Diodes

The spectrum of a Fabry-Perot cavity, as discussed in Chap. 4, is described by the equation:

$$\frac{I_t}{I_i} = \frac{1}{1 + F \, Sin^2(\phi/2)}. \tag{12.78}$$

where F is the finesse, which is given by:

$$F = \frac{4R}{(1-R)^2}. \tag{12.79}$$

Here, R is the reflectivity of the cavity facets, I_t is the transmitted intensity, I_i is the incident intensity, and ϕ is the phase difference between successive reflections.

Figure 12.23 demonstrates how the output beam profile of the laser diode is influenced by the gain spectrum and the resonant modes of the cavity. We consider two scenarios: (1) assume the reflectivity of the cavity facets to be $R = 0.9$, as shown in Fig. 12.23b, and (2) assume $R = 0.32$, as shown Fig. 12.23e. The net gain of the cavity above the threshold is depicted in Fig. 12.23a, d. The output optical power spectra of the laser diode for these cases are presented in Fig. 12.23c, f, respectively. In the high reflectivity scenario ($R = 0.9$), the longitudinal modes are very sharp, resulting in distinctly distinguishable modes in the laser output. Conversely, in the low reflectivity case for the GaAs cavity ($R = 0.32$), the modes are not as sharply separated. Achieving high reflectivity in a laser diode typically requires the end facets to be coated with a highly reflective material or structures as

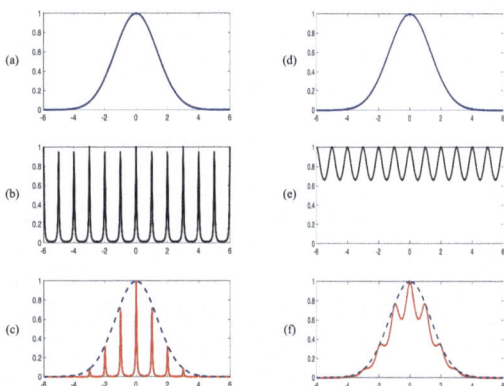

Fig. 12.23 The plots illustrate how the output optical power of a laser diode is influenced by the net gain above the threshold condition combined with the response of the cavity modes. Panels (**a**) and (**d**) show the gain spectrum of the active region of the LD. Panel (**b**) depicts the Fabry-Perot longitudinal modes for a reflectivity $R = 0.9$, whereas panel (**e**) presents these modes for $R = 0.32$. The corresponding output spectra of the laser diode are represented by the red line in panels (**c**) and (**f**) for $R = 0.9$ and $R = 0.32$, respectively. The $x\text{-}axis$ scale is arbitrary, and the amplitudes of the curves are normalized

will be discussed later which will be another level of complexity in the fabrication of the LD. Laser diodes that operate a simple cavity as above are referred to as Fabry-Perot laser diodes (FPLDs).

12.5.4 Laser Diode Output characteristics

The laser diode function converts electrical signals into optical signals to be used, for example, in fiber communication systems. Therefore, the relationship between the injected current (i) and output optical power (P_{opt}) is a major characteristic of the laser diode.

We have thus far discussed the transparency current density (J_T), which is the current density that makes the gain coefficient to be equal to zero. Additionally, the condition for lasing is that the gain should equal the total loss of the cavity, i.e.,

$$\gamma_{pth} = \alpha_r, \tag{12.80}$$

where the γ_{pt} threshold peak gain that satisfies the lasing condition. Additionally, we can rewrite the peak gain coefficient, Eq. (12.57), for the threshold condition using Δn_{th} as the threshold carrier density; then,

$$\gamma_{pth} = \alpha_a \left(\frac{\Delta n_{th}}{\Delta n_T} - 1 \right) = \alpha_a \left(\frac{J_{th}}{J_T} - 1 \right) = \alpha_r, \tag{12.81}$$

Consequently, the threshold current density is given by:

$$J_{th} = J_T \left(1 + \frac{\alpha_r}{\alpha_a} \right). \tag{12.82}$$

Similarly, the threshold current is given by:

$$i_{th} = i_T \left(1 + \frac{\alpha_r}{\alpha_a} \right). \tag{12.83}$$

It is clear that the threshold current is larger than the transparency current, typically $J_{th} \cong 1.1 J_T$. Importantly, the transparency current density is proportional to the active region thickness (d); hence, reducing d results in a smaller J_T.

If we include the confinement factor (Γ) of the optical waveguide into the threshold current expression, we obtain:

$$i_{th} = i_T \left(1 + \frac{\alpha_r}{\Gamma \alpha_a} \right). \tag{12.84}$$

12.5.5 Double Heterostructure Optical Confinement

Double heterostructure optical confinement is a method used in semiconductor lasers and light-emitting diodes (LEDs) to increase their efficiency and performance by confining both the charge carriers (electrons and holes) and the light within a specific region of the device. This is achieved by using a layered structure of different semiconductor materials, each with distinct bandgaps, forming two heterojunctions.

Layers with larger bandgaps act as barriers for the charge carriers. Electrons and holes are injected into the active region and are confined there because they cannot easily pass into the surrounding layers. This increases the likelihood of recombination (when electrons and holes combine to emit light) within the active region. The difference in refractive indices between the active region and the surrounding layers creates an optical waveguide. This confines the light within the active region, reducing optical losses and improving the intensity and directionality of the emitted light.

The optical confinement factor Γ is inversely proportional to the size of the waveguide width (w), whereas the transparency current density is proportional to the thickness (d). Therefore, there is an optimum thickness that results in minimizing J_T. The optimum thickness value is usually between 0.1 and 0.2 μm.

Example An $AlGaAs - GaAs - AlGaAs$ laser diode with the following parameters:
$\alpha_r = 60\,\text{cm}^{-1}$, $\alpha_a = 600\,\text{cm}^{-1}$, and $\Gamma = 0.9$.
Calculate the threshold current in terms of transparency current.

Solution Given

$$i_{th} = i_T \left(1 + \frac{\alpha_r}{\Gamma \alpha_a}\right)$$

Therefore,

$$i_{th} = i_T \left(1 + \frac{60}{0.9 * 600}\right)$$

and

$$i_t = 1.1 i_T$$

Consider the optical power output as a function of the injected current density in a laser diode, as in Fig. 12.24. If we compare this with the output from an LED

Fig. 12.24 The optical power output as a function of current injection in a semiconductor laser. Above the threshold, the presence of a high photon density causes stimulated emission to dominate

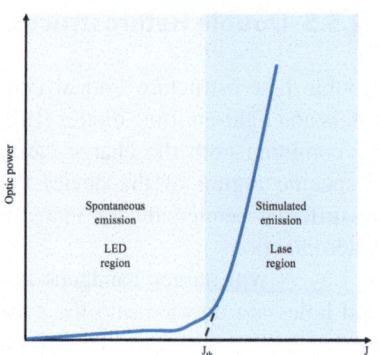

shown in Fig. 12.16, we notice an important difference. The optical power output from a laser diode displays a rather abrupt change in behavior below the "threshold" condition and above this condition. The threshold condition is usually defined as the condition where the cavity gain overcomes the cavity loss for any photon energy, i.e., in the range v_1 and v_2 in Fig. 12.22

When the pn-junction diode making up the semiconductor laser is forward biased, electrons and holes are injected into the active region of the laser. These electrons and holes recombine to emit photons. It is important to identify two distinct regions of operation of the laser. When the forward bias current is small, the number of electrons and holes injected is small. As a result, the gain in the device is too small to overcome the cavity loss. The photons that are emitted are either absorbed in the cavity or lost to the outside. Thus, in this regime, there is no buildup of photons in the cavity. However, as the forward bias increases, more carriers are injected into the device until the threshold condition is eventually satisfied for some photon energy. As a result, the number of photons starts to build-up in the cavity. As the device is further biased beyond the threshold, stimulated emission starts to occur and dominates the spontaneous emission. The light output in the photon mode for which the threshold condition is satisfied becomes very strong. Below the threshold, the device essentially operates as an LED except that there is a higher cavity loss in the laser diode since photons cannot escape from the device due to the mirrors.

Laser diode performance is very sensitive to temperature because of the semiconductor properties. Figure 12.25 illustrates the input-output characteristics for a range of temperatures. The threshold currents clearly increase with temperature. The relationship between the threshold current (i_{th}) and the temperature is given by:

$$i_{th} = i_o e^{T/T_o}, \tag{12.85}$$

where i_o is the characteristic current, T is the temperature in Kelvin, and T_o is the characteristic temperature. For an $AlGaAs/GaAs LD$, $T_o = 140$ K.

12.5 Semiconductor Lasers Diodes

Fig. 12.25 Laser diode threshold current dependence on temperature where $T_3 > T_2 > T_1$

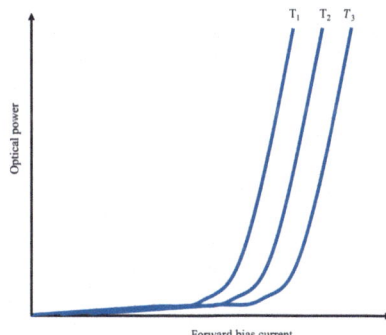

Fig. 12.26 Angular profile of the laser diode beam

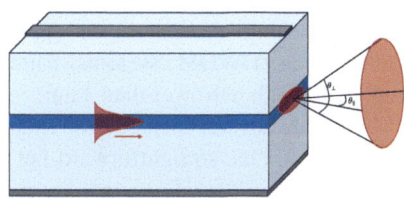

12.5.6 Angular Profile of the Beam

The optical beam emitted from the laser diode is influenced by the shape of the optical mode propagating along the active medium and the dimensions of the active region at the end facet. As depicted in Fig. 12.26, the beam diffracts non-uniformly, forming a cone of light with an elliptical cross section. The beam's vertical and horizontal spread is characterized by the angles where:

$$\theta_\perp > \theta_\parallel. \tag{12.86}$$

Typically, $\theta_\perp \cong 30°$, and $\theta_\parallel \cong 10°$.

12.5.7 Single-Frequency Laser Diode

Figure 12.23c illustrates the emission spectrum of a laser diode, which includes multiple operating frequencies. These frequencies correspond to the resonances dictated by the cavity length l, as defined by Eq. (12.75), rewritten here as:

$$v_q = q \frac{c}{2nl}, \tag{12.87}$$

where q is the mode number, c is the speed of light in free space, n is the refractive index, and l is the cavity length. The separation between these frequencies, known as the free spectral range, is given by Eq. (12.76), rewritten here as:

$$\nu_F = \frac{c}{2nl}. \quad (12.88)$$

For example, for a GaAs cavity with $l = 300\,\mu\text{m}$ and $n = 3.6$, the free spectral range is:

$$\nu_F = \frac{3 \times 10^8}{2 \times 3.6 \times 300 \times 10^{-6}} = 1.4 \times 10^{11}\,\text{Hz},$$

which is significantly smaller than the typical bandwidth of the emitted spectrum, potentially reaching 10^{13} Hz. Consequently, hundreds of resonant modes can be emitted, posing challenges for applications such as dense wavelength-division multiplexing (DWDM) systems, which require lasers with single-mode operation and a linewidth narrower than 1 nm.

There are several methods for enabling laser diodes (LDs) to operate as single-mode lasers. One straightforward but generally impractical approach is to reduce the cavity length (l). For example, reducing l to $3\,\mu\text{m}$ increases the free spectral range ν_F to 1.4×10^{13} Hz. This configuration is likely to support only one mode within the gain bandwidth. However, the short cavity length leads to such a small gain that it cannot surpass the inherent losses, making stable single-mode operation unfeasible. To address this challenge, alternative methods have been developed to facilitate single-mode operation in laser diodes, including the following methods:

1. External Cavity Laser (ECL),
2. Distributed Feedback (DFB) Laser,
3. Distributed Bragg Reflector (DBR) Laser.

These approaches involve implementing a mechanism that significantly increases the cavity's loss at specific, narrow-frequency ranges, thereby enhancing the amplification of a single resonant mode. This concept is schematically represented in Fig. 12.27.

Fig. 12.27 A single-mode selection in which the gain is less than the losses except at one resonant mode.

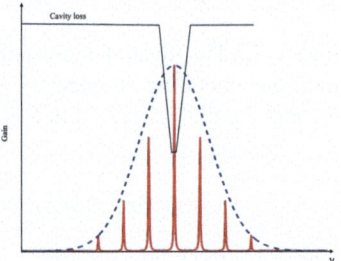

Fig. 12.28 External cavity laser utilizing an Fabry-Perot etalon

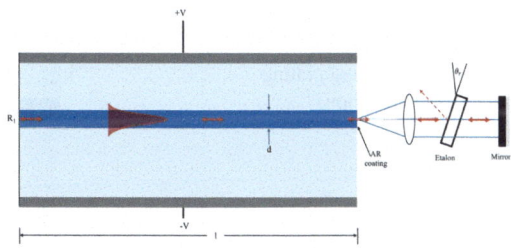

12.5.7.1 External Cavity Laser (ECL)

An external cavity laser (ECL) integrates two Fabry-Perot cavities with distinct thicknesses and is designed to match a single resonant frequency across an extensive free spectral range. This section explores one method for achieving such mode selection. Figure 12.28 depicts this configuration: the laser cavity's end facet is coated with an antireflective (AR) layer, which allows light to exit the laser without reflecting back into the active region. The emitted light first encounters a collimating lens, then travels through a tilted Fabry-Perot etalon, and finally reflects off a mirror. In the absence of the etalon, the laser cavity extends from the active region's left side facet to the mirror, forming a very long cavity that supports numerous modes due to the available gain.

Introduced in Chap. 6, the etalon consists of two highly reflective surfaces separated by a narrow gap with thickness t. Due to this small gap, the etalon has a large free spectral range and a narrow linewidth in its transmission. As illustrated in Fig. 12.28, the tilted etalon selectively allows one specific frequency of the light generated by the laser to pass through and similarly controls the reflection from the mirror. This selected frequency is the only one that resonates within the laser cavity, and thus, it predominates in the laser's output, emerging from the left facet of the cavity. Adjusting the angle θ_r enables the selection of different resonant modes, allowing the laser to emit various single frequencies. The purpose of the tilted etalon in an external cavity laser is to selectively filter out all the resonant modes except one. It reflects unwanted modes and allows only the desired mode to pass through the etalon in both directions, resulting in a single-mode output from the laser.

However, this configuration poses practical challenges for integration with other photonic components, making it less desirable for compact and integrated photonic systems.

12.5.7.2 Distributed Feedback (DFB) Laser Diode

Because of the broad gain bandwidth of diode lasers, it is necessary to incorporate a frequency-selective element in the cavity to achieve single-wavelength operation under a wide range of operating conditions. The method of choice is to use a Bragg reflecting grating, which reflects only in a relatively narrow wavelength band, as the feedback element. The two configurations in which a Bragg reflector is typically used are the distributed Bragg reflector (DBR) laser, in which the grating is external to the gain region, and the distributed feedback (DFB) laser, in which the grating

Fig. 12.29 Schematic of a distributed feedback (DFB) laser utilizing a Bragg grating along the active regions

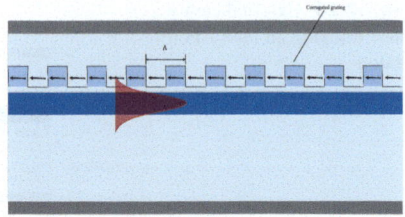

is formed in the gain region. Although single-line emission has been achieved with both DBR and DFB lasers, it was decided to concentrate on the DFB structure, mainly because it seemed to yield the best performance on the basis of published information. The basic idea of the distributed feedback (DFB) laser is illustrated in Fig. 12.29 which shows the usual Fabry-Perot type of feedback (reflection at the cleaved laser ends) and grating feedback (the return of radiation by scattering from the grating protrusions). This back-scattering process is distributed along the cavity (which explains the origin of the name). The relationship between the grating period Λ and the wavelength at which the light is reflected by the grating is given by the expression:

$$\Lambda = m \frac{\lambda_o}{2n_{eff}}, \qquad (12.89)$$

where n_{eff} is the effective refractive index of the waveguide, λ_o is the vacuum wavelength, and m is an integer denoting the order of the grating. As the wave propagates in the active region, the tail of the modes (the evanescent wave) is being reflected at the grating.

In a distributed feedback laser, the grating period (Λ), which refers to the spacing between the periodic structures within the laser cavity, plays a crucial role in determining the emitted wavelength of the laser. DFB lasers utilize a wavelength-dependent reflector to emit light at a single wavelength, achieved through a grating structure that causes constructive interference for a specific wavelength, resulting in feedback and lasing at that wavelength. Any changes in the grating period lead to corresponding changes in the reflected wavelength. Specifically, decreasing the grating period results in a shorter reflected wavelength, whereas increasing the grating period leads to a longer reflected wavelength. Hence, the laser can be tuned to emit light at different wavelengths within its gain spectrum by adjusting the grating period. This tunability enables precise control over the emitted wavelength, making DFB lasers highly suitable for various applications in telecommunications and optical sensing.

12.5.7.3 Distributed Bragg Reflector (DBR) Laser Diodes

Distributed Bragg reflector (DBR) laser diodes use a Bragg grating to achieve single-mode operation, similar to DFB lasers, but with the grating located outside the gain medium, acting like the two mirrors of a resonator. This configuration is

12.6 Vertical Cavity Surface-Emitting Lasers (VCSEL)

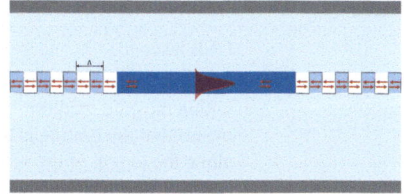

Fig. 12.30 Schematic of a distributed Bragg reflector (DBR) laser diode

illustrated in Fig. 12.30. The period of the Bragg reflectors is designed to allow only one wavelength to reflect back and forth in the active region, where it is amplified and then emitted by the laser diode. The properties of the Bragg grating period are similar to those described for the DFB laser, determining the specific wavelength that is reflected and ultimately emitted.

Table 12.3 provides a general comparison between the DFB and DBR lasers.

The Bragg grating for both DFB and DBR lasers can vary in form and placement. For simplicity, we have illustrated a square-wave grating; however, in many practical designs, a sinusoidal grating is used. The sinusoidal grating offers better wavelength selectivity and effectively suppresses the side modes of the laser cavity.

12.6 Vertical Cavity Surface-Emitting Lasers (VCSEL)

Vertical cavity surface-emitting lasers (VCSELs) represent a distinct class of semiconductor lasers that emit light perpendicular to the surface of the fabricated wafer, in contrast to edge-emitting lasers that emit light parallel to the wafer surface. This unique emission direction is a result of the VCSEL structure, which consists of a vertically oriented cavity formed by two distributed Bragg reflectors (DBRs) on either side of an active region. VCSELs offer several advantages over traditional edge-emitting lasers, including the following:

1. Efficient Manufacturing: VCSELs can be tested and characterized at the wafer level before they are separated into individual devices, significantly reducing production costs and improving yield.
2. Low Threshold Currents: Due to their small active region and efficient cavity design, VCSELs typically exhibit low threshold currents, which contribute to lower power consumption and heat generation.
3. High Modulation Speeds: VCSELs are capable of high-speed modulation, making them suitable for high-bandwidth communication systems, such as data centers and short-reach optical interconnects.
4. Scalability: VCSEL arrays can be easily fabricated, enabling high-density integration for applications requiring multiple light sources, such as imaging and sensing.

The basic structure of a VCSEL includes an active region sandwiched between two DBRs, which provide high reflectivity and form the optical cavity. The active region

Table 12.3 Comparison of distributed feedback and distributed Bragg reflector lasers

	DFB laser	DBR laser
Grating location	The Bragg grating is integrated within the gain medium. This means that the grating is distributed along the length of the active region, providing feedback directly within the region where amplification occurs	The Bragg grating is located outside the gain medium. The grating acts like the mirrors of a resonator, reflecting the specific wavelength back into the active region where it is amplified
Mode selection	The integrated grating in DFB lasers allows for very precise mode selection within the gain medium. The feedback mechanism is inherently distributed, which can lead to more stable single-mode operation	DBR lasers rely on external gratings for feedback, which can still effectively select a single mode but may be slightly less stable compared to the distributed feedback in DFB lasers due to the separation of the grating and gain regions
Complexity and fabrication	The fabrication of DFB lasers is generally more complex because the grating needs to be accurately integrated within the gain medium. This can lead to higher manufacturing costs and more stringent process control requirements	DBR lasers are somewhat simpler to fabricate as the grating is external to the gain medium. This separation can ease the manufacturing process, although precise alignment of the grating with the active region is still critical
Performance	DFB lasers typically exhibit very stable single-mode operation due to the distributed nature of the feedback. They can provide excellent performance in terms of linewidth, stability, and spectral purity, making them suitable for high-precision applications	DBR lasers also provide stable single-mode operation but might be slightly less stable compared to DFB lasers. However, they still offer good performance and are used in many practical applications where precise control over the emission wavelength is required
Applications	Due to their superior stability and performance, DFB lasers are widely used in telecommunications, spectroscopy, and other applications requiring high precision and narrow linewidths	DBR lasers are also used in telecommunications and sensing applications but might be preferred in scenarios where the fabrication simplicity and cost-effectiveness are more critical than the absolute performance characteristics
Typical cavity length	300–500 μm	500–1000 μm

typically comprises quantum wells that facilitate efficient light generation through the recombination of electrons and holes. The DBRs are composed of alternating layers of materials with different refractive indices, creating a mirror with high reflectivity at the desired emission wavelength.

12.6 Vertical Cavity Surface-Emitting Lasers (VCSEL)

Example Calculate the length of the active region that provides a gain, $\gamma = 70\,\text{cm}^{-1}$) and assume the absorption loss, $\alpha_s = 40\,\text{cm}^{-1}$ and let $R_1 = R_2 = R$.

Determine the length of the active region (l) for: $R = 0.32, 0.9, 0.99$, and 0.98.

Solution For the laser to operate, the gain must equal the loss, i.e.,

$$\gamma = \alpha_s + \frac{1}{2l}\left[\ln\left(\frac{1}{R_1 R_2}\right)\right].$$

Therefore, the length of the active region is given by:

$$l = \frac{1}{\gamma - \alpha_s}\left[\ln\left(\frac{1}{R_1 R_2}\right)\right],$$

$$l = \frac{1}{\gamma - \alpha_s}\left[\ln\left(\frac{1}{R}\right)\right], \qquad (12.90)$$

$$l = \frac{1}{70 - 40}\ln\left(\frac{1}{R}\right).$$

The length of the active cavity l for the various reflectances R are listed in the following table:

R	l (μm)
0.32	380
0.8	74
0.3	3.5
0.99	0.3
0.998	0.1

To have a VCSEL, the active region must be small and grow as a layer of semiconductor. The previous example shows that this is possible if we can grow Bragg reflecting layers with $R \geq 0.99$. Figure 12.31 illustrates a vertical cavity surface-emitting laser diode.

The Bragg grating reflectors are made of a large number (N_L) of alternating layers of high (n_h) and low (n_l) refractive index materials, such that each layer is a

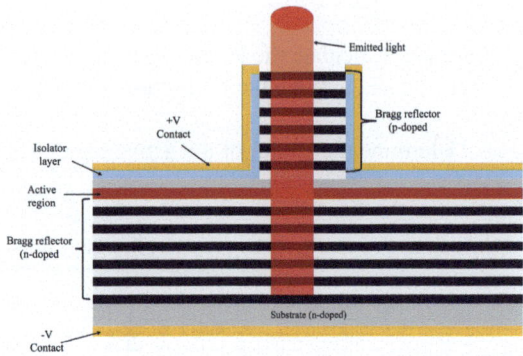

Fig. 12.31 Schematic of a double heterostructure vertical cavity surface-emitting laser (VCSEL) utilizing Bragg reflectors

quarter-wavelength thickness (t_h and t_l). The thicknesses of the layers are given by:

$$t_h = \frac{\lambda_o}{4n_h}, \tag{12.91}$$

and

$$t_l = \frac{\lambda_0}{4n_l}. \tag{12.92}$$

The period of the Bragg reflector is:

$$\Lambda = t_h + t_l. \tag{12.93}$$

The Bragg grating reflectance (R_B) is a function of the number of stacks and the wavelength. Each stack is made of one high-index layer and a low-index layer. Consider a stack is made of a layer of $AlGaAs$ ($n_l = 3.2$) and a layer of $GaAs$ ($n_h = 3.6$). For a wavelength of operation $\lambda_o = 1\,\mu m$, the layer thicknesses will be:

$$t_l = \frac{1 \times 10^{-6}}{4 \times 3.2} = 78.1\,\text{nm},$$

and

$$t_h = \frac{800 \times 10^{-9}}{4 \times 3.6} = 69.4\,\text{nm}.$$

Therefore, the Bragg grating period is $\Lambda = 78.1 + 69.4 = 147.5$ nm.

Figure 12.32 illustrates Bragg reflector stacks and the Reflectance R_B as a function of wavelength and for 4, 6, 8, and 16 Bragg periods. The reflectance will reach almost 100% for the 16 periods. In this case, the size of the Bragg reflector is

12.7 Tunable Laser Diodes (TLDs)

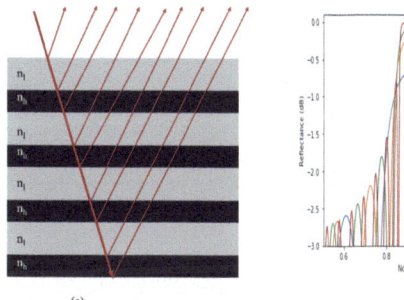

Fig. 12.32 (a) Bragg reflector stacks and (b) the reflectances of the stacks as a function of wavelength plotted for various number of periods

2.36 μm. Therefore, by growing 16 stacks on each side of the active region to create a vertical cavity that is tuned to the modes to be emitted by the VCSEL.

The cavity length in these lasers can be $l = 1$ μm, which results in a free spectral range given by:

$$\nu_F = \frac{c}{2nl} = \frac{3 \times 10^8}{2 \times 3.5 \times 10^{-6}} = 4.3 \times 10^{15} \, \text{Hz}.$$

This value is much larger than the bandwidth of the gain medium. Therefore, the laser will operate in a single mode, as there will be only one resonant frequency within the gain spectral width.

12.7 Tunable Laser Diodes (TLDs)

Unlike fixed-wavelength laser diodes, TLDs can adjust their output wavelength over a specified range. This tunability is achieved through various mechanisms, such as temperature control, current injection, or mechanical adjustments. The operation of tunable laser diodes is based on semiconductor materials, typically involving quantum well or quantum dot structures that enable precise control over electronic and optical properties.

The core of a TLD comprises an active region where electron-hole recombination generates photons, and a feedback mechanism, usually a distributed Bragg reflector (DBR) or distributed feedback (DFB) structure, to select and stabilize the desired wavelength.

In this section, we introduce tunable laser diodes based on a distributed Bragg reflector (DBR) where its period is controlled by current injection. Consider Fig. 12.33, the upper electrode is segmented in two pieces. The first one carries current i_1 that passes through the active region generating the gain, and the second electrode carries current i_2 that passes through the Bragg grating. The injected current that passes through the Bragg grating changes the effective refractive index

Fig. 12.33 Tunable laser diodes based on a distributed Bragg reflector

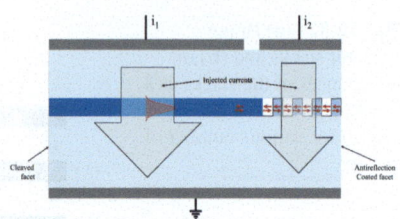

and, in turn, changes the reflected frequency of the Bragg grating ν_B as given by:

$$\nu_B = \frac{c}{2\Lambda n_{eff}}. \tag{12.94}$$

Therefore, current i_1 controls the power emitted by the laser, whereas current i_2 controls the emitted wavelength of the laser. These lasers can be tuned between 5 and 10 nm.

There are other methods, such as external cavity lasers (ECLs), that can provide tunability up to 50 nm. These can use an external tilted reflection grating, replacing the etalon and mirror in Fig. 12.28. In such an implementation, tilting the grating changes the resonant frequency of the cavity.

12.8 Quantum Well Laser Diodes (QLED)

Quantum well laser diodes represent a significant advancement in the field of semiconductor lasers, offering enhanced performance characteristics when compared to conventional laser diodes. These devices utilize quantum wells, which are thin layers of semiconductor material with thicknesses on the order of nanometers, to confine charge carriers (electrons and holes) in a dimensionally restricted region. This quantum confinement leads to unique electronic and optical properties that improve the efficiency and functionality of the laser.

Key features of quantum well laser diodes:

1. **Enhanced Carrier Confinement:** The quantum well structure confines electrons and holes in a potential well, leading to increased carrier density in the active region. This results in a higher probability of radiative recombination, which is essential for efficient light emission.
2. **Lower Threshold Current**: Due to the enhanced carrier confinement and increased radiative recombination rate, quantum well laser diodes typically exhibit lower threshold currents than their bulk counterparts. This makes them more energy-efficient and suitable for low-power applications.
3. **Improved Temperature Stability**: Quantum well lasers show improved temperature stability because the density of states in a quantum well is less sensitive to

12.8 Quantum Well Laser Diodes (QLED)

temperature variations. This leads to more stable performance across a range of operating temperatures.

4. **Wavelength Tunability**: The emission wavelength of quantum well lasers can be precisely controlled by adjusting the thickness of the quantum well and the material composition. This tunability makes them versatile for various applications, including telecommunications, data storage, and medical devices.

In quantum well lasers, the active region where electron-hole recombination takes place is only approximately 10 nm thick. As discussed in Chap. 10, the thin active region results in only a few available quantum states for occupation by charge carriers. The density of states becomes valid only in discrete energy states. Figure 12.34a illustrates the $E - k$ diagram for a quantum well semiconductor, whereas Fig. 12.34b, c illustrate the density of states as a function of energy.

As shown in Chap. 10, as the thickness of the quantum well (d_z) decreases around the de Broglie wavelength, the number of quantum states decreases and becomes discretized. As the thickness approaches the lattice constant, only one available quantum state exists. Figure 12.35a shows the energy band diagram as a function of z, with the width of the quantum well-being w_z. Figure 12.35b plots the energy levels of the quantum states as a function of the thickness d_z. In this illustrative example, if $d_z \leq 4$ nm, there exists only one available quantum state. If this is the quantum well of a laser, then the quantum well laser diode (QLED) will have a single mode.

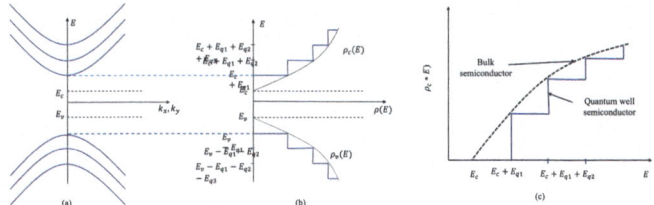

Fig. 12.34 Quantum well structure (a) $E-k$ diagram, (b) and (c) the density of states as a function of energy

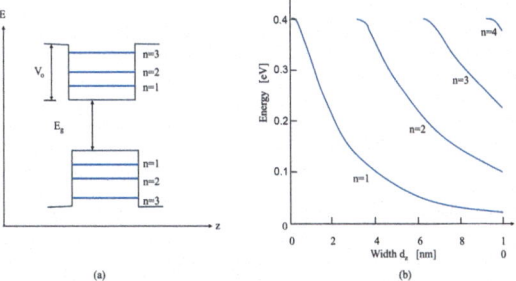

Fig. 12.35 (a) Energy band diagram for a quantum well structure illustrating the available quantum states and (b) the effect of the width d_x on the energy of the available quantum states

Quantum well laser diodes have a low threshold current because:

$$\Delta n_T = \frac{J_T \tau}{qd} \tag{12.95}$$

$$J_T = \frac{\Delta n_T q d}{\tau} \tag{12.96}$$

$$\gamma = \alpha_a \left(\frac{J}{J_T} - 1 \right) \tag{12.97}$$

Since d is small for QW, the transparency current density J_T becomes small. The gain coefficient is given by:

$$\gamma = \frac{\lambda^2}{8\pi \tau} \rho(\nu) f_g(\nu), \tag{12.98}$$

and the threshold current is given by:

$$i_{th} = i_T \left(1 + \frac{\alpha_r}{\Gamma \alpha_a}\right). \tag{12.99}$$

Γ is smaller for quantum wells due to the small dimensions of the core of the waveguide, ($\Gamma \approx 0.4$) compared to bulk semiconductors ($\Gamma \approx 0.8$–0.9), but the gain is much larger since $\alpha_r = \Gamma \gamma$, resulting in a low threshold current. Quantum well laser diodes have a threshold current that is less sensitive to temperature because the discrete nature of the quantum states prevents interband (phonon) transitions. The threshold current is given by:

$$i_{th}(T) = i_o e^{T/T_o}, \tag{12.100}$$

where $T_o = 400$ K for quantum wells, while $T_o = 140$ K for bulk semiconductor lasers. The internal quantum efficiency (η_i) is also less sensitive to temperature for the same reason.

Figure 12.36 illustrates the gain coefficient (γ) for bulk and quantum well semiconductors. The gain coefficient for the quantum well follows the step functions demonstrated by the density of states.

12.9 Summary

This chapter delves into the principles, structures, and applications of semiconductor light sources, focusing on light-emitting diodes (LEDs), laser diodes (LDs), and optical amplifiers. It begins with the theory of light emission in semiconductor materials, discussing the operation of pn-junctions under unbiased, forward-biased, and reverse-biased conditions. Homojunction and heterojunction LEDs are examined in

Fig. 12.36 Gain coefficient for bulk and quantum well semiconductors as a function of energy

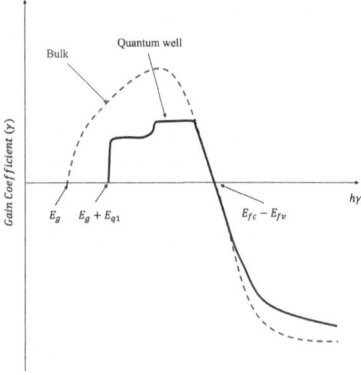

detail, emphasizing advancements in photon emission efficiency through improved carrier confinement and reduced recombination losses.

A significant portion of the chapter is dedicated to laser diodes, starting with their fundamental operation based on stimulated emission. The integration of a semiconductor optical amplifier (SOA) with an optical feedback system forms the core of laser diode functionality. Key components of LDs, such as gain media, electrical pumping mechanisms, and resonant optical cavities, are discussed alongside their role in generating coherent light. Fabry-Perot cavities and the dynamics of photon confinement and amplification within laser structures are explored in depth. LD characteristics, including threshold currents, spectral linewidths, and enhanced efficiency through heterostructures, are also covered comprehensively.

Various types of laser diodes are presented, each designed for specific applications. Distributed feedback (DFB) lasers use a periodic grating within the cavity for wavelength-selective feedback, ensuring single-mode operation, making them indispensable for optical communications. Vertical cavity surface-emitting lasers (VCSELs) emit light perpendicular to the wafer surface, offering a compact and scalable design suited for high-speed data transmission and sensing. Quantum well and quantum do lasers improve carrier confinement and reduce threshold currents through nanostructures, with quantum dot lasers providing exceptional temperature stability and narrower linewidths. Heterostructure lasers, constructed with materials of varying bandgaps, achieve superior carrier and optical confinement, significantly outperforming homojunction lasers. Tunable lasers, which allow wavelength adjustment by modifying the refractive index or cavity length, are crucial for dense wavelength-division multiplexing (DWDM) systems.

12.10 Problems

1. Calculate the built-in potential of a pn-junction at room temperature (300 K) with doping concentrations $N_A = 10^{18}$ cm^{-3} and $N_D = 10^{16}$ cm^{-3}. Assume intrinsic carrier concentration $ni = 1.5 \times 10^{10}$ cm^{-3}.

2. Determine the width of the depletion region for a pn-junction with uniform doping ($N_A = 10^{17}$ cm^{-3}, $N_D = 10^{15}$ cm^{-3}) and a built-in potential of 0.7 V. Use the permittivity of the material as $\varepsilon = 11.7\varepsilon_0$.
3. For a forward-biased pn-junction diode, calculate the reduction in the potential barrier when a voltage of 0.3 V is applied. Explain its effect on carrier injection across the junction.
4. An LED has a radiative recombination lifetime of 100 ns and a non-radiative recombination lifetime of 50 ns. Calculate the internal quantum efficiency (η_i).
5. A $GaAs$ LED with a bandgap energy of 1.42 eV emits photons with an internal quantum efficiency of 0.8. If a current of 50 mA is applied, calculate the optical power output at the junction.
6. Determine the peak wavelength and full-width at half-maximum (FWHM) for a $GaAs$ LED operating at 300 K. Use the bandgap energy of 1.42 eV.
7. For a Fabry-Perot laser diode, calculate the threshold current if the transparency carrier concentration is 1.2×10^{18} cm^{-3} and the active region dimensions are $200\,\mu$m $\times$ $10\,\mu$m $\times$ $1\,\mu$m. Assume the carrier lifetime $\tau = 2$ ns.
8. A semiconductor optical amplifier (SOA) has a gain coefficient of 300 cm^{-1}. Calculate the total gain of the device if its length is $500\,\mu$m. Express the result in dB.
9. An LED has a refractive index of $n = 3.5$, and the surrounding medium has a refractive index of $n_o = 1.0$. Calculate the fraction of photons that escape the surface, considering total internal reflection.
10. Discuss the mechanism of wavelength tuning in a tunable laser and its importance for dense wavelength-division multiplexing ($DWDM$) systems. Provide an example of a practical scenario where tunable lasers are indispensable.

Bibliography

1. Sze, S. M., & Ng, K. K. (2006). Physics of Semiconductor Devices (3rd ed.). Wiley.
2. Kasap, S. O. (2020). Principles of Electronic Materials and Devices (4th ed.). McGraw-Hill.
3. Coldren, L. A., Corzine, S. W., & Mašanović, M. L. (2012). Diode Lasers and Photonic Integrated Circuits (2nd ed.). Wiley.
4. Wilson, J., & Hawkes, J. F. B. (1998). Optoelectronics: An Introduction (3rd ed.). Prentice Hall.
5. Bhattacharya, P. (1996). Semiconductor Optoelectronic Devices (2nd ed.). Pearson.
6. Schubert, E. F. (2006). Light-Emitting Diodes (2nd ed.). Cambridge University Press.
7. Chuang, S. L. (2009). Physics of Photonic Devices (2nd ed.). Wiley.
8. Kressel, H., & Butler, J. K. (1977). Semiconductor Lasers and Heterojunction LEDs. Academic Press.
9. Agrawal, G. P., & Dutta, N. K. (1993). Semiconductor Lasers. Springer.
10. Saleh, B. E. A., & Teich, M. C. (2019). Fundamentals of Photonics (3rd ed.). Wiley.

Optical Detectors   13

Photodetectors are devices that convert photons into electrons, i.e., convert optical signals into electrical signals. The physical mechanism of photodetectors is the absorption of photons, which changes the electric properties, such as the generation of a photocurrent in a photoconductor or a photovoltage in a photovoltaic detector. The performance of a photodetector depends on the optical absorption process, the carrier transport, and the interaction with the circuit system.

For example optical absorptions, such as interband processes in a direct semiconductor, the general theory for the absorption spectrum was presented in Chap. 10. The interband absorption creates electron-hole pairs. The carrier transport of these electrons and holes after generation depends on the design of the photodetectors.

There are several types of photodetectors, each designed to meet specific requirements. Among the most commonly used devices are photodiodes, photomultiplier tubes (PMTs), and charge-coupled devices (CCDs). Photodiodes are particularly notable for for their fast response time, making them ideal for applications requiring high-speed detection.

Photodetectors can be categorized into the following types on the basis of their light-sensing mechanisms:

1. **Photoemissive Devices:** These devices, such as photomultiplier tubes, convert photons directly into electrons.
2. **Photoconductive Devices:** In these devices, the electrical conductivity changes in response to incident photons. Examples include photoconductors, phototransistors, and photodiodes.
3. **Pyroelectric Devices:** In these devices the energy conversion involves the generation of heat that increases the temperature of the device, which changes its polarization and hence its relative permittivity

The choice of photodetector depends on various factors, including the wavelength range of the incident light, the required sensitivity, and the speed of response.

In this chapter, we study photodiodes using p-n junctions and p-i-n structures and avalanche photodiodes. Our focus is on the understanding of the physical processes of the carrier generation and transport. We derive the essential parameters of photodiodes, such as the responsivity, quantum efficiency, rise time, and noise current.

13.1 Photoconductors

Consider a uniform p-type bulk semiconductor illuminated with a uniform optical beam, as shown in Fig. 13.1. When photons are absorbed by the semiconductor, mobile charge carriers are generated, ideally forming an electron-hole pair for every absorbed photon. The electrical conductivity of the material increases in proportion to the photon flux Φ. An electric field applied to the material by an external voltage source causes the electrons and holes to be transported. This in turn results in a measurable electric current in the circuit, as illustrated in Fig. 13.1. Photoconductive detectors operate by measuring either the photocurrent i_p, which is proportional to the photon flux Φ, or the voltage drop V across a load resistor R placed in series with the circuit.

The total electron and hole carrier concentrations, n and p, deviate from their thermal equilibrium values, n_o and p_o, by the excess carrier concentrations, Δn and Δp, respectively, due to the optical excitation of the carriers, where:

$$n = n_o + \Delta n \text{ and } p = p_o + \Delta p. \tag{13.1}$$

Since $p_o \gg n_o$ in this extrinsic p-type semiconductor, the net thermal recombination rate can, assuming low injection levels, (Δn and $\Delta p \ll p_o$) be expressed as:

$$R = \frac{\Delta n}{\tau}. \tag{13.2}$$

where τ is the recombination lifetime. The recombination rate is proportional to the photon flux such that:

Fig. 13.1 Schematic of a photoconductor detector

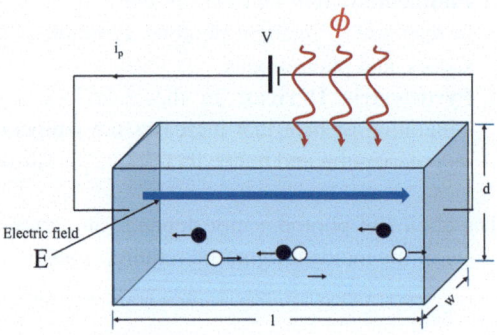

13.1 Photoconductors

$$R = \frac{\eta \Phi}{lwd}, \qquad (13.3)$$

hence,

$$\Delta n = \frac{\eta \tau_n \Phi}{lwd} \qquad (13.4)$$

where η is the quantum efficiency (i.e., the percentage of the photons that are absorbed and generate electron-hole pairs) and lwd is the volume of the sample. The excess carriers generated (Δn) by the electrons will change the conductivity σ. The change in the conductivity of the device is given by:

$$\Delta \sigma = q \Delta n (\mu_e + \mu_h). \qquad (13.5)$$

Therefore,

$$\Delta \sigma = \frac{\eta \tau q \Delta n (\mu_e + \mu_h)}{lA} \Phi, \qquad (13.6)$$

where $A = wd$ is the surface area of the device and where μ_n and μ_h are the electron and hole mobilities, respectively. Equation (13.6) shows that the change in conductivity is linearly proportional to the photon flux illuminating the device.

The photon-generated current density is given by:

$$J_{ph} = \Delta \sigma E, \qquad (13.7)$$

where E is the electric field generated by the applied V across the device as shown in Fig. 13.1. The photocurrent will be given by:

$$i_{ph} = J_{ph} A. \qquad (13.8)$$

Since $v_e = \mu_e E$ and $v_h = \mu_h E$, therefore the current is given by:

$$i_{ph} = \left(\frac{\eta \tau q (v_e + v_h)}{l} \right) \Phi. \qquad (13.9)$$

Therefore, the photocurrent in a photoconductive detector is directly proportional to the photon flux. Therefore,

$$i_{ph} \approx \eta q \tau \frac{v_e}{l} \Phi \approx \eta q \frac{\tau}{t_e} \Phi, \qquad (13.10)$$

where t_e is the electron transit time across the detector.

Example Given:

Quantum efficiency, $\eta = 0.8$; electron charge, $q = 1.6 \times 10^{-19}\,C$; recombination lifetime, $\tau = 10^{-6}$ s; electron mobility, $\mu_e = 1350\,\text{cm}^2/\text{V·s}$; applied voltage, $V = 5$ V; distance between electrodes (thickness of the sample), $l = 0.1$ cm; and photon flux, $\Phi = 10^{16}$ photons/cm^{2}s.

Calculate the photocurrent i_p.

Solution First, calculate the electric field E:

$$E = V/l = 5/0.1 = 50\,\text{V/cm}$$

Now calculate the electron velocity v_e:

$$v_e = \mu_e E = 1350 \times 50 = 67500\,\text{cm/s}$$

Now, using the following formula, the photocurrent can be calculated as:

$$i_{ph} \approx \eta q \tau \frac{v_e}{l} \Phi$$

Substitute the given values:

$$i_{ph} \approx 0.8 \times 1.6 \times 10^{-19} \times 10^{-6} \times \frac{67500}{0.1} \times 0^{16}$$

$$i_{ph} \approx 8.64 \times 10^{-7}\,A = 86.4\,\mu A.$$

Therefore, the photocurrent i_p is approximately 86.4 micro Amps.

Example An optical beam with power $P_{opt} = 5$ mW at wavelength 500 nm illuminating a photoconductor detector with dimensions $l = 1$ mm, $d = 1$ mm, $and\, w = 10$ mm. Determine the photocurrent and the responsivity of the detector by using the same parameters from the privies example.

Solution The relationship between the optical power and photon flux is given by:

$$P = \Phi A h\nu,$$

where A is the surface area of the detector.

(continued)

Therefore,

$$\Phi = \frac{P}{Ah\nu} = \frac{P\lambda}{wdhc},$$

$$\Phi = \frac{5 \times 10^{-3} \times 500 \times 10^{-9}}{10 \times 10^{-3} \times 10^{-3} \times 1.626 \times 10^{-34} \times 3 \times 10^8},$$

$$\Phi = 5.125 \times 10^{21} \text{ photons/m}^2.\text{s}.$$

Let us use the parameters from the previous example but ensure that the units are consistent. Let us change the units for the photon flux to be $photons/cm^2.s$; therefore,

$$\Phi = (5.125 \times 10^{21}) \times 10^{-4} = 5.12 \times 10^{17} \text{ photons/cm}^2.\text{s}.$$

The photocurrent will be:

$$i_{ph} \approx \eta q \tau \frac{v_e}{l}\Phi = 0.8 \times 1.6 \times 10^{-19} \times 10^{-6} \times \frac{67500}{0.1} \times 5.12 \times 0^{17}$$

$$i_{ph} = 44.2 \times 10^{-6} \text{ A} = 44.6\,\mu\text{A}.$$

The responsivity of the detector is:

$$Responsivty = \frac{i_{ph}}{P_{opt}} = \frac{44.2 \times 10^{-6}}{5 \times 10^{-3}} = 8.85\,\text{mW/A}.$$

13.2 The pn Photodiode

In the previous chapter, we introduced pn-junction diodes as light sources when they are forward biased. In this case, the forward biased injection current causes electron-hole recombination, resulting in the emission of photons if the recombinations are radiative. Photodiodes operate differently: illuminating photon flux creates electron-hole pairs which, because of the reverse bias voltage, are swept across the junction, resulting in a photocurrent.

Figure 13.2 schematically shows a pn-junction diode under reverse bias. To understand the pn-junction photodetector, we need to consider the energy band diagram. Figure 13.2a shows the pn-junction reverse biased by V_r, which increases the built-in voltage to $V_o + V_r$. The electric field E in the depletion region is found by the integration of the net space charge density ρ_{net} in Fig. 13.2b across w subject to

Fig. 13.2 (a) A schematic diagram of a reverse biased pn-junction photodiode. (b) Energy band diagram for a reverse biased pn-junction, (c) net space charge across the diode in the depletion region, and (d) the electric field in the depletion region

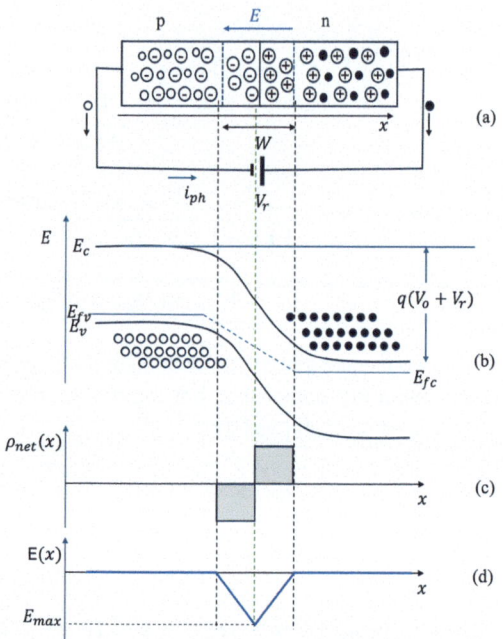

a voltage difference of $V_o + V_r$, i.e., V_r. The field only exists in the depletion region and is not uniform. It varies across the depletion region as shown in Fig. 13.2d, where it is at the maximum at the junction. The regions outside the SCL are the neutral regions in which there are majority carriers on both sides of the depletion region. The electric field in the SCL also increases. The Fermi levels E_{fc} and E_{fv} on the n- and p-sides are separated by V_r, creating a potential hill. The potential hill, that is, the change in E_c from the E_c on the n-side to that on the p-side in the SCL, is very steep due to the large field in the SCL. The absorption of a photon in the SCL creates an electron-hole pair (EHP). The electron and hole become separated and drift due to the field. The drift corresponds to the electron rolling down the energy hill (along E_c) toward the n-side, whereas the hole rolls down the energy hill toward the p-side. (Remember that the hole energy increases in the downward direction.) The drift creates a photocurrent i_{ph}, which is detected in the external circuit and lasts for the duration of the drift of the electron and hole.

Photogeneration within a diffusion length of the SCL also generates a photocurrent, as illustrated in Fig. 13.2b. The photogenerated electron diffuses to the SCL, where the internal field then drifts the electron over to the n-side. The drift creates the photocurrent. As the electron drifts in the SCL toward the neutral n-side, an electron flows in the external circuit toward the positive terminal of the battery; this electron comes from the n-side. The photogenerated hole on the p-side is neutralized by the flow of an electron from the external circuit into the p-side. The photocurrent due to photogeneration in the neutral region is weaker than that due to photogeneration in the depletion region; in the latter, the field separates and drifts the carriers immediately.

13.3 Current-Voltage Characteristic of a pn Photodiodes

The current-voltage relationship for a photodiode under no light illumination behaves as a standard pn-junction diode, described by the following expression:

$$I = I_o(e^{qV/k_BT} - 1), \qquad (13.11)$$

where I_o is the reverse saturation current, which is very small. Figure 13.3 shows a typical plot of the current as a function of voltage for a p-n photodiode. Figure 13.3a illustrates a photodiode circuit diagram where the voltage source can vary to either forward or reverse bias the diode. Figure 13.3b shows the I–V curve for a dark and an illuminated photodiode. When the photodiode is illuminated, the generated photocurrent (i_{ph}) flows in the opposite direction of I, shitting the I–V curve downward by i_{ph}. Consequently, the curve intersects the vertical axis at $I = -i_{ph}$ and crosses the vertical axis at $I = 0$ and $V = V_{oc}$, where V_{oc} is the open-circuit voltage for an illuminated photodiode. This is the open circuit voltage that would be generated when the photodiode is illuminated in an open circuit, without any current

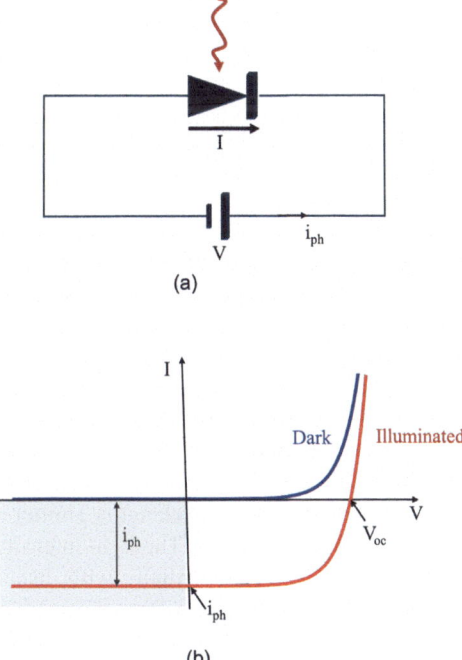

Fig. 13.3 (a) Photodiode circuit diagram and (b) the I–V characteristic curves for a dark (blue line) and illuminated (red line) photodiode

in the external circuit. However, within the diode itself, there are two currents that cancel each other exactly; one due to photogeneration (i_{ph}), and the other due to the diode current (I) caused by V_{oc}, which acts like a forward bias across the junction.

The regions of the pn-junction I–V characteristics that are bound by the positive V and negative I axes represent a photovoltaic mode, as in the case of solar cells. There is no applied bias, and the light generates a photocurrent and a voltage across the device. When the pn-junction is shorted, the current is i_{ph}, and when it is in an open circuit, the voltage is V_{oc}. If there is a load resistance R connected to the pn-junction, we need to draw a load-line construction to find the operation (see Sect. 13.3). The region bound by the negative V-$axis$ and the negative I-$axis$ represents a reverse biased photodiode mode of operation, shown as a gray region; this is the most common mode of operation for the detection of light.

13.4 Absorption and Photodiode Materials

Photodiodes are devices that convert photons into electrons through the process of photogeneration, which involves the creation of electron-hole pairs. For this generation to occur, the photon energy must be at least equal to the bandgap energy E_g of the semiconductor material, enabling an electron to be excited from the valence band (VB) to the conduction band (CB). The upper cutoff wavelength λ_g for a photo-generative absorption is determined by the bandgap energy E_g of the semiconductor, as given by:

$$\lambda_g(\mu m) = \frac{1.24}{E_g(eV)}. \tag{13.12}$$

This cutoff wavelength defines the maximum wavelength the material can absorb; otherwise, the photon energy will be less than the bandgap energy. For example, Si has an energy gap $E_g = 1.12\,\text{eV}$, resulting in a cutoff wavelength. Thus, silicon cannot be used for optical telecommunications operates at wavelengths 1.3 and 1.55 μm. On the other hand, germanium (Ge) has $E_g = 0.66\,\text{eV}$, corresponding to $\lambda_g = 1.87\,\mu m$, which makes it suitable for such applications. Therefore, different semiconductor materials can be used for different applications across the wavelength spectrum.

Table 13.1 lists some typical bandgap energies and the corresponding cutoff wavelengths of various photodiode semiconductor materials.

The incident photons on the photodiode with wavelengths longer than the cutoff wavelength (λ_g) will pass through the material without being absorbed. In contrast, incident photons with wavelengths shorter than λ_g are absorbed as they travel through the semiconductor. The light intensity, which is proportional to the number of photons, decays exponentially with distance into the semiconductor. The light intensity $I(x)$ at a distance x from the semiconductor surface is given by:

$$I(x) = I_o e^{-\alpha x} \tag{13.13}$$

13.4 Absorption and Photodiode Materials

Table 13.1 Bandgap energy Eg at 300 K, cutoff wavelength λ_g, and type of bandgap (D Direct and I Indirect) for some photodetector materials

Semiconductor	Bandgap (eV)	Bandgap Type	Cutoff wavelength λ_g (nm)
$InAs$	0.38	D	3263
Ge	0.66	I	1879
$In_{0.53}Ga_{0.47}As$	0.75	D	1.65
$In_{0.7}Ga_{0.3}As_{0.47}P_{0.53}$	0.89	D	1.4
Si	1.12	I	1107
InP	1.35	D	919
$GaAs$	1.42	D	873
$AlAs$	2.16	I	574
GaP	2.26	I	549
AlP	2.45	I	506

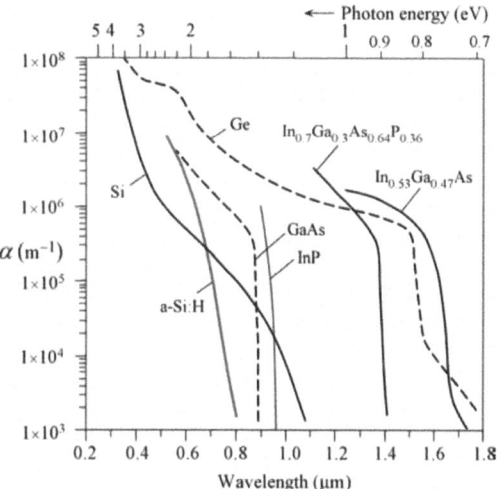

Fig. 13.4 Absorption coefficient (α) versus wavelength for some of the semiconductors listed in Table 13.1

where I_o is the intensity of the incident radiation and α is the absorption coefficient of the material. Most of the photon absorption (63%) occurs over a distance $1/\alpha$, known as the penetration or absorption depth δ. Figure 13.4 shows the α as a function of λ for various semiconductors.

In semiconductors with direct bandgaps, the electrons excited by photon absorption transition directly from the valence band to the conduction band without requiring phonon assistance, as there is no need for a change in momentum. In contrast, in indirect bandgap semiconductors, the transition of excited electrons involves a change in both energy and momentum, where the phonon momentum is given by:

$$Photon\ momentum = \hbar k_{CB} - \hbar k_{VB}. \tag{13.14}$$

The absorption process in indirect bandgap semiconductors is therefore dependent on lattice vibrations (phonons), which are influenced by temperature. For semiconductors with indirect bandgaps such as silicon (Si) and germanium (Ge), at wavelengths much shorter than λ_g, the photon energies are significantly greater than E_g. For example, at these wavelengths, the transition between their direct bandgaps influences the absorption coefficient..

Example For a germanium detector operating at wavelengths $\lambda = 0.8, 1.3,$ and $1.7\,\mu$m. Determine the penetration depth and plot $I(x)$ for these wavelengths as a function of x. Use plot in Fig. 13.4 to determine α at these wavelengths.

Solution The following are the approximate absorption coefficients for the given wavelengths and the corresponding penetration depths:

Wavelength (μm)	Absorption coefficient (m^{-1})	Penetration depth
0.8	4×10^6	$0.25\,\mu$m
1.3	7×10^5	$1.4\,\mu$m
1.7	4×10^3	0.25 mm

Figure 13.5 shows the intensity plots for $I(x)/I_o$ as a function of the distance x.

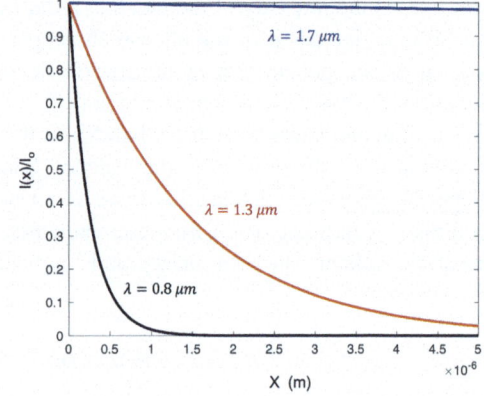

Fig. 13.5 Absorption in Ge for light beams with wavelength and absorption coefficients given un the example

13.5 Quantum Efficiency and Responsivity

The conversion of photons illuminating the photodiode into electron-hole pairs determines the quantum efficiency (QE) (η) of the device which is defined as:

$$\eta = \frac{\text{Number of collected electrons at detector terminals}}{\text{Number of incident photons}}. \quad (13.15)$$

The collected electrons constitute the photocurrent (i_{ph}), and the number of photons illuminating the photodiode is given by the photon flux (ϕ). Therefore, the quantum efficiency is expressed as:

$$\eta = \frac{i_{ph}/q}{\phi}, \quad (13.16)$$

where

$$\phi = \frac{P_{opt}}{h\nu}. \quad (13.17)$$

Thus, the quantum efficiency can be rewritten as:

$$\eta = \frac{i_{ph}/q}{P_{opt}/h\nu} = \frac{i_{ph}h\nu}{P_{opt}q}. \quad (13.18)$$

The responsivity R of a photodiode characterizes its performance in terms of the photocurrent generated (i_{ph}) per incident optical power (P_{opt}) at a given wavelength, i.e.,

$$\mathcal{R} = \frac{\text{Output Photocurrent}}{\text{Incident optical power}} = \frac{i_{ph}}{P_{opt}} \text{ A/W}. \quad (13.19)$$

Using the definition of the quantum efficiency, substituting Eq. (13.18) into Eq. (13.19) yields

$$\mathcal{R} = \eta \frac{q}{h\nu} = \eta \frac{\lambda \,(\mu m)}{1.24 \,(eV)}. \quad (13.20)$$

Therefore, the responsivity is a function of the material and wavelength. Figure 13.6 shows typical responsivity curves for Si, Ge and $InGaAs$ as a function of wavelengths.

Consider the simplified photodiode structure shown in Fig. 13.7 with an annular front surface. Let us follow what happens to the light incident on the front surface of the device:

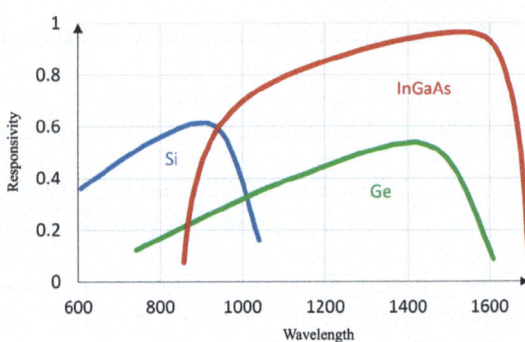

Fig. 13.6 Typical responsivity vs. wavelength for Si, Ge, and $InGaAs$

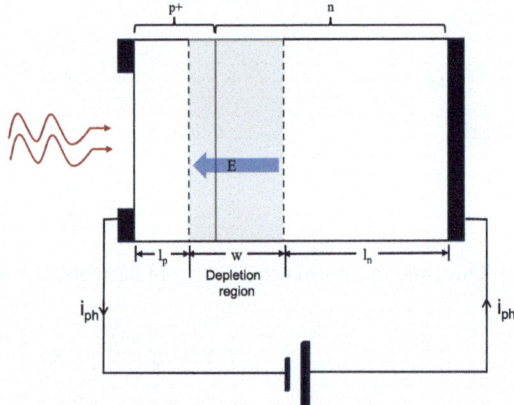

Fig. 13.7 A schematic of a reverse biased photodiode

1. Some of the photons are reflected by the air-semiconductor interface,
2. Some of the photons are absorbed as they travel through the $p+$ material before reaching the depletion region without contributing to the photocurrent
3. A fraction (ξ) of the photons that reach the depletion region can be absorbed creating electron-hole pairs that contribute to the photocurrent.

Considering these factors in determining the total quantum efficiency of the photodiode yields:

$$\eta = (1 - R)(1 - e^{-\alpha w})\xi \qquad (13.21)$$

where w is the width of the depletion region, R is reflectance at the surface of the photodiode, and ξ is the fractional number of absorbed photons that generate electron-hole pairs contributing to the photocurrent. Additionally, ξ is maximized by applying a bias voltage and by reducing the number of defects in the material.

Maximizing the quantum efficiency (η) depends on the device's design. Applying an antireflection coating to the device surface reduces the reflectance, resulting in $R = 0$. Increasing the factor αw can be achieved by increasing the thickness (w)

13.5 Quantum Efficiency and Responsivity

of the depletion region and by choosing the proper semiconductor mater material to increase α for the wavelength of operation.

Example Determine the thickness of the depletion region w to make the absorption in the depletion region to be 0.9 for a semiconductor with an absorption coefficient $\alpha = 20{,}000 \, \text{cm}^{-1}$.

Solution The absorption in the depletion region is:

$$1 - e^{-\alpha w} = 0.9$$

$$e^{-\alpha w} = 0.1$$

Taking the natural logarithm of both sides to get:

$$\alpha w = ln(10)$$

$$w = \frac{2.3}{20{,}000} = 1.15 \, \mu\text{m}.$$

Example Consider a silicon photodiode with a peak responsivity of 0.6 at wavelength $\lambda = 840 \, \text{nm}$. Determine the quantum efficiency at peak responsivity. If the photosensitive device area is $1 \, \text{mm}^2$, what would be the light intensity corresponding to a photocurrent of $20 \, \text{mA}$ at the peak responsivity?

Solution From Eq. (13.20) the responsivity is given by:

$$\mathcal{R} = \eta \frac{\lambda \, (\mu m)}{1.24}$$

therefore, the quantum efficiency is:

$$\eta = \frac{1.24 \, \mathcal{R}}{\lambda} = \frac{1.24 \times 0.6}{0.84} = 0.89$$

Since the responsivity is also given by Eq. (13.19) as:

$$\mathcal{R} = \frac{i_{ph}}{P_{opt}}.$$

(continued)

Therefore, the optical power is given by:

$$P_{opt} = \frac{i_{ph}}{\mathcal{R}} = \frac{20 \times 10^{-3}}{0.79} = 25.3\,\text{mW}.$$

The intensity I is given by:

$$I = P_{opt} \times \text{photosensative area}$$

Therefore, the intensity is:

$$I = 25.3 \times 10^{-3} \times 10^{-3} = 25.3\,\mu\text{W/mm}^2.$$

It is clear from the previous example that increasing the width of the depletion region results in higher quantum efficiency. Additionally, decreasing the distance the photons travel through the semiconductor before reaching the depletion region reduces the losses, thereby increasing the quantum efficiency. Both of these can be achieved by altering the photodiode design.

13.5.1 p-n Photodiode Impulse Response

The detection of dynamic signals with a photodiode depends on its response time, which is the duration for the detector output to change in response to changes in the input light intensity. This response time determines the upper limit of the data rates and the bandwidth of the system. The EHPs generated in the depletion layer require time to reach the device electrodes to form the photocurrent. The transit time of the charge carriers depends on the width of the depletion layer, the applied voltage, and the electron and hole mobilities.

The photocurrent generated by electrons (i_e) and by holes (i_h) is expressed as:

$$i_{ph}(t) = i_h(t) + i_e(t), \tag{13.22}$$

and according to Ramo's theorem:

$$i_{ph}(t) = \frac{qv_h(t)}{w} + \frac{qv_e(t)}{w} \tag{13.23}$$

where v_e and v_h are the drift velocities for the electrons and the holes, respectively, which are given by:

$$v_e = \mu_e E, \tag{13.24}$$

13.5 Quantum Efficiency and Responsivity

and

$$v_h = \mu_h E, \qquad (13.25)$$

where μ_e and μ_h are the electron and hole mobilities, respectively, and E is the built-in electric field. Thus, the photocurrent is expressed as:

$$i(t) = \frac{q}{w}(\mu_e + \mu_h)E. \qquad (13.26)$$

13.5.1.1 Impulse Response Based on Transit Time

To qualitatively determine the impulse response of a photodiode based on EHP transit time, consider a photon absorbed at location $x = L$ in the detector shown in Fig. 13.8a. This photon is absorbed and generates an electron-hole pair, where the hole travels a distance L to reach the collection electrode and the electron travels a distance w-L to reach the electrode connected to the positive terminal of the voltage source. The electron and the hole drift in opposite directions with respective drift velocities v_e and v_h. The transit time of a carrier is the duration it takes for a carrier to drift from its generation point to the collecting electrode. The electron and hole

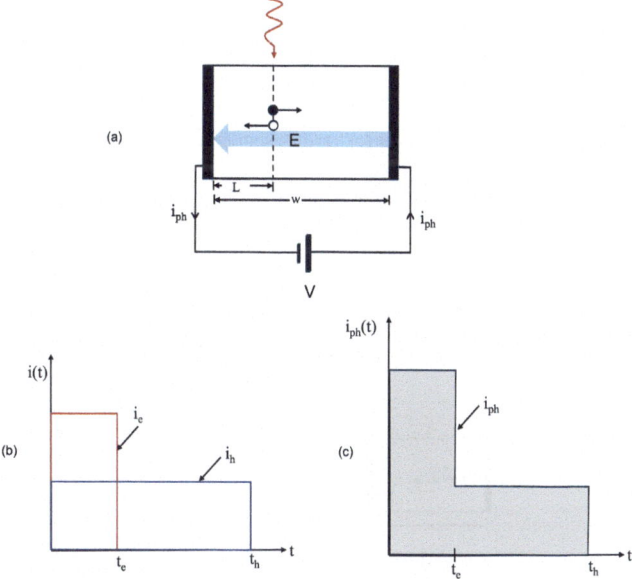

Fig. 13.8 (a) A photodetector illuminated with a single photon at $x = L$, the electron-hole pair generated drifts across the detector to reach the respected collecting electrodes, (b) a plot of the electron and hole currents, and (c) a plot of the total photocurrent generated by the electron and the hole drift currents

transit times t_e and th_h are given by:

$$t_e = \frac{w-L}{v_e}, \text{ and } t_h = \frac{L}{v_h}. \tag{13.27}$$

According to Table 13.2, electron mobility is significantly higher than hole mobility, implying:

$$t_h \gg t_e. \tag{13.28}$$

First, consider only the drifting electron. Suppose that the external photocurrent due to the motion of this electron is $i_e(t)$. The electron is acted on by the force qE of the electric field. This current continues to flow as long as the electron is drifting in the detector. It lasts for a duration t_e at the end of which the electron reaches the battery. Thus, although the electron has been photogenerated instantly, the external photocurrent is not instantaneous and has a time spread. The electron and hole currents are plotted as a function of time in Fig. 13.8b, and the resultant photocurrent which is the sum of both currents is shown in Fig. 13.8c. The curve in Fig. 13.8c represents the impulse response of the detector.

Next, consider that a very large number of photons falling across the detector as shown in Fig. 13.9a. At $x = 0$, an electron-hole pair was generated by the absorption of a photon. The hole immediately reaches the electrode while the electron drifts a distance w before reaching the collecting electrode. Conversely, for an electron-hole pair generated at $x = w$, the electron immediately reaches the collecting electrode

Table 13.2 Electron and hole mobilities for various semiconductors

Semiconductor	Electron mobility (μ_e)	Hole mobility (μ_h)	μ_e/μ_h
Si	1500	450	3.3
Ge	3900	1900	2.1
GaAs	850	400	2.125
InP	4600	150	31
InGaAsP	14,000	400	35

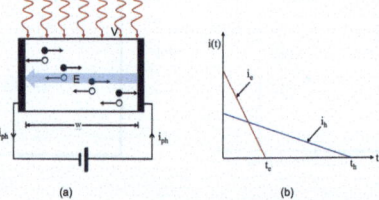

Fig. 13.9 (a) A photodetector illuminated with a very large number of photons across the device surface, the electron-hole pairs generated drift across the detector to reach the respected collecting electrodes, and (b) a plot of the electron and hole currents as a function of time

13.5 Quantum Efficiency and Responsivity

and the whole drifts a distance w before reaching the collecting electrode at $x = 0$. The electron and hole currents are represented in Fig. 13.9b.

Example For a $GaAs$ with the following parameters, determine the electron and hole currents and their transit times, given the following:
$V = 5\,\mathrm{V}$, $w = 5\,\mu\mathrm{m}$, and $L = 25\,\mu\mathrm{m}$.

Solution The electron current is given by:

$$i_e = \frac{q\mu_e E}{w} = \frac{q\mu_e (V/w)}{w} = \frac{q\mu_e V}{w^2} = \frac{1.6 \times 10^{-19} \times 850 \times 5}{(50 \times 10^{-6})^2} = 0.27\,\mu\mathrm{A},$$

and the hole current is given by:

$$i_h = \frac{q\mu_h V}{w^2} = \frac{1.6 \times 10^{-19} \times 400 \times 5}{(50 \times 10^{-6})^2} = 0.13\,\mu\mathrm{A}.$$

Therefore, the photocurrent is given by:

$$i_{ph} = 0.27 + .13 = 0.4\,\mu\mathrm{A}.$$

Therefore, the electron and hole transit times are given by:

$$t_e = \frac{w - L}{v_e} = \frac{w - L}{\mu_e(V/w)} = \frac{50 \times 10^{-6} - 25 \times 10^{-6}}{850 \times 5/(50 \times 10^{-6})} = 0.29\,\mathrm{ps},$$

and

$$t_h = 0.62\,\mathrm{ps}.$$

The response time thus far has focused on the impulse response on the basis of the carriers transit time. To reduce the response time, $t_e = w/v_e$, which can be achieved increase the drift velocity and reducing the junction width. The drift velocity can be increased by increasing the built-in electric field, $E = V/d$, but this linear relationship does not continue and the drift velocity reaches a peak for some value of the electric field.

13.5.1.2 Impulse Response Based on Diode Induced Capacitance

Another factor that contributes to the impulse response of the photodiode is the $R_L C_{pd}$ time-constant of the circuit, where C_{pd} is the photodiode junction capacitance. Figure 13.10a shows a typical photodiode circuit. The load resistance R_L determines the output voltage of the receiver circuit. The bandwidth of the

Fig. 13.10 (a) A typical photodiode receiver circuit and (b) photodiode equivalent circuit

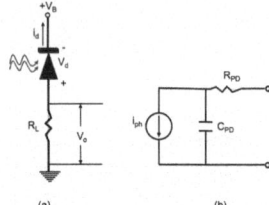

receiver circuit is given by:

$$f_c = \frac{1}{2\pi R_L C_{pd}}. \tag{13.29}$$

The capacitance of a p-n junction can be evaluated by determining how the charge on either side of the junction changes in response to a changing diode voltage. The junction capacitance is given by:

$$C_{pd} = \frac{\epsilon_o \epsilon_r A}{w}, \tag{13.30}$$

where A is the area of the junction and ϵ_r is the permittivity of the semiconductor.

A typical photodiode equivalent circuit is shown in Fig. 13.10b. The typical values for the junction capacitance is 0.2 pF, and the device resistance is 20 Oms. This results in a time constant of 4 ps, and $f_c = 4$ Ghz.

13.6 p-i-n Photodiodes

As discussed in the previous section, the simple pn-junction photodiode suffer from several drawbacks including decreased responsivity. The narrow size of the depletion layer, typically a few few microns thick, allows photons to pass through it without being absorbed, especially at longer wavelengths, where the absorption coefficient is lower at these wavelengths and the penetration depth exceeds the depletion layer thickness. Consequently, many photons are absorbed outside of the depletion layer, where there is no induced electric field exists to separate and drift the EHPs. Furthermore, the narrow depletion layer results in higher junction capacitance, limiting the photodiode's ability to detect high-frequency signals. These limitations are significantly mitigated in p-i-n (p–intrinsic–n) photodiodes.

Figure 13.11a illustrates the structure a $p^+ -$ intrinsic $- n+$ (p-i-n photodiode. The intrinsic layer is very lightly doped, either as a p- or n-$type$, and it is much wider than the other regions, typically 5 to 50 μm depending on the particular application. When the structure is formed, holes diffuse from the p^+ region and electrons from the n^+ region into the intrinsic i-$layer$, where they recombine. This process leaves behind a thin layer of exposed negatively charged acceptor ions on the p^+ region and a thin layer of exposed positively charged donor ions in the n^+

13.6 p-i-n Photodiodes

Fig. 13.11 (a) p-i-n photodiode in reverse bias, (b) space charge density across the device, and (c) induced electric field

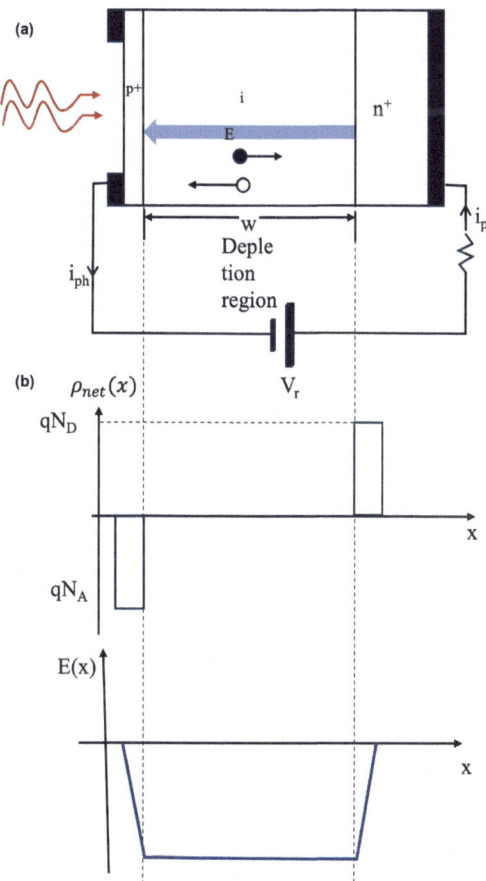

region as shown in Fig. 13.11b. These charges are separated by the *i-layer* of width w. Creating a uniform built-in electric field $E_r + E_o$ across the *i-layer* as illustrated in Fig. 13.11c.

When a reverse bias voltage V_r is applied, it enhances the electric field in the depletion region:

$$E = E_o + E_r = E_o + \frac{V_r}{w} \approx \frac{V_r}{w}, \qquad (13.31)$$

since most of the reverse bias voltage V_r drops across the depletion layer.

The p-i-n photodiode design allows for maximum photon absorption within the wide *i-layer*. Photogenerated EHPs in the *i-layer* are efficiently separated and drifted by the electric field E toward the n^+- and p^+-sides, respectively, as illustrated in Fig. 13.11a–d. As the photogenerated carriers drift through the *i-layer*, they generate external photocurrent, which can be detected as a voltage across a load resistance R, as shown in Fig. 13.11a.

Table 13.3 Generic parameters for p-i-n photodiodes

Parameter	Units	Silicon (Si)	Germanium (Ge)	InGaAs
Wavelength range	nm	400–1100	800–1600	1100–1700
Peak wavelength	nm	800	1550	1550
Responsivity	A/W	0.7–0.9	0.4–0.5	0.75–0.9
Rise time	ns	0.5–1	0.1–0.5	0.05–0.5
Dark current	nA	1–10	50–500	0.5–2
Bandwidth	GHz	0.3–0.7	0.5–3	1–2
Bias voltage	V	5	5–10	5

A significant advantage of the p-i-n photodiode is its ability to absorb a wider spectral range within the i-$layer$, where photogeneration occurs. This design improves the photodiode responsivity R, which can be further optimized by adjusting the width of the i-$layer$. Both the Si and $InGaAs$ p-i-n photodiodes cover a wide range of wavelengths, from approximately 300 nm to 1700 nm m, making them suitable for various applications.

Table 13.3 summarizes the typical performance parameters for p-i-n photodiodes.

13.7 Avalanche Photodiodes (APD)

Avalanche photodiodes (APDs) are widely used in optical communications because of their high speed and internal gain. A simplified schematic diagram of a Si reach-through APD is shown in Fig. 13.12a. The n^+-side is thin, and it is the side that is illuminated through a window. There are three p-type layers of different doping levels next to the n^+-layer to suitably modify the field distribution across the diode. The first is a thin p-type layer and the second is a thick, lightly p-type doped (almost intrinsic) layer, called the π-layer, and the third is a heavily doped p^+-layer. The diode is reverse biased to increase the fields in the depletion regions. The net space charge distribution across the diode caused by exposed dopant ions is detected in Fig. 13.12b. Under zero bias, the depletion layer in the p-region (between n^+ and p) does not normally extend across this layer. However, when a sufficient reverse bias is applied, the depletion region in the p-layer widens and extends to reach through the p-layer (and hence the name reach-through APD). The field extends from the exposed positively charged donors in the thin depletion layer on the n^+-side all the way to the exposed negatively charged acceptors in the thin depletion layer on the p^+-side.

The electric field is given by the integration of the net space charge density ρ_{net} across the diode subject to an applied voltage V_r across the device. The variation in the field across the diode is shown in Fig. 13.12c. The field lines start at positive ions and end at negative ions which exist through the p and p^+-layers. The electric field E reaches its maximum at the n^+-p junction and then decreases slowly through the p-layer. Through the p-layer, it decreases only slightly as the net space charge

Fig. 13.12 (a) Schematic illustration of an avalanche photodiode (APD) structure and the identification of the absorption and multiplication regions. (b) Net space charge density across the photodiode. (c) Electric field across the diode

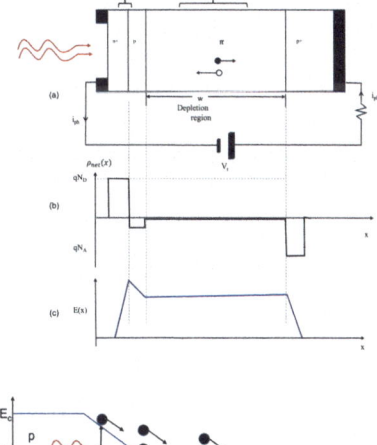

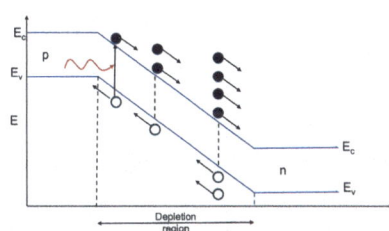

Fig. 13.13 The process of impact ionization

density (ρ_{net}) is minimal. The field vanishes to zero at the end of the narrow depletion layer on the p^+-side.

13.7.1 Photon Absorption and Avalanche Multiplication

Photon absorption and subsequent electron-hole pair (EHP) generation primarily occur in the long p-layer. The nearly uniform field here separates the electron-hole pairs and drifts them at velocities near saturation toward the n^+- and p^+-sides, respectively. When the drifting electrons reach the p-layer, they experience even greater fields and therefore acquire sufficient kinetic energy (greater than E_g) to impact-ionize some of the Si covalent bonds and release EHPs. We can visualize the impact ionization process as shown in Fig. 13.12d, where an electron entering the avalanche region (width w) gains energy from the field as it "drifts" in the opposite direction to the field, and its energy (which is its kinetic energy) increases with respect to E_c. Eventually, the energy gained from the field is sufficient to excite an electron across the bandgap E_g as illustrated in Fig. 13.13. These impact-ionization-generated carriers are called secondary carriers. These secondary EHPs themselves can also be accelerated by the high fields in this region to sufficiently large kinetic energies to further cause impact ionization and release more EHPs, leading to an avalanche of impact ionization processes. Thus, from a single electron entering the p-layer, one can generate many EHPs, all of which contribute to the photocurrent. The APD possesses an internal gain mechanism in that single-photon absorption leads the generation of many EHPs being generated. Therefore, the photocurrent in the APD in the presence of avalanche multiplication corresponds to an effective quantum efficiency in excess of unity.

13.7.2 Ionization Coefficients and Ratios

The ionization coefficients for electrons and holes, α_e and α_h, are related to the average distance between consecutive ionizations, L_{impact}, as:

$$L_{impact} \approx \frac{1}{\alpha_e} \text{ and } \frac{1}{\alpha_h}. \tag{13.32}$$

Both coefficients increase with higher electric fields in the depletion layer and with higher device temperatures. The ionization ratio k_A is defined as:

$$k_A = \frac{\alpha_e}{\alpha_h}. \tag{13.33}$$

When $\alpha_h \ll \alpha_e$, ionization due to holes is negligible compared to that of electrons. For silicon at $E = 10^5$ V/cm^2 the ionization ratio $k_A = 30$.

13.7.3 APD Gain

For a single-carrier (electron) multiplication ($\alpha_h = 0$), the ionization ratio $K_A = \alpha_e$.

Let $J_e(x)$ is the electric-current density at position x in the depletion region, as shown in Fig. 13.14. The rate of change of the electron current density is:

$$\frac{d}{dx} J_e(x) = \alpha_e J_e(x). \tag{13.34}$$

Solving this first-order differential equation gives:

$$J_e(x) = J_e(0) e^{\alpha_e x}. \tag{13.35}$$

Fig. 13.14 Gain across the APD

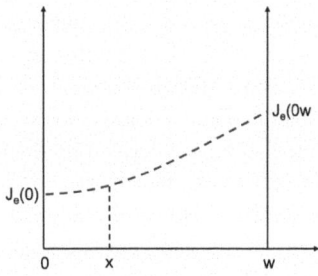

The gain for a depletion region of width w is given by:

$$G = \frac{J_e(w)}{J_e(0)} = e^{\alpha_e w}. \tag{13.36}$$

For both electron and hole multiplication, the total current density satisfies:

$$\frac{dJ}{dx} = \alpha_e J_e(x) + \alpha_h J_n(x), \tag{13.37}$$

where:

$$\frac{dJ_e}{dx} = -\frac{dJ_h}{dx} \tag{13.38}$$

This implies that the sum of the current densities remains constant across the region:

$$J_e(x) + J_h(x) = J_e(w). \tag{13.39}$$

Since no holes are injected at $x = w$, $J_h(w) = 0$, the rate of change of $J(x)$ becomes:

$$\frac{dJ_e(x)}{dx} = \alpha_e J_e(x) + \alpha_h [J_e(w) - J_e(x)] = (\alpha_e - \alpha_h) J_e(x) + \alpha_h J_e(w). \tag{13.40}$$

Therefore, the APD gain is defined as:

$$M = \frac{J_e(w)}{J_e(0)}. \tag{13.41}$$

When $\alpha_e \neq \alpha_h$, we obtain the gain to be:

$$M = \frac{1 - k_A}{e^{-[(1-k_A)\alpha_e w]} - k_A}. \tag{13.42}$$

13.7.4 APD Rise Time

The APD rise time, τ τ, accounts for carrier transit and multiplication times. It is given by:

$$\tau = \frac{w_d}{V_e} + \frac{w_d}{V_h} + \tau_m, \tag{13.43}$$

where w_d is the width of the absorption region, w_m is the width of the multiplication region, and τ_m is the multiplication region's contribution:

$$\tau_m \cong \frac{GK_A w_m}{V_e} + \frac{w_m}{V_h}. \qquad (13.44)$$

Example Determine the APD rise time for the following parameters:

$w_d = 50\,\mu m\,\mu m = 0.5\,\mu m$, $V_h = 10^2$ cm/s, $V_e = 5 \times 10^6$ cm/s, $G = 100$, $k = 0.1$

Solution

$$\tau_m = \frac{100 \times 0.1 \times 0.5 \times 10^{-6}}{10^5} + \frac{0.5 \times 10^{-6}}{5 \times 10^4} = 20\,\text{ps}$$

Therefore the rise time is (Table 13.4):

$$\tau = \frac{50 \times 10^{-6}}{10^5} + \frac{50 \times 10^{-6}}{5 \times 10^4} + 20 \times 10^{-12}$$

$$\tau = 1.07\,\text{ns}$$

13.8 Optical Receivers

In an optical receiver, the photodiode is typically followed by a trans-impedance amplifier (TIA) which converts the small photocurrent into a measurable voltage signal. The design of the TIA is critical as it determines the receiver's noise performance and bandwidth, both of which significantly affect the overall sensitivity and reliability of the communication system.

Table 13.4 Typical parameters for avalanche photodiodes

Parameter	Units	Silicon (Si)	Germanium (Ge)	InGaAs
Wavelength range	nm	400–1100	800–1600	1100–1700
Peak wavelength	nm	800	1550	16,000
Avalanche gain		200–400	50–500	10–50
Rise time	ns	0.1–2	0.5–0.8	0.1–0.5
Dark current	nA	0.1–1	20–500	10–50
Bandwidth	GHz	100–400	2–10	20–250
Bias voltage	V	150–400	20–40	20–30

13.8 Optical Receivers

Receiver noise is a crucial factor that impacts the signal-to-noise ratio (SNR) and determines the minimum detectable signal and the reliability of data transmission.

The following are the definitions of important quantities we need to consider in the discussion of optical receivers:

1. Signal-to-noise ratio (SNR) which is defined as:

$$SNR = \frac{P_{signal}}{P_{noise}}, \quad (13.45)$$

where P_{signal} and P_{noise} represent the power of the signal and noise, respectively.

2. Minimum detectable signal is the signal that yields:

$$SNR = 1 \quad (13.46)$$

3. Bit error rate (BER) is the probability of error per bit in a digital optical receiver, usually 10^{-9} to 10^{-12}.

4. Receiver sensitivity is the signal corresponding to a specific SNR_o that can be between:

$$10 < SNR_o < 1000 \quad (13.47)$$

and

$$10\,dB < SNR_o < 30\,dB. \quad (13.48)$$

13.8.1 Noise in Optical Receivers

Noise in optical receivers arises from multiple sources, including thermal noise, shot noise, and photon noise. Since noise is statistical in nature, its effects are quantified using time-averaged currents and variances.

Thermal Noise Thermal noise, also known as Johnson-Nyqvist noise, is generated by the random motion of electrons in the resistive components of the receiver circuit. This noise is proportional to the temperature and resistance and is a fundamental limit to the performance of electronic circuits. The thermal noise in a resistor (R_L) is given by:

$$<i_{th}>^2 = \frac{4k_B T B}{R_L} \quad (13.49)$$

where T is the temperature in Kelvin, k_B is the Boltzmann constant, and B is the bandwidth of the receiver in $bits/second$.

Shot Noise hot noise is another significant source of noise in optical receivers, originating from the discrete nature of the charge carriers (electrons and holes) generated by the photodiode. It is inherent in the process of photodetection and is directly proportional to the average photocurrent. The randomness in the arrival of photons causes fluctuations in the photocurrent, contributing to the overall noise. The shot noise current is given by:

$$<i_{sh}>^2 = 2qIB \tag{13.50}$$

where $I = i_{ph} + I_d$, and I_d is the dark current.

Photon Noise Photon noise is caused by random arrival of photons on the photodetector which can be expressed as:

$$<n> = \phi\tau \tag{13.51}$$

where ϕ is the photon flux and τ is the pulse width. It obeys a Poison distribution with variance is given by:

$$\sigma_n^2 = <n> \tag{13.52}$$

so the photon number SNR is

$$SNR = \frac{<n>^2}{\sigma_n^2} = <n>, \tag{13.53}$$

Example Determine the minimum detectable photon number $<n> = 1$ for $\tau = 1\,\mu s$ and $\lambda = 1.24\,\mu m$.

Solution The minimum detectable power is:

$$P_{opt} = n \times h\nu/\tau = 0.16\,\text{pw}$$

The receiver sensitivity for $SNR = 10^3 = 30\,\text{dB}$ is 1000 photons. If $\tau = 10\,\text{ns}$, then 10^{11} photons/s

$$P_{opt} = 16\,\text{nW}$$

The total noice current including thermal and shot boise currents is given by:

$$\sigma^2 = <(\Delta I)^2> = 2q(I_{ph} + I_d)B + \frac{4k_BTB}{R_L}. \tag{13.54}$$

13.8 Optical Receivers

Hence for p-i-n receivers the signal-to-noise ratio is given by:

$$SNR = \frac{I_{ph}^2}{\sigma^2}, \quad (13.55)$$

$$SNR = \frac{\text{Re}^2 P_{in}^2}{2q(I_{ph} + I_d)B + 4k_B T B/R_L} \quad (13.56)$$

where Re is the responsivity of the photodiode and is given by:

$$\text{Re} = \eta \frac{1.24}{\lambda (\mu m)}. \quad (13.57)$$

All these sources of noise are present in receiver systems; however, under certain condition, one of these sources dominate. We explore these scenarios next.

13.8.2 Thermal Noise Limited Case

In most practical cases, thermal noise dominates the receiver performance neglecting shot noise:

$$SNR = \frac{R_L \text{R}^2 P_{in}^2}{4k_B T B} \quad (13.58)$$

Therefore the thermal noise increases by the product of P_{in} and R_L; thus, receivers are usually designed with high input impedance.

The effect of thermal noise is quantified through the **Noise-Equivalent-Power (NEP)**, which is defined as the minimum optical power per unit bandwidth required to produce a signal-noise-ratio of unity, i.e.,

$$SNR = 1.$$

Therefore, the NEP is expressed as:

$$NEP = \frac{P_{in}}{\sqrt{B}} = \left(\frac{4K_B T}{B R_L}\right)^{1/2}, \quad (13.59)$$

simplifying yields:

$$NEP = \frac{h\nu}{\eta q}\left(\frac{4K_B T}{B R_L}\right)^{1/2}. \quad (13.60)$$

Typical values for NEP are in the range of 1 to 10 pw/$\sqrt{\text{Hz}}$.

13.8.3 Shot Noise Limited Case

Consider the following case where the shot-noise is much greater than thermal-noise, i.e.,

$$\sigma_{sh}^2 \gg \sigma_{th}^2. \tag{13.61}$$

For a p-i-n photodiode the the signal-to-noise ratio is:

$$SNR = \frac{\Re P_{in}}{2qB}, \tag{13.62}$$

substituting for the responsivity yields the following:

$$SNR = \frac{\eta P_{in}}{2h\nu B} \tag{13.63}$$

Therefore, the SNR increases linearly with P_{in}, which depends on quantum efficiency (η). Let the number of photons of the input signal represented by N_p, which are contained in "1" bit.

For a pulse bit duration $\tau = 1/B_d$, where B_d is bit rate; hence, the photon energy for one bit is given by:

$$E_p = \frac{P_{in}}{B_d} = N_p h\nu, \tag{13.64}$$

therefore,

$$P_{in} = N_p h\nu B_d. \tag{13.65}$$

The bandwidth of the system is related to the bit rate as:

$$B = \frac{B_d}{2}. \tag{13.66}$$

Therefore, the signal-to-noise ration is:

$$SNR = \frac{\eta N_p h\nu (2B)}{2h\nu B} = \eta N_p \tag{13.67}$$

in the shot noise limit a SNR of 20 dB can be realized if $N_p = 100$ and $\eta \simeq 1$.

13.8 Optical Receivers

Example For 1.55 µm receiver operating at 10 Gb/s and $N_p = 100$, calculate the input power.

Solution

$$P_{in} = \frac{N_p h \nu B}{\lambda} = \frac{100 \times h \times 3 \times 10^8 \times 10^{10}}{1.55 \times 10^{-6}}$$

$$P_{in} \approx 130 \, \text{nW}$$

Optical receivers that employ avalanche photodiodes generally produce a larger signal-to-noise ratio for the same input optical power because of their gain (M). The improvement is due to the internal gain from the avalanche photodiode. The photocurrent is given by:

$$I_{ph} = M \times R \times P_{in}, \tag{13.68}$$

where R is the responsivity.

Shot Noise Enhancement Thermal noise remains the same for avalanche photodiode receivers. The avalanche photodiode gain results in a second source of electron-hole pairs generation through the impact ionization process; thus, the formula for the shot noise becomes:

$$\sigma_{sh}^2 = 2qM^2(RP_{in} + I_d)BF_A, \tag{13.69}$$

where F_A is the excess noise factor of the APD, which is defined as:

$$F_A = k_A M + (1 - k_A)(2 - 1/M), \tag{13.70}$$

where k_A is the impact ionization factor:

$$k_A = \begin{cases} \dfrac{\alpha_n}{\alpha_e} & for\ \alpha_n < \alpha_e \\ \dfrac{\alpha_e}{\alpha_n} & for\ \alpha_n > \alpha_e \end{cases}. \tag{13.71}$$

Therefore,

$$0 < k_A < 1 \tag{13.72}$$

and

$$2 < F_A < M. \tag{13.73}$$

The SNR for an avalanche photodiode is:

$$SNR = \frac{I_{pn}^2}{\sigma_{sh}^2 + \sigma_{th}^2} = \frac{(M\,\mathrm{Re}\,P_{in})^2}{2qM^2 F_A(\mathrm{Re}\,p_{in} + I_d)B + \frac{4k_B T B}{R_L}}. \tag{13.74}$$

In the thermal noise limit case, the SNR becomes:

$$SNR = \frac{R_L \,\mathrm{Re}^2\, M^2 P_{in}^2}{4k_B T B}, \tag{13.75}$$

whereas in the shot-noise limit case, the SNR becomes:

$$SNR = \frac{\mathrm{Re}\,P_{in}}{2q F_A B} = \frac{\eta P_{in}}{2h\nu F_A B}. \tag{13.76}$$

13.8.4 Optimum APD Gain

For a given P_{in}, the SNR will be maximum for an optimum gain M_{opt}, which satisfies the following relationship:

$$k_A M_{opt}^3 + (1 - k_A) M_{opt} = \frac{4k_B T B}{q R_L (\mathrm{Re}\,P_{in} + I_d)}, \tag{13.77}$$

simplifying yields the following:

$$M_{opt} \simeq \left[\frac{4k_B T B}{R_A q R_L (\mathrm{Re}\,P_{in} + I_d)}\right]^{1/3}. \tag{13.78}$$

Example Determine the signal-to-noise ratio for the modulated signal given by:

$$P_{in} = P_o[1 + mf(t)],$$

where m is the modulation index.

(continued)

13.8 Optical Receivers

Solution The photocurrent detected by the photodiode is given by:

$$I_{pn} = \text{Re } P_o[1 + mf(t)].$$

The average photocurrent is:

$$I_{pn} = R_o P_o.$$

The small signal component is given by:

$$i_s(t) = I_{ph} + mf(o).$$

Therefore,

$$SNR = \frac{I_{ph}}{2qB} m^2 < f^2(t) > .$$

13.8.5 Receiver Sensitivity

A receiver is more sensitive if it achieves the same performance with less optical power. The performance criterion for digital systems is the bit error rate (BER), as shown in Fig. 13.15. To detect the received signal and distinguish between 1 and 0 bits, the current of choice that determines the bits is given by I_D. For bit 1, the condition is:

$$I > I_D, \quad (13.79)$$

and for bit0, the condition is:

$$I < I_D. \quad (13.80)$$

Fig. 13.15 Digital signal detection and noise distribution

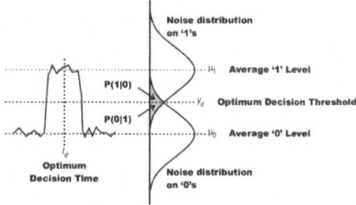

The bit error rate is defined by:

$$BER = p(1)p(0/1) + p(0)p(1/0), \tag{13.81}$$

where $p(1)$ and $p(0)$ are the probabilities of the occurrence of bit 1 and 0, respectively. Whereas, $p(0/1)$ is the probability of detecting 1, while the bit is 0, and $p(1/0)$ is the probability of detecting 0, while the bit is 1.

Typically, the occurrence of bits 1 and 0 is the same, i.e.,

$$p(1) = p(0) = 1/2, \tag{13.82}$$

therefore,

$$BER = \frac{1}{2}[p(0/1) + p(1/0)]. \tag{13.83}$$

Substituting the typical values for $p(0/1)$ and $p(1/0)$ in Eq. (13.83), we obtain:

$$BER = \frac{1}{4}\left[erfc\left(\frac{I_1 - I_D}{\sigma_1\sqrt{2}}\right) + erfc\left(\frac{I_D - I_o}{\sigma_o\sqrt{2}}\right)\right], \tag{13.84}$$

where $erfc(x)$ is the error function defined as

$$erfc(x) = \frac{2}{\sqrt{\pi}} \int_x^\infty e^{-y^2} dy, \tag{13.85}$$

where σ_1 and σ_0 are the variances for bits 1 and 0, respectively. Therefore, the BER depends on the choice of I_D. The minimum BER occurs when the I_D is chosen such that

$$\frac{(I_D - I_o)^2}{2\sigma_o^2} = \frac{(I_1 - I_D)^2}{2\sigma_1^2} + \ln\left(\frac{\sigma_1}{\sigma_o}\right), \tag{13.86}$$

therefore, the I_D is given by:

$$I_D = \frac{\sigma_o I_1 + \sigma_1 I_o}{\sigma_o + \sigma_1}. \tag{13.87}$$

When $\sigma_o = \sigma_1$, then

$$I_D = \left(\frac{I_1 + I_o}{2}\right). \tag{13.88}$$

13.8 Optical Receivers

Fig. 13.16 BER plotted as a function of the Q-factor using Eq. (13.90)

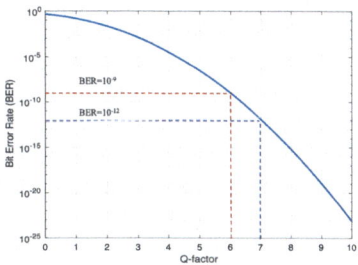

This is the case for most p-i-n receivers because thermal noise dominates. The factor Q is defined as:

$$Q = \frac{I_D - I_o}{\sigma_o} = \frac{I_1 - I_D}{\sigma_1}. \tag{13.89}$$

The BER will be given by the following for the optimum case:

$$BER = \frac{1}{2} erfc\left(\frac{Q}{\sqrt{2}}\right) \simeq \frac{e^{-Q^2/2}}{Q\sqrt{2\pi}} \tag{13.90}$$

For example, a $BER = 10^{-12}$, the Q factor much satisfy the condition $Q \geq 7$; whereas, for a $BER = 10^{-9}$, the condition should be $Q \geq 6$, as shown in Fig. 13.16.

13.8.6 Amplifier Noise Figure

There are other sources of noise generated by components such as transistors, resistors, and other components, which refer to the amplifier noise figure F_A. It is is usually added to the thermal noise such that:

$$\sigma_{th}^2 = \frac{(4k_B T F_n B)}{R_L}. \tag{13.91}$$

Therefore, the total current is given by:

$$I = I_{ph} + I_{sh} + I_{th}. \tag{13.92}$$

The variance for the total noise signal is given by:

$$\sigma^2 = (\Delta I)^2 = 2q(I_d + I_{ph})BM^2 F_A + \frac{4k_B T B F_n}{R_L}. \tag{13.93}$$

Since the SNR is

$$SNR = \frac{I_{ph}^2}{\sigma_{sh}^2 + \sigma_{th}^2}, \tag{13.94}$$

therefore, by substituting Eq. (13.93) into Eq. (13.94), then the SNR can be expressed as follows:

$$SNR = \frac{(M \operatorname{Re} P_{in})^2}{2qM^2 F_A (\operatorname{Re} P_{in} + I_d) B + 4kTBF_n/R_L}. \tag{13.95}$$

13.8.7 Minimum Receiver Power

The minimum power for a receiver needs to operate with a BER below a specified value. Let P_0 and P_1 be the optical power for bits 0 and 1. However,

$$I_1 = M \operatorname{Re} P_1 = 2M \operatorname{Re} \overline{P}_{rec}, \tag{13.96}$$

where $\overline{P}_{rec}$ is the average received power, which is given by:

$$\overline{P}_{rec} = \frac{P_1 + P_0}{2}. \tag{13.97}$$

The root mean square (RMS) noise currents are expressed as:

$$\sigma_1 = (\sigma_{sh}^2 + \sigma_T^2)^{1/2}, \text{ and } \sigma_0 = \sigma_{th}. \tag{13.98}$$

The shot noise is expressed as:

$$\sigma_{sh}^2 = 2qM^2 F_a \operatorname{Re}(2\overline{P}_{rec})B \tag{13.99}$$

and the thermal noise is given by:

$$\sigma_{th}^2 = \frac{2K_B T B F_n}{R_L}. \tag{13.100}$$

The Q factor is defined as:

$$Q = \frac{I_1}{\sigma_1 + \sigma_0} = \frac{2M \operatorname{Re} \overline{P}_{rec}}{(\sigma_{sh}^2 + \sigma_{th}^2)^{1/2} + \sigma_{th}} \tag{13.101}$$

and the BER as determined by the Q is given by:

$$BER \cong \frac{e^{-Q^2/2}}{Q\sqrt{2\pi}}, \qquad (13.102)$$

and the average received signal is expressed as:

$$\overline{P}_{rec} = \frac{Q}{R}\left(qF_A QB + \frac{\sigma_{th}}{M}\right) \qquad (13.103)$$

for a p-i-n, $M = 1$; therefore, the thermal noise (σ_{th}) dominates the received signal.

13.9 Summary

In this chapter, we explored the fundamental principles and various types of optical detectors, focusing on their physical mechanisms, performance characteristics, and practical applications. Photodetectors, such as photodiodes, photoconductors, and avalanche photodiodes (APDs), were analyzed in detail, highlighting their roles in converting optical signals into electrical signals. We delved into the parameters governing their performance, including the responsivity, quantum efficiency, noise characteristics, and impulse response.

The design and operation of photodiodes, particularly p-n and p-i-n structures, were discussed, emphasizing the impact of material properties, junction width, and bias voltage on their efficiency and speed. APDs were presented as advanced devices with internal gain, allowing for enhanced sensitivity under low-light conditions, and their noise characteristics and gain mechanisms were thoroughly examined.

Finally, the chapter addressed the integration of optical detectors within receiver systems, focusing on the signal-to-noise ratio (SNR), noise sources, and receiver sensitivity. The interplay between thermal and shot noise in determining system performance was also covered, with a discussion on optimizing receiver designs for specific applications.

This comprehensive overview provides the foundation for understanding and selecting optical detectors for various technological applications, paving the way for innovations in optical communications, imaging systems, and beyond.

13.10 Problems

1. A photomultiplier tube (PMT) has a photoemission efficiency of 25% and is illuminated by a 500 nm light source with a power of 1 mW. Calculate the rate of emitted electrons.

2. A photoconductive device has $\eta = 0.6$, $q = 1.6 \times 10^{-19}$ C, and $\tau = 2$ μs. If the device is illuminated by 10^{16} photons/cm^2/s, calculate the photogenerated current for an electric field of 50 V/cm.
3. Determine the open-circuit voltage of a reverse-biased silicon photodiode illuminated by light that generates a photocurrent of 50 μA. Assume a reverse saturation current of 5 nA and temperature $T = 300$ K.
4. For a germanium photodiode operating at 1.3 μm, determine the absorption depth if the absorption coefficient is 7×10^5 m^{-1}.
5. A photodiode detects 8×10^{12} photons per second with a light power of 1 mW at 800 nm. Calculate the quantum efficiency.
6. Compute the responsivity of a photodiode with $\eta = 90$ at a wavelength of 1.55 μm.
7. An avalanche photodiode has a gain of 50 and detects 10^{12} photons per second at 1.55 μm. If the quantum efficiency is 80%, calculate the photocurrent.
8. An APD has a dark current of 1 nA, gain $M = 100$, and operates at a bandwidth of 10 MHz. Compute the shot noise current.
9. A p-i-n photodiode has $w = 50$ μm and $V_r = 10$ V. Calculate the transit time for electrons and holes, assuming mobilities $\mu_e = 1500$ cm^2/V.s and $\mu_h = 500$ cm^2/V.s.
10. An optical receiver operates at 10 Gbps with a p-i-n photodiode having $\eta = 0.85$ and $SNR = 10$. Determine the minimum detectable power.

Bibliography

1. Sze, S.M., and Ng, K.K. (2007) Physics of Semiconductor Devices. 3rd Edition, Wiley.
2. Rogers, A.J., and Pennock, S.T. (2018) Handbook of Optoelectronics. CRC Press.
3. Jain, F.C., and Kapoor, A. (2001) 'Avalanche Photodiodes: Design and Applications.' Journal of Lightwave Technology, vol. 19, no. 8.
4. Rogalski, A. (2011) Infrared Detectors. 2nd Edition, CRC Press.
5. Madhukar, A., et al. (2000) 'Quantum Efficiency and Responsivity of Photodiodes.' Journal of Applied Physics, vol. 45.
6. Smith, R.G. (1971) 'Noise in Avalanche Photodiodes.' Bell System Technical Journal, vol. 50, pp. 1055–1104.
7. Capasso, F. (1991) 'Bandgap Engineering for Photodetectors.' Physics Today, vol. 44, no. 8, pp. 34–42.
8. Rogalski, A. (2019) Advanced Infrared Photodetectors. Springer.
9. Saleh, B.E.A., and Teich, M.C. (2019) Fundamentals of Photonics. 3rd Edition, Wiley.
10. Yariv, A., and Yeh, P. (2007) Photonics: Optical Electronics in Modern Communications. 6th Edition, Oxford University Press.

Optical Modulators 14

Optical modulators are crucial devices used for controlling and manipulating light properties, primarily to modulate various aspects of light waves. They enable the modification of optical wave characteristics such as the intensity, phase, polarization, and frequency of light signals. Optical modulation provides the means to control an optical wave or encode information onto a carrier optical wave, while the corresponding process is known as demodulation. They are an essential component in optical communication and optical signal processing systems.

14.1 Modulation Methods and Format

Optical modulation can be categorized into different schemes based on the specific optical-field parameter being manipulated. These categories include phase modulation, frequency modulation, polarization modulation, and amplitude modulation. Additionally, optical modulation can be classified as either analog modulation or digital modulation, depending on whether the information is encoded in analog or digital form.

Optical modulation can be categorized into two main types: direct (internal) modulation and external modulation. Direct modulation involves the direct manipulation of an optical source, typically a light-emitting diode (LED) or a laser diode (LD), without the use of a separate optical modulator. On the other hand, external modulation entails the alteration of an optical wave emitted by a light source by employing a separate optical modulator to change one or more of its characteristics.

As discussed in Chap. 3, an optical wave propagating along the z-axis can be represented by the electrical field $\bar{E}(z, t)$ as follows:

$$\bar{E}(z, t) = \bar{E}_o \cos[\omega t - kz],$$

where $\bar{E}_o$ is the vectorial amplitude, which determines the amplitude of the electric field and its polarization, and where ω is the radial frequency and k is the wavenumber. If we represent the polarization by a vector $\bar{p}$, then we can express the electric field of the planewave as:

$$\bar{E}(z,t) = \bar{p}|\bar{E}_o|cos(\omega t = kz + \varphi).$$

The phase of the planewave is given by:

$$\phi = (\omega t - kz + \varphi),$$

where φ is the initial phase of the wave. The modulation of the optical wave is achieved by controlling one of the wave's parameters, such as the amplitude ($|\bar{E}_o|$), frequency (ω), phase (ϕ), and/or polarization ($\bar{p}$).

The optical wave can serve as a carrier for a modulating signal, which can be either analog or digital. In the case of analog signal modulation, common formats include amplitude modulation (AM), phase modulation (PM), and frequency modulation (FM). For digital modulation, formats such as amplitude shift keying (ASK), phase shift keying (PSK), and frequency shift keying (FSK) are employed.

The modulation of light beams can be achieved by using a modulating signal to control the drive current of the light source such as in the case of directly modulating a light emitting diode (LED) or a laser diode (LD) as shown in the subsequent sections. High speed optical data communication mostly uses devices (external modulators) to modulate a laser beam. These modulators use the modulating signal to control some of their material properties.

14.2 Direct Modulation of Laser Diodes

The optical power of laser diodes is directly proportional to the drive current for the range of currents between the threshold current I_{th} and the saturation current. The modulation of the optical output power from a laser diode is achieved by modulating the LD drive current. The modulation can be from an analog or digital signal.

14.2.1 Modulation of an Analog Signal

We first consider small-signal AC modulation as shown in Fig. 14.1a. A DC current I_1 is applied to bias the LD at the middle of the linear $I - P$ curve yielding a steady-state optical output power P_{o1}. The LD is then modulated by a sinusoidal signal:

$$i_{in}(\omega) = I_m sin(\omega t) \quad (14.1)$$

14.2 Direct Modulation of Laser Diodes

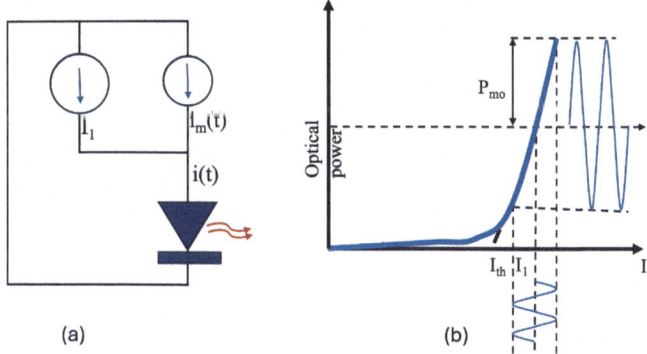

Fig. 14.1 (a) Laser diode modulation drive circuit including the DC bias current I_1 and the modulating drive current i_{in} and (b) the input current and output optic power characteristics

superimposed on I_1 as in Fig. 14.1a, where I_m is the maximum amplitude of the signal and ω is the modulation frequency. The modulation of the diode input current around I_1 results in an increase in the output optical power varying around its DC value of P_{o1} represented by:

$$P_{mo}(\omega) = P_m sin(\omega t) \tag{14.2}$$

as shown in Fig. 14.1b, where P_m is the maximum amplitude of the modulated output optical power. For simplicity, we are dropping any phase shift that might be between the input and output signals.

In the linear region of the $I - P$ curve, the output optical power is directly proportional to the input current. The frequencies of the modulating signal ω and that of the output signal are expected to be the same. The frequency response of the LD depends on the characteristics of the device as discussed in Chap. 12. Consequently, the output optical power ($P_{mo}(\omega)$) depends on the modulation frequency. Let the output optical power at DC be $P_{mo}(0)$; therefore, the ratio $P_{mo}(\omega)/P_{mo}(0)$ is expected to remain constant for low frequencies until it reaches the upper limits of the LD frequency response. At high frequencies, this ratio exhibits a prominent peak at a certain frequency called the relaxation oscillation frequency f_r, as marked in Fig. 14.2. This relaxation oscillation frequency depends on the LD characteristics and is given by

$$f_r = \frac{1}{2\pi \sqrt{\tau \tau_{ph}}} \left(\frac{I_1}{I_{th}} - 1 \right)^{1/2} \tag{14.3}$$

where τ is the effective carrier recombination lifetime, which represents the radiative and nonradiative recombinations; τ_{ph} is the photon cavity lifetime, which is the average time it takes for a photon to be lost from the laser diode cavity; I_1 is the operating diode bias current; and I_{th} is the threshold current of the LD. The

Fig. 14.2 The frequency dependence of $P_{mo}(2\pi f)/P_{mo}(0)$ [1]

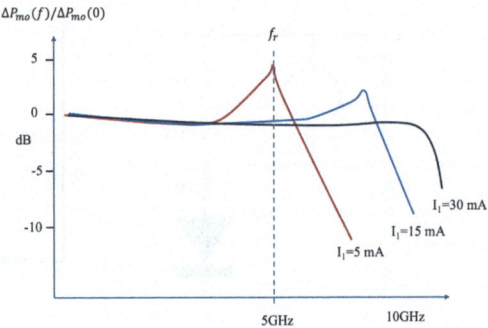

characteristic times τ and τ_{ph} are typically in the nano- and picosecond range so that for the case when I_1 is twice I_{th}, f_r is typically a few GHz. Therefore, the relaxation oscillation frequency f_r increases as a function of the bias current. Laser diodes are usually modulated below f_r.

The electron injection and photon emission concentrations are coupled through the stimulated emission process. The phase difference ϕ between the output and the input signals arises from the delay in obtaining the photon output while keeping up with the electron injection. This delay associated with the recombination process, represented by τ, and the number of times for the photons to exit the cavity, represented by τ_{ph}. As a result, the radiation output lags behind the input current as shown in Fig. 14.1b.

14.2.2 Modulation of a Digital Signal

The modulation of digital signals can be achieved similarly to the analog signal by a bias current around the threshold current. Consider the input current as a step function between $I_1 < I_{th}$ and $I_2 > I_{th}$, as shown in Fig. 14.3. As discussed in the previous section, the output optical power will be lag in the input current by time t_d. Assume that I_1 is initially very small. The lasing radiation output begins only after a delay time td and reaches a steady-state value after a few damped oscillations as illustrated in Fig. 14.4a. The damped relaxation oscillations in the output in Fig. 14.4b have the frequency f_r.

Under direct modulation, LDs are generally the obvious choice in high-speed optical communications, given also their narrow spectral width. There is also a disadvantage to directly modulating LDs by the diode current. The modulation process modifies the refractive index of the gain medium and hence causes the radiation frequency to shift, called frequency chirping, during the modulation period, over the time of a digital bit. The reason is that the refractive index n of the active region depends on the injected electron concentration, which is controlled by the drive current. The frequency chirp can be eliminated by operating the LD as a CW laser and using a high-speed external modulator as discussed in the following sections.

Fig. 14.3 Laser diode direct modulation of a digital signal

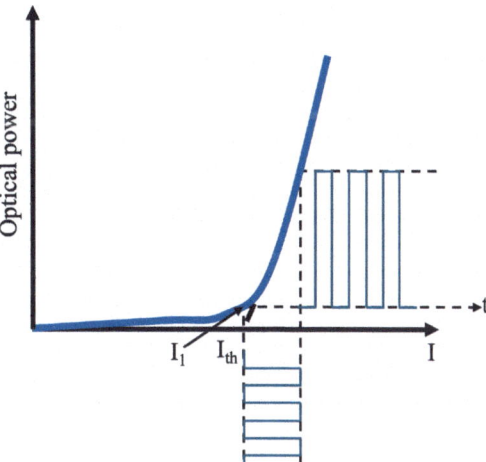

Fig. 14.4 (**b**) Laser diode emitted optical power for (**a**) an input step function drive current

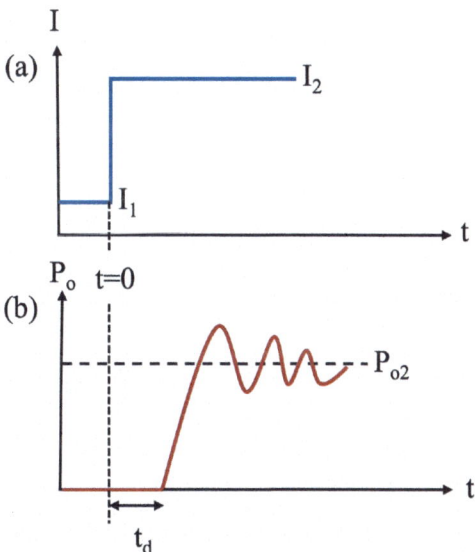

14.3 Indirect (External) Optical Modulation

Indirect or external optical modulation is a critical technique used to control the intensity, phase, or frequency of a laser beam without directly modulating the laser source itself. This method employs external modulators to impose the desired modulation onto the optical signal after it has been generated by the laser. External modulation offers several advantages over direct modulation, including higher modulation speeds, reduced chirp, and the ability to maintain a stable and continuous-wave operation of the laser source.

In this section, we will explore the principles, types, and applications of external optical modulators. We discuss electrooptic modulators (EOMs), electroabsorption modulators (EAMs), and other relevant devices, highlighting their operational mechanisms, performance characteristics, and typical applications.

14.3.1 Electrooptic Effects

When a strong electric field is applied to a material, the refractive index of the material changes. If the change is linearly proportional to the electric field, the effect is called Pockels effect. If it is proportional to the square of the field, then, it is called the **Kerr** effect.

The relationship between the refractive index $n(E)$ and the electric field E can be expressed as:

$$n(E) = n + a_1 E + \frac{1}{2} a_2 E^2 + \ldots, \tag{14.4}$$

where n is the refractive index when no electric field is applied (i.e., $E = 0$), a_1 represents the linear coefficient, and a_2 represents the quadratic coefficient. These coefficients are defined as:

$$a_1 = \frac{dn}{dE}\Big|_{E=0}, \tag{14.5}$$

and

$$a_2 = \frac{d^2 n}{dE^2}\Big|_{E=0}. \tag{14.6}$$

To simplify, we introduce the **Pockels coefficient** r and the **Kerr coefficient** s:

$$r = -\frac{2a_1}{n^3}, \tag{14.7}$$

and

$$s = -\frac{2a_2}{n_3}. \tag{14.8}$$

Thus, the refractive index as a function of the electric field can be expressed as:

$$n(E) = n - \frac{1}{2} r n^3 E - \frac{1}{2} s n^3 E^2 + \ldots \tag{14.9}$$

14.3 Indirect (External) Optical Modulation

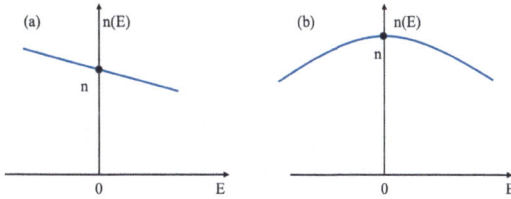

Fig. 14.5 Dependence of the refractive index on the electric field: (**a**) Pockels medium and (**b**) Kerr medium

The effective refractive index in a Pockels medium can be approximated by:

$$n(E) \approx n - \frac{1}{2}rn^3 E, \tag{14.10}$$

while the effective refractive index in a Kerr medium can be approximated as:

$$n(E) \approx n - \frac{1}{2}sn^3 E^2. \tag{14.11}$$

Figure 14.5 depicts the effective refractive index in **Pockels** and **Kerr** materials as a function of changes in the electric field.

The Pockels coefficient r typical ranges from 10^{-12} to 10^{-10} m/V, while the Kerr coefficient s varies depending on the material. For crystals, s ranges from 10^{-18} to 10^{-14} m^2/V^2, and for liquids, it ranges from 10^{-22} to 1^{-19} m^2/V^2.

The Pockels effect is observed in non-center-symmetric materials, such as lithium niobate ($LiNbO_3$) and $GaAs$, which lack a center of symmetry, allowing a linear relationship with the electric field. This effect is widely used in electrooptic modulators, which control the phase and amplitude of light in various optical communication and photonic systems.

On the other hand, the Kerr effect is present in all materials to some extent, as it arises from the intrinsic nonlinear optical properties of the material. This phenomenon is significant in applications such as all-optical switching, where the refractive index changes because the electric field can be used to control light with light.

14.3.2 Electrooptic Phase Modulator

A beam of light traveling across a Pockels cell of length L undergoes a phase shift $\phi = n(E)k_o$, where k_o is the wavenumber in free space and $k_O = 2\pi/\lambda_o$. Since the refractive index $n(E)$ is a function of the electric field, thus, the phase shift ϕ is a function of the electric field (E) and the thickness L of the Pockels cell. Therefore, for a fixed thickness of the Pockels cell, the phase shift can be varied by changing the electric field applied across the cell as:

$$\phi(E) = n(E)k_o L = \left(\frac{2\pi L}{\lambda_o}\right) n(E) \tag{14.12}$$

where λ_o is the wavelength of the optic beam in free space. Let the phase shift of the optic beam undergoes by traveling across the cell without the application of an electric field is given by:

$$\phi_o = \left(\frac{2\pi n L}{\lambda_o}\right). \tag{14.13}$$

Therefore, the beam phase shift after applying an electric field E is:

$$\phi = \phi_o - \frac{1}{2}rn^3\left(\frac{2\pi L}{\lambda_o}\right)E = \phi_o - \frac{\pi rn^3 L}{\lambda_o}E, \tag{14.14}$$

where r is the Pockels coefficient. Applying an electric field across the Pockels cell causes the phase shift to decrease linearly, as shown in Fig. 14.6. An important value is the electric field that produces a π phase shaft, known as the **half-wave electrical field** E_π. Since the electric field is generated by applying voltage, the **half-wave voltage** is denoted as $\mathbf{V}_\pi$, as shown in Fig. 14.6. Substituting for the phase shift π in Eq. (14.4) results in:

$$\pi = \frac{\pi rn^3 L}{\lambda_o}E_\pi, \tag{14.15}$$

$$E_\pi = \frac{\lambda_o}{rn^3 L}. \tag{14.16}$$

The electric field can be applied along the direction of propagation of the optical beam (longitudinal modulator) or perpendicular to the direction of propagation of the optical beam (transverse modulator). These two configurations are shown in Fig. 14.7.

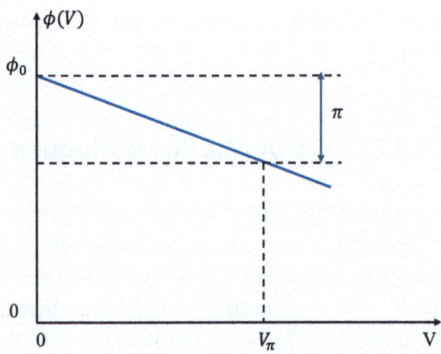

Fig. 14.6 Phase shift of a Pockels cell as a function of the voltage applied across the cell

14.3 Indirect (External) Optical Modulation

Fig. 14.7 Configuration of (**a**) the longitudinal phase modulator and (**b**) the transverse phase modulator. The voltage is applied across the Pockels cell via transparent electrodes

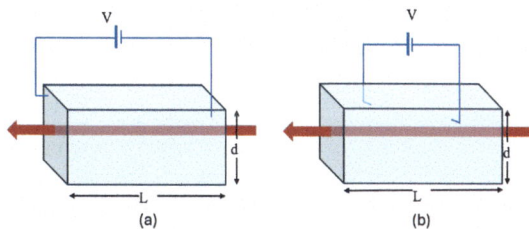

In the **longitudinal phase modulator**, the applied voltage V induces an electric field across the modulator of thickness d given by:

$$E = \frac{V}{d}. \tag{14.17}$$

Therefore, the phase shift is of the phase modulator is represented by:

$$\phi = \phi_o - \frac{\pi r n^3 V L}{d \lambda_o}, \tag{14.18}$$

and the half-wave voltage is given by:

$$V_\pi = \frac{d}{L} \frac{\lambda_o}{r n^3}, \tag{14.19}$$

and the total phase shift of the optical beam as it propagates along the phase modulator is given by:

$$\phi = \phi_o - \pi \frac{V}{V_\pi}. \tag{14.20}$$

In the **transverse phase modulator**, the applied voltage V induces an electric field across the modulator of thickness d given by:

$$E = \frac{V}{L}. \tag{14.21}$$

Therefore, the phase shift is of the phase modulator is represented by:

$$\phi = \phi_o - \frac{\pi r n^3 V}{\lambda_o}, \tag{14.22}$$

and the half-wave voltage is given by:

$$V_\pi = \frac{\lambda_o}{r n^3}, \tag{14.23}$$

and the total phase shift of the optical beam as it propagates along the phase modulator is given by:

$$\phi = \phi_o - \pi \frac{V}{V_\pi}. \qquad (14.24)$$

Therefore, we can modulate the phase of an optical wave by varying the voltage V applied across a material through which the light passes. The parameter V_π is an important characteristic of the modulator. It depends on the material properties (n and r), the wavelength λ_o, and the aspect ratio d/L. The value of the electrooptic coefficient r depends on the direction of propagation and the applied field since the crystal is, in general, anisotropic. The typical values of the half-wave voltage are in the range of 1 to a few kilovolts for longitudinal modulators and hundreds of volts for transverse modulators.

The speed at which an electrooptic modulator operates is limited by electrical capacitive effects and the transit time of the light through the Pockels cell. If the electric field $E(t)$ varies significantly within the light transit time T, the traveling optical wave will be subjected to different electric fields as it traverses the crystal. The modulated phase at a given time t is then be proportional to the average electric field $E(t)$ at times from $t - T$ to t. As a result, the transit-time-limited modulation bandwidth is $\approx 1/T$. If the velocity of the traveling electrical wave matches that of the optical wave, the transit time effects can be eliminated. Commercial modulators in the forms shown in Fig. 14.7 generally operate at several hundred MHz, but modulation speeds of several GHz are possible.

Electrooptic modulators can also be constructed as integrated-optical devices, which operate at higher speeds and lower voltages than do bulk devices. An optical waveguide is fabricated in an electrooptic substrate, such as lithium niobate ($LiNbO_3$), and an electric field is applied to the waveguide using electrodes, as shown in Fig. 14.8. Because the configuration is transverse and the width of the waveguide is much smaller than its length ($d \ll L$), the half-wave voltage can be as small as a few volts. These modulators can be operated at speeds in excess of 100 GHz. Light can be conveniently coupled into, and out of, the modulator by the use of optical fibers.

An important factor that affects the design and size of the modulator which is given by the product $V_\pi L$ as demonstrated by Eq. (14.29):

$$V_\pi L = \frac{d\lambda_o}{rn^3}.$$

Fig. 14.8 Integrated optical waveguide phase modulator

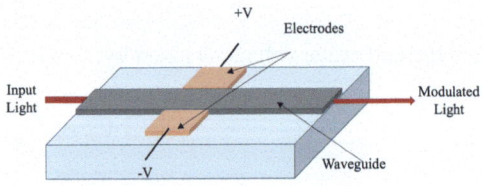

This factor depends on the material properties and wavelength of the light. Typical values are as follows:

- Lithium niobate: $V_\pi L = 14 V$ cm,
- Silicon (depletion-mode): $V_\pi L = 4\ V$ cm
- Silicon (MOS-capacitor): $V_\pi L = 0.4 V$ cm.

14.3.3 Electrooptic Intensity Modulator

Optical phase modulators can be incorporated into an interferometric setup, such as a **Mach-Zehnder interferometer** (*MZI*), to create an intensity modulator. In an optical interferometer, the incoming light is split into two paths, undergoes phase shifts in each path, and is then recombined at the output. The relative phase difference between the two paths determines the interference pattern and, consequently, the output intensity.

- When the phase shift between the two waves is 0°, maximum constructive interference occurs, resulting in the highest output intensity.
- When the phase shift is 180°, maximum destructive interference occurs, producing the lowest output intensity.

A **Mach-Zehnder modulator** (*MZM*) operates by controlling the relative phase difference between the two paths using a modulation voltage. This is achieved via the electrooptic effect, where the refractive index of the material changes in response to the applied electric field. By varying the modulation voltage, the phase difference between the two paths is altered, modulating the output intensity to produce the desired signal.

Consider the interferometer setup illustrated in Fig. 14.9. An optical phase modulator is placed in one arm of the interferometer, and the incoming light is split equally between the two arms. The phase modulator introduces a phase shift between the light beams in the two paths. By changing the voltage applied to the phase modulator, the transmittance $T(V)$ of the interferometer can be controlled, as shown in Fig. 14.9.

If the beam splitters divide the optical power equally, the intensity transmitted through one output port of the interferometer I_o is related to the incident intensity I_i, as:

$$I_o = \frac{1}{2}I_i + \frac{1}{2}I_i cos(\phi) = I_i cos^2(\phi/2). \qquad (14.25)$$

Therefore, the transmittance of the interferometer using Eq. (14.24) is given by:

$$T(V) = cos^2\left(\frac{\phi_o}{2} - \frac{\pi}{2}\frac{V}{V_\pi}\right). \qquad (14.26)$$

Fig. 14.9 (a) A Mach-Zehnder intensity modulator and (b) the transmittance of the modulator as a function of the applied voltage

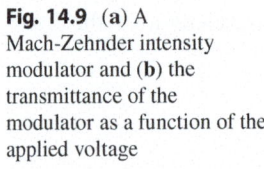

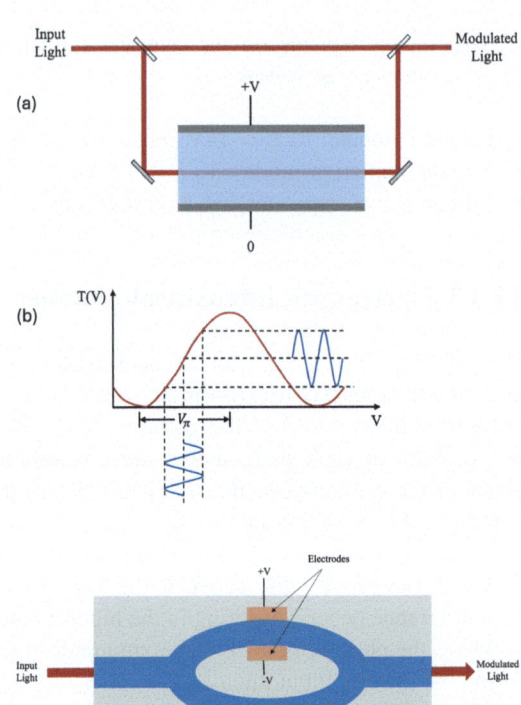

Fig. 14.10 Integrated Mach-Zehnder intensity modulator

This function is plotted in Fig. 14.9b for an arbitrary value of ϕ_o. The device may be operated as a linear intensity modulator by adjusting the optical path difference so that $\phi_o = \pi/2$ and operating in the nearly linear region around $T = 0.5$. Alternatively, the optical path difference may be adjusted so that ϕ_o is a multiple of 2π. In this case $T(0) = 1$, and $T(V_\pi) = 0$, so that the modulator switches the light on and off as V is switched between 0 and V_π.

A Mach-Zehnder intensity modulator can be constructed in the form of an integrated-optical device. Waveguides are placed on a substrate in the geometry shown in Fig. 14.10. The beam splitters are implemented with the use of a Y waveguide. The optical input and output may be coupled in and out by optical fibers. Commercially available integrated-optical modulators generally operate at speeds of a few GHz, but modulation speeds exceeding 100 GHz have been achieved.

Furthermore, we can use two-phase modulators to make an amplitude modulator using the layout shown in Fig. 14.11. The input electrical field $E_{in}(t)$ is split in two using a 3 dB coupler. Each of the each arm passes through a phase modulator that is driven by a voltage $V(t)$. The outputs of the phase modulators are combined by a 3 dB coupler to for the output of the amplitude modulator. This configuration is referred to as a push-pull amplitude modulator.

Fig. 14.11 A dual-drive push-pull optical amplitude modulator

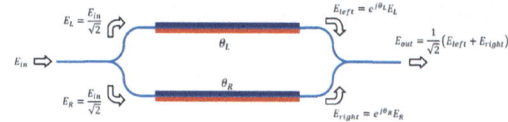

Let

$$v_2(t) = -\frac{V(t)}{2} = -v_1(t), \quad (14.27)$$

therefore,

$$E_{out}(t) = \frac{E_{in}}{2}\left(e^{j\left(\frac{\pi v(t)}{2V_\pi}\right)} + e^{-j\left(\frac{\pi v(t)}{2V_\pi}\right)}\right), \quad (14.28)$$

simplifying yields to:

$$E_{out}(t) = E_{in}\cos\left(\frac{\pi v(t)}{2V_\pi}\right). \quad (14.29)$$

Push-pull amplitude modulators are typically implemented in a $LiNbO_3$, but silicon-on-insulator (SOI)-based modulators have been introduced. In a properly designed push-pull-driven Si modulator, the contributions of both the phase shifter and the MZI transfer function to the overall third-order nonlinearity can cancel each other and be lower than that of a conventional modulator with an ideal linear phase shifter.

Mach-Zehnder modulators (MZMs) have two general configurations: single-drive and a dual-drive. A single-drive MZM uses a phase shifter in one arm, which requires the application of V_π. In this case, only a single high-speed input signal is needed, but there is generally some chirp in the output signal. On the other hand, a dual-drive MZM drives both arms in a differential/push-pull manner. This ideally results in no chirp at the output and only needs to apply $\pm V_\pi/2$ on the two arms to achieve the maximum extinction ratio.

14.4 Electroabsorption Modulators

Electroabsorption modulators (EAMs) enable the modulation of light signals by leveraging the electroabsorption effect, where the absorption of light in a semiconductor material changes under an applied electric field. EAMS are quantum well modulators. The absorption spectrum shift process allows EAMs to control the intensity of light passing through them, effectively encoding information onto a light wave.

Fig. 14.12 Franz-Keldysh effect

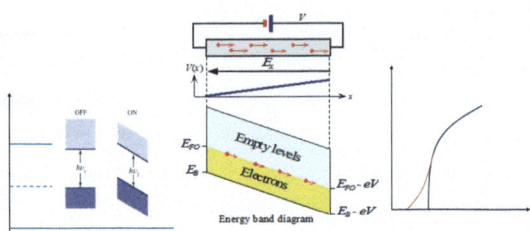

Fig. 14.13 Energy bands tilt caused by the application of an electric field in a quantum well structure

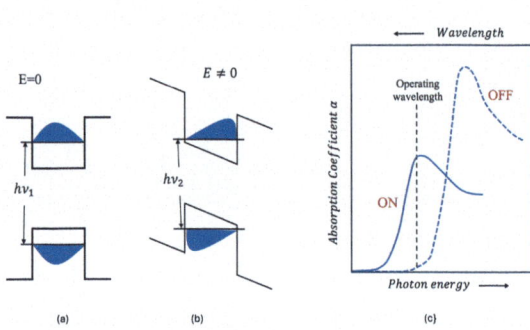

14.4.1 Franz-Keldysh Effect

Consider the energy band diagram for a bulk semiconductor material shown in Fig. 14.12a. A photon with $h\nu$ greater than the energy gap E_g can be absorbed by an electron causing it to transit from the valence band to the conduction band. When a voltage V is applied across the semiconductor, as shown in Fig. 14.12b, an electric field is created causing electrons to move from the negative to the positive terminal. This applied voltage V creates a potential energy qV, where q is the charge of an electron, resulting in a tilt of the energy bands, causing the conduction, Fermi, and valence energies to change across the device, as shown in Fig. 14.12b. As a result, the energy bands become tilted, reducing the separation energy to less than E_g. This shift causes the absorption edge to move to lower energy, or longer wavelengths, as illustrated in Fig. 14.12c. This phenomenon is known as the Franz-Keldysh effect. Additionally, the applied electric field broadens and ultimately causes the disappearance of the exciton absorption peaks.

14.4.2 Quantum-Confined Stark Effect

In bulk semiconductors, electrons and holes are not confined by potential wells. However, in quantum well structures, the electron and hole wave functions are confined by a potential well, as shown in Fig. 14.13. This principle is referred to as quantum-confinement. When an electric field (E) is applied to a quantum-well device, Fig. 14.13a, the energy bands tilt as shown in Fig. 14.13b. Photons

14.4 Electroabsorption Modulators

with energy $h\nu_2 < E_g$ can be absorbed causing the transition of electrons to the conduction band, whereas without an electric field the photon energy required for absorption is $h\nu_1 \geq E_g$. This causes the absorption spectrum edge from $h\nu_1$ to $h\nu_2$. Therefore, the application of an electric field to quantum well structures shifts the absorption spectrum to longer wavelengths, Fig. 14.13c. This phenomenon is known as the **quantum-confined Stark effect** ($QCSE$).

The core principle behind EAMs is the quantum-confined Stark effect (QCSE). When an electric field is applied to a semiconductor material, it alters the electronic band structure, leading to a shift in the absorption spectrum. Specifically, the bandgap energy of the material decreases, causing a shift in the absorption edge to longer wavelengths. This change modulates the intensity of the transmitted light. EAMs are typically made from compound semiconductor materials such as indium phosphide (InP) and gallium arsenide ($GaAs$). These materials were chosen for their favorable electronic and optical properties, which are essential for efficient modulation. The design of an EAM involves a waveguide structure where the light signal travels and an electrode structure that applies an electric field, as shown in Fig. 14.14.

Compared with electrooptic modulators, which operate by changing of the refractive index in response to an externally applied electric field, electroabsorption modulators typically operate at greater speeds and at lower voltages. Since they can be integrated on a single chip with semiconductor light sources, they are convenient for use in optical fiber communication systems. Additionally, they exhibit less chirp than directly modulated laser diodes

The electroabsorption effect is more pronounced in semiconductor multi-quantum-well (MQW) structures. An electric field applied in the plane of a quantum well gives rise to the quantum-confined Stark effect, including a shift of the absorption edge to a longer wavelength. The QCSE results in the energy difference between the conduction- and valence-band energy levels decreasing with increasing electric field ($h\nu_2 < h\nu_1$), and the band tilt causes the locations of the wave functions to shift toward the edges of the well, as shown in Fig. 14.14

As a result of these MQW characteristics, the wavelength shift of the absorption peak is greater, and the absorption edge is more abrupt, than that in bulk semiconductors. Electroabsorption modulators based on the QCSE have excellent characteristics, including high speeds, large extinction ratios, low drive voltages, and low chirp. The simplest transmission implementation directs light through

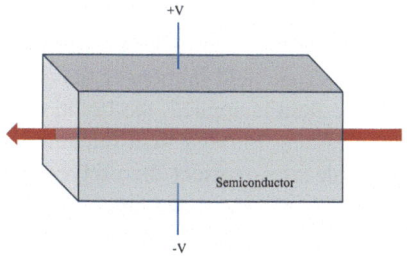

Fig. 14.14
Electro-absorption modulator integrated waveguide device

an intrinsic MQW structure sandwiched between p and n regions across which a voltage is applied. Switching is accomplished by simply turning the voltage on and off. A device of this sort can also be fabricated in a waveguide configuration and can be integrated with a DFB laser on a single chip.

EAMs offer several advantages, including high-speed operation, compact size, and integration capability with other photonic components. However, they also face challenges such as temperature sensitivity and relatively high insertion loss compared with other types of modulators, such as e Mach-Zehnder modulators. Innovations in integration techniques enable seamless incorporation of EAMs into photonic integrated circuits (PICs), which are vital for next-generation optical networks.

14.5 Liquid Crystal Optical Modulators

Liquid crystal optical modulators (LCOMs) are vital components of modern optical systems, playing an essential role in a wide range of applications, from display technology to advanced communication systems. These devices leverage the unique properties of liquid crystals—substances that exhibit both liquid and crystalline characteristics—to modulate light in various ways. By altering the orientation of liquid crystal molecules in response to an applied electric field, LCOMs can control the phase, amplitude, or polarization of the light passing through them.

Liquid crystals (LCs) are a state of matter that have properties between those of conventional liquids and solid crystals. The most common type used in optical modulators is the nematic liquid crystal, characterized by the long-range orientational order of its rod-like molecules. When an electric field is applied, these molecules reorient themselves, causing changes in the optical properties of the material.

There are several types of liquid crystal optical modulators, each designed for specific functions and applications:

1. **Phase Modulators**: These devices alter the phase of the incoming light wave. They are commonly used in adaptive optics, interferometry, and holography.
2. **Amplitude Modulators**: By controlling the alignment of the liquid crystal molecules, these modulators can vary the intensity of light. They are often used in display technologies and optical switching.
3. **Polarization Modulators**: These modulators change the polarization state of light, which is crucial in various imaging and sensing applications.

The operation of an LCOM relies on the electrooptic effect, where the application of an electric field influences the optical properties of the liquid crystal. When no electric field is applied, the liquid crystal molecules are in a relaxed state, and the light passing through than remains unaffected. When an electric field is applied, the molecules reorient themselves, changing the optical path length, intensity, or polarization of the transmitted light.

14.5.1 Liquid Crystal Intensity Modulators

In the nematic phase, the rod-like molecules of the liquid crystal are oriented parallel to each other but without positional order. This means that while the molecules align in a common direction, they do not form a regular lattice as they do in solid crystals. The average orientation of the molecules is described by a unit vector called the "director." The direction of the director can be influenced by external factors such as electric or magnetic fields, surface interactions, or temperature changes. Despite the orientational order, nematic liquid crystals maintain fluidity similar to conventional liquids. This allows them to flow and conform to the shape of their container while preserving their ordered structure. In nematic liquid crystals, the orientations of the molecules tend to be the same, but their positions are completely random.

Nematic liquid crystals are birefringent, having different refractive indices along and perpendicular to the director. This property enables the modulation of the light phase and polarization when passing through the liquid crystal. When an electric field is applied, the orientation of the liquid crystal molecules can be reoriented, changing optical properties of the material. This electrooptic effect is the basis for many applications of nematic liquid crystals, including displays and modulators.

The orientation of the LC molecules depends on the strength of the electric field $E = V/d$. There is a critical Fréedericksz threshold voltage V_c below which the electric field has no effect on the orientation of the molecules. For voltage $V > V_c$, the angle of the orientation θ becomes proportional to $V - V_c$ as given by the following relationship:

$$\theta = 0 \; for \; V \leq V_c,$$

and

$$\theta = \frac{\pi}{2} - 2tan^{-1}\left(e^{-\left(\frac{V-V_c}{V_o}\right)}\right), \; for \; V > V_c,$$

where V is the applied voltage, V_c is the critical voltage at which the tilting process begins, and V_o a constant. When $V - V_c = V_o$, $\theta = 50°$; as $V - V_c$ increases beyond V_o, θ approaches $90°$ when $(V - V_c)/V_o \approx 4$, as indicated in Fig. 14.15.

When the electric field is removed, the orientation of the molecules near the glass plates is reasserted, and the molecules return back to their original orientations, in planes parallel to the plates.

Consider a thin film of liquid crystal sandwiched between two glass plates coated with transparent electrodes. The electrodes are patterned such that the LC molecules will be oriented such that the director on both electrodes is in the same direction, as shown in Fig. 14.16. The molecular rods allow the light beam polarized along the rods passes through while it block all other polarizations. The molecules reorient themselves with a continuous twist to match the orientations of the molecules at the electrodes. The light entering the device with polarization parallel to the input electrode will be oriented such that it exits parallel to the polarization of the incident

Fig. 14.15 Liquid crystal molecules tilt angle θ as a function of the applied voltage

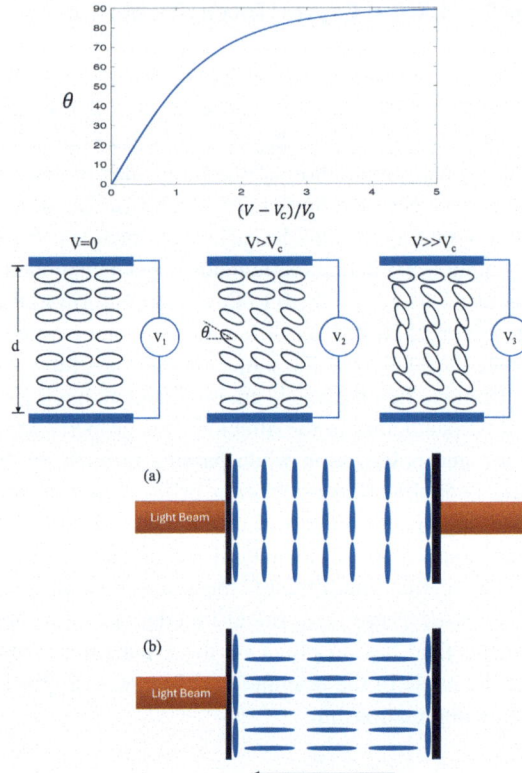

Fig. 14.16 Liquid crystal intensity modulator: (**a**) no electric field applied and light beam passes through the "ON" state, and (**b**) electric field E applied across the device and the light beam does not pass through the "OFF" state

beam. Therefore, in this case, the device is transparent to the incident beam. This is the "ON" state for this LC modulator, Fig. 14.16a. On the other hand, when an electric field is applied parallel to the transmission path, the orientation of the LC molecules away from the electrodes realign themselves in the direction of the electric field, as shown in Fig. 14.16b. As a result, the light is blocked, and the device becomes opaque representing the "OFF" state of the modulator.

Liquid crystals are relatively slow, with response timed depending on the thickness of the liquid crystal layer, the viscosity of the material, the temperature, and the applied drive voltage. The rise time is on the order of tens of milliseconds if the operating voltage is near the critical voltage V_c, but decreases to a few milliseconds at higher voltages. This switching speed is not suitable for communication applications but is suitable for other applications such as displays.

14.5.2 Liquid Crystal Displays

Liquid crystal displays (LCDs) have become an integral part of modern technology and are ubiquitous in everything from smartphones and televisions to computer

14.5 Liquid Crystal Optical Modulators

monitors and digital signage. LCDs leverage the unique properties of liquid crystals to create images and text by modulating light.

LCDs are essentially a two-dimensional array of liquid crystal intensity modulators. By applying a variable voltage, the transmittance of these elements can be modified to transmit several levels of intensity. Each element of these liquid crystal modulators represents a pixel of the displayed image.

Structure of Color LCDs A color LCD consists of several layers, each playing a crucial role in the display's operation:

1. **Backlight Unit**: This layer provides the necessary illumination for the display. Modern LCDs typically use LED backlights, which offer high efficiency and brightness.
2. **Polarizing Filters**: There are two polarizing filters, one at the front and one at the back of the liquid crystal layer. These filters are oriented perpendicular to each other. The light from the backlight passes through the first polarizer, becoming polarized light.
3. **Liquid Crystal Layer**: The liquid crystal layer is sandwiched between two glass substrates with transparent electrodes. This layer contains liquid crystals that can twist and untwist in response to an applied electric field, thus controlling the passage of light. When an electric field is applied, the orientation of the liquid crystal molecules changes, modifying the light phase and polarization.
4. **Color Filter Array**: Each pixel is divided into three subpixels, with each subpixel covered by a red, green, or blue filter. By controlling the intensity of light passing through each subpixel, the display can produce a full spectrum of colors through additive color mixing.
5. **Transparent Electrodes**: These electrodes are typically made of indium tin oxide (ITO) and are used to apply an electric field across the liquid crystal layer. The electrodes are patterned to correspond to each pixel and subpixel.
6. **Alignment Layers**: These layers are applied to the surfaces of the glass substrates in contact with the liquid crystal layer. They help align the liquid crystal molecules in a specific direction when no electric field is applied.

Operation of Color LCDs The operation of an LCD involves the following steps.

The light emitted from the backlight passes through the first polarizing filter, becoming polarized. The polarized light then enters the liquid crystal layer, where the degree of twist of the liquid crystal molecules is determined by the voltage applied to the transparent electrodes. This twisting alters the light's polarization. The modulated light subsequently passes through the color filter array, which separates it into red, green, and blue components. The light then encounters the second polarizing filter, often referred to as the analyzer. If the light's polarization matches the orientation of this filter, it passes through; otherwise, it is blocked. The alignment of the liquid crystal molecules, which is controlled by the applied voltage, determines the degree to which light passes through. By controlling the voltage applied to each subpixel, the LCD can adjust the intensity of red, green, and blue light passing through each pixel. This modulation allows the display to produce a wide range of colors and shades, creating the desired image.

In summary, color LCDs use an array of liquid crystal modulators, backlight, polarizing filters, color filters, and transparent electrodes to control light and create images. The precise modulation of light at each pixel and subpixel enables the display to produce high-quality, full-color images for various applications.

14.6 Microring Optical Modulators

Microring optical modulators are based on microring resonators, as introduced in Chap. 6. They are constructed using a ring-shaped waveguide made from materials with high refractive indices, such as silicon or silicon nitride. These rings are typically very small in size, often on the order of few micrometers, allowing for integration into photonic integrated circuits ($PICs$). The specific operation of a microring modulator depends on its design and the type of modulation it is intended for.

Microring resonators have resonant frequencies depending on the size of the ring and its effective index of refraction. Changing one of these parameters causes a shift in the resonant frequency. Whereas changing the size of the ring is not an option but changing the effective index of refraction is possible. These devices operate by changing the effective refractive index of the ring resonator through various mechanisms such as free carrier injection, thermo-optic effects, or electrooptic effects, which in turn affect the phase or intensity of light passing through the ring.

A schematic of the electrooptic modulator is shown in Fig. 14.17. A microring modulator consists of a ring resonator coupled to a single waveguide. The transmission of the waveguide is highly sensitive to the signal wavelength and is greatly reduced at wavelengths in which the ring circumference corresponds to an integer number of guided wavelengths. By tuning the effective index of the

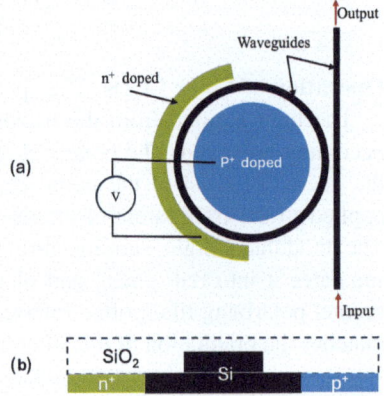

Fig. 14.17 (a) Schematic layout of the ring resonator-based modulator and (b) the cross section of the ring

ring waveguide, the resonant wavelengths are modified, which strongly modulates the transmitted signal. The effective index of the ring is modulated electrically by injecting electrons and holes using a $p-i-n$ junction embedded in the ring resonator. Figure 14.17(inset) shows the cross section of this rib waveguide. It consists of a strip waveguide formed on a 50 nm-thick slab layer. Because the thickness of the slab is much smaller than the wavelength propagating in the device ($\sim$1.5 μm), the mode profile of this waveguide is very close to that of the silicon strip waveguide. The highly doped p- and n-regions are doped around the ring using ohmic contacts are deposited on the doped regions. To minimize absorption losses, the doped regions are formed approximately 1 μm away from the ring resonator, ensuring that the overlap of the resonating mode with the doped regions is minimal.

The $p-i-n$ ring resonator is typically fabricated on a silicon-on-insulator (SOI) substrate with a 3 μm thick buried oxide layer. Both the waveguide coupling to the ring and that forming the ring have a width of 450 nm and a height of 250 nm. The diameter of the ring is 12 μm, and the spacing between the ring and the straight waveguide is 200 nm. These dimensions illustrate the modulator structure of a typical fabricated devices and are dependent on the specific application intended.

There are numerous examples of microring resonator optical modulators that are based on one or several microring resonators. Many devices have been demonstrated to haves exceeding 100 GBPs. The great advantage of these optical modulators is their compact size and their integrability with other photonic devices built on SOI technology.

14.7 Summary

In this chapter, we examined the principles, mechanisms, and applications of optical modulators, highlighting their significance in controlling and manipulating light for communication and processing systems. Key concepts of modulation methods, including amplitude, phase, polarization, and frequency modulation, were introduced, along with their classification into direct and external techniques.

Direct modulation of laser diodes was analyzed, emphasizing its practicality for high-speed communication, while discussing challenges such as frequency chirp. The chapter then explored external modulation techniques, with a focus on electrooptic and electroabsorption modulators, including their underlying mechanisms like the Pockels effect, Kerr effect, and quantum-confined Stark effect. These modulators' advantages, such as high-speed operation and integration capability, were contrasted with limitations like insertion loss.

Liquid crystal optical modulators and displays demonstrated the versatility of liquid crystal technology for intensity, phase, and polarization modulation, whereas microring modulators highlighted advances in compact, high-speed solutions enabled by integrated photonics.

This comprehensive overview of optical modulators equips the reader with a foundational understanding of their design principles and applications, emphasizing their role in advancing modern photonic technologies.

14.8 Problems

1. A laser diode is biased at $I_1 = 50$ mA, with a threshold current of $I_{th} = 20$ mA. If a sinusoidal current with amplitude $I_m = 10$ mA is superimposed, calculate the resulting modulated optical output power.
2. Calculate the relaxation oscillation frequency (f_r) of a laser diode with $\tau = 2$ ns, $\tau_{ph} = 1$ ps, $I_1 = 2I_{th}$, and $I_{th} = 1$ mA.
3. A Pockels cell with $r = 10^{-12}$ m/V, $n = 2.3$, $L = 1$ cm, and $\lambda_o = 1550$ nm is used. Determine the phase shift introduced by an applied voltage of $V = 200$ V.
4. An MZI modulator operates at $V_\pi = 4$ V. Calculate the output intensity I_{out} if the applied voltage is $V = 2$ V and the incident intensity is $I = 10$ mW.
5. For a bulk semiconductor with an energy gap $E_g = 1.2$ eV, determine the shift in the absorption edge when an electric field $E = 10^7$ V/m is applied, considering the Franz-Keldysh effect.
6. A quantum well device has $E_g = 1.5$ eV and a redshift in absorption edge of 0.1 eV under an electric field $E = 5 \times 10^7$ V/m. Calculate the QCSE coefficient.
7. For a nematic liquid crystal device with $V_c = 1$ V, $V_o = 0.5$ V, and applied voltage $V = 2$ V, determine the tilt angle θ of the LC molecules.
8. A microring modulator has a radius $R = 5$ μm and effective refractive index $n_{eff} = 2$. Calculate the free spectral range (FSR) for a wavelength of 1550 nm.
9. A microring resonator has a linewidth $\Delta\lambda = 0.02$ nm at a resonance wavelength $\lambda = 1550$ nm. Calculate the quality factor (Q) of the resonator.
10. An EAM has an extinction ratio of 30 dB and operates at a drive voltage range from 0 V to 2 V. Calculate the corresponding transmission coefficients for the "ON" and "OFF" states.

Bibliography

1. Bhattacharya, P. (1993), Semiconductor Optoelectronic Devices. 2nd Edition, Prentice-Hall.
2. Yariv, A., and Yeh, P. (2007), Photonics: Optical Electronics in Modern Communications. 6th Edition, Oxford University Press.
3. Agrawal, G.P. (2010), Fiber-Optic Communication Systems. 4th Edition, Wiley.
4. Chuang, S.L. (2009), Physics of Photonic Devices. 2nd Edition, Wiley.
5. Jiang, H., and Lipson, M. (2012), 'High-Speed Microring Modulators.' Nature Photonics, vol. 6, pp. 369–373.
6. Takahashi, H., et al. (2005), 'High-Speed Electroabsorption Modulators for Optical Fiber Communication.' Journal of Lightwave Technology, vol. 23, no. 1, pp. 23–30.
7. Reed, G.T., and Knights, A.P. (2004), Silicon Photonics: An Introduction. Wiley.
8. Saleh, B.E.A., and Teich, M.C. (2019), Fundamentals of Photonics. 3rd Edition, Wiley.

Optical Fiber Communication Systems 15

15.1 Introduction

Optical fiber communication systems have become the cornerstone of modern telecommunications over the past four decades. As the demand for high-speed, high-capacity data transmission continues to grow exponentially, these systems have become increasingly essential. Harnessing the power of light, optical communication systems enable the transmission of information over vast distances with unparalleled speed and minimal loss, forming the backbone of the global Internet infrastructure.

This chapter presents the fundamental principles behind optical communication, focusing on the critical components comprising these systems, building on concepts introduced in earlier chapters of this book, such as light generation, modulation, and detection as well as how it propagates through optical fibers. We will introduce additional components, such as connectors, splicers, and fiber Bragg gratings, which play crucial roles in deploying optical networks. We will also demonstrate how to integrate these components into a functional optical communication system.

Additionally, the chapter covers key topics such as dense-wavelength-division multiplexing, optical amplifiers, and the various challenges involved in maintaining signal integrity over long distances. We will also introduce the principle of designing optical communication links, including time- and power-budget calculations.

By the end of this chapter, students will gain a comprehensive understanding of how optical communication systems and their impact on the future of global communication. These systems are poised to remain at the heart of how we stay connected in the modern world.

15.2 Historical Development

The evolution of light wave systems in fiber-optic communication began in the mid-1970s and has seen significant advancements over several generations. The transmission wavelengths used in these generations spanned several bands, as shown in Fig. 15.1. The first generation used GaAs semiconductor lasers operating near 0.8 μm, offering bit rates of 45 Mbps and repeater spacings of up to 10 km. The second generation, developed in the early 1980s, moved to 1.3 μm wavelengths, reducing fiber loss and increasing the repeater spacing to 50 km, with bit rates reaching 1.7 Gbps.

The third generation, introduced around 1990, shifted to 1.55 μm wavelengths, where fiber losses were minimized. These systems can achieve bit rates of up to 2.5 Gbps and use dispersion-shifted fibers to manage pulse spreading. However, signal regeneration was still required every 60–70 km. The advent of optical amplification and wavelength-division multiplexing (WDM) in the fourth generation, starting in approximately 1992, drastically increased system capacity, allowing for bit rates up to 10 Tbps by 2001, with long-distance transmission capabilities.

The fifth generation focused on extending the WDM wavelength range beyond the conventional C-band to include the L- and S-bands, and on improving spectral efficiency via advanced modulation formats. This generation has achieved remarkable transmission capacities, such as 10 Tbps over 10,000 km, by employing high-density WDM channels and sophisticated modulation techniques. Current DWDM systems can employ more than 60 wavelengths.

Overall, fiber-optic communication technology has matured rapidly, with continuous improvements in capacity, efficiency, and distance, making it a critical component of global telecommunications.

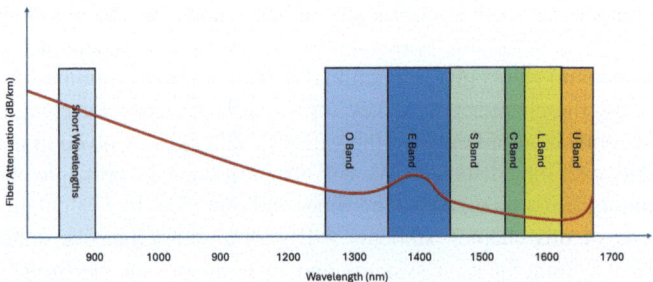

Fig. 15.1 Sketch of the fiber attenuation and optical fiber communication system wavelength transmission bands

15.3 Basic Principles of Communication Systems

Communication can be broadly defined as the transmission of information from one point to another. Over the past 150 years, this has been achieved primarily through the use of electromagnetic waves. In this process, information is superimposed (modulated) onto a high-frequency carrier wave, which is then transmitted to the desired destination. Upon arrival, the original information signal is extracted through a process known as demodulation. Carrier waves can take various forms, including radio waves, ultra-high-frequency (UHF) waves, microwaves, or millimeter waves. Additionally, communication can be achieved using electromagnetic carriers selected from the optical range of frequencies.

A basic communication system consists of three key components (Fig. 15.2):

- **Transmitter**: Modulates the information signal onto an electromagnetic carrier.
- **Communication Channel**: The medium through which the modulated signal is transmitted. This can be a wired medium such as a twisted copper cable, coaxial cable, or optical fiber.
- **Receiver**: Demodulates the information signal from the carrier.

Optical fiber networks can be utilized to transmit both analog and digital data. For analog data networks, the quality of signal integrity is typically measured by the signal-to-noise ratio (SNR), while for digital data networks, it is assessed by the bit error rate (BER). Figure 15.3 provides an illustration of a point-to-point fiber network.

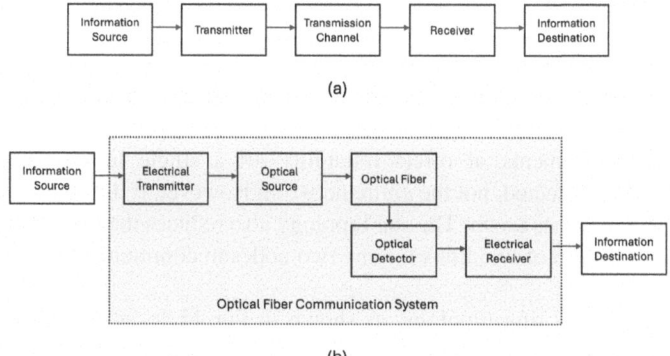

Fig. 15.2 Schematic block diagram of (**a**) an electrical communication system and (**b**) an optical fiber communication system

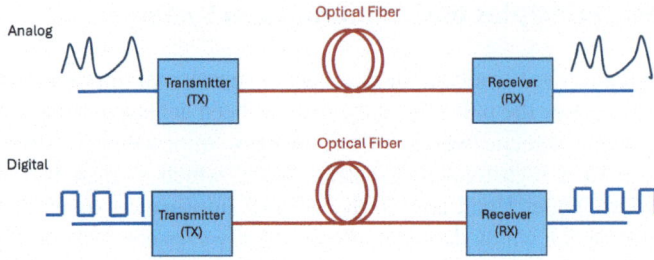

Fig. 15.3 Point-to-point optical fiber network

15.4 Communication Network Topologies

In general, communication networks operate on the principle that each node in the network can send and receive information, with links or channels facilitating communication between these nodes. The network's performance is heavily influenced by its topology or the arrangement of these nodes and links. Various topologies have different advantages and limitations:

Bus Topology In a bus topology, as shown in Fig. 15.4a, all nodes share a common communication channel. When a node broadcasts a message, all nodes receive it, but only the intended recipient responds. This topology is simple and requires minimal cabling, making it scalable—adding a new node only requires adding a single link. However, its drawbacks include a large network span (the maximum signal travel distance) and vulnerability to failures; a single link failure can bring down the entire network.

Star Topology A star topology, as shown in Fig. 15.4b, connects all nodes to a central hub, which manages communication. While this setup increases complexity and cabling requirements, it offers reliability—if a single link fails, only the associated node is affected, not the entire network. However, if the central hub fails, the whole network goes down. The star topology also reduces the network diameter, requiring a maximum of two links for any two nodes to communicate.

Ring Topology In a ring topology, as shown in Fig. 15.4c, nodes are connected in a circular arrangement, distributing the communication load more evenly. This topology offers fault tolerance; if one link fails, communication can continue in the opposite direction. However, adding new nodes is challenging, and any change can disrupt the network.

Mesh Topology The mesh topology, as shown in Fig. 15.4d, is the most interconnected, with each node connected to every other node. This setup provides

15.4 Communication Network Topologies

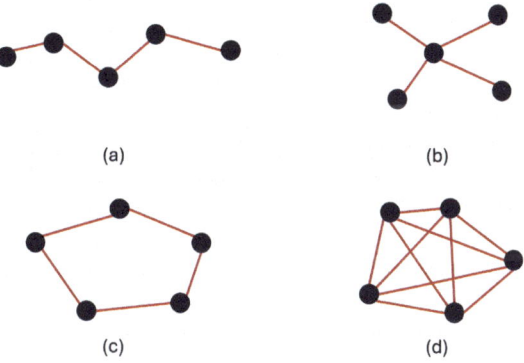

Fig. 15.4 Schematic of communication networks topologies

maximum fault tolerance and the shortest possible network diameter, but it requires extensive cabling and complex routing decisions, especially as the network grows.

In practice, the network topology is chosen based on specific needs and constraints. Smaller networks, such as those in homes or small offices, often use bus or star topologies, while larger networks may combine multiple topologies to meet different requirements. For example, a city's communication infrastructure might use a star topology to connect individual users to local hubs, a ring topology to connect those hubs, and a mesh topology to link cities together. In fiber-optic networks, the nodes consist of optical transmitters and receivers, connected by optical fibers. These connections are made by components such as optical couplers, which will be discussed in more detail in subsequent sections.

Communication networks can be categorized on the basis of their geographical span and can be divided into local area networks (LANs), metropolitan area networks (MANs), and wide area networks (WANs).

LANs cover small geographical areas such as university campuses or corporate buildings and are usually owned and maintained by a single organization. They typically use Ethernet protocols, and their reach depends on the communication medium, with fiber optics extending the range more than copper cables or Wi-Fi.

MANs span larger areas, such as cities or districts, and are often shared by multiple organizations. They interconnect LANs and use fiber optics for higher data rates and longer distances, with protocols such as the asynchronous transfer mode (ATM) and evolving Ethernet standards.

WANs cover vast areas, including multiple cities, countries, or continents. They connect smaller networks like MANs and LANs and rely heavily on fiber optics and advanced multiplexing techniques, such as wavelength-division multiplexing (WDM), due to their extensive range and high data demands. WANs use a variety of topologies and protocols, such as ATM and SONET to manage their complex infrastructures.

15.5 Optical Network Components

Optical fiber network design and implementation require a large assembly of active and passive components. These include optical fibers (Chap. 9), light sources and detectors, (Chaps. 12 and 13), and optical modulators (Chap. 14). In this section, we introduce a number of components that are required to connect these components together such as connectors, splices, cables, optical filters, switches, and optical amplifiers.

15.5.1 Wavelength-Division Multiplexers/Demultiplexers

Wavelength-division multiplexers (WDM) and demultiplexers (WDMD) are passive components that combine multiple wavelengths of light on a fiber at the transmitter side and separate the wavelengths at the receiver, respectively. Figure 15.5 shows a schematic of both of these devices.

The implementation of WDM devices can be achieved using the property of diffraction gratings in separating different wavelengths as a function of the diffraction angle. In implementation, we demonstrate two architectures, one using a reflecting diffraction grating (Littrow) and the second using an arrayed waveguide grating (AWG).

Littrow Diffraction Grating Wavelength Multiplexer/Demultiplexer A diffraction grating reflects light in specific directions according to the grating constant, the angle at which the light is incident on the grating, and the optical wavelength. The Littrow device employs a single lens and a separate plane reflective diffraction grating. In a Littrow-mounted grating, the blaze angle of the grating is such that the incident and reflected light beams follow virtually the same path, as illustrated in Fig. 15.6, thereby maximizing the grating efficiency and minimizing lens astigmatism. For a given center wavelength λ, the blaze angle is set such that:

$$\theta_B = sin^{-1}\left(\frac{\lambda}{2\Lambda}\right), \qquad (15.1)$$

where Λ is the grating period (refer to Chap. 3 for further details). Schematic diagrams of Littrow-type grating multiplexer and demultiplexer employing a GRIN-rod lens are provided in Fig. 15.6. The GRIN-rod lens configuration offers advantageous such as compactness and ease of alignment. In Fig. 15.6a, multiple input fibers and

Fig. 15.5 Schematic of wavelength-division multiplexer and demultiplexer

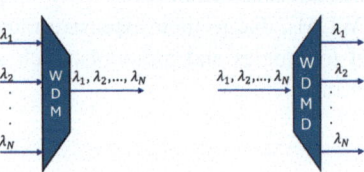

15.5 Optical Network Components

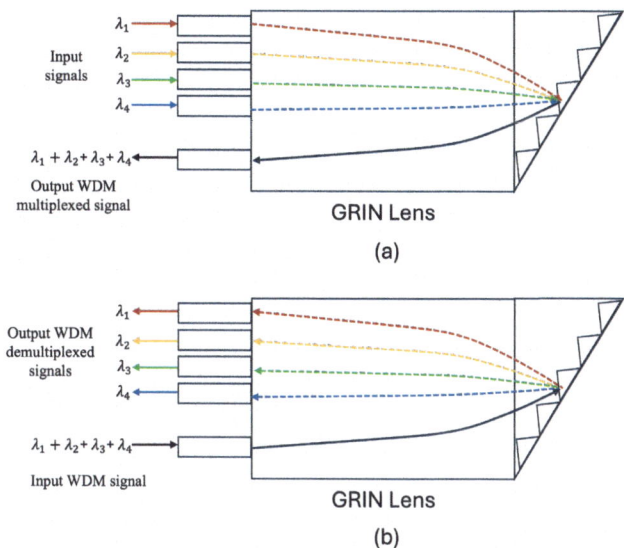

Fig. 15.6 (a) Reflective diffraction grating based wavelength-division multiplexer and (b) demultiplexer

a single output fiber are arranged on the focal plane of the lens. For a quarter pitch GRIN-rod lens, this focal plane coincides with the fibers end faces. Input signals wavelengths are reflected by the grating and consequently focused on the output fiber. An offset angle required for proper operation can be introduced by inserting a prism (or glass wedge) between the lens and the grating.

Figure 15.6b illustrates a wavelength-division demultiplexer. Here, the input WDM signal is focused by the GRIN-lens on the grating. Upon reflection, the grating disperses the light angularly based on the optical wavelengths. The dispersed wavelengths then pass back through the lens and are focused onto different output fibers, each corresponding to a specific wavelength.

This design highlights the Littrow configuration's efficiency in wavelength multiplexing and demultiplexing applications, leveraging the GRIN-rod lens for compact, high-performance optical systems.

Arrayed Waveguide Grating Multiplexer An arrayed waveguide grating (AWG) multiplexer is a device that utilizes the grating property of spreading light into its spectrum and is commonly used for multiplexing and demultiplexing optical signals, as shown in Fig. 15.7. The AWG consists of several components designed to split and combine wavelengths for optical communication:

1. **Input/Output Waveguides**: These guide the optical signal into and out of the AWG.

Fig. 15.7 Schematic of an arrayed-waveguide-grating multiplexer/demultiplexer [John Senior,...]

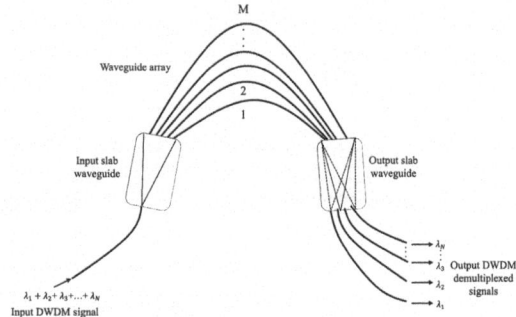

2. **Arrayed Waveguides:** A set of waveguides with varying lengths (represented by ΔL) that introduce phase shifts between optical signals of different wavelengths.
3. **Focusing Slab Waveguides:** These act as multimode interference couplers that are responsible for splitting and combining signals at both the input and output ends of the device.

The core functionality of multiplexing/demultiplexing is achieved through the two focusing slab waveguides, each of which acts as either a multimode interference coupler or a free space propagation region. When a dense wavelength-division multiplexed (DWDM) signal is coupled into the input waveguide, it propagates through the input slab waveguide, where the optical signal is split into the arrayed waveguides—typically more than 64 channels.

The optical signals then travel through the arrayed waveguides, each of which has a different path length. As a result, the optical wavefronts arrive at the input ports of the second slab waveguide with a phase difference. At the output, the slab waveguide acts as a combiner, making the AWG function as a wavelength demultiplexer. The signals from each arrayed waveguide interfere with one another in the output slab waveguide, and due to constructive interference, each individual wavelength from the original DWDM signal is coupled into a specific output waveguide, as depicted in Fig. 15.7.

The optical signal corresponding to the central wavelength (λ_c) is focused into the central output waveguide in the image plane. When the wavelength shifts to $\lambda_c + \Delta\lambda$, the phase of the signals in the waveguides changes progressively from the lower to the upper channel. This causes a slight tilt in the phase front at the output aperture, which focuses the beam at a different position in the image plane. Channel spacings of 100, 50, *and* 25 GHz in the L- and C-bands are common in commercial AWG devices.

15.5.2 Optical Add/Drop Multiplexers

Optical add/drop multiplexers (OADMs) are passive devices used in wavelength-division multiplexing (WDM) systems to selectively extract (drop) or insert (add) specific wavelength channels from an optical fiber carrying multiple channels, each at a different wavelength. This operation allows the dynamic management of data traffic, enabling a network to route certain wavelengths to different destinations without interrupting the other wavelengths.

In a WDM network, a single fiber transmits multiple wavelength channels simultaneously. There are instances where a specific wavelength needs to be redirected (dropped) to a separate network branch or to serve a specific hub or customer. Conversely, there may be situations where a wavelength needs to be inserted (added) into the WDM stream from a different part of the network. OADMs allow this process to occur seamlessly, without affecting the other wavelengths traveling along the fiber.

Figure 15.8 illustrates an optical add/drop multiplexer that employs a fiber Bragg grating (FBG) to reflect the selected wavelength. This system uses two optical circulators to manage the direction of the optical signals. A three-port optical circulator functions as follows:

- The light fed into Port 1 exits through Port 2.
- The light input into Port 2 exits through Port 3.
- The light input into Port 3 exits through Port 1.

In this setup, the WDM fiber carrying multiple wavelength channels is connected to Port 1 of the first circulator, and the signals exit through Port 2. The specific wavelength to be dropped is reflected by the FBG and routed back into Port 2 of the first circulator, exiting via Port 3. The remaining WDM channels continue forward into Port 1 of the second circulator and exit through Port 2. If a wavelength needs to be added, it is fed into Port 3 of the second circulator. The signal then exits from Port 1, where it encounters the FBG. The FBG reflects the new wavelength, integrating it with the existing WDM channels, which pass through the FBG. The combined signal then exits through Port 3 of the second circulator, thereby merging the added wavelength into the WDM stream.

As an example, in Fig. 15.8, wavelength λ_3 is dropped in the first circulator and routed to a different destination, and a new wavelength is added into the system via the second circulator, effectively integrating it into the WDM fiber.

The OADM provides many benefits to the WDM network, including the following:

- **Non-disruptive:** The ability to add or drop a specific wavelength without interrupting the rest of the traffic ensures high flexibility in optical networks.

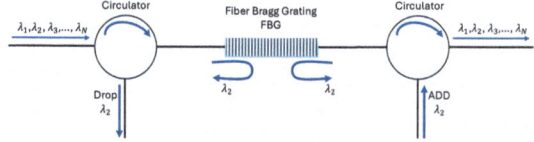

Fig. 15.8 An optical add/drop multiplexer (OADM) using an FBG reflective filter

- **Dynamic Channel Management:** OADMs allow networks to efficiently manage bandwidth by enabling or disabling specific wavelengths as needed.
- **Scalability:** As networks grow, OADMs provide a scalable solution for adding or dropping channels without the need for additional fiber infrastructure.

By using FBGs and optical circulators, OADMs provide a precise and efficient way to manage individual wavelength channels, making them a key component in modern optical networks.

15.5.3 Optical Cross-Connect Switches

Optical Cross-Connect (OXC) switches are crucial devices used in optical networks to manage and route optical signals without the need to convert them into electrical signals. These switches enable the dynamic connection of optical fibers from different parts of a network, allowing signals to be rerouted based on changing traffic demands or fault management.

In a typical optical network, multiple fibers carry many wavelength division multiplexed channels. The purpose of an OXC is to establish paths between different input and output fibers by switching entire wavelengths or groups of wavelengths between fibers. This enables flexible routing and network reconfiguration, which is especially important in large, scalable networks.

Key functions of optical Cross-Connect switches:

1. **Routing Optical Signals:** OXC switches manage the routing of signals by dynamically configuring optical paths. These paths can be established or reconfigured based on network demands, helping balance traffic loads or rerouting around network failures.
2. **Wavelength Switching:** In wavelength-selective OXC switches, each wavelength channel of a WDM signal can be individually switched to a different output port. This allows the network to manage specific wavelength channels efficiently, making it more flexible and capable of handling complex traffic patterns.
3. **Non-blocking Architecture:** OXC switches are often designed with a non-blocking architecture, meaning that any input signal can be switched to any output port without interfering with other signals. This enables multiple channels to be routed simultaneously, optimizing bandwidth usage.

Components of Optical Cross-Connect Switches:

1. **Input/Output Ports:** OXCs have multiple input and output fiber ports. Each input fiber may carry a WDM signal, and each output fiber can be connected to a different part of the network.

2. **Optical Switching Fabric:** This is the core of the OXC, where the actual signal routing occurs. It can be implemented using various technologies, including the following:
   (a) **Micro-Electro-Mechanical Systems (MEMS)**: Tiny mirrors are used to physically direct light beams to different output fibers.
   (b) **Wavelength Selective Switches (WSS):** Enables selective switching of individual wavelength channels.
   (c) **Liquid Crystal on Silicon (LCoS)**: A technology that can steer light by modulating its phase, enabling efficient routing of optical signals.
3. **Control Plane:** The control plane is responsible for managing the configuration of the switching fabric. It can be programmed to route signals dynamically based on predefined rules or real-time network conditions.

All optical cross-connect switches have many advantages, including:

- **Transparency**: Since OXCs work in the optical domain, they are protocol- and bit-rate-agnostic, meaning that they can switch signals regardless of the underlying data format or speed.
- **Scalability**: OXCs can handle a large number of wavelength channels, making them highly scalable as networks grow.
- **Low Latency**: By avoiding optical-electrical-optical conversions, OXCs introduce minimal latency, which is crucial for high-performance networks.

OXCs are typically used in the backbone of large telecom networks where multiple fiber paths converge, and there is a need for dynamic and efficient traffic management. In the case of network failures, OXCs can quickly reroute traffic to avoid disrupted paths, ensuring reliable service. OXCs help optimize traffic distribution by managing the routing of high-volume data streams across different optical paths.

Optical cross-connect switches are fundamental in modern optical networks, offering efficient, dynamic, and scalable solutions for routing optical signals. By enabling flexible reconfiguration of wavelength channels and managing high-capacity WDM signals, OXCs help optimize network performance while minimizing latency and power consumption.

15.6 Optical Fiber Amplifiers

Optical fiber amplifiers are crucial components in optical communication networks that increase the strength of optical signals without the need to convert them into electrical signals. By amplifying optical signals directly in the fiber, these devices overcome the attenuation (signal loss) that naturally occurs as light travels long distances through the optical fiber. Two common types of fiber amplifiers used in modern optical systems are erbium-doped fiber amplifiers (EDFAs) and Raman amplifiers.

15.6.1 Erbium-Doped Fiber Amplifiers (EDFAs)

Erbium-doped fiber amplifiers (EDFAs) are among the most widely used optical amplifiers in wavelength-division multiplexing systems. EDFAs operate using a 10 to 30 meter length of optical fiber that has been doped with erbium ions (Er^{3+}). When a signal in the 1.55 μm wavelength range (the operating wavelength for many WDM systems) passes through the erbium-doped fiber, the erbium ions are excited by an external pump laser. These excited ions then transfer their energy to the signal, amplifying it.

Figure 15.9 illustrates a typical configuration of an EDFA. A pump laser, typically operating at 980 nm or 1480 nm, excites the erbium ions within the fiber. The input optical signal (approximately 1550 nm) triggers the release of energy from the erbium ions, resulting in signal amplification. The amount of signal gain can be significant, often ranging from 20 to 40 dB, making EDFAs highly efficient for long-haul optical communication.

The active medium in an EDFA consists of a nominal 10 to 30 meter length of optical fiber lightly doped (typically around 1000 parts per million by weight) with a rare-earth element, such as erbium (Er^{3+}), ytterbium (Yb), neodymium (Nd), or praseodymium (Pr). The host fiber material can be standard silica, fluoride-based glass, or multicomponent glass. The operating regions of these amplifiers depend on the host material and the specific doping elements.

For long-haul telecommunication applications, the most commonly used material is silica fiber doped with erbium, referred to as an erbium-doped fiber amplifier. Initially, EDFA operation was limited to the C-band (1530–1560 nm) because the erbium atoms exhibit a particularly high gain coefficient in this region, which is why it is called the conventional band (C-band). Beyond the C-band, the erbium gain decreases significantly, with only approximately 20% of the peak gain observed in the L-band. However, advancements in erbium-doped fiber designs and the use of high-power pump lasers operating at different wavelengths have extended the operation of EDFAs into the L-band. Figure 15.10 illustrates a typical gain spectrum of an EDFA. To ensure a uniform gain across the C- and L-bands, a gain flattening filter (GFF) is used to flatten the gain spectrum.

Some of the advantages of EDFAs include the following:

1. **Wide Operating Band:** EDFAs operate efficiently in the C-band (1530–1565 nm) and L-band (1565–1625 nm), which are commonly used for DWDM systems.

Fig. 15.9 Schematic of an erbium-doped fiber (EDF) amplifier

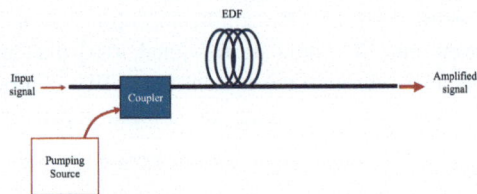

15.6 Optical Fiber Amplifiers

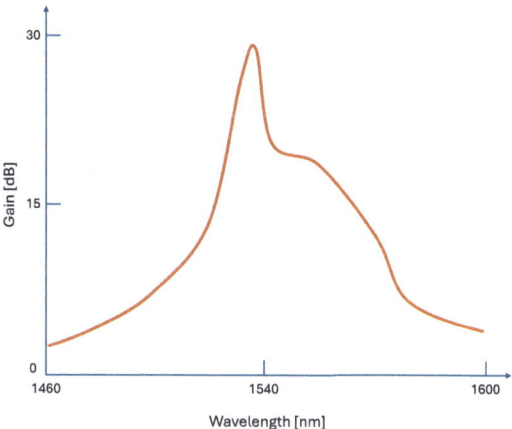

Fig. 15.10 Schematic for an erbium-doped amplifier (EDFA) spectrum

2. **Low Noise:** EDFAs introduce relatively low noise to the amplified signal, making them highly suitable for long-distance transmission.
3. **High Output Power:** EDFAs provide high output power, which makes them ideal for high-capacity WDM systems.
4. **Transparency:** EDFAs can amplify multiple WDM channels simultaneously, without requiring separation or conversion of individual channels.

15.6.2 Raman Optical Amplifiers

Raman optical amplifiers are another type of optical amplifier that utilizes the nonlinear interaction between the optical signal and the fiber medium itself to amplify light. Unlike erbium-doped fiber amplifiers, Raman amplifiers do not rely on rare-earth-doped fibers. Instead, they depend on the stimulated Raman scattering (SRS) effect, which occurs when a high-power pump laser interacts with the optical fiber.

Raman amplification is based on the stimulated Raman scattering (SRS) effect, which occurs because of the interaction between an optical energy field and the vibration modes within the material's lattice structure. In this process, an atom first absorbs a photon at one energy level and then emits another photon at a lower energy level (i.e., a longer wavelength). The difference in energy between the absorbed and emitted photons is transferred to a phonon, which represents a vibration mode in the material. This energy transfer results in a wavelength shift of about 80 to 100 nm, commonly known as the Stokes shift.

Figure 15.11 demonstrates the Stokes shift and the resulting Raman gain spectrum produced by a pump laser operating at 1453 nm. In this example, a signal at 1550 nm (97 nm away from the pump wavelength) is amplified. This process transfers optical energy from the powerful pump laser to the weaker transmission signal, which is 80 to 100 nm higher than the pump wavelength. For example,

Fig. 15.11 Stokes shift and the resulting Raman gain spectrum produced by a pump laser operating at 1453 nm. In this example, a signal at 1550 nm

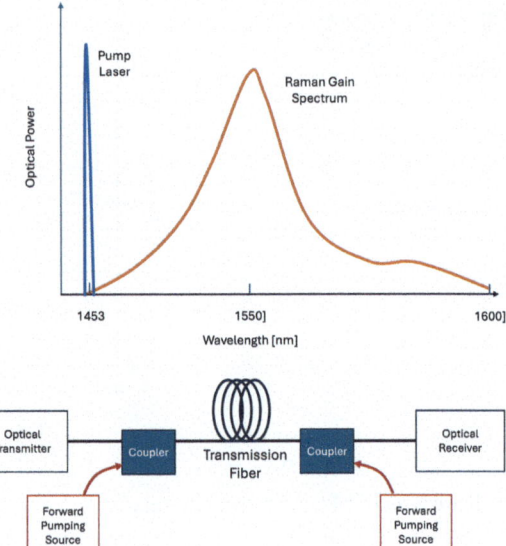

Fig. 15.12 Typical configuration of a Raman optical amplifier with co- and counter-directional coupling

pumping at 1450 nm results in signal amplification at approximately 1530 to 1550 nm. Due to the molecular structure of optical fiber (usually glass), multiple vibration modes exist, resulting in an optical gain region roughly 30 nm wide. To achieve a flat and wideband gain spectrum, several pump lasers operating at different wavelengths are often used together.

Figure 15.12 illustrates a typical configuration of a Raman optical amplifier. In a Raman amplifier, the amplification occurs over the length of the transmission fiber, as opposed to a separate doped fiber, which allows the optical signal to be amplified while it propagates through the same fiber that carries the data. A pump laser, typically in the 1450–1500 nm range, is injected into the fiber, and through the SRS process, energy is transferred from the pump to the optical signal, amplifying it. Raman amplifiers provide a more flexible gain spectrum compared to EDFAs, as the amplification band depends on the pump wavelength used. Raman amplifiers can be tuned to amplify different wavelength regions by adjusting the pump laser wavelength. Since the amplification occurs throughout the transmission fiber, Raman amplifiers reduce signal loss across the entire fiber span, improving overall system performance. Raman amplifiers can also be used alongside EDFAs to extend the amplification bandwidth and reduce signal attenuation in long-haul networks. Unlike EDFAs, Raman amplifiers are not limited to the C- and L-bands, making them more versatile for certain systems.

Traditionally, the signal and pump beams travel in opposite directions, a method known as counter-directional pumping, though pumping can also be performed co-directionally with the signal. In counter-directional pumping, amplification predominantly occurs in the last 20 to 40 km of the transmission fiber, which is why Raman amplification is often referred to as distributed amplification. However,

15.7 Single-Channel Networks

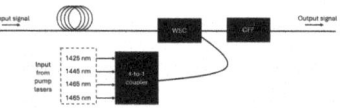

Fig. 15.13 Raman amplifier with four pumping laser sources in counter-directional pumping. WSC is a wavelength selective filter

due to noise considerations, the practical gain of distributed Raman amplifiers is typically limited to less than 20 dB.

For amplifying signals in the C-band and L-band, pump lasers with high output powers in the 1400 to 1500 nm range are used. These lasers, which provide fiber launch powers of up to 300 mW, are widely available. Figure 15.13 illustrates the setup of a typical Raman amplification system. In this setup, a pump combiner multiplexes the outputs of four pump lasers operating at different wavelengths (e.g., 1425, 1445, 1465, *and* 1485 nm) into a single fiber. The combined pump power is injected into the transmission fiber in a counter-propagating direction through a broadband WDM coupler. To ensure uniform gain across different wavelengths, a gain-flattening filter (GFF) is employed.

The following table compares EDFA and Raman amplifiers:

Features	EDFA	Raman
Doping medium	Erbium-doped fiber	Transmission fiber (no doping)
Pump wavelength	980 nm or 1480 nm	1450 nm or higher
Gain medium	External erbium-doped fiber	The transmission fiber itself
Gain bandwidth	C- and L-band (1530–1625 nm)	Adjustable depending on pump
Amplification location	In the erbium-doped fiber	Distributed along the transmission fiber
Noise figure	Low	Slightly higher than the EDFA
Application	Long-haul WDM networks	Long-haul, ultra-long-haul, distributed amplification

15.7 Single-Channel Networks

The simplest optical fiber network is based on a single wavelength link. This is applicable to local connections such as point-to-point links, local area network (LAN), and distribution networks.

15.7.1 Point-to-Point Links

Point-to-point links are the simplest network configuration for transporting information between two locations on a network, as illustrated in Fig. 15.14. The

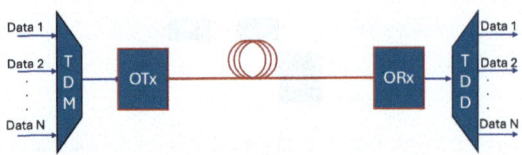

Fig. 15.14 A block diagram of a point-to-point optical link, where OTx represents the optical transmitter and ORx represents the optical receiver

Fig. 15.15 Block diagrams for (**a**) an optical transmitter and (**b**) and optical receiver

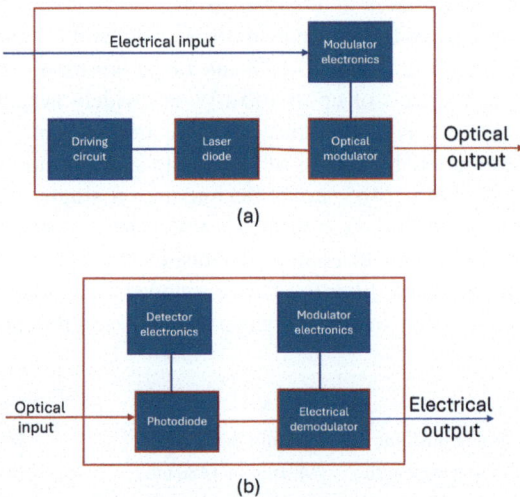

information to be transmitted—whether data, voice, or video—is electronically time-division multiplexed (TDM). This multiplexed signal is then used to modulate an optical carrier generated by a laser diode, which is subsequently transmitted over an optical fiber.

At the receiver end, the optical signal is detected and demodulated by an optical receiver (ORx), and the electrical signal is time-division demultiplexed (TDD) to retrieve the original data.

Figure 15.15 depicts the block diagrams of the optical transmitter (OTx) and optical receiver (ORx). Depending on the data rate and span of the fiber link, the light source may be a laser diode (LD) or a light-emitting diode (LED).

The configuration of the link depends on the distance between the transmitter and receiver. For shorter connections, a simple fiber link is sufficient. However, when the link length exceeds a certain threshold—typically within the range of 30 to 70 km, depending on the operating wavelength—it becomes essential to compensate for fiber losses, as the optical signal may weaken to an unreliable level. In such cases, optical fiber amplifiers can be employed to enhance the optical signal. Reliable operation over spans of thousands of kilometers can be achieved using optical fiber amplifiers placed every 40 to 50 km. Notably, that periodically amplified light wave

15.7 Single-Channel Networks

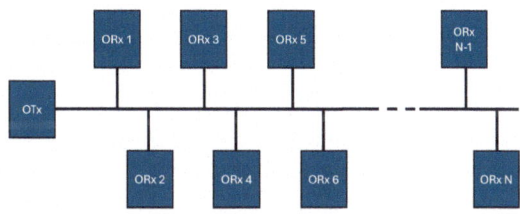

Fig. 15.16 Optical fiber bus-network with N-taps

systems are often constrained by fiber dispersion unless dispersion-compensation techniques, as discussed in Chap. 7, are applied.

15.7.2 Bus Network

The optical bus network is a distribution framework where optical signals are transported via a fiber link, allowing customers to connect to the bus through taps. In this setup, a single fiber cable carries a multichannel optical signal across the service area. Distribution occurs using optical taps, which divert a small fraction of the optical power to each subscriber, Fig. 15.16.

However, a significant drawback of the bus topology is that signal loss increases exponentially with the number of taps, thereby limiting the number of subscribers that can be served by a single optical bus. Even when fiber losses are disregarded, the power available at the Nth tap can be expressed by the following formula:

$$P_N = P_T C \left[(1 - \delta)(1 - C)\right]^{N-1}, \tag{15.2}$$

where P_T is the transmitted power, C is the fraction of power coupled out at each tap, and δ accounts for insertion losses, which are assumed to be the same at each tap. The power from the optical transmitter as coupled into the fiber bus is multiplied by the fraction C.

Example An optical bus with the following parameters:
$\delta = 0.05$, $C = 0.05$, $P_T = 1$ mW, and $P_N = 0.1\,\mu$W.
Determine the number of customers N using the following formula:
$P_N = P_T C \left[(1 - \delta)(1 - C)\right]^{N-1}$.

Solution

$$0.1 \times 10^{-6} = 0.05 \times 10^{-3} \left[(1 - 0.05)(1 - 0.05)\right]^{N-1}$$

(continued)

By taking the logarithm of both sides, we obtain:

$$N = 1 + \frac{log\left(0.1 \times 10^{-6}/0.05 \times 10^{-3}\right)}{log\left[(1-0.05)(1-0.05)\right]} \approx 61.$$

Therefore, the 64 customers can be served by this bus-network.

15.8 Optical Access Network

An optical access network refers to the segment of telecommunications infrastructure that connects end users (residential or business customers) to the service provider's central office or a metropolitan area network. These networks utilize optical fibers, which offer high bandwidth and reliability, making them ideal for supporting modern communication demands such as high-speed Internet, voice, video streaming, and data services.

There are several topologies of optical access networks. The most widely used is the passive optical networks (PON). A PON uses unpowered optical splitters to share a single optical fiber among multiple users educing the need for expensive active components and thus making it more cost-effective. The most common PON architectures are shown in Fig. 15.17. A typical PON network consists of optical line terminal (OLT), located in the central office (CO), which connects to the metropolitan area network (MAN) through one of its nodes. The signal received by the OLT is time-division multiplexed (TDM) data. The optical fiber network from the optical line terminal (OLT) at a CO to the optical network units (ONUs) at customer sites is completely passive. In a TDM-PON, a passive power splitter serves as the remote terminal. The same signal from the OLT is broadcast to different ONUs by the power splitter, with signals for different ONUs multiplexed in the time domain. Each ONU recognizes its own data through the address labels embedded in the signal.

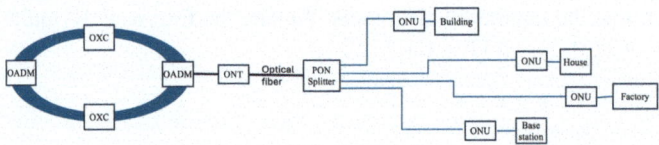

Fig. 15.17 Schematic of a passive optical network (PON) system architecture

Bidirectional traffic between the OLT and ONU can be implemented via several methods:

1. **Conventional Two-Fiber Approach:** In this method, two separate fibers are used for bidirectional communications. This approach is also called space division duplex or two-fiber approach. One fiber for upstream (ONU to OLT) and one for downstream (OLT to ONU) signals. The method is straightforward and does not require separation of signals in the time, frequency, or wavelength domains. It commonly uses a 1.3 μm wavelength for both upstream and downstream transmission, as low-cost Fabry-Perot (FP) lasers are readily available at this wavelength. However, despite its simplicity, the two-fiber approach is more expensive than single-fiber solutions due to the additional fiber, splitter management, and higher capital and operational costs.
2. **One-Fiber Single-Wavelength Full Duplex**: This method uses a single optical fiber for both upstream and downstream communication. A simple 3 dB one-to-two directional coupler is placed at the OLT and ONU to separate the upstream and downstream signals. The main disadvantage of this approach is that the 3 dB couplers introduce about 3.5 dB of signal loss at each end of the transmission link, which negatively impacts the system's power budget.
3. **Time-Division Duplex (TDD)**: In the time-division duplex method, the OLT and ONU take turns using the fiber for upstream and downstream transmissions, alternating in a "ping-pong" fashion. Similar to the one-fiber single-wavelength approach, directional couplers are used at both the OLT and ONUs to separate the optical signals. However, TDD eliminates the need for multiple fibers and avoids the significant signal loss associated with 3 dB couplers.

15.9 Wavelength-Division Multiplexed (WDM) Networks

In WDM link is illustrated in Fig. 15.18. The signal fed in the link is connected to a series of optical transmitters (OTx), each employing fixed or tunable laser sources operating at different wavelengths. The outputs of these optical transmitters

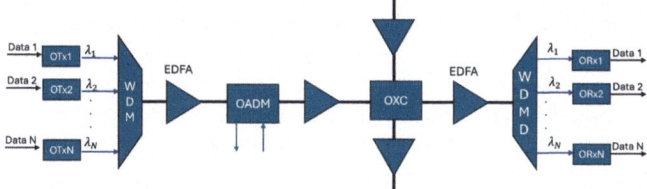

Fig. 15.18 Block diagram for an example of a wavelength-division multiplexed network identifying key optical components. The spacing between the components may be in tens of kilometers

are fed into a wavelength-division multiplexers to combine the independent light signals from the transmitters onto a single fiber. Along the path of the link between the transmitters and receivers, there may be optical amplifiers, optical add/drop multiplexers ($OADM$)s for inserting or subtracting individual wavelengths along the path, optical-cross connects ($OXCs$), and other devices to enhance the link performance. At the end of the link, there is an optical-division demultiplexer for separating the wavelengths into independent signal streams and an array of tunable optical receivers (ORx). A major application of WDM technology is to increase the capacity of long-haul networks. Owing to the large amount of traffic carried on these long links, high-performance wideband components are required. Metro WDM links have a different set of applications that require lower-cost narrowband components. One important point is that WDM-based networks are bit rate- and protocol-independent, so they can carry various types of traffic at different speeds concurrently.

15.9.1 WDM Channel Spacing and Data Rates

In wavelength-division multiplexing systems, multiple data signals are transmitted simultaneously over a single optical fiber by assigning each signal a unique wavelength. This technique increases the capacity of the fiber without requiring additional physical infrastructure. Two key factors in WDM systems are channel spacing and data rates.

WDM Channel Spacing: Channel spacing refers to the difference in wavelength (or frequency) between two adjacent WDM channels. Proper spacing ensures that channels do not interfere with each other, minimizing cross talk and inter-channel interference. The channel spacing depends on the type of WDM system:

- **CWDM (Coarse Wavelength-Division Multiplexing):**

CWDM has larger channel spacings, typically 20 nm apart, corresponding to around 2.5 THz in frequency. CWDM typically operates in the range of 1270 nm to 1610 nm, using up to 18 channels. The wider channel spacing in CWDM allows for simpler and cheaper components because of reduced demands on wavelength accuracy and temperature control, but it limits the number of channels that can be used.

- **DWDM (Dense Wavelength-Division Multiplexing):**

DWDM systems have much tighter channel spacing, typically 0.8 nm (100 GHz), 0.4 nm (50 GHz), or even 0.2 nm (25 GHz). DWDM operates primarily in the **C-band** (1530–1565 nm) or **L-band** (1565–1625 nm), allowing for more channels (often 40 to 120 or even higher). Tighter spacing in DWDM systems requires more precise laser sources and filters, making the system more complex and expensive

15.9 Wavelength-Division Multiplexed (WDM) Networks

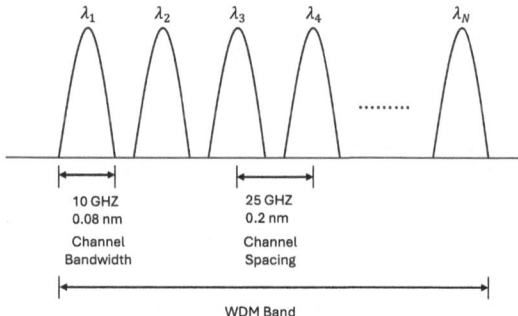

Fig. 15.19 Schematic for a DWDM multiplexing and channel bandwidth and channel spacing for the 25 GHz standard

but allowing much higher capacity over the same fiber. Figure 15.19 illustrates the DWDM multiplexing and channel bandwidth and channel spacing for the 25 GHz standard.

Data Rates in WDM Systems: Data rates in WDM systems depend on several factors, including the modulation format used, the bandwidth available per channel, and the efficiency of the coding techniques.

The data rate per channel is influenced by the modulation format. Common modulation formats used in WDM systems include the following:

- **On-Off Keying (OOK):** A simple format where a binary bit (1 or 0) is represented by the presence or absence of light. It supports data rates of up to 10 Gbps per channel.
- **Differential Phase Shift Keying (DPSK):** This format encodes data in the phase of the light wave and is more efficient than OOK. It allows data rates of 40 Gbps or higher per channel.
- **Quadrature Amplitude Modulation (QAM):** A more complex modulation format that encodes multiple bits per symbol by varying both the amplitude and phase of the light. Formats such as 16-QAM or 64-QAM can achieve data rates of 100 Gbps or more per channel.
- **Coherent Detection:** Advanced coherent systems combined with high-order modulation formats (e.g., QAM) and sophisticated digital signal processing (DSP) allow data rates of 400 Gbps to 1 Tbps per channel.

Example Determine the total rates for the following WDM links:

1. CWDM link where the number of channel is 18 using OOK modulation format

(continued)

2. DWDM link where the number of channels is 80 using:
   (a) 100 GHz channel spacing with OOK modulation format
   (b) 25 GHz channel spacing with 16-QAM modulation format

Solution

1. CWDM link where the number of channel is 18 using OOK modulation format:
   Total Data Rate = 18 × 10 Gbps = 180 Gbps
2. DWDM link where the number of channels is 80 using:
   (a) 100 GHz channel spacing with OOK modulation format:
       Total Data Rate = 80 × 10 Gbps = 800 Gbps
   (b) 25 GHz channel spacing with 16-QAM modulation format:
       Total Data Rate = 80 × 100 Gbps = 8,000 Gbps = 8 Tbps

15.9.2 Network Transport Protocols and Standards

SONET (Synchronous Optical Network) and SDH (Synchronous Digital Hierarchy) are two closely related standards for transmitting digital data over optical fibers. They are widely used in telecommunications networks, particularly for transporting large amounts of data over long distances. Both SONET and SDH define how data is multiplexed, framed, and synchronized, enabling efficient and high-capacity transport of voice, video, and data.

SONET and SDH systems are hierarchical, meaning that they support a range of standard transmission rates, each of which is a multiple of the base rate. The Synchronous Transport Signal (STS) and Optical Carrier (OC) standards are:

1. STS-1- OC-1: 51.84 Mbps (basic rate)
2. STS-3- OC-3: 155.52 Mbps (3 × STS-1)
3. STS-12- OC-12: 622.08 Mbps
4. STS-48- OC-48: 2.488 Gbps
5. STS-192- OC-192: 9.953 Gbps
6. STS-768- OC-768: 39.813 Gbps

15.9.3 Switching and Routing

In optical networks, circuit switching and packet switching are two fundamental techniques used for routing data across the network. Each approach has its own advantages and applications, depending on the requirements for data transmission, latency, and efficiency.

15.10 Long-Haul Optical Fiber Networks

Circuit Switching in Optical Networks Circuit switching establishes a dedicated communication path (or circuit) between the source and the destination for the entire duration of the communication session. In optical networks, this means reserving a wavelength or a specific fiber path between nodes before transmitting data.

Circuit switching provides a fixed path (wavelength or fiber) established for the entire duration of the session and guarantees bandwidth and low latency, which are crucial for real-time communications.

Packet Switching in Optical Networks In packet switching, data is divided into small packets, which are independently routed through the network based on the destination address contained in each packet. There is no dedicated path; instead, the network dynamically routes each packet across the available paths.

Unlike circuit switching, packet switching does not require a pre-established path. Data packets are routed individually, and each packet can potentially take a different route to the destination. Packet switching has the advantage sharing network resources and dynamic resource allocation.

15.10 Long-Haul Optical Fiber Networks

Long-haul optical fiber networks refer to optical fiber communication systems designed to transmit data over very long distances, typically hundreds to thousands of kilometers. These networks are used to connect geographically distant locations, such as between cities, countries, or continents, using either terrestrial (land-based) or submarine (undersea) fiber cables. Long-haul networks utilize CWDM or DWDM technologies to provide the high data rates needed.

15.10.1 Types of Long-Haul Optical Networks

Terrestrial Long-Haul Networks These are land-based optical fiber networks that connect distant locations within and between countries or continents. They typically span thousands of kilometers and rely on amplifiers and regenerators to maintain signal quality over long distances.

These networks are typically deployed over land, spanning cities, countries, or continents. They are routed through data centers, network hubs, and telecom exchanges. Since these links may span hundreds or thousands of kilometers, they employ optical amplifiers that are spaced approximately every 80 to 100 km. Depending on the fiber type and network design. Regenerators may be used for extremely long distances where signal quality degrades significantly. Terrestrial long-haul networks can be dynamically reconfigured via reconfigurable optical add multiplexers (ROADMs) to adapt to changing traffic patterns or restore service after failure.

Submarine (Undersea) Long-Haul Networks These networks connect continents and islands using undersea fiber-optic cables laid on the seabed. Submarine cables are much longer than terrestrial links and face unique challenges such as undersea pressure, cable damage, and difficult maintenance.

Submarine cables are laid on the ocean floor, often spanning thousands of kilometers to connect continents. These networks are critical for global communications, as they carry the vast majority (over 99%) of intercontinental Internet traffic. Unlike terrestrial networks that use amplifiers on land, submarine networks rely on repeaters, which are placed along the cable (typically every 50–100 km) to optically amplify and regenerate the optical signal as it traverses long distances under the ocean. These networks are point-to-point links with no branching out data on the route under the sea. The repeaters and optical amplifiers require electric power which is feed at the landing sites and carried on the same fiber-optic cables.

The following table illustrates some examples of optical submarine long-haul networks.

Network name	WDM channels	Length (km)	Capacity (Gbps)
TAT-14	16	15.400	640
SEA-ME-WE 3	48	39,000	960
FLAG	60	28,00	4,800
Asia-America Gateway	96	20,000	1,920
Gateway India-ME-WE	96	13,000	3,840
MAREA cable	64	7,800	160,000

The highest capacity submarine optical link currently in use is the MAREA cable, which spans the Atlantic Ocean, connecting the United States and Spain. This cable, a joint project by Microsoft, Facebook, and Telxius, has a capacity of up to 160 Tbpa. This extraordinary capacity allows it to handle massive amounts of Internet and data traffic, making it one of the most advanced and fastest undersea cables in operation. The MAREA cable was designed with eight fiber pairs. Each fiber pair supports advanced dense wavelength-division multiplexing (DWDM) technology. Each fiber pair is optimized for up to 26.2 Tbps.

15.11 Optical Link Design

The design of optical fiber communication systems must consider the unique properties of glass fiber as a transmission medium. The primary design criteria for a given application, whether digital or analog transmission techniques, are the required transmission distance and the rate of information transfer. In optical fiber links these criteria are closely tied to the fiber's key transmission characteristics: optical attenuation and dispersion. These factors, along with the limitations imposed

15.11 Optical Link Design

by terminal equipment, ultimately determine the maximum distance permissible between the optical fiber transmitter and receiver. For long-haul telecommunications, where terminal equipment must be spaced farther apart than this maximum distance, optical amplifiers are inserted at regular intervals to maintain signal strength.

The design of optical links involves performing a power-budget calculation to take account of fiber attenuation, optical component losses, and optical amplifier gain. Another crucial aspect of the link design is the rise-time budget calculation, which accounts for fiber dispersion and determines the link's data-carrying capacity. These calculations are dependent on the choice of optical components.

15.11.1 Component Choices

The initial step in system design is selecting the operating wavelength, whether in the shorter wavelength range (0.8–0.9 µm) or the longer wavelengths in the C- and L-bands (1450–1650 nm). This choice dictates the selection of transmitter and receiver components. The major component choices include the following:

1. **Optical source**: Laser or LED, considering parameters such as optical power launched into the fiber, the source's rise time, and spectral width.
2. **Optical fiber:** Single-mode or multimode fibers, on the basis on their attenuation and dispersion characteristics.
3. **Optical amplifier:** Options include erbium-doped fiber amplifiers (EDFAs) and Raman fiber amplifiers, with considerations for gain and bandwidth.
4. **Optical receiver**: Photodiode types such as p-i-n or avalanche photodiodes, considering factors such receiver sensitivity and dark current.

The critical design parameters are the span length and data rate of the optical link.

15.11.2 Power-Budget Analysis

Power-budget calculations are essential for determining the feasibility and performance of an optical link by ensuring that sufficient optical power reaches the receiver for reliable communication. This process involves accounting for all the losses encountered along the optical path, including fiber attenuation, connector and splice losses, and the insertion losses of other components such as optical amplifiers.

The power budget can be expressed as the difference between the transmitted optical power and the minimum required receiver sensitivity. The power available at the receiver must be greater than or equal to the minimum receiver sensitivity for the link to function properly. The calculation generally follows this structure expressed as:

$$P_{budget} = P_{Tx} - P_{Rx\text{-}min} \tag{15.3}$$

where:

P_{Tx} is the optical power launched into the fiber by the transmitter.

$P_{Rx\text{-}min}$ is the minimum optical power required by the receiver for error-free performance.

The following losses must be considered when determining the available power at the receiver:

1. **Fiber attenuation**: Optical power decreases as it propagates through the fiber due to inherent material absorption and scattering. This is typically expressed in dB/km and is wavelength-dependent.
2. **Connector and splice losses:** Each connector or splice introduces a small amount of loss, typically between 0.1 and 0.5 dB per connection or splice.
3. **Component insertion losses:** Components such as optical amplifiers, couplers, and splitters may introduce additional losses that must be factored into the budget.
4. **Link margin**: A margin (often 3 to 6 dB) is usually added to account for unexpected losses due to aging components, temperature variations, or environmental factors.

The overall power-budget equation can thus be written as:

$$P_{link} = P_{Tx} - (L_{fiber} + L_{connectors} + L_{splices} + L_{components} + L_{margin}) \tag{15.4}$$

where:

L_{fiber} is the total fiber attenuation (dB), $L_{connectors}$ and $L_{splices}$ are the respective losses for connectors and splices, $L_{components}$ accounts for losses due to additional optical components, and L_{margin} is the system margin to ensure reliable operation.

If the resulting P_{link} exceeds the receiver sensitivity, the link is viable. Otherwise, measures such as using optical amplifiers or selecting lower-loss components may be needed to improve the link performance.

Example An edge-emitting LED operating at a wavelength of 1.3 μm launches −22 dBm of optical power into a single-mode fiber pigtail. The pigtail is connected to a single-mode fiber link, which exhibits an attenuation of 0.4 dB/km at this wavelength. In addition, the splice losses on the link provide an average loss of 0.05 dB/km. The transmission rate of the system is 280 Mbit/s, and the sensitivity of the p–i–n photodiode receiver is −35 dBm. The link requires an allowance of a safety margin of 6 dB is specified. If the

(continued)

connector losses at the LED transmitter and p–i–n photodiode receiver are each 1 dB, calculate the link length over which the link will operate.

Solution **Step 1**: Known Parameters

- Transmitted power: $P_{Tx} = -22\,\text{dBm}$
- Receiver sensitivity: $P_{Rx} = -35\,\text{dBm}$
- Fiber attenuation coefficient: $\alpha = 0.4\,\text{dB/km}$
- Splice loss: $L_{\text{splice}} = 0.05\,\text{dB/km}$
- Connector losses at transmitter and receiver: $L_{\text{connector}} = 1\,\text{dB}$ *each*
- Link margin: $L_{\text{margin}} = 6\,\text{dB}$

Step 2: Total Loss Calculation} The total system losses include:

$$L_{\text{total}} = L_{\text{connector (tx)}} + L_{\text{connector (rx)}} + L_{margin}$$

$$L_{total} = 1\,\text{dB} + 1\ \text{dB} + 6\ \text{dB} = 9\ \text{dB}.$$

Step 3: Available Power Budget
The available power budget for the fiber and splice losses is:

$$P_{budget} = P_{Tx} - P_{Rx} - L_{\text{total}}$$

$$P_{budget} = (-22\,\text{dBm}) - (-35\,\text{dBm}) - 8\,\text{dB} = 5\,\text{dB}$$

Step 4: Fiber and splice losses per km}
The combined loss per kilometer due to fiber attenuation and splicing is as follows:

$$L_{\text{km}} = \alpha + L_{\text{splice}}$$

$$L_{\text{km}} = 0.4\,\text{dB/km} + 0.05\,\text{dB/km} = 0.45\,\text{dB/km}$$

Step 5: The maximum link length is given by:

$$D = \frac{\text{Power Budget}}{L_{\text{km}}}$$

$$D = \frac{3.5\,\text{dB}}{0.45\,\text{dB/km}} \approx 7.78\,\text{km}$$

The maximum link length over which the link will operate is approximately 7.78 km.

Example Long-haul single-mode optical fiber system operating at a wavelength of 1.55 μm with the following parameters:

- Mean power launched from the laser transmitter into the fiber is −3 dBm
- Optical fiber attenuation is 0.4 dB/km
- Splice loss is 0.1 d/km
- Connector losses at the transmitter and receiver is 1 dB each
- APD receiver sensitivity: when operating at 50 Mbps is −50 dBm, and when operating at 400 Mbps is −45 dBm
- Required link margin is 6 dB.

Determine the following:

(a) The maximum possible link length when operating at 50 Mbps.
(b) The maximum possible link length when operating at 400 Mbps.

Solution (a) When the system is operating at 50 Mbps the optical power budget is:

$$P_{budget} = P_{Tx} - P_{Rx} = -3\,\text{dBm} - (-50\,\text{dBm}) = 47\,\text{dBm}$$

The total losses are given by:

$$L_{total} = (\alpha_{fiber} + \alpha_{splice}) \times D + 2 \times \alpha_{connector} + L_{margin} = 47\,\text{dBm}$$

$$(0.4 + 0.1) \times D + 2 + 6 = 47$$

Therefore, the maximum possible link length is:

$$D = \frac{39}{0.5} = 78\,\text{km}$$

(b) When the system is operating at 50 Mbps the optical power budget is:

$$P_{budget} = P_{Tx} - P_{Rx} = -3\,\text{dBm} - (-45\,\text{dBm}) = 42\,\text{dBm}$$

Total losses are

$$L_{total} = (\alpha_{fiber} + \alpha_{splice}) \times D + 2 \times \alpha_{connector} + L_{margin} = 42\,\text{dBm}$$

$$(0.4 + 0.1) \times D + 2 + 6 = 42$$

(continued)

15.11 Optical Link Design

Therefore, the maximum possible link length is

$$D = \frac{34}{0.5} = 68 \, \text{km}$$

15.11.3 Rise-Time Budget Analysis

The rise-time budget is essential to ensure that the system operates efficiently at the intended bit rate. Even if the bandwidth of individual components exceeds the bit rate, the total system might not function at the required speed. The concept of rise time helps allocate bandwidth among different components.

The rise time T_r of a linear system is the time it takes for the system's response to increase from 10% to 90% of its final value after an abrupt input change. The rise-time budget is crucial in determining the data-carrying capacity of an optical link, as it ensures the system meets the required bit rate without signal distortion. A key focus is minimizing intersymbol interference (ISI), where signal pulses spread and overlap due to dispersion, leading to errors in data transmission.

In digital communication, rise time refers to the time taken for the optical signal to transition between low and high states, typically measured between 10% and 90% of the signal's amplitude. The total system rise time must be short enough to prevent pulse broadening and ensure accurate data detection.

Relationship Between Rise Time and Bandwidth An inverse relationship exists between bandwidth Δf and rise time T_r for a linear system. For example, in an RC circuit, when an input voltage changes abruptly from 0 to V_o, the output voltage $V_{out}(t)$ evolves as:

$$V_{out}(t) = V_o \left(1 - e^{(-t/RC)}\right) \tag{15.5}$$

where R is resistance and C is capacitance. The rise time in this system is given by:

$$T_r = ln(9) RC \approx 2.2 \, RC \tag{15.6}$$

and

$$\Delta f = \frac{1}{2\pi RC} \tag{15.7}$$

Thus, the relationship between bandwidth and required rise time is:

$$T_r \Delta f \approx 0.35. \tag{15.8}$$

This inverse relationship holds for most linear systems and the product $T_r \Delta f \approx 0.35$ is often used as a guideline in optical system design. However, depending on the digital format, the product can vary. For example,

Return-to-zero (RZ) format:

$$BT_r = 0.35, \tag{15.9}$$

where B is the bit rate.

Non return-to-zero (NRZ) format:

$$BT_r = 0.7. \tag{15.10}$$

The rise time of the communication system must be designed to meet the bit rate by ensuring:

$$T_r \leq \begin{cases} 0.35 & for\ RZ\ format \\ 0.7 & for\ NRZ\ format \end{cases} \tag{15.11}$$

Components Contributing to the System Rise-Time An optical communication system comprises three key components that contribute to the total rise time: the transmitter, fiber, and receiver.

The total system rise time T_{sys} is the square root of the sum of the squares of the individual rise times:

$$T_{sys} = \sqrt{T_{Tx}^2 + T_{fiber}^2 + T_{Rx}^2} \tag{15.12}$$

Transmitter Rise Time T_{Tx}: Determined by the optical source, which is typically shorter for lasers compared to LEDs. The transmitter's rise time must be small enough to support the required data rate.

Fiber Rise Time T fiber T_{fiber}: This factor is influenced primarily by dispersion in the fiber. For single-mode fibers, chromatic dispersion dominates, whereas in multimode fibers, modal dispersion is the primary cause of pulse broadening. The fiber-induced rise time due to chromatic dispersion can be approximated by:

$$T_{fiber} \approx D_m \times \Delta\lambda \times L \tag{15.13}$$

where D_m is the chromatic dispersion coefficient, $\Delta\lambda$ is the spectral width of the source, and L is the fiber length.

Receiver Rise Time T_{Rx}: Depends on the type of photodetector used (e.g., p-i-n or avalanche photodiodes). Faster photodiodes improve the system's ability to detect high-speed signals accurately.

15.11 Optical Link Design

Example Consider a fiber-optic system operating at 1 Gbps over a single-mode fiber with a repeater spacing of 50 km. The transmitter rise time T_{Tx} is specified as 0.25 ns, and the receiver rise time T_{Rx} is 0.35 ns. The source spectral width is $\Delta\lambda = 3$ nm, and the chromatic dispersion coefficient D_m is 2 ps/(km-nm). For a fiber length $L = 50$ km, determine the system rise-time.

Solution The fiber rise time is:

$$T_{fiber} = D_m \times \Delta\lambda \times L = 2 \times 3 \times 50 = 0.3 \text{ ns}$$

Using the total rise-time equation:

$$T_{sys} = \sqrt{0.25^2 + 0.3^2 + 0.35^2} = 0.524 \text{ ns}$$

Therefore,

$$BT_{sys} = 10^9 \times 0.524 \times 10^{-9} = 0.524$$

For the *RZ* format, the criteria is $BT_r = 0.35$, indicating the system cannot operate at 1 Gbps. However, for the *NRZ* format, the system meets the requirement.

Example Make the rise-time budget for an 850 nm, 10 km fiber link designed to operate at 50 Mbps. The LED transmitter and the *Si p-i-n* receiver have rise times of 10 and 15 ns, respectively. The multimode fiber has a core index of 1.46, $\Delta = 0.01$, and $D_m = 80$ ps/(km-nm). The LED spectral width is 50 nm. Can the system be designed to operate with the NRZ format?

Solution The total system rise time including modal dispersion is:

$$T_{sys} = \sqrt{T_{Tx}^2 + T_{Rx}^2 + T_{fiber}^2 + T_{modal}^2},$$

$$T_{fiber} = D_m \times \Delta\lambda \times L = 80 \times 50 \times 10 = 40 \text{ ns}$$

$$T_{modal} = \frac{\Delta \times L}{v} = \frac{0.01 \times 10^4}{3 \times 10^8 / 1.46} = 487 \text{ ns}$$

(continued)

It is clear that the modal dispersion is much larger than the chromatic dispersion. Therefore, the system rise time is:

$$T_{sys} = \sqrt{10^2 + 15^2 + 40^2 + 487^2} = 489 \text{ ns}.$$

The required rise time for 50 Mbps NRZ operation is:

$$T_{required} = \frac{0.7}{50 \times 10^6} = 14 \text{ ns}$$

Therefore, the system cannot operate with NRZ format at 50 Mbps, primarily due to the large modal dispersion in the multimode fiber.

15.12 Summary

This chapter provides an in-depth discussion of fiber-optic communication systems, exploring their evolution, principles, and modern implementations. It begins by emphasizing the importance of optical fiber communication, which forms the backbone of global telecommunications due to its high-speed and long-distance transmission capabilities.

The chapter starts with a discussion on the historical development of optical systems, highlighting several generations of technology advancements. Initially, systems operated at 0.8 μm wavelengths with limited bit rates and repeater spacings. Progressing through the 1980s and 1990s, systems adopted longer wavelengths (1.3 μm and 1.55 μm) and optical amplification, which drastically increased capacity and transmission distance. The latest advancements involve dense wavelength-division multiplexing (DWDM), which can achieve data rates beyond to 10 Tbps over thousands of kilometers.

Key communication principles are outlined, focusing on modulation and demodulation processes, essential components such as transmitters, communication channels, and receivers, as well as the importance of optical fiber for transmitting both analog and digital data.

Network topologies such as buses, stars, rings, and meshes are analyzed in terms of efficiency, reliability, and fault tolerance. These topologies serve different purposes based on network size and configuration.

A significant portion is dedicated to optical network components such as wavelength-division multiplexers/demultiplexers, optical add/drop multiplexers, and optical cross-connect switches, all of which are key for handling WDM systems and routing signals.

Optical fiber amplifiers, with detailed descriptions of erbium-doped fiber amplifiers (EDFAs) and Raman amplifiers were discussed in detail. Their applications are crucial for overcoming signal attenuation over long distances without converting optical signals into electrical ones.

Several types of optical networks have been explored, including single-channel, WDM networks, and long-haul networks, which span large geographical areas and use technologies like DWDM for high data capacities.

Finally, the chapter covers the design of optical links, including critical calculations such as power-budget and rise-time budget analysis. These ensure that enough power reaches the receiver and the system operates efficiently at the required bit rate, accounting for signal losses and dispersion.

In conclusion, this chapter provides a comprehensive guide to the components, design, and operation of modern optical fiber communication systems, which are essential for meeting the growing demands of global data transmission.

15.13 Problems

1. Derive the equation for the blaze angle in a Littrow-mounted diffraction grating. Explain the role of the grating period (Λ) in determining the central wavelength.
2. For a Littrow grating with a period of 1.5 μm, calculate the blaze angle (θ_B) required for a central wavelength of 1550 nm.
3. A transmitter launches -5 dBm of optical power into a fiber with an attenuation of 0.3 dB/km. If the receiver sensitivity is -28 dBm and the link includes two connectors (each with 0.5 dB loss) and a 3 dB margin, calculate the maximum permissible link length.
4. Design a power budget for a 100 km link using an $EDFA$ with 20 dB gain placed at the 50 km mark. The fiber attenuation is 0.25 dB/km, and there are 0.4 dB splice losses every 10 km. The transmitter power is -4 dBm, and the receiver sensitivity is -30 dBm. Assume 3 dB for margin.
5. In a 200 km link, $EDFAs$ with 15 dB gain are placed every 50 km. The fiber attenuation is 0.2 dB/km, and each $EDFA$ introduces 4 dB noise. If the transmitter power is -3 dBm and the receiver sensitivity is -35 dBm, determine if the link operates reliably with a 5 dB margin.
6. For a 30 km fiber link with 0.35 dB/km attenuation and 0.5 dB connector losses, calculate the minimum transmitter power required to meet a receiver sensitivity of -25 dBm, including a 3 dB margin.
7. A communication system has the following rise times: transmitter 0.3 ns, fiber 0.4 ns, and receiver 0.5 ns. Calculate the total system rise time and determine if it supports a bit rate of 2.5 Gbps for NRZ format.
8. For a single-mode fiber link of 20 km, the chromatic dispersion is 3.5 ps/(nm·km), and the spectral width of the source is 5 nm. Calculate the rise time contribution of the fiber.
9. A multimode fiber link has a transmitter rise time of 1 ns, receiver rise time of 2 ns, chromatic dispersion rise time of 5 ns, and modal dispersion rise time of 25 ns. Determine the total system rise time and verify if the system can support a data rate of 100 Mbps with NRZ modulation.

10. Given a chromatic dispersion coefficient of 2 ps/(nm· km), a source spectral width of 3 nm, and a fiber length of 30 km, calculate the fiber-induced rise time. Include the transmitter rise time (0.2 ns) and receiver rise time (0.3 ns) to compute the total system rise time.

Bibliography

1. Agrawal, G. P. (2012). Fiber-Optic Communication Systems (4th ed.). Wiley.
2. Keiser, G. (2011). Optical Fiber Communications (4th ed.). McGraw-Hill.
3. Senior, J. M., & Jamro, M. Y. (2009). Optical Fiber Communications: Principles and Practice (3rd ed.). Pearson.
4. Mukherjee, B. (2006). Optical WDM Networks. Springer.
5. Simmons, J. M. (2008). Optical Network Design and Planning. Springer.
6. Tsuchiya, Y., Hirano, A., & Kato, S. (2002). Reconfigurable optical add/drop multiplexer using fiber Bragg gratings and optical circulators. IEEE Photonics Technology Letters.
7. Desurvire, E. (1994). Erbium-Doped Fiber Amplifiers: Principles and Applications. Wiley.
8. Dutta, N. K., & Wang, Q. (2006). Raman Amplifiers for Telecommunications 1: Physical Principles. Academic Press.
9. Keiser, G. (2003),Optical Communications Essentials, McGraw-Hill.
10. Islam, M.,(2004), Raman Amplifiers for Telecommunications 1 Physical Principles, Springer.
11. P. S. Henry, (1988), R. A. Linke, and A. H. Gnauck, in Optical Fiber Telecommunications II, S. E. Miller and I. P. Kaminow, Eds., Academic Press.
12. Ramaswami, R., Sivarajan, K. N., & Sasaki, G. H. (2009). Optical Networks: A Practical Perspective (3rd ed.). Morgan Kaufmann.
13. ITU-T Recommendation G.694.1 (2020). Spectral grids for WDM applications: DWDM frequency grid.

The manufacturer's authorised representative in the EU is Springer Nature Customer Service Centre GmbH, Europaplatz 3, 69115 Heidelberg, Germany. If you have any concerns regarding our products, please contact ProductSafety@springernature.com

Printed and bound by CPI Group (UK) Ltd, Croydon, CR0 4YY
26/03/2026
02078967-0005